Current Progress in Medical Mycology

Héctor M. Mora-Montes
Leila M. Lopes-Bezerra
Editors

Current Progress in Medical Mycology

Springer

Editors
Héctor M. Mora-Montes
Universidad de Guanajuato
Guanajuato, Mexico

Leila M. Lopes-Bezerra
Universidade do Estado do Rio de Janeiro
Rio de Janeiro, Brazil

ISBN 978-3-319-64112-6 ISBN 978-3-319-64113-3 (eBook)
DOI 10.1007/978-3-319-64113-3

Library of Congress Control Number: 2017953012

Printed on acid-free paper

This Springer imprint is published by Springer Nature
The registered company is Springer International Publishing AG
The registered company address is: Gewerbestrasse 11, 6330 Cham, Switzerland

Preface

It is estimated that the Kingdom Fungi groups have about five million species, and they significantly contribute to the biological interactions that sustain and shape the different ecosystems on Earth. The heterotrophic nature of fungi forces them to interact with other living organisms, and besides some extremophile species, all organisms, unicellular or multicellular, are susceptible to be a source of nutritive macromolecules for the fungal cell. In most of the cases such interaction is established with death cells and tissues, and fungi play a beneficial role helping to clean the environment and recycling the basic building blocks of biomolecules. However, when the interaction occurs with other organisms this could lead to the establishment of mutualistic, commensalism or parasitism. The latter can potentially affect the health and physiological status of the host, leading to the establishment of a disease. Fungi can parasite other fungi, plants, and animals, including the human being. The number of fungal species that cause infections in humans are just a handful, if compared with the fungal biodiversity currently known. It is important to notice that new emerging species adapted to mammalian parasitism are increasing. After viral infections, mycoses are the most frequent infections in humans and are of outstanding interest because the morbidity and mortality associated to them. Depending on the tissues affected, we can classify mycoses as superficial, subcutaneous and systemic or deep-seated infections; and they can be caused by primary pathogens or opportunistic species. Opportunistic fungal infections are of special interest because they are a consequence of a temporal or permanent immunodeficiency, and are associated to high mortality rates. Despite we have several antifungal drugs to treat fungal infections, the diversity is not enough, as some fungal species are naturally resistant to the therapeutic alternatives, or they develop or acquire drug resistance faster than the development of new drugs to combat these pathogens. This book offers a comprehensive overview of medical mycology.

Written by scientists and physicians, it addresses basic aspects of the causative agents and the clinical manifestations of the most common and relevant human mycoses, as well as their diagnosis and the therapeutic alternatives. Further, it discusses the immune response, virulence factors and the development of modern molecular diagnostic tools.

Guanajuato, Mexico Héctor M. Mora-Montes
Rio de Janeiro, Brazil Leila M. Lopes-Bezerra

Contents

Contributors

Priscila C. Albuquerque, Ph.D. Centro de Desenvolvimento Tecnológico em Saúde (CDTS), Fundação Oswaldo Cruz (Fiocruz), Rio de Janeiro, RJ, Brazil

Rodrigo de Almeida-Paes Laboratório de Micologia, Instituto Nacional de Infectologia Evandro Chagas, Fundação Oswaldo Cruz (Fiocruz), Rio de Janeiro, Brazil

Brenda Alvarado-Sánchez, Ph.D. Unidad Académica Multidisciplinaria Zona Huasteca, Universidad Autónoma de San Luis Potosí, Valles, San Luis Potosi, Mexico

Francisco Antunes, M.D., Ph.D. Instituto de Saúde Ambiental, Faculdade de Medicina de Lisboa, Universidade de Lisboa, Lisboa, Portugal

Roberto Arenas, M.D. Sección de Micología, Hospital General "Manuel Gea González", México, Cd.Mx, Mexico

Bridget M. Barker, Ph.D. Division of Pathogen Genomics, Translational Genomics Research Institute (TGen-North), Flagstaff, AZ, USA

Pathogen and Microbiome Institute, Northern Arizona University, Flagstaff, AZ, USA

Department of Medicine, Valley Fever Center for Excellence, University of Arizona, Tucson, AZ, USA

Alexandro Bonifaz, Ph.D. Department of Mycology and Dermatology Service, Hospital General de México, "Dr. Eduardo Liceaga", Mexico, DF, Mexico

Chelsea Bueter, Ph.D. Divison of Infectious Diseases, College of Medicine, University of Cincinnati, Cincinnati, OH, USA

Beatriz Bustamante, M.D. Instituto de Medicina Tropical Alexander von Humboldt, Universidad Peruana Cayetano Heredia, Lima, Peru

Departamento de Enfermedades Infecciosas, Tropicales y Dermatológicas, Hospital Cayetano Heredia, Lima, Peru

Emma Camacho, Ph.D. Department of Molecular Microbiology and Immunobiology, Johns Hopkins Bloomberg School of Public Health, Johns Hopkins University, Baltimore, MD, USA

Pablo E. Campos, M.D., M.P.H. Investigaciones Medicas en Salud, Lima, Peru

Arunaloke Chakrabarti, M.D. Department of Medical Microbiology, Postgraduate Institute of Medical Education and Research, Chandigarh, India

George S. Deepe, M.D. Divison of Infectious Diseases, College of Medicine, University of Cincinnati, Cincinnati, OH, USA

Esperanza Duarte-Escalante, Ph.D. Departamento de Microbiología y Parasitología, Facultad de Medicina, Universidad Nacional Autónoma de México (UNAM), Mexico, Cd.Mx, Mexico

Fabianno F. Dutra, D.Sc. Laboratório de Inflamação e Imunidade, Departamento de Imunologia, Instituto de Microbiologia, Universidade Federal do Rio de Janeiro (UFRJ), Rio de Janeiro, Brazil

Ahmed Hassan Fahal, MBBS, FRCS, FRCSI, FRCS (Glasg) Department of Surgery, University of Khartoum, Khartoum, Sudan

Peiying Feng, Ph.D. Department of Dermatology, The Third Affiliated Hospital of Sun Yat-Sen University, Guangzhou, China

Fernanda L. Fonseca, D.Sc. Centro de Desenvolvimento Tecnológico em Saúde (CDTS), Fundação Oswaldo Cruz (Fiocruz), Rio de Janeiro, RJ, Brazil

María Guadalupe Frías-De-León, Ph.D. División de Investigación, Hospital Juárez de México, Edificio E. Av. Instituto Politécnico Nacional 5160, México, Cd. Mx, Mexico

Dirección de Investigación, Hospital Regional de Alta Especialidad de Ixtapaluca, Ixtapaluca, Estado de México, Mexico

Beatriz L. Gómez, B.Sc., Ph.D. School of Medicine and Health Sciences, University of El Rosario, Bogotá, Colombia

Corporación para Investigaciones Biológicas (CIB), Medellín, Colombia

Rodrigo Maciel da C. Godinho, D.Sc. Instituto de Microbiologia Paulo de Góes, Universidade Federal do Rio de Janeiro (UFRJ), Rio de Janeiro, Brazil

Jose A. Hidalgo, M.D. Division de Enfermedades Infecciosas, Hospital Guillermo Almenara, Lima, Peru

G.S. de Hoog, Ph.D. Westerdijk Fungal Biodiversity Institute, Utrecht, The Netherlands

Leila M. Lopes-Bezerra, Ph.D. Laboratory of Cellular Mycology and Proteomics, Biology Institute, University of Rio de Janeiro State (UERJ), Rio de Janeiro, Brazil

Jean-Paul Latgé, Ph.D. Department of Mycology, Unité des Aspergillus, Institut Pasteur, Paris, France

Ángel León-Buitimea, Ph.D. Unidad Académica Multidisciplinaria Zona Huasteca, Universidad Autónoma de San Luis Potosí, Valles, San Luis Potosí, Mexico

José R. Macías-Pérez, Ph.D. Unidad Académica Multidisciplinaria Zona Huasteca, Universidad Autónoma de San Luis Potosí, Valles, San Luis Potosí, Mexico

Erick Martínez-Herrera, Ph.D. Dirección de Investigación, Hospital Regional de Alta Especialidad de Ixtapaluca, Carretera Federal México, Estado de México, Mexico

Olga Matos, M.D., M.Sc., Ph.D. Medical Parasitology Unit, Group of Opportunistic Protozoa/HIV and Other Protozoa, Global Health and Tropical Medicine, Instituto de Higiene e Medicina Tropical, Universidade NOVA de Lisboa, Lisboa, Portugal

Héctor M. Mora-Montes, Ph.D. Departamento de Biología, División de Ciencias Naturales y Exactas, Campus Guanajuato, Universidad de Guanajuato, Guanajuato, Gto., Mexico

Ángela Restrepo Moreno, Ph.D. Corporación para Investigaciones Biológicas (CIB), Medellín, Colombia

María J. Navarro-Arias, M.Sc. División de Ciencias Naturales y Exactas, Departamento de Biología, Universidad de Guanajuato, Campus Guanajuato, Guanajuato, Guanajuato, Mexico

Gustavo A. Niño-Vega, Ph.D. Departamento de Biología, División de Ciencias Naturales y Exactas, Universidad de Guanajuato, Guanajuato, Guanajuato, Mexico

Instituto Venezolano de Investigaciones Científicas, Centro de Microbiología y Biología Celular, Caracas, Venezuela

Débora L. Oliveira, Ph.D. Centro de Desenvolvimento Tecnológico em Saúde (CDTS), Fundação Oswaldo Cruz (Fiocruz), Rio de Janeiro, RJ, Brazil

Luis A. Pérez-García, Ph.D. Unidad Académica Multidisciplinaria Zona Huasteca, Universidad Autónoma de San Luis Potosí, Valles, San Luis Potosí, Mexico

Mayra D. Ramírez-Quijas, M.Sc. Departamento de Ingeniería Biomédica, Universidad Politécnica del Bicentenario, Silao de la Victoria, Guanajuato, Mexico

Chad A. Rappleye, Ph.D. Department of Microbiology, Ohio State University, Columbus, OH, USA

María del Rocío Reyes-Montes, Ph.D. Departamento de Microbiología y Parasitología, Facultad de Medicina, Universidad Nacional Autónoma de México (UNAM), México, Cd.Mx, Mexico

Saraí C. Rodríguez-Reyes, Ph.D. Centro Universitario de Ciencias de la Salud, Instituto de Investigación en Ciencias Biomédicas, Universidad de Guadalajara, Guadalajara, Jalisco, Mexico

Marcio L. Rodrigues, D.Sc. Instituto de Microbiologia Paulo de Góes, Universidade Federal do Rio de Janeiro (UFRJ), Rio de Janeiro, Brazil

Centro de Desenvolvimento Tecnológico em Saúde (CDTS), Fundação Oswaldo Cruz (Fiocruz), Rio de Janeiro, RJ, Brazil

Nandini Sethuraman, M.B.B.S., M.D. Department of Microbiology, Apollo Hospitals, Chennai, India

Marcus M. Teixeira, Ph.D. Division of Pathogen Genomics, Translational Genomics Research Institute (TGen-North), Flagstaff, AZ, USA

Pathogen and Microbiome Institute, Northern Arizona University, Flagstaff, AZ, USA

Translational Genomics Research Institute, Northern Arizona Center For Valley Fever Research, Flagstaff, AZ, USA

Angela María Tobón Facultad de Medicina, Centro de Estudios en Salud (CES), Medellín, Colombia

Corporación para Investigaciones Biológicas (CIB), Medellín, Colombia

Ana Luisa Tomás, Ph.D., Student Medical Parasitology Unit, Group of Opportunistic Protozoa/HIV and Other Protozoa, Global Health and Tropical Medicine, Instituto de Higiene e Medicina Tropical, Universidade NOVA de Lisboa, Lisboa, Portugal

Sarah Sze Wah Wong, Ph.D. Department of Mycology, Unité des Aspergillus, Institut Pasteur, Paris, France

Chapter 1
Introduction to Medical Mycology

Arunaloke Chakrabarti and Nandini Sethuraman

Abstract Fungi are ubiquitous and nearly 1.5 million fungal species exist in the universe, but only few fungi were known to be pathogenic to humans. The temperature of 37 °C, low redox potential in tissues and immune barrier prevent majority of fungi from invading human hosts. However with the change of host environment due to co-morbidities, several saprophytic fungi get the opportunity to adapt to human tissue. The impact of modern medical interventions and fungi adapting to this environment are reasons why many fungi are now known to cause invasive human disease. Fungal infection in humans varies from superficial colonization to invasive diseases and allergic manifestations; each group associated with unique risk factors. Impact of the global burden of fungal infections is often overlooked and estimates are often extrapolations of limited available data. There is also wide variation in the geographical niches occupied by fungi across the world for reasons largely unknown. Natural disasters also contribute to the upsurge of invasive fungal infections. The severity of infection closely parallels a suppression of immunity. Invasive fungal infections are important causes of morbidity and mortality of hospitalized patients. Diagnosis of these fungal infections is often a challenge due to poor sensitivity, specificity and long turnaround times of conventional diagnostic methods. Availability of advanced, rapid technologies is sparse in the majority of laboratories in developing countries. Several biomarkers and molecular techniques are being studied to reduce these turnaround times. In spite of availability of new antifungals, therapy is often empirical due to rapid progression of disease and non-availability of early diagnosis. There is no ideal antifungal agent. Limited spectrum, drug-interaction, toxicity and cost of antifungals are limiting factors. A brief description of these challenges is presented in this introductory chapter.

A. Chakrabarti, M.D. (✉)
Department of Medical Microbiology, Postgraduate Institute of Medical Education and Research, Chandigarh, India
e-mail: arunaloke@hotmail.com

N. Sethuraman, M.B.B.S., M.D.
Department of Microbiology, Apollo hospitals, Chennai, India
e-mail: drnandinipgi@gmail.com

H.M. Mora-Montes, L.M. Lopes-Bezerra (eds.), *Current Progress in Medical Mycology*, DOI 10.1007/978-3-319-64113-3_1

1.1 Definition and Structure

Fungi are aerobic, eukaryotic, heterotrophic, spore-bearing, non-motile saprobes that range from microscopic unicellular forms known as yeasts to multicellular moulds to complex forms that give rise to edible mushrooms. True to their widely used description of being ubiquitous, they can be found in virtually any habitat ranging from dry to humid, tropical to temperate, deserts to deep sea and symbionts to parasites. They reproduce by sexual as well as asexual means. The sexual stage may be seen in a few homothallic fungi, but is largely unknown in majority of fungi (mitosporic fungi). A characteristic feature unique to fungi is the presence of a rigid cell wall composed mainly of chitin and glucans that makes them impenetrable by many substances including antibacterial compounds. There are approximately 1.5 million species of fungi of which around 500 species have the propensity to cause disease in humans and are studied under the term 'medical mycology'.

Morphologically, fungi are broadly divided into two groups:

1. Yeasts: The unicellular forms that reproduce by budding, also known for blastospore formation. Some species produce pseudohyphae, which are elongated, jointed yeast cells with no cell-to-cell continuity or communication. The pseudohypha is differentiated from true hyphae by the presence of a constriction at the junction of the two cells and branching at the constricted point. Macroscopically, yeasts produce moist and creamy colonies. The most common medically important yeasts belong to *Candida* and *Cryptococcus*.
2. Moulds: Filamentous, multicellular forms that exhibit complex structures depending on their sexual or asexual mode of reproduction. The basic unit of a mould is a hypha, which is a filamentous structure that may or may not be divided by septa. A group of hyphae is known as mycelium. Macroscopically mycelia exist both basally (vegetative) and aerially. Asexually reproducing fungi form fruiting bodies that arise from the vegetative mycelia as conidiophores that end up forming conidia. The method of conidiation, structure and arrangement of conidia are useful tools for morphological identification of many moulds. Sexually reproducing fungi produce additional spore sacs denoted by various names, all of which help in morphological identification of fungi.

   Moulds, which are broad ribbon like and lack septa or sparsely septate, are the agents that cause mucormycosis and entomophthoramycosis. They reproduce sexually as well as asexually by way of zygospore and sporangiospore formation respectively. Moulds with septate hyphae are classified under *Ascomycota* and *Basidiomycota*. The majority of the medically important fungi belong to *Ascomycota* with few exceptions. The evolutionary advantage of septations is that even if a part of the hypha is damaged, the rest of the hypha survives as against primitive aseptate hyphae which are continuous and are completely damaged.
3. Dimorphic fungi: Apart from the above two broad categories, certain fungi exist as yeasts in human tissues and specialized media at 37 °C, whereas in the environment at room temperature, exist as moulds. These are phenotypically classified as dimorphic fungi (Fig. 1.1).

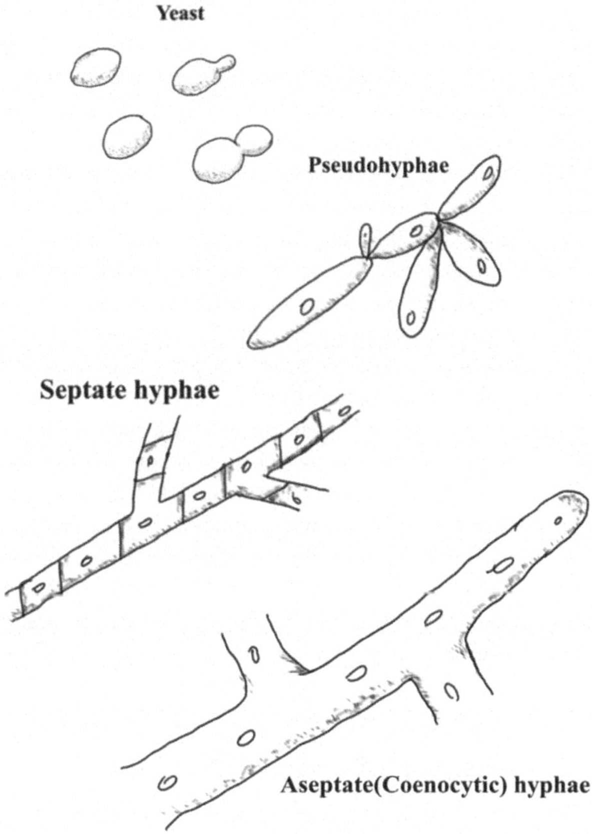

Fig. 1.1 Fungal cell morphology

1.2 Emergence of Challenge Due to Fungal Infections

The progress in modern medicine closely parallels the emergence of invasive fungal infections. Historically, with the exception of dermatophytes and pathogenic dimorphic fungi, few other fungi were known to be pathogenic in immunocompetent individuals. The scenario changed over the last few decades, with the entry of AIDS, immunosuppressive and immunomodulatory therapy, solid and stem cell transplants, frequent use of broad-spectrum antibiotics, aging population staying longer in hospital with many co-morbidities, extensive abdominal surgeries and use of invasive devices and lifestyle diseases such as diabetes. Fungi could access these patients easily. Many new saprophytic fungi could adjust to earlier known hostile human tissue environments and possibly acquired virulence factors while passing through predator soil amoebae [1]. This has led to tremendous rise in number of opportunistic fungal infections.

We face several challenges while managing opportunistic fungal infections: (1) patients do not have characteristic signs or symptoms suggesting fungal infections; (2) localization of infection often requires sophisticated radiological and laboratory support; (3) awareness about invasive fungal infections among clinicians is limited to specialized units only; (4) appropriate sampling of infected tissues often requires invasive procedures that may be difficult in the immunocompromised cohort; (5) conventional diagnosis like direct microscopy, culture and histopathology have poor sensitivity and high turnaround time; (6) although there is transformational change in diagnostics over the last decade (MALDI-TOF, sequencing, beta glucan assay, galactomannan, etc.), the availability of these advanced tests is patchy in developing countries; (7) there are limited antifungal drug target sites as fungi are eukaryotic and there is non-availability of many antifungal drugs in developing countries; (8) the majority of antifungal agents are fungistatic and cannot get rid of fungi from tissue in absence of host immunity in immunosuppressed patients; (9) antifungal drug dosing needs strict titration as therapeutic window is often narrow with most of the available antifungal agents; (10) antifungal drugs like azoles have plenty of drug interaction issues and requires regular therapeutic drug monitoring; (11) toxicity of antifungal drugs; (12) the intrinsic antifungal resistance in many species and recent development of acquired resistance; (13) the emergence of multi-drug resistant superbugs like *Candida auris* across globe; (14) the cost of antifungal therapy and poor antifungal stewardship programs which leads to heavy economic burden while managing these patients.

1.3 Spectrum of Fungal Infections

1.3.1 Superficial Mycoses

1. Cutaneous—affecting skin and mucous membranes. Cutaneous infections are usually caused by dermatophytes, *Candida* and *Malassezia* species (Pityriasis versicolor) in healthy and compromised individuals. The infection occurs in moist folds of the skin in tropical climates and is treatable with long-term topical and occasional systemic antifungals.
2. Subcutaneous tissue—These are commonly acquired during traumatic implantation (implantation mycoses). Common diseases include sporotrichosis, chromoblastomycosis, phaeohyphomycosis and mycetoma. The lesions are often disfiguring and multiple in individuals who are immunosuppressed. Sporotrichosis is caused by a dimorphic fungus- *Sporothrix schenckii* sensu *lato*. Chromoblastomycosis and phaeohyphomycosis are caused by melanized fungi. Mycetoma can be either bacterial (actinomycotic) or fungal (eumycotic) origin. Rhinosporidiosis caused by *Rhinosporidium seeberi* and lacaziosis (lobomycosis) caused by *Lacazia loboi* are also grouped under subcutaneous fungal infections due to similarity of infection, though the causative agents are no longer classified under Kingdom Fungi.

1.3.2 Invasive/Opportunistic Mycoses

In general, it is believed that invasive fungal infections or opportunistic mycoses occur when patient's immunity is compromised. Both innate and acquired immunity play important roles against fungi. When cell-mediated immunity is compromised (as in HIV infection), infections like cryptococcosis, endemic fungal infections are prevalent. On the other hand during the fall of innate immunity or neutropenia (patients with malignancies undergoing chemotherapy, transplant recipients, etc.), invasive candidiasis, aspergillosis, mucormycosis and several other mould infections prevail. These infections may present in the form of fungaemia, invasive pulmonary infections, central nervous system infections, disseminated infections, etc.

Many saprophytic fungi (considered non-pathogenic) are also reported to cause human infection even in apparently healthy hosts. These new fungal diseases revealed many new immunodeficiencies. The deficiencies like CARD9 or STAT1in signal transduction pathway lead to frequent deep dermatophytosis, central nervous system candidiasis, disseminated phaeohyphomycosis, aspergillosis and mucormycosis (primary immunodeficiency). The use of anti-TNF-α agents has increased due to the dramatic response in autoimmune disorders, but it leads to increased occurrence of endemic mycoses, aspergillosis, pneumocystosis and cryptococcosis (acquired immunodeficiency).

1.3.3 Endemic Mycoses

These are a group of infections caused by a distinct set of fungi which have restricted geographical endemicity and common morphological characteristics of exhibiting thermal dimorphism, i.e. the ability to switch from yeast or spherule form in the tissue phase at 37 °C to mould form at room temperature. They also share the ability to cause invasive infections even in immunocompetent individuals. This group includes blastomycosis, coccidioidomycosis, histoplasmosis, paracoccidioidomycosis and penicilliosis. A new endemic mycosis, emmonsiosis has emerged in AIDS patients of South Africa and sporadic cases have been reported from Italy, India, China and Germany [2, 3]. With the application of molecular tools, the taxonomy of some of these agents has altered or new cryptic species have been explored. The agent causing penicilliosis is now called *Talaromyces marneffei* and many cryptic species have been identified under *Blastomyces* and *Sporothrix*.

1.3.4 Allergic Manifestations

Allergic mycoses occur as a result of hypersensitivity to many fungi. The systemic manifestations range from fungal asthma including allergic bronchopulmonary aspergillosis and severe asthma with fungal sensitization, allergic alveolitis, allergic fungal rhinosinusitis and localized or systemic allergy to many fungi.

1.4 Classification

Fungi were originally classified along with plants in kingdom *Plantae* when Carolus Linnaeus in 1735 classified all living material in two kingdoms. As the understanding of their distinct nature developed, they were placed separately as a fifth kingdom by Whittaker in 1969 [4], on the basis of their mode of reproduction, heterotrophic nutrition, saprobic (living on dead organic matter) existence, cellular structure and lack of motility [4]. With the current level of understanding of taxonomy and molecular phylogeny, we know that over 1.5 million fungal species exist on the planet. The fungi have been placed as a separate kingdom *Fungi* under superkingdom *Eukaryota* in the Catalogue of Life database, which classifies all life into seven kingdoms [5]. The main characteristic of this Kingdom is the presence of chitin (also found in arthropods) along with beta-glucans (also found in plants as beta 1,4 glucans) in their cell wall. Evolutionarily related parasites belonging to *Oomycota* have been separated and placed in kingdom *Chromista*. A simplified, clinically useful classification of fungi is presented in Table 1.1.

Table 1.1 Clinically useful classification of medically important fungi

Type	Prominent members
Yeasts	*Candida* spp. *Cryptococcus* spp.
Hyaline fungi	*Aspergillus* spp. *Fusarium* spp. *Penicillium* spp. (other than *Talaromyces marneffei*) *Paecilomyces* spp. *Scedosporium* spp.
Black/Dematiaceous fungi • Chromoblastomycosis • Phaeohyphomycosis (subcutaneous and systemic)	*Cladophialophora bantiana* *Phialophora* spp. *Fonsecaea* spp. *Bipolaris* spp. *Curvularia* spp. *Alternaria* spp. *Ochroconis gallapova*
Dimorphic fungi	*Histoplasma capsulatum* *Blastomyces dermatitidis* *Coccidioides* spp. *Paracoccidioides* spp. *Talaromyces marneffei* *Sporothrix* spp.
Coelomycetes (also cause phaeohyphomycosis)	*Pyrenochaeta* spp. *Phoma* spp. *Natrassia mangifera* *Colletotrichum* spp.
Fungus-like organisms	*Pneumocystis jirovecii* (placed in *Fungi*) *Pythium insidiosum* (placed in *Chromista*) *Rhinosporidium seeberi* *Lacazia loboi*

One-Fungus One-Name Concept Fungi with known sexual stage have two names, one for anamorph (asexual) and other for teleomorph (sexual) stage. For example, the anamorph *Cryptococcus neoformans* is also known as *Filobasidiella neoformans*, which is its teleomorph stage. This was convenient as long as microscopy was the basis of classification, though it remained confusing. With the advent of genotypic methods, the dual nomenclature of a single organism is neither necessary nor relevant. The Amsterdam Declaration on Fungal Nomenclature recommended the abolition of Article 59 of the Code of Botanical Nomenclature, the provision that sanctioned multiple names for the same fungus, and a new Code of Nomenclature for Algae, Fungi and Plants was created from 1 January 2013, which does not permit dual nomenclature, advocating the one-fungus one-name concept [6]. Anatomical classification of fungi into *Coelomycetes*, *Hyphomyctes*, Zygomycetes have all become redundant. Another advantage of the phylogenetic approach is that close relatives come together even if they may be morphologically quite different. This may be useful for predictions of pathogenicity or antifungal susceptibility. Conversely, distant relationships are expected to predict large differences in clinically relevant parameters. For example all *Candida* species have been grouped together due to morphological and physiological similarities but have huge evolutionary distance responsible for varying antifungal susceptibilities [7]. However, approximately 2000 genus and 10,000 species names need to be reassessed, though not all names will necessarily change. Hopefully medically important species names will be conserved like *Cryptococcus neoformans*. But it will take many years, cost huge money, as all databases would require validation and updating. Many gaps still exist in the understanding of the 'real' phylogenetic tree of the Fungi, and the nomenclature adaptation will be a continuous process with more understanding of taxonomy. Newer techniques like whole genome sequencing have made the classification of fungi at more stable platform.

1.5 Epidemiology of Fungal Infections

1.5.1 Global Burden of Disease

Burden of fungal infections on human health and economy is often overlooked as is exemplified by the fact that the World Health Organization (WHO) has no program on fungal infections. Estimates of global burden are often extrapolations of data available from limited single centre studies or sentinel surveillance sites. Superficial fungal infections constitute the major share of all fungal infections. It is known that about 25% of the worldwide population suffers from fungal infections of the skin and nails at any given point of their lifetime [8]. These infections refer to conditions commonly known as athletes' foot, ringworm, onychomycosis, etc. caused by dermatophytes and *Candida* species. Superficial acute and chronic mucosal infections of the gastrointestinal and genital tracts are also common. It is known that about 50–75% of all women in the reproductive age group suffer from vulvovaginal candidiasis

(VVC) and about 492 million women have recurrent VVC (RVVC) in their lifetime [9, 10]. With the advent of HIV, the numbers of oral and oesophageal candidiasis have gone up remarkably. While these conditions lead to personal and social stigmata, they are usually manageable in the long term with appropriate therapy.

The challenge due to invasive fungal infections is gaining more importance. The global burden of top ten deep seated fungal infections is given in Table 1.2 (modified from [11, 12]). The top five killers among fungi are *Candida*, *Aspergillus*, *Cryptococcus*, *Mucorales* and *Pneumocystis jiroveci* with an average mortality associated with these conditions being around 50%. The predominant pathogen in fungal sepsis is *Candida* and is responsible for 80% of all nosocomial fungal infections and 15% of nosocomial infections overall [13]. With the advent of fluconazole prophylaxis, the rise in incidence of candidemia has been checked considerably in the USA. However non-albicans *Candida* species with inherent resistance to fluconazole are on the rise. This is especially true for *C. glabrata* and *C. krusei* [14]. In many regions of the world, *C. albicans* candidemia still predominates, but the number of cases with non-albicans *Candida* is increasing. *C. tropicalis* or *C. parapsilosis* are the second leading cause rather than *C. glabrata* in developing countries. In tropical region of Asia, *C. tropicalis* is the most common agent and *C. albicans* has dropped to third or fourth in frequency. The true burden of fungal infections is not known and underestimated in developing countries. There are several conditions that allow these infections to thrive in tropical climate: the large economically deprived populations with malnutrition, limited access to health care, improper antibiotic stewardship practices and steroid use and over the counter availability of these drugs add to this problem. Natural disasters like tsunami, tornado, earthquake, etc., predispose to large land upheavals leading to high spore counts in the atmosphere and implantation of saprophytic fungi from soil due to injuries during disasters.

1.5.2 Predisposing Factors for Fungal Infections

The common risk factors for most invasive fungal infections include:

- Neutropenia
- Cytotoxic chemotherapy
- Immunosuppressive therapy for autoimmune disorders
- Haematological malignancies
- Cancer chemotherapy
- HIV—AIDS
- Haematological and solid organ transplantation
- Extremes of age
- Diabetes mellitus
- Extensive trauma and burns
- Prolonged antibiotic therapy
- Total parenteral nutrition
- Central vascular access devices

Table 1.2 Morbidity and mortality of top ten significant fungal infections (adapted from Brown et al. [11] and Vallabhaneni et al. [12])

Opportunistic mycosis (organism)	Location	Life-threatening new infections per year	Mortality (%)	Comments
Candidemia (*Candida* species)	Worldwide	>400,000	46–75	Incidence of 4–14 per 100,000 population
Aspergillosis (*Aspergillus* species)	Worldwide	>200,000	30–95	Annual incidence in transplant recipients is 0.65%, second only to candidiasis
Mucormycosis (*Rhizopus, Lichtheimia, Rhizomucor, Mucor, Apophysomyces*, etc.)	Worldwide	>10,000	30–90	Mucormycosis in India is very high in number among the diabetics, which is predicted at 70 times the rate in other countries
Pneumocystosis (*Pneumocystis jirovecii*)	Worldwide	>400,000	20–30	Opportunistic infections mainly in HIV but also in transplant recipients, malignancies, etc.
Cryptococcosis (*Cryptococcus* species)	Worldwide	>1,000,000	20–70	Prevalence is high in HIV infection with 700,000 cases annually in sub-Saharan Africa. *C. gattii* commonly infects immunocompetent hosts
Endemic mycosis				
Histoplasmosis (*Histoplasma capsulatum*)	Worldwide distribution, the endemicity demarcates well in Midwestern United States	~25,000	28–50	African histoplasmosis (*H. capsulatum var. duboisii*) is restricted to Africa and Madagascar
Coccidioidomycosis (*Coccidioides immitis, C. posadasii*)	Southwestern USA, Mexico, some parts of Central and South America	~25,000	<1–70	The numbers increase during earthquake in endemic zone
Blastomycosis (*Blastomyces dermatitidis*)	Midwestern and Atlantic USA	~3000	<2–68	Also reported in Canada. Incidence of 0.3–41 cases per 100,000 population. In Asia, reported from India
Paracoccidioidomycosis (*Paracoccidioides brasiliensis*)	South of Mexico to Argentina, high number in Brazil	~3000	5–27	Incidence of 0.2–2 per 100,000 population in endemic area
Penicilliosis (*Talaromyces marneffei*)	Southeast Asia	>8000	2–75	The number of cases decreased after introduction of anti-retroviral therapy

1.5.3 Geographical Distribution

The distribution of majority of opportunistic fungal infections is independent of geography as they are present ubiquitously in the environment or are commensals in humans. However, due to socioeconomic diversity and better control of lifestyle related diseases like diabetes in the developed world, some infections are more common in Southeast Asia, Africa and Latin America. The incidence of mucormycosis is highest in India especially in uncontrolled diabetic patients and has been reported at 1.6 cases per 1000 consecutive diabetics [15]. Human pythiosis due to opportunistic oomycete *Pythium insidiosum* remains almost exclusive to Thailand and neighboring Southeast Asian countries [16]. Trichosporonosis in Japan and south Asian countries, scedosporiosis in Australia and Spain, fusariosis in Brazil, phaeohyphomycosis due to *Rhinocladiella mackenziei* in the Middle East Asian region are prevalent opportunistic fungal infections in the respective regions [17–19]. Neglected tropical disease like mycetoma is prevalent in Sudan, India and a few Latin American countries possibly due to life style and existence of causative agents in the environment [20]. Endemic mycoses, as the name suggest are endemic with restricted geographic distribution (see Table 1.2).

1.5.4 Community Acquired Fungal Infections

Apart from endemic fungi and superficial infections, community acquired fungal infections are most commonly seen in immunocompromised individuals and are frequently opportunistic in nature. One of the common opportunistic infections seen mainly in HIV patients is cryptococcal meningitis. This is common among developing countries of Africa and Southeast Asia which parallels the burden of HIV. *Pneumocystis jirovecii* is a pathogen that causes opportunistic pneumonia in HIV infected individuals. Community acquired candidemia is rare. However according to the SENTRY surveillance data, about a third of the candidemia in the US is community acquired. This is attributable to the increasing preference for home based health care in developed countries [21]. When immunosuppressed patients undertake gardening without protection, they may acquire *Aspergillus* and many other saprophytic fungi from environment. Construction activities and natural disasters predispose individuals to aspergillosis, mucormycosis and invasive mould infections especially when the individual is immunosuppressed.

1.5.5 Nosocomial Fungal Infections

Though other than dermatophytes and *Malassezia* infection, fungal infections are considered non-contagious, nosocomial transmission of fungal infections is not rare. *Candida* and related yeasts may spread from the hands of health-care providers to the patients. Several outbreaks have been reported in the hospitals across the world.

In a single outbreak in India 379 babies suffered from *Pichia anomala* fungaemia [22, 23]. Analysis of the outbreak revealed that the transmission of this agent from the index case to the subsequent cases occurred through the hands of a resident doctor. Fungi from environment also get access to the patients due to breach of physical barrier (intravascular catheters) or through respiratory route. The most recent nosocomial outbreak involving contaminated methylprednisolone injections, which were directly inoculated intraspinally, led to an outbreak of fungal meningitis claiming over two dozen lives in USA. The agent identified in this outbreak was *Exserohilum rostratum*, an extremely rare black fungus causing fungal meningitis [24, 25].

1.6 Fungal Infection Immunity and Pathogenesis

Fungi are present ubiquitously in the environment. Some may reside on human body as resident or transient commensals. In order to co-exist, the human immune response strikes a balance between invading fungi and host resistance. When the balance is lost, fungal infections occur. Both innate and adaptive immunity play a role in protection against fungal infections.

1.6.1 Innate Immunity

Primary immunity to fungal infections is provided by barriers—both skin and mucosal barriers of respiratory, gastrointestinal and genitourinary epithelia, which are continually in contact with fungi [26]. Cells of innate immunity such as monocytes, macrophages, neutrophils and even epithelial and endothelial cells contribute to fungal clearance by phagocytosis aided by opsonic recognition of cell wall components by defensins, collectins and complement. Neutropenia is one of the most common underlying factors for development of invasive candidiasis, aspergillosis and mucormycosis. The functional neutropenia seen in diabetic ketoacidosis underlies development of mucormycosis prevalent in India.

Host cells express pathogen recognition receptors (PRRs) that specifically recognize pathogen associated molecular patterns (PAMPs) and initiate downstream signalling pathways. The most important PRRs in antifungal immunity are toll-like receptors (TLRs) and C-type lectin receptors (CLRs). Of these, Dectin 1, TLR 2 and 4 play important role in fungal infections (Table 1.4).

1.6.2 Adaptive Immunity

Although both cellular and humoral arms are involved in immune response to fungi, it is the T cell-mediated immunity that is mainly protective (Table 1.3). The capacity of dendritic cells (DCs) to initiate different adaptive antifungal immune responses

Table 1.3 T cell Subsets in Antifungal Immunity

T cell subset	Function	Key cytokines	Susceptibility in deficiency
Th1	Protective	IFNγ	Overwhelming opportunistic fungal infections
Th2	Dampens Th1, promotes fungal allergy and relapses, role in antibody production	IL-4, IL-5, IL-13	Th2 reactivity- allergic broncho-pulmonary aspergillosis, allergic fungal rhinosinusitis
Th17	Promotes Th1 and depresses Th2	IL-17A, IL-22	Th17 hyperactivity-role in chronic fungal infections
Treg	Regulation of other Tcell subsets	TGFβ, IL-10	Fungal persistence to immunosuppression

depends on balanced activation of DC subsets. Inflammatory DCs initiate antifungal Th17 and Th2 cell responses through signalling pathways involving the TLR adaptor MYD88, whereas tolerogenic DCs activate Th1 and regulatory T (TReg) cell differentiation through TRIF (TIR domain-containing adaptor protein inducing IFNβ). The multiple, functionally distinct, receptor signalling pathways in DCs ultimately affect the balance between CD4+ effector T cells and Treg cells and thus are likely to be exploited by fungi to enable them to establish commensalism or infection.

1.6.3 Fungal Virulence

Several mechanisms are involved in fungal pathogenesis. Fungi produce several virulence factors involved in adhesion, tissue invasion and evasion of immune response. The tough fungal cell wall composed of various polysaccharides has many roles to play (Table 1.4) (Fig. 1.2). Primarily it protects the organism from external environmental stresses and also resists turgor pressure generated inside the cell during hyphal growth. Melanin of many pigmented fungi is involved in evasion of immune response by preventing opsonization and binding of C3 to the fungal cell wall, as well as recognition by PRRs [27]. Chitin blocks dectin 1 and the subsequent cytokine production [28].

Tissue invasion is assisted by formation of pseudohyphae and true hyphae in tissues. Elaboration of phospholipases, extracellular proteases such as secreted aspartyl proteinases (SAPs) and other hydrolytic enzymes are important especially for *Candida* species [29]. Rapid growth, angioinvasion and heat tolerance are important features for *Aspergillus* and mucormycosis agents. Cot III (invasin) has been recognized as important virulence marker of *Mucorales*, which act as a ligand for endothelial cell glucose-regulated protein 78 (GRP78) receptor causing angioinvasion. Other specific virulence mechanisms are shown in Table 1.5.

Table 1.4 Important PAMPs and their effects on host immunity

Pathogen associated molecular pattern (PAMP)	Pattern recognition receptor (PRR)	Cytokine signalling/ effect	Cytokines activated	Fungal infection related to deficiency
β-glucans	Dectin-1	CARD9 pathway and RAF pathway	NLRP3 inflammasome-IL-1β, IL-18, ROS	Chronic mucocutaneous candidiasis, invasive aspergillosis
High mannose structures—hyphae > yeasts	Dectin-2	CARD 9 pathway	NLRP3 inflammasome-IL-1β, IL-18, ROS	*C. albicans*, recurrent vulvovaginal candidiasis
Damaged cells, polysaccharides of *Malassezia, Candida*	Mincle	CARD 9 pathway	NLRP3 inflammasome-IL-1β, IL-18, ROS	*C. albicans*, recurrent vulvovaginal candidiasis
N-linked mannans	DC-SIGN	DC endocytic pathway	IL-10, activation of T cell subsets	
N-linked mannans, α-glucans, chitin	Mannose receptor	DC endocytic pathway	Activation of T cell subsets including TH17	*cryptococcosis*
Zymosan, phospholipomannan, *O*-linked mannans and fungal DNA	TLRs- 2,4 and 9	MyD88 pathway	Antigen presentation by DCs and Tcell activation	TLR4- pulmonary aspergillosis, candidemia TLR9- ABPA

1.6.4 Tissue Reaction

Tissue reaction to various fungi depends on the infecting fungus as well as the immune status of the host [30]. It ranges from suppurative changes in acute infections to granulomatous and fibrotic changes in subacute to chronic infections (Table 1.5). Broadly, yeasts or moulds can be visualized on histopathology. Some useful pointers are shown in Table 1.5.

1.7 Diagnosis of Fungal Infections

Superficial infections can be diagnosed easily in the outpatient department as they often produce characteristic lesions. A Wood's lamp examination helps in diagnosis of dermatophytosis and a potassium hydroxide (KOH) mount of skin scrapings helps to identify the hyphal forms. Calcofluor white fluorescent stain improves sensitivity of direct microscopy. The challenge lies in the diagnosis of deep invasive infections whose clinical manifestations are non-specific and require a high index of clinical suspicion. The setting of infection, the immune status of the host, residence,

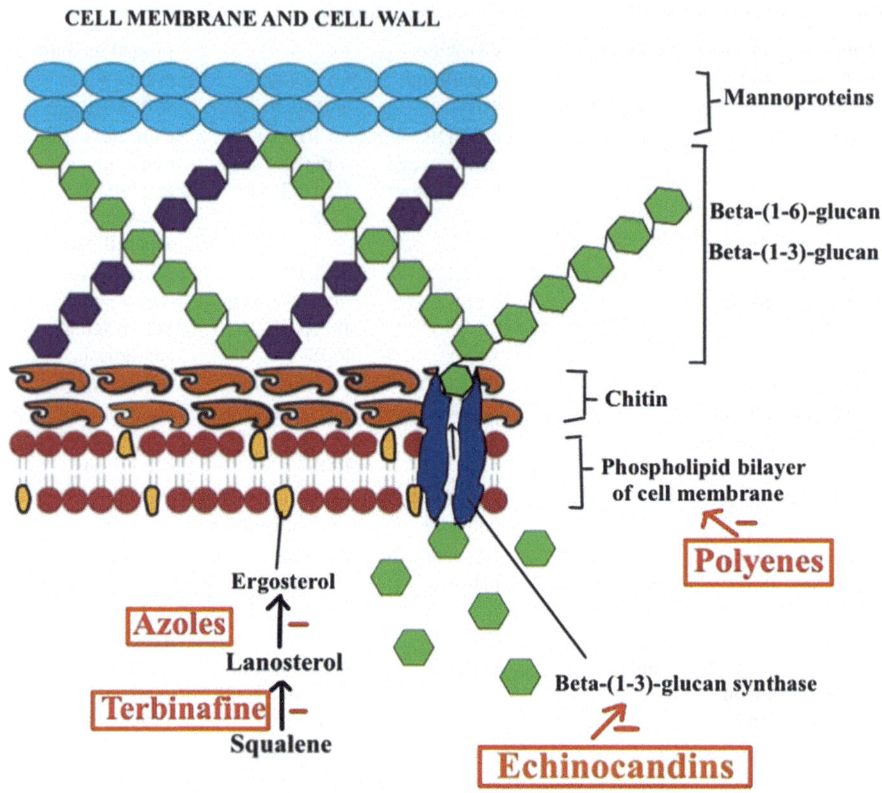

Fig. 1.2 Schematic diagram of the fungal cell wall

travel and occupational history are clinical pointers to the type of infection. Available diagnostic tests are far from perfect and the gold standard tests lack sensitivity and are of long turn-around times. Sample collection from deep sites of infection is also difficult due to thrombocytopenia in these patients. Though European Organization for Research and Treatment of Cancer/Invasive Fungal Infections Cooperative Group and the National Institute of Allergy and Infectious Diseases Mycoses Study Group (EORTC/MSG) a consensus definition has been laid down criteria for proven, probable and possible invasive fungal infections to serve as case definitions in clinical trials [31]. Many clinicians use the same criteria to diagnose invasive fungal infections in their routine clinical practice. Briefly, a proven IFD is defined as histopathological, cytopathological or direct microscopic evidence or culture of yeasts or moulds in a specimen derived from a sterile site. In the absence of these diagnostic criteria, other predisposing host factors, clinical clues and indirect evidences in the form of imaging, serological tests and biomarkers are used to define a probable or possible IFD.

Table 1.5 Pathology of common fungal infections

Organism	Clinical features	Virulence	Host tissue reaction	Morphology on H&E, PAS and GMS
Yeasts				
Candida	Superficial infections to invasive health care associated disease	Adherence and biofilm formation, tissue invasion using secreted aspartyl proteinases, phospholipases. Yeast to hyphal switch	Suppurative inflammation, angioinvasion, necrotizing vasculitis	Small yeasts of 3–5 μm size with or without pseudohyphae and occasionally true hyphae
Cryptococcus	Pneumonia, pleural effusion, chronic meningitis, cryptococcomas	Polysaccharide (glucuronoxylomannan) capsule evades immune response, melanin	Granulomas in lung and skin, various inflammatory reactions depending on immune status, mucinous and occasionally fibrotic granuloma	Narrow based budding yeasts of various sizes between 4 and 10 μm with thick capsule
Dimorphic fungi				
Histoplasma	Acute pneumonia, chronic pneumonia, mediastinitis, disseminated disease	Dimorphism, endotoxin-like substance, collagenases and elastases, siderophores and ability to neutralize acidic pH of macrophages	Nodules with vascular necrosis and lympho-histiocytic vasculitis to granulomatous inflammation with abundant intracellular yeasts	Small yeasts (2–4 μm) with narrow-based budding grouped in clusters inside macrophages
Blastomyces	Acute and chronic pneumonia or disseminated and cutaneous disease	Dimorphism, Blastomyces adhesion (BAD1), α-glucan	Mixed suppurative and granulomatous inflammation	Broad-based budding yeasts (10–15 μm)
Coccidioides	Acute and chronic pneumonia or disseminated and cutaneous disease	Dimorphism	Mixed suppurative (including eosinophils) and granulomatous inflammation with a rim of lymphocytes, Splendore-Hoeppli pheromenon likely	Spherules with multiple endospores of 10–100 μm
Paracoccidioides	Acute and chronic pneumonia or disseminated and cutaneous disease	Dimorphism, α-glucan, chitin	Mixed suppurative and epitheloid granulomatous inflammation with or without fibrosis	Multiple-budding yeast forms, 4–60 μm with pilot wheel appearance

(continued)

Table 1.5 (continued)

Organism	Clinical features	Virulence	Host tissue reaction	Morphology on H&E, PAS and GMS
Talaromyces (Penicillium) marneffei	Cutaneous to disseminated disease	Dimorphism, acid phosphatases, survival in acidic pH of macrophages	Suppurative inflammation with infected macrophages and granulomas	Small yeasts (2–4 μm) with transverse septations inside macrophages
Sporothrix	Cutaneous, lympho-cutaneous and occasionally disseminated disease	Cell wall components	Mixed suppurative (including eosinophils) and granulomatous inflammation, Splendore-Hoeppliphenomenon, asteroid bodies, epidermis with pseudoepitheliomatous hyperplasia	Asteroid bodies surround small round, oval to cigar shaped yeast cells 2 to 6 μm in size with narrow based budding
Moulds				
Aspergillus	Allergic bronchopulmonary aspergillosis and allergic fungal rhinosinusitis	Thermal tolerance, siderophore production, ability to survive in oxygen deprived environments, elaboration of proteases	Eosionphilic mucus, Curshmann's spirals, Charcot-Leyden crystals	Hyaline, septated hyphae with acute angle branching
	Chronic and necrotising pulmonary aspergillosis		Fungus ball with surrounding fibrosis, necrosis, granulation tissue, granulomatous inflammation	
	Invasive disease		Angioinvasion by hyphae with consequent necrosis or hemorrhage of surrounding tissue	
Fusarium, Scedosporium	Locally invasive and disseminated disease	Thermal tolerance, proteases	Angioinvasion by hyphae with consequent necrosis or hemorrhage of surrounding tissue	Hyaline, septated hyphae with acute angle branching[a]
Mucorales	Cutaneous, rhinocerebral, pulmonary, gastrointestinal and disseminated	Thermal tolerance, siderophore production, rapid growth	Angioinvasion by hyphae with consequent necrosis or hemorrhage of surrounding tissue; inflammation is suppurative more than granulomatous	Hyaline, aseptate to sparsely septate ribbon-like hyphae with right-angle branching

Entomophthorales	Muco-cutaneous, subcutaneous and occasionally gastrointestinal tract disease	Proteases and lipases	Fibrosis, granulation tissue, mixed eosinophilic and granulomatous inflammation; Splendore-Hoeppli phenomenon	Hyaline, aseptate to sparsely septate ribbon-like hyphae with right-angle branching
Dematiaceous fungi such as *Cladophialophora*, *Fonsecaea*, *Madurella*, *Curvularia*, etc.	Superficial and deep subcutaneous infections and invasive disease	Melanin	Mixed suppurative and granulomatous inflammation with pseudo-epitheliomatous hyperplasia and draining sinuses	Pigmented irregular hyphae and yeast-like swellings with septations
Others				
Pneumocystis	Atypical pneumonia to occasional disseminated infections	not clearly known	Minimal reaction; rarely fibrosis or granuloma, foamy eosinophilic exudates	Round cysts with GMS staining only the wall, seen in groups
Rhinosporidium seeberi	Nose, nasopharynx	not clearly known	Granulomatous inflammation with fibrosis	Immature trophocytes are spherical measuring 10-100μm whereas mature sporangia are 100-350 μm in diameter, containing many spores
Lacazia loboi	Cutaneous, disseminated disease	not clearly known	Granulomatous inflammation constituted by a dense histiocytic infiltrate with multiple epithelioid and multinucleated giant cells with accentuated fibrosis. Asteroid bodies may be seen	Lemon-shaped budding yeast-like organisms measuring 6–12 μm

[a]At times, *Fusarium* and *Paecilomyces* exhibit adventitious sporulation that may be seen in histologic sections and blood culture positive broths as yeast-like structures which promote their dissemination in blood

1.7.1 Radiological Imaging

Imaging plays a crucial role in early detection of signs attributable to fungal infections. Imaging of lungs, paranasal sinuses, central nervous system (CNS) provide clues for possible fungal infections. Chest radiographs may be normal or non-specific in non-neutropenic patients with pulmonary aspergillosis, however the presence of nodules may be suggestive [32]. *Pneumocystis* pneumonia can seem normal at presentation, followed by typical bilateral, perihilar, diffuse granular, or hazy (ground-glass) opacification, which becomes dense with worsening infection, progressing to areas of consolidation [33].

The computed tomography (CT), multidetector CT (MDCT) and high-resolution CT (HRCT) are preferred imaging techniques. In neutropenic patients, the characteristic halo sign is transitory and an early feature of invasive aspergillosis but more specifically nodules greater than 1 cm suggest fungal infection over bacterial or viral [34]. CT-Angiography may improve the specificity for fungal infections. Air crescent sign appears late in invasive aspergillosis. Reverse halo sign or multiple (>10) nodules suggest mucormycosis. CT and MRI are also crucial for diagnosis of invasive sinusitis and CNS infection. Recently antibody guided PET-CT/MR provides better sensitivity and specificity of fungal diagnosis. While radiology provides room for suspicion of IFD, the changes are still non-specific and need confirmation with mycological diagnosis.

1.7.2 Conventional and Culture Based Diagnosis

Diagnosis of IFDs often requires a biopsy or aspirates from sterile body sites and blood. Haematoxylin and eosin staining is useful for evaluating tissue reaction whereas fungus specific stains such as Gomorri's or Grocott'smethenamine silver (GMS), periodic acid Schiff (PAS) or Giemsa are often required for morphological diagnosis of infecting fungi (Table 1.5).Often, the biopsy material may land only in the histopathology laboratory in 10% formol saline, as the clinician had not suspected a fungal infection. In such situations, an attempt may be made to extract DNA from tissue and identify the pathogen by sequencing of the PCR amplified DNA product.

For samples from non-sterile sites, it is difficult to separate colonizers from true infections. This is especially true for genitourinary and respiratory specimens where *Candida* and occasionally filamentous fungi may be present as commensals. There is no accepted cut-off for candiduria on quantitative cultures although several have been proposed [35]. Conventional blood culture yields for candidemia are typically under 50%. Automated blood cultures have greatly improved this scenario with the availability of specialized containers for fungal blood cultures increasing yield [36]. The volume of blood cultured determines sensitivity. At least 10 ml of blood should be collected from central and peripheral line separately to improve sensitivity and to

Table 1.6 Identification of fungi from positive cultures

Method	Time required	Comment
Non-commercial conventional	3–4 days	Most laboratories in developing countries perform this technique
Commercial phenotypic	1–3 days	Accurate identification depends on database
PNA-FISH	90 min	Mainly *C. albicans, C. glabrata*
Sequencing—ITS barcoding and other conserved sequences	1 day	Best method for identification, but technically challenging
MALDI-TOF	Few minutes	Future for identification

establish central line related infections. Lysis-centrifugation (Isolator) tubes increase yield at the cost of introduction of possible contamination during processing. However, sensitivity of blood culture still remains around 50–60%. Moreover, the average time to detection for these systems ranges from 14 to 38 h and may take up to 72 h (Table 1.6). This is unacceptable considering the rate at which mortality increases with delay in start of therapy for candidemia. Recent technique T2 magnetic resonance enabled nano-particle-mediated rapid diagnosis is promising, as it can detect candidemia within 3–4 h compared to 2.6 days by BACTEC or BacT/ALERT [37]. However, this technique can only detect five common *Candida* species and cannot perform antifungal susceptibility testing.

1.7.3 Sepsis Biomarkers

Commonly used biomarkers of sepsis may be used to distinguish candidemia and bacterial sepsis, as the levels do not rise in fungal infections. A low procalcitonin (PCT) value may be interpreted as high likelihood for IFI in critically ill patients with clinical symptoms of sepsis, though some clinicians may not agree for the same. Fungaemia, even severe sepsis or septic shock, does not necessarily elicit a substantial increase in serum PCT. In a meta-analysis of utility of PCT for diagnosis of fungal infections, it was found that he PCT value was markedly lower in the fungal infection group (range, 0.69–1.23) than in the bacteraemia group (range, 4.18–12.9) at the onset of fever [38]. The same analysis found C-reactive protein to be neither sensitive nor specific for fungal infections.

1.7.4 Serology

Several serological tests have been validated and introduced into routine clinical practice. They are useful compared to culture in that the turnaround time is few hours, and they are positive weeks before culture positivity. The β-D-glucan (BDG) assay is a useful adjunct to culture for diagnosis of most of the invasive fungal infections including *Pneumocystis*. The BDG is a major cell wall component of fungi found in sera of patients with IFDs. It is negative in mucormycosis and yeast stage of blastomycosis. It may also be negative in cryptococcosis. BDG can be measured in serum prior to clinical symptoms and facilitates early initiation of antifungal therapy. Serial measurements are associated with improved sensitivities and predict response to therapy. However cut-offs for measuring BDG still need evaluation and standardization. It is associated with false positivity especially in patients with haematological malignancies who may be heavily colonized by many fungi [39]. Dialysis membranes, *Pseudomonas* bacteraemia among other bacteraemias may also lead to a false positive BDG. BDG in respiratory samples, though appears more sensitive, requires standardization. The test is expensive and can be performed by highly skilled personnel in specialized laboratories only.

Galactomannan (GM) measurement is another useful tool for early detection of invasive aspergillosis (Table 1.7). It is US FDA approved for serum and brochoalveolar lavage (BAL) samples for the presumptive diagnosis of invasive aspergillosis. It has good positive and negative predictive values in haematology–oncology units where patients are frequently neutropenic. Sensitivity is lower in non-neutropenic patients. GM test also suffers from limitations of cost and high turnaround time. Other available serological techniques are presented in Table 1.7.

1.7.5 Molecular Detection

Molecular detection directly from samples or cultures is challenging mainly due to the thick cell wall of fungi. Extraction of nucleic acids and intrinsic proteins often require specialized harsh methods for cell wall lysis, distinct from viruses or bacteria. The fungal burden may be less per gram of tissue or fluid and thereby lower DNA quantities, requiring larger volumes of samples for DNA extraction. Environmental contamination can lead to false positive amplification of fungi. PCR is also hindered by presence of inhibitors such as heparin, haemoglobin and lactoferrin. Contamination is often difficult to avoid as about 20% of commercially available blood collection tubes may have fungal DNA [41]. Also there is lack of standardization as to which component of blood to use for maximal fungal recovery. European *Aspergillus* PCR Initiative standardized a test protocol in a multicentric study. It recommends large volume of sample should be used with small elution volume, and mechanical breakage rather than enzymatic disruption should be practiced for extraction of DNA. Plasma is best sample though serum and whole blood may also be used [42].

As fungi are eukaryotic, target selection for PCR amplification needs to be distinct from human genomic DNA and preferably in multiple copies due to low fungal burden. The usual target for nucleic acid amplification are 18S, 28S and internal transcribed spacer (ITS) region. Nested PCRs are preferred, as they increase specificity of target amplified. However there is a chance of introducing contamination. Use of panfungal primers with post-amplification processing for identification by

Table 1.7 Serological diagnosis of fungal infections

Technique	Antigen/Antibody	Organisms	Sensitivity (%)	Specificity (%)	Comments
Latex agglutination	Glucuronoxylomannan antigen detection	*Cryptococcus*	97–100	97–100	False positives are seen in trichosporonosis and *Capnocytophaga* infections
ELISA	Galactomannan antigen	*Aspergillus*	72–90	84–92	Detectable GM precedes clinical infection. BAL GM precedes serum GM. Useful in monitoring response to therapy
	Antigen	*Cryptococcus*	97–100	97–100	
		Histoplasma	87–94	88–92	Antigenuria more sensitive than antigenaemia
		Blastomyces	90–93	90–93	
		Candida (mannan)	58–60	90–95	
	Antibodies	*Coccidioides, Blastomyces, Histoplasma, Candida* (antimannan)	88–100	95–99	
		Sporothrix [40]	89	82	*In-house* ELISA using purified cell wall antigen SsCBF (*S. schenckii* ConA binding fraction)
Immunodiffusion	Antibodies	*Histoplasma* and *Blastomyces*	60–70	100	
Complement fixation	Antibodies	*Histoplasma, Coccidioides*and *Blastomyces*	45–70	100	*Histoplasma* antibodies persist for years after infection. *Coccidioides* antibody titre used to monitor treatment response

(continued)

Table 1.7 (continued)

Technique	Antigen/Antibody	Organisms	Sensitivity (%)	Specificity (%)	Comments
Lateral flow assay	Antigens	*Cryptococcus*	97–100	97–100	More sensitive than presently used tests, can be used in field, takes less time
		Aspergillus	48–100	100	Still presents lot of variability, needs standardization, respiratory samples may be better
Immunofluorescence	Antigen	*Candida* germ tube assay (CAGTA)	77–89	91–100	
Beta-glucan test	All fungi except *Mucorales* and *Cryptococcus*	BDG assay	77–81	87–94	Apart from serum, shown to be highly specific in CSF for fungal meningitis and in BAL for pneumocystosis

sequencing or hybridization provides highly specific results. Quantitative real time PCRs are required for differentiation of colonization and infection especially in case of *Pneumocystis jirovecii* pneumonia.

A variety of other molecular methods have been studied. Fluorescence in situ hybridization (FISH) is a technique that uses fluorescent probes to identify target areas on the genomes of microbial pathogens in human samples, which can then be detected by fluorescence microscopy. This method has been proven to be accurate for the identification of *Candida* spp. infections from blood culture bottles. Nucleic acid sequence-based amplification (NASBA) is a method that amplifies mRNA using an RNA polymerase and is isothermal. A rapid, affordable loop-mediated isothermal amplification (LAMP) assay has been validated for *Histoplasma* with high sensitivity and specificity in urine specimens [43].

MALDI-TOF MS, is based on mass spectrometry to identify the intrinsic ribosomal protein fingerprints of different microorganisms. By direct comparison of the spectral pattern of the organism with databases of known patterns from different microorganisms, it is possible to identify the microbe at the genus, species and even strain level. It is a rapid and cost effective method in comparison with nucleic acid based techniques.

1.8 Treatment

Superficial infections are usually treatable with long-term topical and oral antifungals in patients with extensive or recurrent disease. Disseminated or invasive infections require systemic antifungal therapy. Delay in initiation of therapy is associated with significant mortality risks. According to one study, there is a 20% increase in mortality if empiric therapy is initiated beyond 12 h after the first blood culture sample is drawn [44]. Choice of empiric therapy is also challenging as the spectrum of agents in particular cohort is changing and several species acquire resistance or have intrinsic resistance profiles. Currently available systemic and important topical antifungals are described in detail in Chap. 2. Strategies for therapy based on available evidence of fungal infection are presented in Table 1.8.

Choice of antifungal therapy depends not only on the species but also formulations available, toxicity profiles and pharmacokinetic parameters. Lipid formulations of amphotericin B allow higher dosage of antifungal agent with lesser toxicity, but impose significantly higher costs. Therapeutic drug monitoring is recommended especially for voriconazole, itraconazole and posaconazole to ensure adequate plasma levels for efficacy and to avoid toxicity, and for flucytosine to monitor toxicity.

Antifungal resistance in *Candida* and mycelial fungi is an emerging problem. Fluconazole resistance in *Candida albicans*, though rare, is increasing in certain reports. The problem is more severe for non-albicans *Candida* species where fluconazole resistance has been reported between 5 and 65% [46]. This is a serious problem especially in countries where fluconazole is the only drug available to treat

Table 1.8 Strategies of antifungal therapy

Strategy of therapy	Definition	Recommended population	Drugs used	Comments/Pitfalls
Prophylaxis	Generally in patients at high risk (>10%) of invasive fungal infections before developing signs and symptoms	Patients with haematological cancer, recipients of haematopoietic stem cell transplantation and other immunosuppressed states; intraabdominal candidiasis especially in patients with acute pancreatitis or repeated bowel leak	The IDSA recommends fluconazole prophylaxis when invasive candidiasis rate exceeds 5% in the ICU [45]. Fluconazole for patients with AML, MDS and recipients of SCT.	Fluconazole prophylaxis leads to emergence of non-albicans *Candida* species such as *C. krusei* and *C. glabrata* with high-level resistance to fluconazole.
			Voriconazole or posaconazole for invasive aspergillosis. Posaconazole recommended in haematological malignancies and myelo-dysplastic syndrome	Voriconazole treatment may lead to breakthrough mucormycosis
Empiric	Patients at high risk of IFI with established clinical signs and symptoms of infection, but without a known source	Patients with acute leukaemia or myelodysplastic syndrome during remission induction chemotherapy with prolonged neutropenia	L-AmB at 3 mg/kg OD i.v. if candidemia suspected, caspofungin at 70 mg OD loading dose and 50 mg OD maintenance or voriconazole if aspergillosis suspected	
Pre-emptive	Patients at high risk of IFI with radiographic signs and/or biomarker positivity suggesting IFI, without definitive histopathological and/or cultural pathogen identification	Candida colonization scores, fungal beta-D glucan	Echinocandins	
		Positive serum/BAL galactomannan/beta-D glucan, halo sign or radiological changes in chest	Voriconazole	
Targeted	Diagnostic criteria permit pathogen identification	Proven invasive fungal disease by microscopy/culture/histopathology	Specific therapy based on identification and susceptibility	Therapy based on histopathological report may also lead to inappropriate therapy for example of specimens showing hyaline septate hyphae presumed to be aspergillosis which may in fact turn out to be invasive fusariosis or scedosporiosis

Candida infections. Resistance to echinocandins, the most frequently used drugs for invasive candidiasis occurs to a much lesser degree, although 6–10% resistance in *C. glabrata* is seen in United States and Denmark. Azole resistance in *Aspergillus fumigatus* has recently gained importance especially in patients with chronic exposure to azoles such as for chronic pulmonary aspergillosis and allergic bronchopulmonary aspergillois or due to agricultural use of azoles. The rates are much higher in UK and Netherlands (around 10–15%), but around 0–5% in rest of the world.

Antifungal susceptibility testing of fungi is technically challenging and there is a lack of availability of interpretive breakpoints for majority of moulds and non-*Candida* yeast. Microbroth dilution MIC is the CLSI or EUCAST recommended method of choice for most organism-drug combinations, although disk diffusion can be done for few yeasts with azoles and echinocandin. Automated commercial systems [Sensititre Yeast one (Trek Diagnostic System, USA); ASTY micro-dilution panel (Kyokuto Pharamaceutical Industrial Co. Ltd, Japan); Fungitest, (Biorad); ATB Fungus 2 (bioMérieux), VITEK 2 (bioMérieux)] are available for performing susceptibility to yeasts. The commercial systems are based on agar diffusion principle or colorimetric method. The tests have reasonably good agreement (≥92%) with standard micro-broth dilution method for itraconazole, flucytosine, amphotericin B against yeasts, but not so good for fluconazole (82%), voriconazole (85%), caspofungin [46]. Categorical agreement is lowest for *C. glabrata* and *C. tropicalis,* and significant variation may be seen in the end point obtained.

1.9 Fungal Vaccines

While effective vaccines have been developed for viral and bacterial infections, no vaccine has received formal approval to use against fungal infections, though many researchers are working in this field. With improved understanding of the host–fungus relationships, and the spectacular advances in genomics and proteomics, considerable progress has been made in fungal vaccine research. Some vaccines induce the generation of directly fungicidal antibodies; others are protective in animals carrying major risk factors for fungal infections, such as CD4+ T cell-deficiency or neutropenia. A common immunological pathway studied includes activation of cell-based, pro-inflammatory, Th1 or Th17 immune response to improve phagocytic killing of fungi. Despite the advancement of research, the lack of complete understanding of the market potential for such vaccines has resulted in the non-availability of capital to develop these vaccines. Continued education about the economic importance of vaccines for invasive fungal infections, combined with continued development of well-defined antigens and effective adjuvants with a track record of safety, should enable such vaccines to enter clinical testing in the coming decade.

Acknowledgements The authors would like to thank Dr. Ajay Padmanaban for the illustrations.

References

1. Casadevall A (2012) Amoeba provide insight into the origin of virulence in pathogenic fungi. Adv Exp Med Biol 710:1–10
2. Kenyon C et al (2013) A dimorphic fungus causing disseminated infection in South Africa. N Engl J Med 369(15):1416–1424
3. Malik R et al (2016) Disseminated *Emmonsia pasteuriana* infection in India: a case report and a review. Mycoses 59(2):127–132
4. Whittaker RH (1969) New concepts of kingdoms of organisms. Science 163(3863):150–160
5. Ruggiero MA et al (2015) A higher level classification of all living organisms. PLoS One 10(4):1–60
6. De Hoog GS et al (2015) Name changes in medically important fungi and their implications for clinical practice. J Clin Microbiol 53(4):1056–1062
7. Schmalreck AF et al (2014) Phylogenetic relationships matter: antifungal susceptibility among clinically relevant yeasts. Antimicrob Agents Chemother 58(3):1575–1585
8. Havlickova B, Czaika VA, Friedrich M (2008) Epidemiological trends in skin mycoses worldwide. Mycoses 51(suppl 4):2–15
9. Sobel JD (2016) Recurrent vulvovaginal candidiasis. Am J Obstet Gynecol 214(1):15–21
10. Sobel JD (2007) Vulvovaginal candidosis. Lancet 369(9577):1961–1971
11. Brown GD et al (2012) Hidden killers: human fungal infections. Sci Transl Med 4(165):165rv13
12. Vallabhaneni S et al (2016) The global burden of fungal diseases. Infect Dis Clin N Am 30(1):1–11
13. Pfaller MA, Diekema DJ (2007) Epidemiology of invasive candidiasis: a persistent public health problem. Clin Microbiol Rev 20(1):133–163
14. Trick WE et al (2002) Secular trend of hospital-acquired candidemia among intensive care unit patients in the United States during 1989-1999. Clin Infect Dis 35(5):627–630
15. Bhansali A et al (2004) Presentation and outcome of rhino-orbital-cerebral mucormycosis in patients with diabetes. Postgrad Med J 80(949):670–674
16. Supabandhu J et al (2008) Isolation and identification of the human pathogen *Pythium insidiosum* from environmental samples collected in Thai agricultural areas. Med Mycol 46(1):41–52
17. Al-Tawfiq JA, Boukhamseen A (2011) Cerebral phaeohyphomycosis due to *Rhinocladiella mackenziei* (formerly *Ramichloridium mackenziei*): case presentation and literature review. J Infect Public Health 4(2):96–102
18. Nucci M et al (2013) Increased incidence of invasive fusariosis with cutaneous portal of entry, Brazil. Emerg Infect Dis 19(10):1567–1572
19. Slavin MA (2011) Scedosporiosis. Mycoses 54:26
20. Fahal AH, Sabaa AH (2010) Mycetoma in children in Sudan. Trans R Soc Trop Med Hyg 104(2):117–121
21. Pfaller MA et al (2011) *Candida* bloodstream infections: comparison of species distributions and antifungal resistance patterns in community-onset and nosocomial isolates in the SENTRY antimicrobial surveillance program, 2008-2009. Antimicrob Agents Chemother 55(2):561–566
22. Chakrabarti A et al (2001) Outbreak of *Pichia anomala* infection in the pediatric service of a tertiary-care center in northern India. J Clin Microbiol 39(5):1702–1706
23. Repetto EC, Giacomazzi CG, Castelli F (2012) Hospital-related outbreaks due to rare fungal pathogens: a review of the literature from 1990 to June 2011. Eur J Clin Microbiol Infect Dis 31(11):2897–2904
24. Kainer MA et al (2012) Fungal infections associated with contaminated methylprednisolone in Tennessee. N Engl J Med 367(23):2194–2203
25. Litvintseva AP et al (2014) Whole-genome analysis of *Exserohilum rostratum* from an outbreak of fungal meningitis and other infections. J Clin Microbiol 52(9):3216–3222
26. Romani L (2011) Immunity to fungal infections. Nat Rev Immunol 11(4):275–288

27. Aimanianda V et al (2009) Surface hydrophobin prevents immune recognition of airborne fungal spores. Nature 460(7259):1117–1121
28. Mora-Montes HM et al (2011) Recognition and blocking of innate immunity cells by *Candida albicans* chitin. Infect Immun 79(5):1961–1970
29. Oksuz S et al (2007) Phospholipase and proteinase activities in different *Candida* species isolated from anatomically distinct sites of healthy adults. Jpn J Infect Dis 60(5):280–283
30. Guarner J, Brandt ME (2011) Histopathologic diagnosis of fungal infections in the 21st century. Clin Microbiol Rev 24(2):247–280
31. De Pauw B et al (2008) Revised definitions of invasive fungal disease from the European Organization for Research and Treatment of Cancer/Invasive Fungal Infections Cooperative Group and the National Institute of Allergy and Infectious Diseases Mycoses Study Group (EORTC/MSG). Clin Infect Dis 46(12):1813–1821
32. Leung AN et al (1999) Pulmonary infections after bone marrow transplantation: clinical and radiographic findings. Radiology 210(3):699–710
33. Thomas CF, Limper AH (2004) Pneumocystis pneumonia. N Engl J Med 350(24):2487–2498
34. Bruno C et al (2007) Comparison of CT features of *Aspergillus* and bacterial pneumonia in severely neutropenic patients. J Thorac Imaging 22(2):160–165
35. Kauffman CA (2005) Candiduria. Clin Infect Dis 41(Suppl 6):S371–S376
36. Arvanitis M et al (2014) Molecular and nonmolecular diagnostic methods for invasive fungal infections. Clin Microbiol Rev 27(3):490–526
37. Neely LA et al (2013) T2 magnetic resonance enables nanoparticle-mediated rapid detection of candidemia in whole blood. Sci Transl Med 5(182):182ra54
38. Dou YH et al (2013) The role of procalcitonin in the identification of invasive fungal infection-a systemic review and meta-analysis. Diagn Microbiol Infect Dis 76(4):464–469
39. Karageorgopoulos DE et al (2011) β-D-glucan assay for the diagnosis of invasive fungal infections: a meta-analysis. Clin Infect Dis 52(6):750–770
40. Bernardes-Engemann AR et al (2015) Validation of a serodiagnostic test for sporotrichosis: a follow-up study of patients related to the Rio de Janeiro zoonotic outbreak. Med Mycol 53(1):28–33
41. Harrison E et al (2010) *Aspergillus* DNA contamination in blood collection tubes. Diagn Microbiol Infect Dis 67(4):392–394
42. Loeffler J et al (2015) Analytical comparison of *in vitro* spiked human serum and plasma for PCR-based detection of *Aspergillus fumigatus* DNA: a study by the European *Aspergillus* PCR initiative. J Clin Microbiol 53(9):2838–2845
43. Scheel CM et al (2014) Development of a loop-mediated isothermal amplification method for detection of *Histoplasma capsulatum* DNA in clinical samples. J Clin Microbiol 52(2):483–488
44. Morrell M, Fraser VJ, Kollef MH (2005) Delaying the empiric treatment of *Candida* bloodstream infection until positive blood culture results are obtained: a potential risk factor for hospital mortality. Antimicrob Agents Chemother 49(9):3640–3645
45. Pappas PG et al (2016) Clinical practice guideline for the management of candidiasis: 2016 update by the infectious diseases society of America. Clin Infect Dis 62(4):e1–50
46. Albataineh MT et al (2016) Update from the laboratory: clinical identification and susceptibility testing of fungi and trends in antifungal resistance. Infect Dis Clin N Am 30(1):13–35

Chapter 2
Antifungal Drugs

Beatriz Bustamante, Jose A. Hidalgo, and Pablo E. Campos

Abstract Fungal infections have increased globally due to the increment of the size of population at risk for fungal infection, which is a consequence of the increased use of immunosuppressive drugs and invasive techniques for advanced life support and extended life expectancy among other reasons. Although invasive fungal infections currently are a significant cause of mortality among critically ill patients, development and approval of new systemic antifungal drugs have not occurred at the same rate as the increase in the number of fungal infections. Only one new class of systemic antifungal drugs, Echinocandins, has been included in the antifungal armamentarium in the last 20 years.

The purpose of this chapter is to review the systemic antifungal drugs currently in use, including new insights on pharmacologic and pharmacokinetics properties, clinical indications, adverse events, and resistance mechanisms. Resistance to antifungal drugs is particularly important because it has increased for every drug, including the echinocandins class. New formulations of triazol drugs and combination therapy is also highlighted.

B. Bustamante, M.D. (✉)
Instituto de Medicina Tropical Alexander von Humboldt, Universidad Peruana Cayetano Heredia, Lima, Peru

Departamento de Enfermedades Infecciosas, Tropicales y Dermatológicas, Hospital Cayetano Heredia, Lima, Peru
e-mail: ana.bustamante@upch.pe

J.A. Hidalgo, M.D.
Division de Enfermedades Infecciosas, Hospital Guillermo Almenara, Lima, Peru
e-mail: pepinhidalgo@yahoo.com

P.E. Campos, M.D., M.P.H
Investigaciones Medicas en Salud, Lima, Peru
e-mail: pabloe.campos@gmail.com

H.M. Mora-Montes, L.M. Lopes-Bezerra (eds.), *Current Progress in Medical Mycology*, DOI 10.1007/978-3-319-64113-3_2

2.1 Polyenes (Amphotericin B Deoxycholate and Its Lipid-Associated Formulations)

2.1.1 Amphotericin B Deoxicholate

Amphotericin B (Amph B), derives from *Streptomyces nodosus*, belongs to a group of polyene macrolides characterized by a macrocyclic ring of lactone. Due to its low water solubility, oral bioavailability is low. This drug can be administered intravenous, intrathecal, intraarticular, intravesical, and in surgical sites. This is one of the oldest antifungal drugs but is still used in the fungal therapy because of its broad activity spectrum.

Although amph B is fungicidal in vitro, it may be fungicidal or fungistatic in vivo depending on the concentration of the drug achieved in body fluids and the susceptibility of the fungus [1].

2.1.1.1 Chemical Structure

Amph B is a heptaene macrolide compound comprising of seven conjugated double bonds within the main ring. This drug has an amino sugar called mycosamine, which in the pyranose form is linked to the hydroxyl group at C-19 of the macrolactone ring of amph B through a glycoside side chain, and a free carboxyl group on the macrocycle. Amph B has a molecular formula of $C_{47}H_{73}NO_{17}$ and a molecular weight of 924.09 [2, 3] (Fig. 2.1).

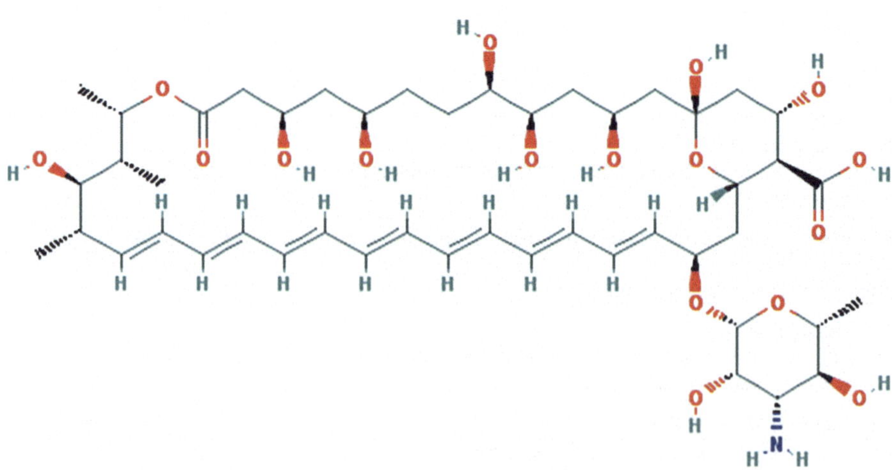

Fig. 2.1 Chemical structure of Amphotericin B deoxicholate. Source: Reference [4]

2.1.1.2 Mechanism of Action

Amph B exerts its antifungal action by binding to ergosterol in the fungal cytoplastic membrane resulting in the formation of pores that causes an increase in their permeability with leakage of small molecules from the cytoplasm, leading to fungal death. This drug may have other effects as an oxidative damage to fungal cell or by immunomodulatory properties on the host cells [5–7].

2.1.1.3 Pharmacokinetics and Pharmacodynamics

Amph B is not appreciably absorbed when taken orally or subsequent to aerosol administration. Then, the intravenous formulation should be used for fungal systemic diseases. After IV administration, this drug is highly bound (>90%) to plasma proteins mainly to albumin, is weakly dialyzable and it is taken up by reticuloendothelial organs, especially the liver, spleen and lung [8–10].

The activity of amph B depends on its concentration and prolonged post-antifungal effect [11]. Its elimination is biphasic. Initially, it is quickly removed with a half-life of about 24 h, while the second elimination phase has a half-life of up to 15 days [12]. In vitro evaluations have shown evident concentration-dependent killing and maximal antifungal activity at concentrations exceeding the MIC by two- to tenfold [11, 13]. In vivo time-kill studies with amph B against different *Candida* species have also shown an improved rate and extent of killing with increasing drug concentrations [14, 15].

Additionally, an in vivo study using a rabbit model of invasive pulmonary aspergillosis to evaluate the pharmacokinetics and pharmacodynamics of amph B deoxycholate, amph B lipid complex (ABLC) and liposomal amph B (L-AmB) found that all formulations of amph B induced a dose-dependent reduction in markers of lung injury and circulating fungus-related biomarkers. Dosing of L-AmB of 3 mg/kg/day predicts complete suppression of galactomannan and (1, 3)-D-glucan levels in the majority of patients [16].

The amph B concentration in urine is low, finding only 3% of the dose, and its penetration in cerebrospinal fluid is limited (2–4% of serum concentrations) [17]. Amph B goes through minimally into the vitreous humor and normal amniotic fluid. In the peritoneal fluid, pleura, or joint, fewer than 50% of the serum levels are obtained [18]. Serum level of the drug is not influenced by hepatic or renal function, or by hemodialysis or peritoneal dialysis. Even in anuric and nephrectomized patients, the drug serum levels and its elimination are just the same as in healthy patients. It seems that amph B is eliminated principally by metabolic conversion or by the bile [9, 19]. However, the metabolites are not yet well known.

2.1.1.4 Spectrum of Activities and Resistance

Amph B has activity against a wide range of fungi including most yeasts, and hyaline and dematiaceous molds. Among the yeast, *Candida* isolates resistant to polyenes are still infrequent. *C. lusitaniae* and *Trichosporum beigelii* are two fungi that have intrinsic resistance, and recently some *C. krusei* and *C. glabrata* strains with high MICs to amph B has been reported [20–22]. In the group of molds, *Aspergillus terreus, Scedosporium apiospermum, Scedosporium prolificans,* and *Fusarium* species are usually resistant to amph B [23].

Although breakpoints for polyenes are not available, most microbiologist use a MIC of ≥1 μg/mL to determine if an isolate is not susceptible to amph B. Reduction or absence of ergosterol in the fungal cell membrane has been associated with resistance. These alterations could be due to mutations in genes coding some of the enzymes involved in the synthesis of ergosterol. The mechanisms involved in the resistance of *C. albicans* isolates to the amph B include a double loss in function of both ERG3 (C-5 sterol desaturase) and ERG11 (lanosterol 14α demethylase) [24, 25]. It has also been identified isolates amph B resistant lacking ERG2, encoding C-8 sterol isomerase, and ERG6, encoding C-24 sterol methyltransferase [26–28].

2.1.1.5 Clinical Uses

Despite the emergence of new antifungal drugs in the last decades, amph B deoxycholate or its lipid formulations are still recommended as primary treatment for several severe or refractory fungal infections or as alternative therapy for other forms of these diseases. According to the Infectious Diseases Society of America (IDSA) guidelines for the management of candidiasis [29], amph B deoxycholate is recommended for the treatment of neonates with disseminated candidiasis or candidiasis in central nervous system, for patients with asymptomatic candiduria who undergo urologic procedures, and for symptomatic candida cystitis or symptomatic ascending candida pyelonephritis due to fluconazole-resistant *C. glabrata* or to *C. krusei*. This drug is recommended as alternative in the treatment for fluconazole-refractory oropharyngeal/esophageal candidiasis. Amph B deoxycholate or lipid formulations are also indicated for the treatment of cryptococcal meningitis, mucormycosis, moderately severe to severe forms of blastomycosis and histoplasmosis [30–33].

Aerosolized amph B, in their different formulations, is utilized as prophylaxis (either alone or in combination with systemic antifungals) in patients at highest risk of invasive fungal infections. A recent meta-analysis presented evidence supporting the concept that the prophylactic use of aerosolized amph B effectively reduces the incidence of invasive pulmonary aspergillosis among high-risk patients [34]. The current IDSA guidelines for the diagnosis and management of aspergillosis stated that aerosolized formulations of amph B may be considered for prophylaxis in patients with prolonged neutropenia and in lung transplant recipients (weak strength recommendation; low-quality evidence) [35].

Because of the low intraocular levels attained with systemic administration, intravitreal injection of amph B is useful to reach high confined antifungal activity for the treatment of severe macular involvement and vitritis. The IDSA guidelines suggest treatment with systemic antifungal drug plus local amph B deoxycholate, 5–10 μg/0.1 mL sterile water, for *Candida* chorioretinitis without vitritis and with macular involvement, and for *Candida* chorioretinitis with vitritis [29]. In addition, IDSA guidelines recommend intraocular amph B with partial vitrectomy as primary treatment of *Aspergillus* endophthalmitis and keratitis [34]. There are controversial reports on toxicity of intravitreal injection of amph B deoxycholate in humans, some of them have described low toxicity and others have reported toxic uveitis [36–41].

There is not a good evidence to recommend bladder irrigation with amph B deoxycholate for treatment of cystitis but, when this infection is due to fluconazole-resistant species, such as *C. glabrata* and *C. krusei*, this method of delivering the antifungal drug directly to the affected site could be used [29, 42].

2.1.1.6 Dosing

Dosing of amph B deoxycholate in adults with normal renal and hepatic function is 0.5–1 mg/kg daily for candidemia, other invasive candidiasis and for endemic dimorphic fungal infections; 1–1.5 mg/kg daily for invasive aspergillosis; 0.3 mg/kg daily for esophageal and oropharyngeal candidiasis; 0.7 mg/kg daily for empiric treatment of febrile neutropenia; and 0.7–1 mg/kg daily (usually with 5-flucytosine) for cryptococcal meningitis in the induction phase of the therapy.

2.1.1.7 Adverse Events and Toxicity

Acute toxicity related to amph B administration is due to stimulation by the antifungal drug of inflammatory cytokine production from innate immune cells via an interaction that requires CD14 and Toll-like receptors [43]. The most frequent acute effects are nausea, vomiting, rigors, fever, hypertension/hypotension, and hypoxia. Other acute adverse events with rare presentation are ventricular arrhythmias, bradycardia, and severe hypertension [44–46]. In addition, two cases with fatal leukoencephalopathy associated with the intravenous administration of the amph B has been reported [47].

Chronic toxicity is referred commonly to nephrotoxicity, but there are other events such as hypomagnesemia, hypokalemia, hyperphosphatemia, anemia, and rare cases of hyperbilirubinemia [48–51]. Reports of anemia of any degree of severity ranges from 33 to 63% in different studies [52–54].

The incidence of nephrotoxicity due to amph B is high, ranging from 49 to 65% [55, 56]. This toxicity is result of renal vasoconstriction producing reduction in glomerular filtration rate, and of direct effect on tubular epithelial cell membranes forming pores. The modification in the permeability of the tubular cell membrane will allow the back diffusion of hydrogen ions thereby weakening the acid elimination [57]. There are some factors that influence the nephrotoxicity of amph B such as cumulative dose, average daily dose, abnormal baseline renal function, concomitant nephrotoxic drugs, and patient's risk category (bone marrow transplant, solid organ transplantation) [58]. Nephrotoxicity is cumulative and dose dependent but it is reversible after amph B treatment is stopped [58, 59]. This toxicity can be decreased by hydration and electrolyte supplementation before amph B infusion [60–63]. In the case of aerosolized amph B, the most common side effects are cough, bad taste, nausea and vomiting; most of them mild or moderate severity [64–67].

2.1.1.8 Contraindication

Amph B is contraindicated in those patients who have known hypersensitivity to this antifungal drug, and because the increment risk of nephrotoxicity and hearing impairment, its use is contraindicated with the simultaneous administration of the following drugs: amikacin, cidofovir, cyclosporine, ioversol, neomycin, streptozocin, and tacrolimus. Amph B should be discontinued before iodinated contrast media administration [68].

2.1.1.9 Drug Interactions

Corticosteroids, thiazide, loop diuretics, and neuromuscular blockers increase risk of amph B-induced hypokalaemia. Amph B can also increase the risk of digoxin toxicity [69]. Use of alternative drugs or monitoring of amph B associated adverse events is indicated when some of the above drugs needs to be used. Administration of amph B with most of the antiretrovirals does not cause interactions, but concomitant or sequential use with tenofovir increases the risk of nephrotoxicity [70]. In the same way, concomitant use of zidovudine and amph B results in increased risk of anemia and neutropenia [71].

2.1.1.10 Use in Special Population

Amph B and their lipid formulations do not need dose adjustment for patients with decreased renal function or in patients receiving hemodialysis or continuous renal replacement therapy or for those with moderate or severe hepatic disease.

Children In pediatric population, pharmacokinetic (PK)/pharmacodynamic (PD) indices of amph B are not validated; therefore, optimal dosing of this drug has not been defined. Amph B pharmacokinetic is very variable in neonates, which may lead to treatment failure or toxicity, and the lack of maturity of the blood-brain barrier in premature infants can be the basis of a better penetration of the antifungal to the CSF [72]. Compared with adults, amph B has a lower volume of distribution and faster clearance in children [72–76]. Even though most of PK studies in children encourage a dosage of 0.5–1 mg/kg/day, a population PK analysis suggested that younger children receiving 1 mg/kg/day might be underexposed, while older children may be overdosed at the same dose [75]. Another study showed that amph B doses of 0.25–1 mg/kg daily to infants causes lower serum concentrations compared with older children and adults [73, 76].

The preferred amph B deoxycholate pediatric dosing is 1–1.5 mg/kg/day. The infusion adverse events and nephrotoxicity due to amph B in children are similar to adults, but fever has been rarely described in infants [77, 78]. Children usually can tolerate higher doses than adults [79].

Pregnancy and Lactation Amph B is considered as category B (B: animal studies no risk, but human studies not adequate, or animal toxicity but human studies no risk) by the US FDA [80]. ABLC has been evaluated in animals without having found harms to the fetus [81]. Although, there are not well-conducted evaluations of amph B use in pregnancy, its use in pregnant women has been described repeatedly without confirmation of teratogenesis in their neonates [82–96].

This is the only systemic antifungal drug that is safe to use during the pregnancy and lactation period, although there are not published data in the last condition. Due to the properties of amph B such as its large molecular size, poor absorption, and high protein binding ability, breast milk amounts are probable insignificant [97]. There is limited information on the use of ABLC and L-AmB in pregnancy [98, 99].

It is known that amph B crosses the placenta and reaches fetal tissues where can persist some weeks after the mother has stopped the drug or given birth. This characteristic possibly would be due to placental deposit or deferred removal by the fetal kidneys [100].

2.1.2 Lipid Formulations of Amphotericin B

Lipid formulations of amph B have higher hydrophobicity, lower nephrotoxicity, and are distributed more efficiently to the sites of fungal infection or inflammation than amph B deoxycholate. The two lipid formulations available in most countries are liposomal amphotericin B (L-AmB) and amphotericin B lipid complex (ABLC), commercially available under the name AmBisome® (Gilead Sciences Inc., Foster City, CA, USA) and Abelcet (Enzon Pharmaceuticals Inc., Bridgewater, NJ, USA; Sigma Tau, Gaithersburg, MD; and Cephalon Limited, Welwin Garden City, UK) respectively. A third lipid formulation, amphotericin B colloidal dispersion (ABCD), is no longer commercially available.

The lipid compositions and particle size are different between the amph B lipid formulations, producing distinct pharmacokinetic parameters and tissue distribution. Then, the lipid formulations of amph B cannot be used interchangeably to treat patients [101–103].

ABLC consists of amph B complexed to two phospholipids (l-α-dimyristoylphosphatidylcholine and l-α-dimyristoylphosphatidylglycerol in a 7:3 molar ratio) in a 1:1 drug-to-lipid molar ratio to form a ribbon-like structure with a diameter of 1600–11,000 nm, making it the largest of the lipid formulations [104]. ABLC is taken by macrophages and becomes sequestered in the liver and spleen. It has lower circulating amph B serum concentrations, when it is compared with the amph B deoxycholate, but larger volume of distribution and clearance. Lung concentration of ABLC is higher than the obtained with L-AmB or amph B deoxycholate.

By contrast, molecules of L-AmB are smaller and no captured by the mononuclear phagocyte system. L-AmB consists of amph B integrated into the lipid bilayer of small unilamellar liposomes, which are composed of hydrogenated soy phosphatidylcholine, cholesterol and distearoyl phosphatidylglycerol [105]. After a dose of L-AmB, it reaches a higher peak plasma level (Cmax) than amph B deoxycholate and a larger area under the concentration–time curve. The highest amph B concentrations with the L-AmB administration are found in the liver and in the spleen, followed by kidneys and lungs. Levels in myocardium and brain tissues are low [106].

The ABLC mechanism of action requires that the fungal lipase, a heat labile extracellular product, produces a lipid breakdown within the ribbon-like material with subsequent discharge of amph B into the tissues [107]. The suggested mechanism of action of L-AmB states it binds to the fungal cell wall and disintegrates itself, discharging amph B that binds to ergosterol in the fungal cell membrane [108, 109]. Table 2.1 summarizes the characteristics of the different amphotericin B formulations.

Dissimilar results have been reported in relation to in vitro susceptibility of amph B deoxycholate and lipid formulations. Montagna et al. found great correlation between the in vitro activities of amph B deoxycholate and L-AmB against 604 clinical yeast isolates [125], while Johnson et al. found that MIC 50 and MIC 90 of ABLC were the same to or lower than those of amph B deoxycholate when they were tested against 190 isolates from different fungal species, including *Aspergillus fumigatus*, *Candida* spp., and *Cryptococcus neoformans* [126]. In the later study, the L-AmB was the least active of the different amphotericin B preparations tested, showing 2–4 dilutions higher in their MICs than those of amph B deoxycholate. Carrillo-Muñoz, studying the in-vitro susceptibilities of 120 clinical isolates of yeasts (including different species of *Candida*, *Rhodotorula rubra*, *Trichosporon* spp., *Cryptococcus laurentii* and *C. neoformans*), found no statistical significance among MICs of amph B deoxycholate, ABLC, L-AmB and other antifungal drugs when evaluating all strains. However, there were differences in the activity of these drugs for individual species. Amph B deoxycholate and ABLC had high MIC values for an isolate of *C. laurentii*, while L-AmB, had moderately low MIC for the same isolate. ABLC and L-AmB had higher mean MICs against *Trichosporon* spp., and L-AmB had high MIC90 values for *C. glabrata* than amph B deoxycholate [127].

In addition, the manufacturer of AmBisome® (L-AmB) claims that it has in vitro activity equivalent to amph B deoxycholate against the following fungi: *Aspergillus* species, *Candida* species, *C. neoformans*, *Fusarium* species and *Blastomyces dermatitidis* [69].

Experimental studies carried out in animals demonstrated that it is necessary higher doses of the two lipid formulations to achieve the same or greater antifungal effect than the obtained by the amph B deoxycholate [123, 128–130]. Although it is also true in the clinical setting, the toxicity, mainly nephrotoxicity, is lower when lipid formulations are used [117, 123, 131–139]. There is some evidence of greater effectiveness of the lipid formulations of amph B in relation to amph B deoxycholate. When used as induction therapy, L-AmB has

Table 2.1 Comparative characteristics of amphotericin B formulations

Characteristics	Amphotericin B deoxycholate	ABLC	L-AmB
Structure	Micelle	Ribbon-like structures	Small spherical unilamellar lipid vesicles
Size (nm)	0.035	1600–11,000	Mean diameter: <100
Composition	Amph B deoxycholate	1:1 molar ratio of amph B/phospholipid (DMPC and DMPG)	HSPC, cholesterol, DSPG and amph B
Mode of action	It binds to ergosterol (fungal membrane) resulting in increased permeability, output of intracytoplasmic compounds and finally in fungal death	Release of amph B from complexes, possibly host macrophage- and phospholipase-mediated	Liposome targeting to fungal cell wall with release of amph B into the fungal cell
Nephrotoxicity	+++	++	+
Infusion-related toxicity	+++	++	+
Pharmacokinetic parameters (mean ± standard deviation)			
Cmax (μg/mL) at steady state	1.1 ± 0.2	1.7 ± 0.8	83 ± 35.2
AUC_{0-24} (μg h/mL)	17.1 ± 5.0	14.0 ± 7.0	555 ± 311
Terminal elimination half-life (h)	91.1 ± 40.9	173.4 ± 78.0	6.8 ± 2.1
Vd (L/kg)	5.0 ± 2.8	131 ± 57.7	0.11 ± 0.08
Clearance (mL/h/kg)	38.0 ± 15.0	436 ± 188.5	11.0 ± 6.0
Distribution			
Protein binding, %	>95	>95	>95
CSF concentration relation to serum, %	0–4	3 (in animal)	<1
Urine concentration relation to serum, %	3–20	<5	4.5
Metabolism	Minor hepatic	Unk	Unk
Elimination	Feces	Unk	Unk
Comparative tissue concentrations of amphotericin formulations			
Liver	++	+++	+++
Spleen	+	+++	+++
Lung	+	++++	+
Kidney	+	±	+
Brain	–	±	±

Source: Data derived from [69, 81, 110–124]

ABLC amphotericin lipid complex, *L-AmB* amphotericin B lipid complex, *HSPC* hydrogenated soy phosphatidylcholine, *DSPG* distearoyl-phophatidylglycerol, *DMPC* dimyristoyl phosphatidylcholine, *DMPG* dimyristoyl phosphatidylglycerol, *Cmax* maximum concentration of drug in serum, *Cmax* peak drug concentration, *AUC* area under the concentration curve, *Vd* volume of distribution

Note: pharmacokinetic parameters after multiple doses of 0.6 mg/kg/day of amph B deoxycholate or 5 mg/kg/day of L-AmB or ABLC demonstrated faster culture conversion in cryptococcal meningitis, and was associated with improved survival and lower toxicity in AIDS patients with moderate to severe disseminated histoplasmosis in comparison to amph B deoxycholate [140, 141]. L-AmB (5 mg/kg/day) also showed superior clinical efficacy to amph B deoxycholate (1 mg/kg/day) in the treatment of neutropenia-associated invasive fungal infections [139].

Any of the lipid formulations of amph B are recommended by the IDSA guidelines for the treatment of non-neutropenic patients with suspected azole- and echinocandin-resistant candidemia, candida suppurative thrombophlebitis, candida endocarditis, and chronic disseminated (hepatosplenic) candidiasis. In addition, they proposed these drugs as an alternative treatment for candidemia in neutropenic and non-neutropenic patients, candida septic arthritis, or osteomyelitis, and as alternative empiric treatment of non-neutropenic patients in the intensive care unit with suspected invasive candidiasis and with intolerance to other antifungal drugs. The same guidelines recommend L-AmB for the initial treatment for central nervous system candidiasis and for candida chorioretinitis without vitritis due to fluconazole/voriconazole-resistant isolates. L-AmB is also suggested as an alternative treatment for central nervous system candidiasis in neonates [29].

For aspergillosis infections, the IDSA guidelines recommend amph B deoxycholate and lipid formulations as alternative treatments for initial or salvage therapy. This document also suggests lipid formulations for refractory and progressive aspergillosis or for empiric and preemptive therapy in allogeneic hematopoietic stem cell transplant recipients and patients treated for acute myelogenous leukemia with prolonged neutropenia who remain persistently febrile despite broad-spectrum antibiotic therapy [35]. The IDSA guidelines also recommend lipid formulations of amph B instead of amph B deoxycholate for patients with cryptococcal meningitis with or predisposed to renal dysfunction and for the treatment of pulmonary, meningitis and disseminated sporotrichosis [30, 142].

2.1.2.1 Dosing

In general, for all fungal infections but CM, the recommended dose of lipid formulations of amphotericin B is 3–5 mg/kg daily for treatment and 1–3 mg/kg daily for prophylaxis. The dose for CM is 3–4 mg/kg per day of L-AmB and 5 mg/kg per day of ABLC, both of them with or without flucytosine, for at least 2 weeks [30, 143]. The pediatric dosing is 5 mg/kg/day of ABLC or L-AmB [144].

2.1.2.2 Adverse Events and Toxicity

In comparison with amph B deoxycholate, rates of infusion related acute reactions are similar with ABLC (60%) but lower with L-AmB (48%) [103, 117]. To reduce infusion-related reactions, it is recommended the use of low-dose hydrocortisone

(1 mg/kg), diphenhydramine (25–50 mg), meperidine (0.5 mg/kg), and nonsteroidal anti-inflammatory agents as premedication [118, 135, 145].

Numerous studies have shown that lipid formulations are less nephrotoxic than amph B deoxycholate. The Barrett et al.'s systematic review found that ABLC and L-AmB decreased all-cause risk of mortality and renal toxicity, compared with amph B deoxycholate [146]. Martino's systematic review also found that ABLC is significantly less nephrotoxic than amph B deoxycholate and can be administered securely to patients with preexisting renal injury [147]. L-AmB used in the treatment of fever in neutropenic patients with cancer showed similar efficacy but significantly less infusion-related reactions and nephrotoxicity than amph B deoxycholate [138]. In comparative trials, adverse events requiring discontinuation of the drug have occurred less frequently when patients received L-AmB (12%), followed by those receiving ABLC (32%) and amph B deoxycholate (44%) [103]. Wade et al. have also reported considerably lower rates of nephrotoxicity, infusion reactions and hypomagnesemia among patients with renal dysfunction and invasive fungal infections receiving L-AmB compared with those receiving ABLC [115].

L-AmB used in 33 consecutive patients at least 65 years old as empirical therapy for the treatment of invasive fungal infections showed equivalent safety and efficacy to those observed in younger patients, but higher incidence of severe hypokalemia when used for extended periods. The incidence of grade III or IV hypokalemia was similar in the older and younger groups [148]. A characteristic triad of acute infusion-related toxicity to L-AmB has been described. The following symptoms alone or in combination of 1 of 3 symptoms can be present: chest pain, dyspnea, and hypoxia; severe abdomen, flank or leg pain; and flushing and urticaria. These reactions occur within the first 5 min of infusion and disappear with administration of diphenhydramine. A multicenter analysis found a mean overall occurrence of 20% (range: 0–100%) of this specific toxicity among 64 centers [149].

2.2 Triazole Antifungal Agents

2.2.1 General Properties

Azoles are a group of antifungals of great importance in the treatment of systemic mycoses that share a common basic chemical structure and mechanism of action, inhibition of membrane sterol synthesis. A major change occurred with the identification of increased antifungal activity of the N substitution in the imidazole ring, which led to the development of current triazoles.

A first generation of triazoles were developed in the eighties and have been in clinical use since then, with limited (itraconazole) or no activity (fluconazole) against filamentous fungi. A second generation of drugs developed in late nineties and more recently is now available (voriconazole, posaconazole, isavuconazole), with improved activity against *Aspergillus* and other filamentous fungi. Clinical studies have shown their effectiveness in the management of multiple fungal infections.

Triazoles are widely used due to their broad spectrum of antifungal activity, availability of both IV and oral formulations, and safety profile. It is expected that in the future newer agents and more indications will be identified [150, 151].

2.2.1.1 Chemical Structure

The imidazole nucleus is a 5-atom heterocyclic structure with 3 C and 2 N (see Fig. 2.2). This structure is present in nitroimidazoles, utilized in antianaerobic and antiparasitic therapy (metronidazole), and azole antifungals (clotrimazole, miconazole, ketoconazole), currently used for local therapy of superficial infections. The substitution of a C for an N atom originates the term "triazoles" (three N atoms in the ring), and it is associated with significant changes in the antifungal activity and pharmacokinetic properties in relation to older azoles (i.e. ketoconazole). The triazoles have a more specific binding to fungal enzymes than to mammal enzymes for sterol synthesis, more potent antifungal activity, and broader spectrum, less metabolic side effects, and better bioavailability and tissue distribution than older azoles.

2.2.1.2 Mechanism of Action

The structural target for triazoles in the fungal cell is the cellular membrane. Action at this level explains part of older azoles limitations, since their activity would not be completely specific for fungal organisms. Older azoles have also some inhibitory activity on steroid metabolism in mammalian cells; for example, ketoconazole was used in the management of primary Cushing syndrome because of its significant inhibition of corticoid synthesis [157].

Triazoles are inhibitors of the enzyme lanosterol 14-α-demethylase, coded by the gene CYP51A, involved in the synthesis of ergosterol. This enzyme catalyzes the oxidative elimination of 14-methyl group from fungal cell lanosterol (monooxigenase P450 activity). Its inhibition causes the accumulation of lanosterol precursors (methyl-esterols) and changes the proper composition of the cell membrane, which produce structural and functional consequences (lanosterol deficiency and decreased membrane fluidity), primarily in cellular reproduction and death. Triazoles block the activity of the enzyme by binding to the active site, and their increased affinity to this site due to the third N atom present in the imidazole ring produces higher antifungal activity [158].

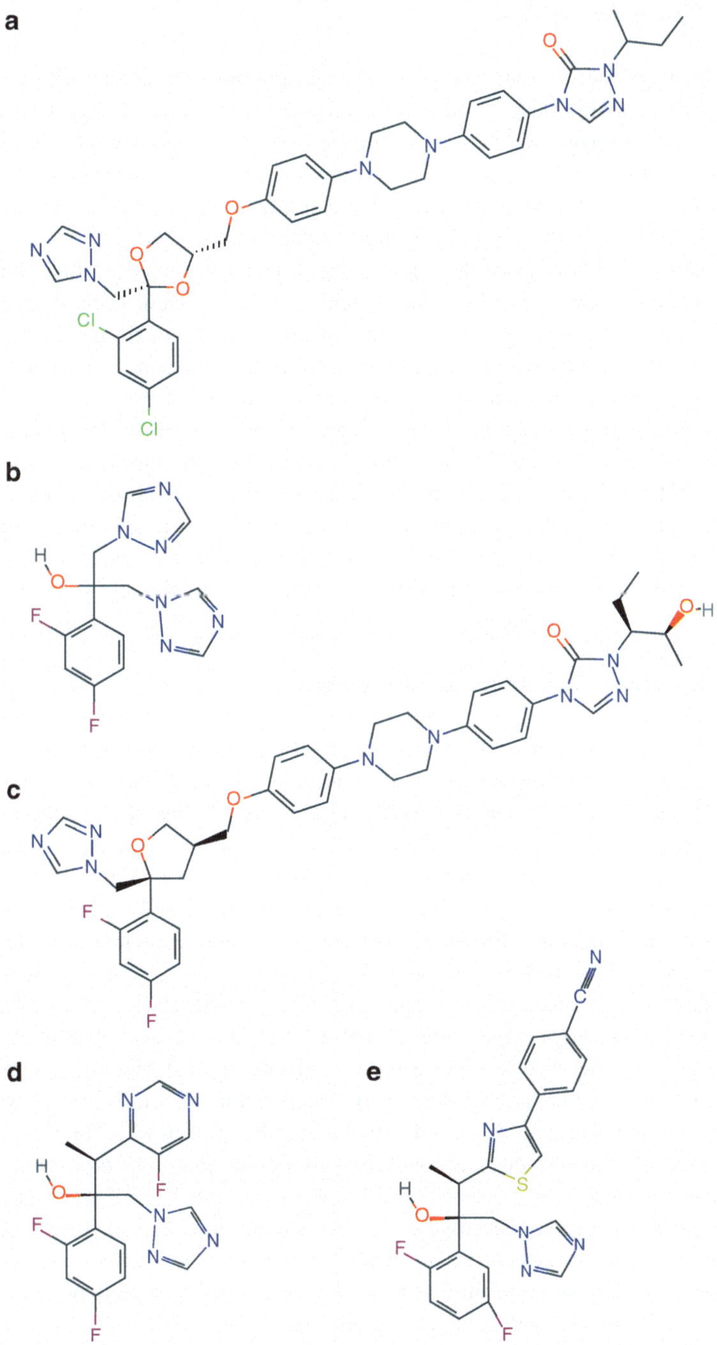

Fig. 2.2 Chemical structure of triazoles. (**a**) Itraconazole, (**b**) Fluconazole, (**c**) Posaconazole, (**d**) Voriconazole, and (**e**) Isavucunazole. Source: References [152–156]

2.2.1.3 Pharmacokinetics

Main pharmacokinetic characteristics of triazoles are displayed in Table 2.2. In summary, these drugs have good oral absorption (with special ingestion requirements for certain formulations of itraconazole and posaconazole), a prolonged half-life (which allows once or twice daily dosing regimens in most cases), good distribution in body tissues with clinical use in different types of invasive disease, and are available in both oral and parenteral formulations [160].

Fluconazole and voriconazole reach higher concentrations in tissues because of their smaller molecular size. Fluconazole shows better concentration in cerebrospinal fluid, and all azoles reach good concentrations in brain tissue. Posaconazole reaches the highest concentration in alveolar cells and voriconazole in bone tissue. Triazoles concentrate similarly well in liver and kidneys [113].

Due to a less predictable absorption, interactions with other drugs, and subject variability of metabolism, therapeutic drug monitoring is recommended to optimize regimens and minimize side effects for itraconazole, voriconazole, and posaconazole [161–163]. Trough levels should be tested within 30 min before patient dosing. The recommended target trough plasma levels for triazoles and their recommended day of testing after initiation of therapy are shown in Table 2.3.

2.2.1.4 Spectrum of Activity and Resistance

All members in the class show good activity against most *Candida* species. Fluconazole is effective against most clinically significant *Candida* sp. and against *Cryptococcus* sp., and to a lesser extent against most dimorphic endemic fungi (*Histoplasma, Blastomyces, Coccidioides*, and *Paracoccidioides* spp). Itraconazole has a broader spectrum that includes *Sporothrix schenckii*, *Aspergillus* spp. and some other filamentous fungi such as dematiaceous fungi and mucorales. Its mold activity is lower than newer triazoles. Voriconazole is active against dose-dependent and fluconazole-resistant *Candida* species (*C. glabrata* and *C. krusei*, respectively). It also shows increased activity against molds, particularly *Aspergillus* spp., *Fusarium* spp., *S.apiospermium,* and dematiaceous fungi. Posaconazole and isavuconazole add to voriconazole spectrum their activity against Mucorales [167, 168].

Resistance to triazoles is well described. Mechanisms of acquisition of resistance include overexpression with increased activity of efflux pumps (ATP binding cassette, ATP and major facilitator superfamily -MFS), and point mutations that cause changes in tridimensional structure or activity of C-14α demethylase [169]. The point mutations cause structural changes in the active site of the demethylase, decreasing the affinity to its ligands. *Candida* species can be intrinsically resistant as is *C. krusei* to fluconazole, show dose-dependent susceptibility as *C. glabrata* versus fluconazole or acquire resistance like *C. albicans* to azoles, largely by changes in the activity of efflux pumps.

Aspergillus resistance to triazoles with mold activity has also been reported in some places. Specific alterations in coding regions of the enzyme in the CYP51A

Table 2.2 Main pharmacologic characteristics of triazoles in clinical use

Characteristics	Itraconazole	Fluconazole	Voriconazole	Posaconazole	Isavuconazole
MW	705.6	306.2	349.3	700.7	814[a]
Half-life (h)	19–22	27–34	6	20–30	56–77
Plasma protein binding (%)	>95	11–22	58	98	>99
Oral bioavailability (%)	55 (capsule with food)	>80	96	See text	98
Volume of distribution (L/kg)	796	46	1.2–4.6	1774	450
Liver failure dose adjustment	Yes	No	Yes	No	No
Metabolism	Liver		Liver	Liver (limited proportion)	Liver
Elimination	Feces	Kidney (60–75% unchanged)	Kidney (2% unchanged)	Feces (mostly unchanged)	Feces
Kidney failure dose adjustment	No	Yes	No (oral)	No	No
Therapeutic drug monitoring	Recommended		Recommended	Recommended	
Most significant toxicities	Nausea, gastrointestinal symptoms	ALT elevation (10%)	Visual disturbances, photosensitivity, confusion	Gastrointestinal symptoms, headache	Gastrointestinal symptoms
Comparative tissue concentrations of triazoles					
Liver	+	+	++	ND	ND
Spleen	+	++	+	ND	ND
Lung	+	+	+	++	ND
Kidney	+	+	+	ND	ND
Brain	+	++	+	+	ND

Source: References [113, 150, 151, 159]
ND No data
[a]MW (Molecular Weight) of isavuconazonium

Table 2.3 Recommended target trough plasma levels for triazoles

Antifungal drug	Target concentration during prophylaxis mg/L	Target concentration during treatment of IFD mg/L	Day of test after initiation of drug
Itraconazole	>0.5	>1	>7 days
Voriconazole*	>1	1–5.5	After 3–5 days
Posaconazole	>0.35 after the first 48 h ≥0.9 after day 7**	>1.8	>7 days

Source: Data from [161, 164–166]
IFD invasive fungal disease
aIrrespective of formulation
bFor the tablet and intravenous formulations TDM is not required when they are used for prophylaxis, as target levels are anticipated to be reached in nearly all the patients

gene have been associated in some cases with itraconazole and posaconazole resistance and with cross-resistance to all triazoles [170].

Identification of triazole resistance is increasingly available with standardization of methods, identification of clinically relevant breakpoints, and consensus from major international institutions (CLSI, EUCAST) in recent years [171].

2.2.1.5 Clinical Utility

Triazoles have a wide spectrum antifungal activity, which includes both yeast and filamentous fungi. Much more limited in the case of the older generation, but with significant increase of activity in the case of the newer products, such as posaconazole and voriconazole.

Fluconazole continues to be of major importance in the management of different clinical presentations of candidiasis in multiple groups of patients [168], and for consolidation treatment of cryptococcosis. Itraconazole is recommended in the management of dimorphic fungi and to some extent in filamentous organisms.

Voriconazole, posaconazole, and isavuconazole have evolving roles in the management of severe fungal infections by filamentous organisms in the most immunosuppressed individuals. Voriconazole is considered a first choice for invasive aspergillosis in most clinical situations. Posaconazole has been used preferentially in antifungal prophylaxis and Mucor infections, and isavuconazole has shown high efficacy in the treatment of aspergillosis and Mucorales [172].

2.2.1.6 Adverse Events and Drug Interactions

Compared to older azoles (i.e. ketoconazole), currently available drugs show much lesser hormonal inhibition, gastrointestinal side effects, and hepatotoxicity. Triazoles as a group are deemed relatively safe drugs.

Itraconazole most commonly causes nausea (10%) and gastrointestinal symptoms (diarrhea, 8%; vomiting, 6%; abdominal discomfort, 6%). Hyperbilirubinemia and liver enzyme elevation is reported in about 5% [173].

Fluconazole used at doses higher than 400 mg/day can cause headache in 2%, anorexia in 3% of patients and transient ALT elevation in 10% of patients.

Reversible visual disturbances (30%), photosensitivity (20%), hallucinations and confusion (15%) have been described with voriconazole use. Recent reports have associated presentation of skin cancer in immune-suppressed patients with use of voriconazole, which needs further clarification [174, 175].

Side effect profile of posaconazole is very similar to that of fluconazole. Experience with isavuconazole in clinical trials has shown that it is largely well tolerated.

Prolongation of the QTc has been observed with triazoles, including posaconazole [176, 177], and this can cause sudden death. The I_{Kr} channel, one of the membrane channels responsible for potassium outflux movement, is inhibited by fluconazole and other triazoles. This inhibition is associated to ventricular repolarization changes that increase vulnerability to cardiac arrhythmias. Patients with other drugs that can cause QTc prolongation, severe bradycardia, hypokalemia, or hypomagnesemia should be cautiously prescribed when used concomitantly with azoles. Contrary to other drugs in the group, isavuconazole has been associated with QTc shortening, of unclear clinical significance.

Triazoles present a significant number of drug interactions [178, 179]. Interactions with immunosuppressants, rifamycins, anticonvulsants, omeprazole, warfarin, statins, and antiretrovirals, amongst others, are particularly important. See Table 2.4 [181].

Interactions of triazole antifungals can be divided into the following categories: modifications of antifungal pharmacokinetics by other drugs, modifications of other drug pharmacokinetics by antifungals, and two-way interactions. The mechanisms involved include azole inhibition of drug metabolizing enzyme cytochrome (CYP) P450 isozymes, such as CYP3A4/5, CYP2C9 and CYP2C19 in varying degrees [182]. Triazoles can also inhibit drug transporter P-glycoprotein (P-gp) within the gastro-intestinal tract and the liver, for which immunosuppressants are substrates.

The degree of these interactions varies greatly, as azole inhibition of relevant enzymes can be dose-dependent and differs in potency and selectivity. Itraconazole and voriconazole are reported to be more potent inhibitors of CYP3A4 than posaconazole and fluconazole. In terms of inhibition of P-gp, itraconazole and posaconazole have the more significant activity [183].

Triazoles are used frequently in transplant patients, either for prevention or treatment of suspected or proven fungal infections, and can interfere with the metabolism and transport of immunosuppressants (i.e. cyclosporine, tacrolimus, sirolimus, and everolimus), which are drugs of narrow therapeutic margin.

Triazoles can increase exposure to immunosuppressant, consequently increasing risk for side effects of these drugs. On the other hand, discontinuation of azoles without dose adjustment of the immunosuppressant drugs may lead to sub-therapeutic immunosuppressant exposure and risk of transplant rejection or graft-versus-host-disease.

Table 2.4 Recommendations for selected drug interactions of triazoles

Drug	Itraconazole	Fluconazole	Voriconazole	Posaconazole	Isavuconazole
Antacids	Avoid				
Atazanavir				Monitor for atazanavir toxicity	
Atorvastatin	Monitor for atorvastatin toxicity				
Carbamazepine			Avoid		Avoid
Cyclosporin A	Monitor for CsA toxicity	Monitor for CsA toxicity	Monitor for CsA toxicity	Monitor for CsA toxicity	Use with caution
Efavirenz	Avoid		Efavirenz dose reduction	Increase dose of posaconazole if need as per therapeutic drug monitoring	
Midazolam	Monitor for midazolam toxicity	Monitor for midazolam toxicity	Monitor for midazolam toxicity	Monitor for midazolam toxicity	
Omeprazole	Avoid capsules, use solution	Omeprazole AUC increased >100%, consider omeprazole dose reduction	Omeprazole AUC increased >100%, consider omeprazole dose reduction		
Phenytoin	Avoid	Monitor for phenytoin toxicity		Avoid	
HIV protease inhibitors (other than atazanavir)	Avoid higher doses of itraconazole		Avoid		
Rifampin	Avoid	Consider fluconazole dose increase	Avoid		Avoid
Sirolimus			Avoid	Avoid	Use with caution
Tacrolimus	Monitor for tacrolimus toxicity	Monitor for tacrolimus toxicity	Monitor for tacrolimus toxicity	Monitor for tacrolimus toxicity	Use with caution

Source: Data from [178–180]

As a potent inductor of CYP enzymes, rifampin significantly increases the metabolism of azoles. Concomitant use of rifampin with itraconazole, voriconazole, or isavuconazole should be avoided, and consideration to increasing fluconazole dosing is required if concomitant use is necessary [184].

In the management of HIV-infected patients, fungal infections are frequent and important complications. Efavirenz should not be coadministered with itraconazole or posaconazole. Voriconazole and itraconazole doses >200 mg/day are not advised in patients receiving protease inhibitors. Posaconazole can increase >100% AUC of atazanavir (boosted or unboosted). Fluconazole use does not generally require adjustment of dosing and can be used with most antiretrovirals [181].

2.2.1.7 Dosing and Administration. Use in Special Populations

Dosing criteria for triazoles are already established for general use in adult populations. However, the newer drugs (posaconazole, isavuconazole) still miss clinical pharmacokinetic data for groups of patients such as neonates, younger children, and pregnant women.

Neonates and Children As more experience is accumulated, use of older triazoles in children is recommended for their routine indications [185, 186]. FDA labeling of voriconazole and posaconazole still restricts their use to older children [187]. In spite of being an older drug, itraconazole has not been developed for pediatric use and does not have formal indications [188]. Recommendations for the use of triazoles in the management of aspergillosis are similar in children and adults, although recognizing that doses of voriconazole are higher in younger children (<12 years) and in younger adolescents with a weight < 40 kg. In these patients, loading dose is 9 mg/kg twice, followed by 4–8 mg/kg (higher dose for invasive molds and more serious infections) [35, 189].

Pregnant Women According with the FDA classification of drugs and risk category in pregnancy, fluconazole status varies with the dose used. It is considered C when a single 150 mg-dose is indicated, but is D for higher doses. This in based on observation of birth defects in five children exposed in utero to fluconazole, and animal experiments showing teratogenic potential. Fluconazole remains contraindicated in pregnancy with the exception of the lower dose. Itraconazole, isavuconazole and posaconazole are currently classified as category C, while voriconazole is in category D. Current consensus is to restrict use of triazoles in pregnancy, in particular in the first trimester [190].

Renal Failure Dose adjustment is recommended for fluconazole. Daily dose should be reduced by 50% with a creatinine clearance lower than 50 mL/min. For end-stage renal disease patients (ESRD) on hemodialysis (HD) dose is normal and goes after dialysis. Parenteral voriconazole preparation with cyclodextrin is not recommended in renal failure due to potential accumulation and toxicity of cyclodextrin. Patients who are already on some form of renal replacement therapy do not have any concern as cyclodextrin is efficiently removed by dialysis. Because cyclodextrin contained in Itraconazole oral solution is metabolized by amylase, patients with renal failure can use it without dose adjustment [191].

Liver Failure Voriconazole has more extensive liver metabolism and in moderate to severe liver failure, its maintenance dose should be reduced to 50% after a regular loading regimen.

Obesity Fluconazole should be dosed by total body weight to achieve AUC/MIC ratios that have been associated with better outcomes. Posaconazole dose should not be corrected for increased body weight. This has still to be evaluated for the newer formulations (tablet, IV solution). Voriconazole dosing using total body weight can reach supratherapeutic concentrations. Its dose should not be changed for increased body weight or BMI in the case of oral voriconazole. Until further studies are performed, use of either the adjusted body weight or ideal body weight when dosing weight-based IV voriconazole could be justified [192].

2.2.2 *Individual Agents*

2.2.2.1 Itraconazole

Itraconazole is available as a 100 mg capsule and an oral suspension in cyclodextrin (100 mg/10 mL). Cyclodextrin makes itraconazole soluble and enhances its absorption. Oral absorption of the capsule depends on food intake, although absorption of the solution is best on an empty stomach. Interestingly, coadministration with a cola beverage increases the AUC of the capsule formulation. Absorption of the capsule is decreased with hypochlorhydria, mucositis, and graft-versus-host intestinal changes, conditions that can be present in AIDS patients or bone marrow transplant recipients.

Therapeutic drug monitoring is useful to adjust proper dosing, taking in consideration the method used for adequate interpretation. Tissue, pus and bronchial secretion concentrations of itraconazole are higher than plasma levels. The drug is metabolized in the liver and excreted in feces; prolonging its half-life in cirrhosis. When administered via oral, minimal amount of active itraconazole is eliminated in urine with most of cyclodextrin (>99%) excreted intact in feces.

Itraconazole most frequent side effects are nausea and abdominal discomfort. Frequency of gastrointestinal symptoms is higher with the solution presentation (osmotic effect). Hypokalemia and edema can be seen with higher doses. Itraconazole is contraindicated in pregnancy and in nursing mothers.

Itraconazole is useful in the treatment of dimorphic organisms (*Blastomyces* spp., *Histoplasma* sp., *Coccidioides* spp., *Paracoccidioides* spp., *Sporothrix* spp.), especially in less severe forms, less immunocompromised individuals or as consolidation treatment. Because activity against *Aspergillus* is lower in comparison to newer agents (second generation), indications of itraconazole in aspergillosis are more limited. It is indicated for the management of allergic bronchopulmonary aspergillosis or as an alternative therapy in aspergilloma [35]. Other uses include phaeohyphomycosis, ringworm, onychomycosis, tinea versicolor, and occasionally candidiasis.

Usual daily dose for cutaneous conditions, including sporotrichosis, is 200 mg. An initial loading dose of 200 mg tid for the first 3 days when used in the treatment of deep mycoses is recommended to ensure adequate serum and tissue levels in the short term. Recommended dose for treatment of invasive mycoses, selected forms of aspergillosis or prophylaxis of invasive aspergillosis is 400 mg daily, divided every 12 h.

2.2.2.2 Fluconazole

Fluconazole is an imidazole analogue to ketoconazole with more specific fungal sterol synthesis inhibition and increased antifungal activity. It has good bioavailability (>90%), which is a significant difference with itraconazole, good distribution in fluids and tissues, long serum half-life (approximately 30 h), and relatively low (11–12%) binding to plasma proteins. Its elimination is mostly renal. The molecule has a second triazole ring that decreases lipophilicity and increases unbound drug in blood.

Fluconazole for systemic use is available in capsules, tablet, powder for oral suspension, and injectable form for IV infusion at a concentration of 2 mg/mL. Tolerance to fluconazole is considered good. Liver toxicity is a concern, especially when higher doses are used for treatment of cryptococcosis or disseminated candidiasis.

Fluconazole is active against yeasts and inactive against molds. It can be used to treat mucosal candidiasis (oro-pharyngeal, esophageal, vaginal), disseminated and invasive candidiasis, cryptococcosis, and systemic dimorphic mycoses (histoplasmosis, coccidioidomycosis, paracoccidioidomycosis, and sporotrichosis). Depending on the disease severity, immunologic status or comorbidities of the patient, and available of other antifungals, fluconazole can be an initial treatment, or a consolidation phase treatment in these indications [193].

Fluconazole has had an important role in the management of serious forms of infection by *Candida* [194], although this role is changing in recent years with newer recommendations favoring use of echinocandins in most seriously ill patients [29, 195, 196]. In the case of invasive candidiasis and candidemia, fluconazole should be considered for patients not critically ill and infected by fluconazole-sensitive organisms, using a loading dose of 800 mg followed by 400 mg daily [29, 195]. In the treatment of mucosal candidiasis, lower doses of fluconazole are appropriate. A single 150 mg is indicated for vulvo-vaginal candidiasis, and daily doses of 100 mg are used for oro-pharyngeal and esophageal candidiasis.

Fluconazole at doses of 800 mg/day combined with amphotericin B is recommended as an alternative regimen for induction treatment in cryptococcal meningitis or disseminated disease in HIV patients when flucytosine is not available. In these patients, doses of 400 or 200 mg daily are used in the consolidation and maintenance phases respectively [30]. Dose of 400 mg daily of fluconazole are recommended as initial antifungal treatment of immunosuppressed and immunocompetent patients with mild-to-moderate pulmonary cryptococcosis [30].

2.2.2.3 Voriconazole

Voriconazole is a second generation triazole that was approved for clinical use in 2002. In voriconazole, the second triazole ring has been replaced with a fluoropyrimidine nucleus, which explains its broader spectrum. The main advantage of Voriconazole over first-generation triazoles is its activity against filamentous fungi, including *Aspergillus* sp., *Fusarium* spp., and *S. apiospermium* [197–199]. Despite its broad-spectrum activity against yeast and molds, voriconazole is not active against Mucorales.

Currently, voriconazole is considered the drug of choice for the treatment of invasive aspergillosis [35, 200]. Reported clinical experience shows some promising results with combination of voriconazole and echinocandins as a rescue regimen [201–204].

Voriconazole is available for oral and parenteral use. It should be taken with empty stomach because food and high content of fat decrease Voriconazole absorption, reducing bioavailability in 20%. Half-life is 6 h, requiring twice daily administration. Parenteral formulation of voriconazole also contains cyclodextrin, which could accumulate in renal failure. Its use in patients with renal impairment should be individualized.

In general, voriconazole is well tolerated. Reversible disturbances in vision (impaired color discrimination, blurring, and photophobia) is reported in about 25–30% of patients and are not seen with other triazoles. Skin rashes, photosensitivity, facial erythema, hallucinations and confusion are other significant side effects. Periostitis has been associated to prolonged use, and serious EKG alterations (torsade de pointes, QTc prolongation) have occurred in patients with predisposing factors to arrhythmia.

For invasive aspergillosis and serious mold infections, a loading dose of 6 mg/kg twice daily is recommended in the first day, followed by 4 mg/kg bid. For the treatment of invasive *Candida* infections dose should be lower, at 3 mg/kg bid. Oral dosing (tablets) is 400 mg bid the first day and then 200 mg bid for persons of >40 kg. For persons under 40 kg, the recommended dose is 200 mg bid for the first day, followed by 100 mg bid. Intake should be 1 h apart from meals.

2.2.2.4 Posaconazole

This triazole has a broad spectrum of antifungal activity, including *Aspergillus* and *Candida* [167]. It can be considered a derivative of itraconazole with structural modifications that enhance its activity and tissue concentration (lipophilic molecule with high concentration in tissues). Oral absorption and bioavailability of posaconazole suspension are difficult to predict, because they are significantly affected by factors like meal ingestion or presence of mucosal lesions in the gastrointestinal tract [205–207]. Posaconazole suspension should be taken with high-fat meals to enhance absorption. Although its half-life is long and suggests the possibility of once daily dosing, AUC is higher with bid or tid dosing, which is recommended for the oral suspension.

The newer delayed-release tablet has better absorption, improved bioavailability, and a fourfold increase in maximum concentration, a threefold increase of the area under the curve, and more prolonged presence in plasma. In contrast to the suspension, the effect of food or drugs that may alter gastric acidity is moderate. Additionally, patient intervariability is reduced. All this favorable pharmacokinetic changes allow for once daily dosing [161, 208, 209].

Posaconazole is usually well tolerated. In clinical trials the most common side effects have been gastrointestinal symptoms including nausea and abdominal pain. Initially available only for oral administration, its indications have been mainly referred to prophylaxis of fungal infection in patients at high risk (prolonged neutropenia, acute myeloid leukemia, post-transplant), in particular for filamentous fungal complications [210, 211]. The oral suspension has been evaluated for refractory aspergillosis at a daily dose of 800 mg with a 42% rate of global response [212]. Currently, a formulation for parenteral administration is also available and undergoing phase III clinical trial in the treatment of invasive aspergillosis, with completion estimated for July 2018 [213].

Posaconazole has good activity against *Mucor* spp. and has a role in the treatment of mucormycosis in combination with surgical procedures, usually after an initial period with amphotericin B [214].

Oral suspension and delayed-release tablet have different dosing regimens. The dose of the suspension is 200 mg tid with food for prophylaxis or 400 mg bid with meals when indicated for treatment [215–217], while the 300 mg tablet is administered once a day. Intravenous dosing of 300 mg daily is approved for prophylaxis.

2.2.2.5 Isavuconazole

It is the newest member of the second-generation triazole antifungal approved by the US FDA [159, 168, 180]. It has been approved for the treatment of both invasive aspergillosis and invasive mucormycosis [218]. It is also under investigation for the treatment of candidemia and invasive candidiasis, cryptococcosis, and dimorphic fungi [219].

Isavuconazole is administered as the hydrosoluble prodrug isavuconazonium, which is available in tablets and for parenteral administration. In preclinical and clinical studies, it has shown significant antifungal potency against a broad range of yeasts, dimorphic fungi, and molds. Isavuconazole has a broad spectrum of antifungal activity, similar to amphotericin B.

Clinical experience so far has revealed that isavuconazole may be associated with less toxicity than voriconazole, even when administered without therapeutic drug monitoring. Additionally, the oral formulation is highly bioavailable and the parenteral presentation is b-cyclodextrin–free (due in large part to the presence of aromatic moieties in the molecule). These are interesting properties that will increase interest on isavuconazole as a new addition to the triazole class of antifungals. Isavuconazole dosing is similar when administered either intravenously or orally. In both cases, the loading dose is 200 mg every 8 h for six times, followed by 200 mg daily.

2.2.3 *Newer and Investigational Agents: Efinaconazole, Albaconazole, Ravuconazole and Others*

A numerous group of newer triazole molecules are currently under different stages of development [220]. Preliminary clinical studies are already available for ravuconazole, albaconazole, and efinaconazole (available as a topical agent).

Ravuconazole is related to fluconazole and voriconazole. It has activity against yeasts (*Candida* spp. and *Cryptococcus* spp.), dimorphic fungi and filamentous organisms (dematecious, mucorales). It is not active in vitro against *Fusarium* spp. Clinical studies for onychomycosis have been reported with a mycological cure rate of 59% and clinical response of 56% [221]. It is still to be determined what potential indications may have and what results are obtained in clinical trials for systemic mycoses.

Albaconazole shows low MICs against *Candida* spp. and has been clinically studied in the treatment of vulvo-vaginal candidiasis and onychomycosis, and experimentally against *S. prolificans*. Its long half-life allows for weekly dosing.

Efinaconazole is a potent antifungal drug against *T. rubrum*, *T. mentagrophytes* and *C. albicans*, approved in 2014 for the treatment of onychomycosis. It also has activity against other species of fungus, including some nondermatophytes molds (*Acremonium* spp., *Fusarium* spp., *Paecilomyces* spp., *Pseudallescheria* spp., *Scopulariopsis* spp., and *Aspergillus* spp.), *Cryptococcus* spp., *Trichosporon* spp., and other species of *Candida* different to *C. albicans* [222].

The list of newer compounds includes RI26638, KP103, T8581, TAK187, FX0685, ZJ522, TAK456, Syn2869, and additional molecular modifications for dioxantriazoles, triazole-quinoxalines, and triazole-benzimidazoles.

The search for newer clinically active compounds might lead to the availability of triazole derivatives with increased antifungal spectrum and effectiveness, as well as better tolerance.

2.3 Echinocandins

Echinocandins are the newest members of the antifungal armamentarium and the first ones targeting the fungal cell wall [223, 224]. Currently, three semi-synthetic echinocandin derivatives have received FDA approval for clinical use: caspofungin (2001), micafungin (2005), and anidulafungin (2006). A fourth compound, the CD101, is under development.

2.3.1 Chemical Structure

Echinocandins are semisynthetic lipopeptides antibiotics, composed of cyclic hexapeptides with modified N-linked acyl lipid side chains [225, 226] (Fig. 2.3).

2.3.2 Mechanism of Action

Echinocandins competitively inhibit the beta-1,3-D-glucan synthesis, a polysaccharide which is an essential component of the fungal cell wall of many fungi. Beta-glucans represent between 30 and 60% of the cell wall mass in yeasts, and its depletion results in fungicidal activity for *Candida* spp. and fungistatic effect for *Aspergillus* spp. [230, 231].This mechanism of action is different from the one of other drugs, allowing a potential use of echinocandins in combination therapy [232], and because the target of echinocandins is unique to fungi, then absent in human cell, these drugs cause less toxicity and have fewer drug–drug interactions. In addition, some evidence from in vitro studies and murine models supports an immunomodulatory effect of echinocandins. They can unmask highly antigenic epitopes and amplify the host immune response [233].

2.3.3 Pharmacokinetics and Pharmacodynamics

Although pharmacokinetic and pharmacodynamic characteristics of echinocandins are similar, they differ in dosing, metabolic elimination pathways, and drug interaction profile. Like other large lipopeptide antibiotics, these drugs are poorly absorbed through the gastrointestinal system and must be administered by intravenous infusion. Due to their long half-life (10–26 h), they are dosed once daily, and because echinocandins are highly bound to plasma proteins, administration of a loading dose is recommended for caspofungin and anidulafungin, although it is not yet clear for micafungin. Also, high binding to plasma protein limit distribution of echinocandins to the cerebrospinal fluid and the eye, making them inadequate treatment for infections of these compartments [234–236].

Echinocandins are primarily eliminated through nonmicrosomal metabolism nonenzymatic degradation to inactive products, and then their urinary concentration is very low [234, 237]. They are not significantly metabolized by the cytochrome P450 enzymes nor are they substrates or inhibitors of P-glycoprotein pumps. As consequence, they have less drug–drug interactions in comparison with others antifungal drugs. However, caspofungin must be used with caution when severely impaired hepatic function is present.

Caspofungin shows a net terminal half-life of 27–50 h, and degrades spontaneously and is metabolized via hydrolysis and N-acetylation to two inactive

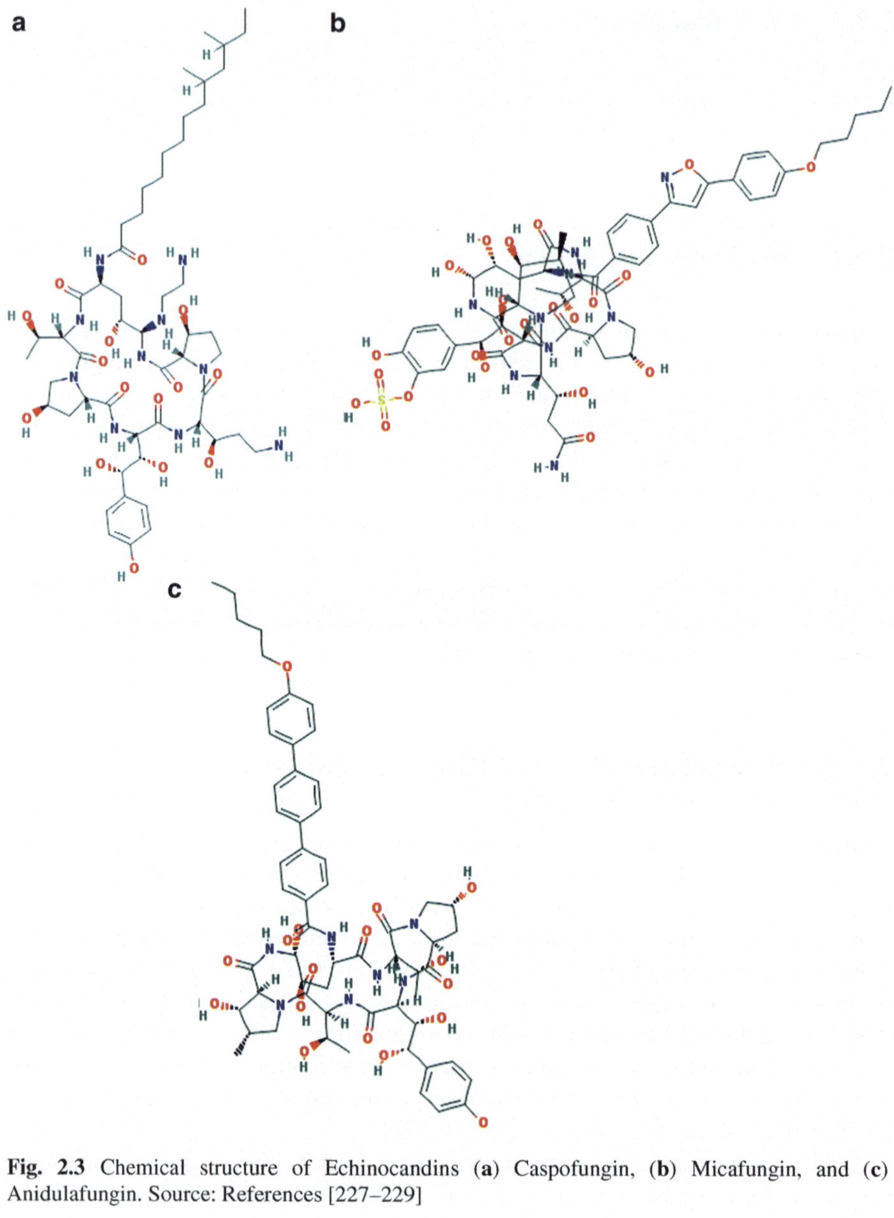

Fig. 2.3 Chemical structure of Echinocandins (**a**) Caspofungin, (**b**) Micafungin, and (**c**) Anidulafungin. Source: References [227–229]

metabolites; micafungin has a terminal half-life of approximately 15 h in adults, and is metabolized hepatically by arylsulfatase, catechol *O*-methyltransferase, and hydroxylation; while anidulafungin shows a terminal half-life of 40–50 h, and is not metabolized but instead eliminated by slow spontaneous degradation. All three echinocandins are nondialyzable, and their breakdown products are excreted predominantly by the fecal route, with only low concentrations of active drugs excreted by urine (less than 2%) [226, 231, 238].

In vitro studies showed that the fungicidal effect of echinocandins against *Candida* spp. is proportional to the maximum plasma drug concentration, that this effect persist after falling of drug concentration below MICs, and that it seems to correlate with the area under time-concentration curve to MIC ratio [237, 239]. However, similar information related to killing or inhibition of Aspergillus spp. is not completely defined yet [239, 240]. In addition, there are not established strategies to conduct therapeutic drug monitoring for echinocandins [237, 239].

2.3.4 Spectrum of Activity and Resistance

Because echinocandins show a similar spectrum of activity, they could be interchangeable specially when treating candidiasis infections. They have potent activity against many Candida spp. (*C. albicans*, *C. glabrata*, *C. dubliniensis*, *C. tropicalis*, and *C. krusei*), and although MICs againts *C. parapsilosis* and *C. guilliermondii* are often higher, they are useful drugs against these candida species [241, 242]. The fungicidal activity against *Candida* spp., including fluconazole-resistant *C. glabrata* and *C. krusei*, is the main advantage of echinocandins [226].

Even though echinocandins inhibit growth of *Aspergillus* species at very low echinocandin levels, their activity against *Aspergillus* spp. is only fungistatic [223, 243–245], this is explained because in *Aspergillus species*, higher activity of cell wall remodeling and beta-glucan synthase is localized in apical and sub-apical branching points. In guinea pig models, echinocandins seem to potentiate the activity of triazoles against *Aspergillus* spp. [219, 246].

Although beta-1,3-D-glucan synthase from Cryptococcus spp. is highly inhibit by caspofungin, echinocandins have not activity against *C. neoformans* and *Cryptococcus gattii*, neither against *Trichosporun* spp. [225, 226]. Echinocandins are not effective drugs to treat mycosis produced by endemic dimorphic fungi (*Blastomyces dermatitidis*, *Histoplasma capsulatum*, and *Coccidioides* spp.), due to their modest activity against the mycelial phase of them. In addition, echinocandins have not significant activity against non-*Aspergillus* molds (Mucorales, *Fusarium* spp., or *Scedosporium* spp.) [226, 247–249], and only modest in vitro activity, without clinical utility, for some phaeohyphomycetes [250, 251]. Echinocandins are effective agents for prophylaxis of *Pneumocystis jirovecci* pneumonia although less effective for established pneumonia in experimental models [252, 253].

In contrast with what happens with amphotericin B and triazoles, activity of echinocandins are not affected by presence of biofilm; echinocandins MICs are minimally affected when tested under biofilm conditions. *C. albicans* inoculum embedded in biofilm is almost completely cleared at the usual echinocandin serum levels [254, 255]. When evaluating activity against *C. tropicalis* biofilm, micafungin showed high activity while liposomal amphotericin B performed poorly [256]. This unique characteristic of echinocandins makes them particularly useful for the treatment of prosthetic device and catheter-associated infections.

Overall resistance to echinocandins of *Candida* spp. has been reported in up to 4%, and results from mutations in conserved regions of the gene-encoding glucan synthase (FK1 and FK2) [257, 258], and resistance to echinocandins has been documented for *C. albicans*, *C. glabrata*, *C. lusitaniae*, *C. tropicalis*, and *C. parapsilosis* [259, 260]. Previous exposure to an echinocandin had been associated with echinocandin resistance on multivariative analysis [254].

Resistance of *C. glabrata* is of particular concern, because it is now reported from around the world, at rates between 3 and 15%, and because isolation of strains with resistance to fluconazole and voriconazole and to echinocandins [259, 261–264]. Among the 162 fluconazole-resistant *C. glabrata* strains isolated between 2006 and 2010 in the US, resistance to any echinocandin was demonstrated in 18 (11%), while there was no resistance to echinocandins among 110 fluconazole-resistant strains isolated between 2001 and 2004 [262]. All the 18 resistant isolates contained an FKS1 or FKS2 mutation.

Resistance to echinocandins is associated with treatment failure and relapse or recurrence if there was an initial response and with higher rates of mortality at days 14 and 30 [261, 265, 266].

In an organ transplant recipient with persistent candidemia, Imbert and colleagues demonstrated that switching from both azole and echinocandin therapy to liposomal amphotericin B, produced that resistant *C. glabrata* isolate lost the FKS2 S663P alteration, regaining full susceptibility to echinocandin, while maintaining their pan-azole resistance. Based on this observation, authors suggest that more restricted use and/or a discontinuous administration of echinocandins may limit the spread of clinical resistance to these drugs [267].

2.3.5 Clinical Uses

Echinocandins are extensively used for prevention and empiric treatment of fungal infection, and for treatment of invasive candidiasis, especially in critically ill and neutropenic patients. The three echinocandins have FDA approval for the treatment of esophageal candidiasis and invasive candidiasis in adults. Micafungin has FDA approval to be used as prophylaxis of *Candida* infections in hematopoietic cell transplanted adults, while caspofungin is approved as empiric treatment for neutropenia febril, and for esophageal candidiasis and invasive candidiasis in children older than 3 months [29, 268, 269]. Echinocandins had demonstrated improved survival when compared to amphotericin B and triazoles in the treatment of candidemia and invasive candidiasis [191, 270, 271] and, similar efficacy to amphotericin

B and fluconazole in the treatment of oropharyngeal or esophageal candidiasis. However, they are not frequently used for these latter indications due to their parenteral-only presentation [272–277].

Although echinocandins are not the choice to treat aspergillosis, they had shown useful for the treatment of refractory aspergillosis, when used in combination with voriconazol or with amphotericin lipid formulations [278–280]. Caspofungin has FDA approval as salvage therapy of invasive aspergillosis, and current IDSA guidelines stated that caspofungin or micafungin can be used to treat aspergillosis in settings in which azole and polyene antifungals are contraindicated [35]. There is also limited evidence supporting the use of echinocandins in combination therapy for the initial treatment of aspergillosis. Association of anidulafungin to voriconazole therapy had shown improved outcome in comparison to monotherapy, although without statistical significance [203, 232, 281].

Because their low urinary excretion rate, echinocandins are not considered for the treatment of UTIs. However, patients with fluconazole-resistant *Candida* spp. or with hepatic injury and fluconazole-sensitive *Candida* spp. have been successfully treated with caspofungin [282].

2.3.6 Adverse Events and Toxicity

Due to the target of echinocandins is absent in human cells, these drugs cause less toxicity. Mostly, echinocandins are well tolerated and their adverse events are mild and similar for all the three drugs currently in use. Serious adverse events requiring treatment discontinuation are fewer with these drugs than with other systemic antifungals. Most common adverse events are gastrointestinal symptoms (diarrhea, nausea, vomiting, abdominal pain, abdominal distention, and constipation), laboratorial alterations (increment of aminotransferases and alkaline phosphatase and bilirubin, hypokalemia, among others) and general disorders and administration site conditions (pyrexia, edema peripheral, Infusion-related reaction, pain at the site of infusion). Table 2.5 summarizes the most frequent adverse reactions, with frequency of at least 5% in any of the groups under evaluation, reported in clinical trials testing echinocandins [283–285].

Asymptomatic elevation of liver enzymes, 5–13% for aminotransferases and 12% for alkaline phosphatase, is less frequent in patients treated with echinocandins in comparison with azoles and amphotericin B. Because hepatitis, hepatomegaly, hyperbilirubinemia, and hepatic failure have been rarely reported, monitoring of hepatic enzymes is recommended when using echinocandins [226, 283–285]. Renal adverse event reported with the use of echinocandins involved mild decrease of serum potassium, reported between 11 and 20% in clinical trials, without significant drug related toxicity observed [226, 283–285]. Occurrence of anemia, neutropenia, and thrombocytopenia have been reported between 6 and 15% in clinical trials, but again hematologic toxicity attributed to echinocandins is infrequent [226, 283–285].

Table 2.5 Adverse reactions[a] in patients who received ECHINOCANDINS in clinical trials[b], incidence 5% or greater in any of the groups under evaluation

	Caspofungin (*n* = 1951)		Micafungin (*n* = 479)		Anidulafungin (*n* = 131[d])	
Adverse reactions[c]	*n*	%	*n*	%	*n*	%
With at least one adverse reaction	**1665**	**85**			**130**	**99**
Investigations	**901**	**46**	**191**	**40**	**66**	**50**
Alanine Aminotransferase Increased	258	13	45	10	7	5
Aspartate Aminotransferase Increased	233	12				
Blood Alkaline Phosphatase Increased	232	12			15	12
Blood Potassium Decreased	220	11			33	25
Blood Bilirubin Increased	117	6				
Urine output decreased			18	4		
White blood cell increased					11	8
Blood creatinine increased					7	5
Hypomagnesemia					15	12
Hypoglycemia					9	7
Hyperkalemia					8	6
Hyperglycemia					8	6
General disorders and administration site conditions	**843**	**43**	**256**	**53**	**70**	**53**
Pyrexia	381	20	103	22	23	18
Chills	192	10				
Edema Peripheral	110	6			14	11
Infusion-related reaction			24	5		
Chest pain					7	5
Gastrointestinal disorders	**754**	**39**	**285**	**60**	**81**	**62**
Diarrhea	273	14	106	22	24	18
Nausea	166	9	91	19	32	24
Vomiting	146	8	91	19	23	18
Abdominal Pain	112	6	76	16	8	6
Abdominal Distension			29	6		
Constipation					11	8
Blood and lymphatic system disorders			**161**	**34**	**34**	**26**
Thrombocytopenia			70	15	8	6
Neutropenia			61	13		
Anemia			63	13	12	9
Febrile neutropenia			23	5		
Leukocytosis					7	5
Infections and infestations	**730**	**37**			**82**	**63**
Pneumonia	115	6			8	6

(continued)

Table 2.5 (continued)

	Caspofungin (n = 1951)		Micafungin (n = 479)		Anidulafungin (n = 131[d])	
Adverse reactions[c]	n	%	n	%	n	%
Bacteremia					23	18
Urinary tract infection					19	15
Sepsis					9	7
Respiratory, thoracic, andmediastinal disorders	**613**	**31**	**194**	**41**	**67**	**51**
Dyspnea					15	12
Pleural effusion					13	10
Respiratory distress					8	6
Cough	111	6			9	7
Epistaxis			45	9		
Cardiac disorders			**97**	**20**		
Tachycardia			47	10		
Renal and urinary disorders			**78**	**16**		
Hematuria			18	4		
Psychiatric disorders			**80**	**17**	**48**	**37**
Anxiety			35	7		
Insomnia					20	15
Confusional state					10	8
Depression					8	6
Skin and subcutaneous tissue disorders	**520**	**27**	**197**	**41**		
Rash	159	8	55	12		
Erythema	98	5				
Pruritus			54	11		
Urticaria			24	5		
Decubitus ulcer					7	5
Nervous system disorders	**412**	**21**				
Headache	193	10			11	8
Vascular disorders	**344**	**18**			**50**	**38**
Hypotension	118	6			19	15
Hypertension					15	12
Deep vein thrombosis					13	10
Musculoskeletal and connective tissue disorders					**27**	**21**
Back pain					7	5

Source: References [283–285]

[a]Defined as an adverse reaction, regardless of causality, while on echinocandins or during the 14-day post-echinocandins follow-up period

[b]Incidence among individuals who received at least 1 dose of trial drug

[c]Within any system organ class, individuals may experience more than 1 adverse event

[d]Patients receiving 100 mg for the treatment of Candidemia/other Candida Infections

Infusion of echinocandins produces several histamine-release symptoms, including rash, pruritus, hypotension, bronchospasm, angioedema, and may be some acute cardiovascular events. Their occurrence is associated with the infusion rate and in most patients is enough to slow it to obtain improvement. In the case of anidulafungin, the infusion rate should not exceed 1.1 mg/min [226, 283–285]. In addition, rare cases of anaphylaxis, erythema multiforme, Stevens–Johnson syndrome, and skin exfoliation have been associated with the use of echinocandins, although a causal relationship has not been established [283–285].

2.3.7 Drug Interactions

Because echinocandins are not significant inhibitors or inducers of the CYP450 enzymatic pathways or p-glycoprotein drug efflux transporters, they have very few drug–drug interactions when compared with other systemic antifungals [226].There are mild interactions of caspofungin with the immunosuppressant tacrolimus and cyclosporine. In the case of tracolimus, standard drug monitoring of tracolimus is recommended. The concomitant use of caspofungin with inducers of hepatic CYP enzymes is expected to reduce the plasma concentration of caspofungin. Then adult patients receiving rifampin, which is a potent inducer of CYP3A4, should receive 70 mg/day and pediatric patients 70 mg/m^2/day of caspofungin. The same dosing should be considered when patient receive other inducers such as efavirenz, nevirapine, phenytoin, dexamethasone, or carbamazepine [283].

There is no drug–drug interaction of micafungin with mycophenolate mofetil, cyclosporine, tacrolimus, prednisolone, fluconazole, and voriconazole.

In the case of nifedipine and itraconazole, the concomitant use of micafungin increment their AUC and C_{max}, while sirolimus AUC was increased but its C_{max} not. It is recommended that patients receiving micafungin with sirolimus, nifedipine or itraconazole should be monitored for these drugs, which dose should be reduced if necessary [284].There is not drug–drug interaction of anidulafungin with cyclosporine, voriconazole, tacrolimus, rifampin, or amphothericin B liposomal [285].

2.3.8 Use in Special Population and Dose Adjustments

Pediatric Caspofungin and micafungin have FDA approval for use in children. Larger doses based on milligrams per kilogram are prescribed for both children and infants because the increased rate of clearance of these drugs among neonates, infants, and younger children compared with adolescents and adults [226, 286]. Caspofungin is considered safe and effective for pediatric patients older than 3 months, having the same indications as adults, with dosing based on body surface area [283]. Micafungin is approved for pediatric patients older than 4 months and is dosed in mg/kg [284].

Pregnancy and Nursing Mothers All three echinocandins are class C agents in the pregnancy category. They should be used only if the potential benefit justifies the risk to the fetus. In animal studies, echinocandins caused embryofetal toxicity, including skeletal changes, increment of abortions and visceral abnormalities. Echinocandins could be detected in the plasma of the fetus, indicating they cross the placental barrier in rats. It is unknown if echinocandins are excreted in human breast milk, but they could be detected in the milk of lactating rats. Again, they should be administered to nursing mothers only if the potential benefit justifies the risk [226, 283–285].

Dose Adjustments As described above, a 70 mg/day dose of caspofungin is recommended when adult patients use rifampin concomitantly, while the pediatric dose is 70 mg/m^2/day. The same dosing should be considered if there is concomitantly use of other inducer of CYP450, such as carbamazepine, dexamethasone, efavirenz, nevirapine, or phenytoin [283–285].There is no need of dose adjustment in presence of renal insufficiency, including patients in hemodialysis or continuous renal replacement therapy [226, 283–285]. In the case of adults with mild hepatic insufficiency, maintenance dose of caspofungina is the same. This should be reduced to 35 mg/day in the case of moderate hepatic insufficiency (Child-Pugh score 7 to 9). There is no recommendation available for dosing caspofungin in adults with severe hepatic insufficiency or pediatric patients with any degree of hepatic insufficiency [283]. There is no need of dose adjustment of micafungin or anidulafungin in presence hepatic insufficiency of any degree [284, 285].

Obesity Because clearance of echinocandins increment with body weight and there is no difference in outcomes of obese and nonobese patients receiving the same dose of caspofungin, it is recommended an increment between 25 and 50% of the daily dose only for patients weighting 75 kg with severe infection [226, 287].

2.3.9 Adult Dosing

The dosing of echinocandins is slightly variable according with the indication. See Table 2.6.

2.3.10 New Echinocandin

Currently, a new echinocandin, named CD101/Bifungina, is under development for topical and weekly IV administration. It exhibits prolonged stability in plasma and aqueous solutions up to 40 °C [288], and has shown in vitro activity against resistant *Candida* spp. and *Aspergillus* spp. strains. There are two phase II studies currently enrolling patients.

Table 2.6 Dosing of echinocandins for adults by indication

Indication	Caspofungin	Micafungin	Anidulafungin
Esophageal candidiasis	Loading dose 70 mg IV, then 50 mg IV daily	150 mg IV daily; no loading dose is required	200 mg IV daily
Invasive candidiasis	Loading dose 70 mg IV, ten 50 mg IV daily	100 mg IV daily; no loading dose is required	Loading dose 200 mg IV, then 100 mg IV
Salvage therapy for invasive aspergillosis[a]	Loading dose 70 mg IV, then 50 mg IV daily. Daily dose can be increased to 70 mg if response is inadequate	150 mg IV daily; no loading dose is required	Loading dose 200 mg IV, then 100 mg IV
Neutropenic fever (empiric therapy)	Loading dose 70 mg IV, then 50 mg IV daily		
Candida prophylaxis in hematopoietic cell transplant recipients		50 mg IV daily; no loading dose is required	

Source: References [283–285]
[a]Only caspofungin has US Food and Drug Administration approval for this indication

2.4 Flucytosine

2.4.1 Chemical Structure

Flucytosine (5-fluorocytosine or 5-FC) is a synthetic nucleoside analogs chemically related to anticancer drugs (fluorouracil and floxuridine). Its molecular formula is C4H4FN3O with a MW of 129.1 (Fig. 2.4).

2.4.2 Mechanism of Action

Flucytosine is transferred into fungal cells by cytosine permeases, where it is converted into 5-fluorouracil and phosphorylated to 5-fluorodeoxyuridine monophosphate. This compound inhibits thymidylate synthase, a crucial enzyme in the synthesis of 2′-deoxythymidine-5′-monophosphate that is an essential precursor for DNA biosynthesis, therefore disturbing DNA synthesis [290]. In addition, the 5-fluorodeoxyuridine monophosphate can be further phosphorylated and be incorporated to RNA, disrupting protein synthesis [291].

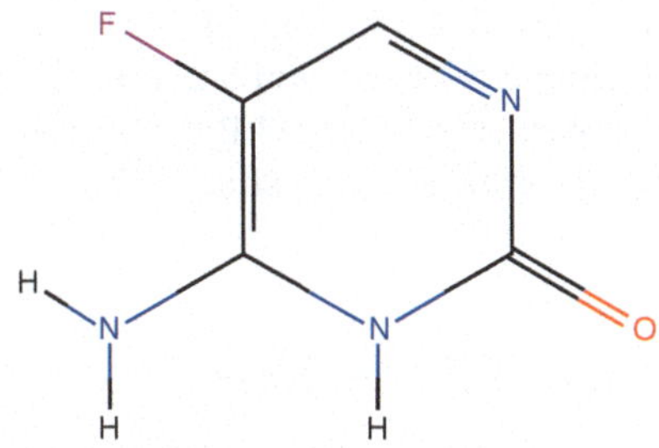

Fig. 2.4 Chemical structure of flucytosine. Source: Reference [289]

2.4.3 Pharmacokinetics and Pharmacodynamics

A feature of the drug is its almost complete and fast absorption after oral administration, having a bioavailability of 76–89% [292]. The AUC is 62 mg·h/L and the maximal concentration is 80 μg/mL [290]. Flucytosine (5-FC) achieves fungistatic levels quickly and distributes extensively throughout the body fluids, including eyes and the cerebrospinal fluid, where it reaches approximately 75% of serum levels.

The 5-FC half-life in humans with normal kidney function is 3–5 h, but it is considerably delayed to 30–250 h in renal insufficiency [293, 294]. Only 2–4% of 5-FC is protein binding, between 80 and 90% is eliminated unchanged in the urine, and the liver metabolizes only a minimal amount. Flucytosine is removed by hemodialysis in 66–75%, but peritoneal dialysis is not as effective as hemodialysis [292, 295].

2.4.4 Spectrum of Activities and Resistance

Flucytosine is active against *C. neoformans* and *Candida* species except *C. krusei*, but isolates of *Aspergillus* species are usually nonsusceptible to 5-FC in vitro. Exist synergy with amphotericin B, which modifies the permeability of the fungal cell membrane allowing greater penetration of 5-FC.

Fungi with primary resistance to 5-FC are rare. A mutation in the *FCY2* gene, which encodes the cytosine permease, affects the absorption of the drug diminishing accumulation of the drug within the cell [296, 297]. Secondary resistance develops during therapy, especially during monotherapy, and it is based on inactivation of enzymes of the pyrimidine pathway. Mutations in the *FCY1* gene that encodes for the cytosine deaminase, or mutation in the *FUR1* gene that encodes for the uracil phosphoribosyl transferase induce acquired resistance by interference in the conversion of 5-FC to 5-fluorouracil, or from 5-fluorouracil to 5-fluorouridine monophosphate respectively [296–300].

Other resistance mechanisms have been suggested for *C. glabrata*. It was found that in the presence of 5-FC the fungal cell wall showed higher resistance to lyticase, suggesting that cell wall alteration occurs in response to 5-FC. Genes CgFPS1 and CgFPS2 of *C. glabrata*, encoding a plasma membrane aquaglyceroporin, are recognized as factors of 5-FC resistance. Both genes facilitate resistance by declining 5-FC accumulation in *C. glabrata* cells. Unlike, the deletion of CgFPS2 and particularly of CgFPS1 was found to improve the susceptibility to 5-FC registered for the parental strain [301].

2.4.5 Clinical Uses

Flucytosine should be used in combination therapy, generally with amphotericin B (Amph B), to decrease development of resistance. This combination is recommended as primary therapy for cryptococcal meningitis, severe pulmonary cryptococcosis and cryptococcocemia [30]. Additionally, 5-FC in combination with Amph B is used for patients with refractory *Candida* infections, such as endocarditis, meningitis, or endophthalmitis and it is also recommended for the treatment of symptomatic ascending *Candida* pyelonephritis due to fluconazole-resistant *C. glabrata* [29].

The ESCMID and ECMM guidelines for the management of rare invasive yeast infections recommend amph B alone or in combination with 5-FC for infections due to *Geotrichum candidum* or *Rhodotorula* spp. They suggest the combination of amph B and 5-FC for infections due to *Saccahromyces cerevisiae,* and the combination of triazole plus echinocandin plus 5-FC to treat cerebral abscess due to dematiaceous fungi when surgery is not possible [30, 302].

2.4.6 Adverse Events and Toxicity

The toxicity to 5-FC is dose-dependent. The most frequent adverse events with this drug are bone marrow depression (leukopenia, anemia, and thrombocytopenia) and gastrointestinal disturbances (nausea, vomiting, diarrhea, abdominal pain, anorexia, dry mouth, and duodenal ulcer) [303–308]. Although bone narrow toxicity can occur with lower serum concentrations of 5-FC, it is more frequent when the concentration is greater than 100 μg/mL [305, 309]. For this reason, it is necessary to monitor the 5-FC serum concentrations to be sure they range between 25 and 100 μg/mL [310].

Less frequently, toxicity occurs in the central nervous system (headache, drowsiness, confusion, vertigo, and hallucinations) or manifest as liver function test abnormalities (jaundice, bilirubin elevation, increased hepatic enzymes, and acute hepatic injury). Colitis is reported infrequently, with toxicity related to local cytotoxic effect on protein synthesis [311–313].

Recently, a study performed in mice suggests that therapy with amph B combined with 5-FC originates a synergistic inflammatory activation in a dose-dependent way in hepatic tissues. Caution when using this antifungal combination is required, particularly for patients with hepatic deficiency [314].

2.4.7 Drug Interactions

Use of clozapine or deferiprone concurrently with 5-FC is not advised. They procainamide. Use of cytosine arabinoside could deactivate the antifungal action of 5-FC by competitive inhibition [290].

It was also noted that drugs which decrease glomerular filtration may extend the half-life of 5-FC [315]. Amph B-associated nephrotoxicity will delay elimination of 5-FC, causing an increase in serum 5-FC concentrations, may increase the risk and severity of bone marrow toxicity. Others agents that can increase the myelotoxic risk and therefore caution should be exercised in their use concomitantly with 5-FC are antineoplastic drugs (cyclophosphamide, doxorubicin, methotrexate, paclitaxel, vinblastine), antiviral agents (gancyclovir, foscarnet), antirretrovirals (zidovudine, lamivudine, didanosine, stavudine), chloramphenicol, dapsone, interferon alfa, linezolid, pentamidine and procainamide.

2.4.8 Use in Special Population

Dose adjustment is necessary in patients with renal dysfunction. In that case, the dose interval has to be extended (see Table 2.7). With a creatinine clearance below 10 mL/min, 5-FC serum levels should be monitored, doing appropriate dose adjustments so not to exceed 80 μg/mL. Because 5-FC is dialyzable, the daily dose must be administered post hemodialysis.

Because of reduced renal function in neonates with a very low birth weight, use of 5-FC in this population should be done with very close monitoring of serum drug levels to avoid large accumulation of 5-FC in plasma [72]. Flucytosine is considered as category C according to the FDA pregnancy category (animal studies show toxicity, human studies inadequate but benefit of use may exceed risk). It is contraindicated during early pregnancy (first trimester) because the drug crosses the human placenta and for its known teratogenic effect in rats and its interference with DNA synthesis in the growing fetus [320, 321]. The delivery of 5-FC in human milk is unknown, and its use during breastfeeding is not recommended. Dosing of 5-FC in obese patients is that for the ideal body weight.

Table 2.7 Flucytosine dosing in adult people according to glomerular filtration rate (standard dose of 100–150 mg/kg/day)

Renal clearance mL/min	Dose	Period (h)
>50	25–37.5 mg/kg	Every 6 h
10–50	25–37.5 mg/kg	Every 12–24 h
<10	25–37.5 mg/kg	Every 24–48 h
Hemodyalisis	25–50 mg/kg	Dose post-dialysis, every 48–72 h
Peritoneal dialysis	0.5–1.0 g	Every 24 h

Source: Data from [316–319]

2.5 Terbinafine

Terbinafine is a drug that belongs to the allylamine group, which includes also the topical antimycotic naftifine. Terbinafine is potent inhibitors of ergosterol biosynthesis, available as tablets, spray, cream, and gel formulations.

2.5.1 *Chemical Structure*

Chemically, terbinafine is (E)-*N*-(6,6-dimethyl-2-hepten-4-ynyl)-*N*-methyl-1- naphthalenemethanamine hydrochloride with a MW of 327.90 (Fig. 2.5).

2.5.2 *Mechanism of Action*

Terbinafine acts as antifungal drug by noncompetitive inhibition of squalene epoxidase, an enzyme that converts the squalene into 2,3-oxidosqualene that then form lanosterol, which is a precursor of ergosterol, an essential constituent of fungal membrane. At that point, the intracellular accumulation of squalene, which is toxic to fungal cells, and a deficiency in ergosterol cause a quick cell death [323, 324].

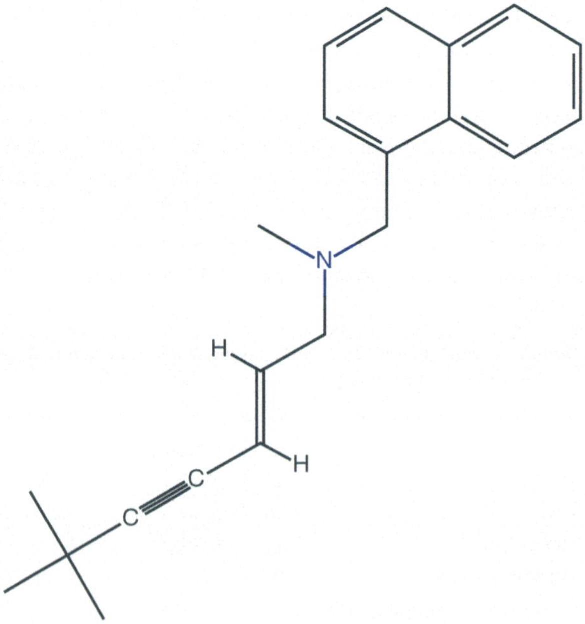

Fig. 2.5 Chemical structure of terbinafine. Source: Reference [322]

2.5.3 Pharmacokinetics and Pharmacodynamics

Terbinafine is well absorbed from the gastrointestinal tract, with a bioavailability ranging from 70 to 85%, which varies discretely when it is taken with foods no requiring dose modification. It has a low affinity for muscle, spleen, and brain tissues, but it has high affinity for the skin and adipose tissues, which are the largest depot for the antifungal drug [325, 326].

Gastric acidity does not seem to influence the absorption. Terbinafine reaches maximal plasma concentrations approximately 2 h after a single dose, getting high concentrations in the adipose tissue, stratum corneum, dermis, epidermis, sebum, nails, and in the hair follicle [327–329]. Penetration of the antifungal drug into the brain ranges between 6 and 43% of the plasma concentration, and is undetectable in CSF [325].

Linear dose-proportional pharmacokinetic has been proven after a range of single doses from 125 to 750 mg of terbinafine [330]. Steady-state plasma levels of terbinafine are attained after 10–14 days of treatment decreasing rapidly after the end of treatment [331, 332]. Plasma half-life ranges from 16 to 26 h after administration of 250 mg of terbinafine in healthy volunteers. A supplementary elimination phase with a half-life of more than 90 h was detected after administration of radiolabelled terbinafine [333–336].

This antifungal drug is extremely lipophilic, 99% protein bound and it has a good penetration in the keratinized tissue, reaching active concentrations in nail in 1–2 weeks. A characteristic of this drug is to remain in the nail in therapeutic concentrations for a long time after discontinuing the drug. Terbinafine is not detected in sweat [328, 329, 335].

Terbinafine is metabolized by the liver. Several cytochrome isoenzymes are involved in the main pathways of terbinafine metabolism. Some of them are CYP2C9, CYP2C8 and CYP1A2 (N-demethylation), CYP3A4 (deamination), CYP2C9, CYP2C8, CYP2C19, and CYP1A2 (alkyl side chain oxidation) and, CYP2C9 and CYP1A2 (dihydrodiol formation). The multiple cytochrome P-450 s implicated in the metabolism of this drug indicates a reduced potential for drug–drug interactions [337, 338]. Nonetheless, terbinafine may inhibit the metabolism of CYP2D6 substrates. It could be important for the interaction with some drugs that are known substrates of CYP2D6 such as amitriptyline, carvedilol, codeine, haloperidol, metoprolol, paroxetine, risperidone, and tramadol. The coadministration of terbinafine with these drugs with could cause a prolonged increase in their plasma levels [339].

Terbinafine is excreted mainly in the urine (80%) and in small amounts in feces (20%) [335]. In children, a dosage adjustment according to bodyweight is recommended because AUC values are significantly higher than in young adults [340].

2.5.4 Spectrum of Activities and Resistance

Terbinafine has potent antifungal activity against a diversity of dermatophytes isolated from patients worldwide, including *Trichophyton tonsurans*, *Microsporum canis*, *M. gypseum*, *T. verrucosum*, *T. violaceum*, *M. audouinii*, *T. rubrum*, *T. interdigitale*, *T. mentagrophytes*, and *Epidermophyton floccosum* [341–343]. Also, in vitro studies have shown that terbinafine is highly active against *Sporothrix brasiliensis*, *S. schenckii*, and *S. globosa* isolated from clinical samples in Brazil [344, 345], and for a wide variety of agents of chromoblastomycosis [346]. Terbinafine is only moderately active against *Madurella mycetomatis* isolates [347].

2.5.5 Clinical Uses

Terbinafine is available in different formulations, tablets, and topical preparations such as cream, jelly, spray, and solution. Topical formulations are used to treat superficial fungal infections due to dermatophytes or Candida species.

Terbinafine is the first line treatment for toenail onychomycosis due to dermatophytes [348, 349]. This antifungal drug presents the greatest rate of mycological cure according to results from a systematic review and a network meta-analysis conducted to compare the relative efficacy of treatments for onychomycosis due to dermatophyte [350]. Oral terbinafine has been also proved effective in treating onychomycosis due to *Scopulariopsis* species [350–352].

Terbinafine is also indicated for tinea capitis caused by *Trichophyton* species in children. A meta-analysis of randomized, controlled trials comparing griseofulvin and terbinafine for the treatment of tinea capitis has shown that terbinafine is superior for tinea capitis caused by *Trichophyton* spp., whereas griseofulvin is superior when *Microsporum* spp. is the etiological agent [353, 354]. For other tinea infections (corporis, cruris, or pedis) without response with cream/gel, oral terbinafine may be used.

For cutaneous and lymphocutaneous sporotrichosis, terbinafine administered at a dosage of 500 mg orally twice daily is considered a safe alternative treatment (A-II) to itraconazole [142]. Based in reports showing successful outcome using combination therapy including terbinafine for treating *Scedesporium* infections, the ESCMID and ECMM joint guidelines proposes the use of caspofungin plus terbinafine as salvage therapy for *Scedosporium* spp. infections in cystic fibrosis patients with lung transplantation (quality of evidence and strength of recommendation: CIII) and voriconazole plus terbinafine for patients with lung infections, osteomyelitis/septic arthritis and for disseminated infection due to *S. prolificans* (quality of evidence and strength of recommendation: BIII) [355–359].

ESCMID also recommends use of an azole or terbinafine plus surgery (AIII) for the treatment of eumycetoma or combination therapy with azoles plus terbinafine or flucytosine for refractory eumycetoma cases (B III). High-dose of terbinafine (1000 mg/day) alone for 24–48 weeks had shown to be clinically effective and well tolerated when used to treat patients with eumycetoma, whose only therapeutic option is surgery. Among 23 eumycetoma patients, 16 out of 20 who completed the study showed improvement or cure [360].

Few case reports about successful outcome of the use of terbinafine plus amphotericin B in the treatment of disseminated fusariosis had been published [361, 362].

Terbinafine can be used as alternative to itraconazole (BIII) in the treatment of chromoblastomycosis, and in combination with itraconazole for cases of refractory disease [302]. Terbinafine dosing for the most common indications in adult and pediatric patients are shown in Tables 2.8 and 2.9.

2.5.6 Adverse Events and Toxicity

Terbinafine is an antifungal drug well tolerated even in people over 65 years, many of whom may be taking antihypertensives, antidiabetics, or lipid-lowering agents concomitantly [363]. Adverse reactions related to terbinafine are usually mild in

Table 2.8 Terbinafine dosing for the most common indications in adult patients

Fungal disease	Doses (mg)	Duration	Comments
Tineas corporis or cruris	250 qd	2–4 weeks	
Tinea pedis	250 qd	2–6 weeks	
Tinea capitis	250 qd	6 weeks	Dosing for patients with more than 40 kg
Onychomycosis	250 qd	6 weeks (fingernails) or 12 weeks (toenails)	Sometimes a treatment for >6 months may be necessary
Sporotrichosis	500 bid	Until 2–4 weeks after clinical cure	Usually for 3–6 months

Source: References [142, 332, 348–350, 363]

Table 2.9 Pediatric dosing according body weight in patients with onychomycosis

Body weight (kg)	Doses (mg)	Comments
10–20	62.5	6 weeks (fingernails) or 12 weeks (toenails)
20–40	125	6 weeks (fingernails) or 12 weeks (toenails)
>40	250	6 weeks (fingernails) or 12 weeks (toenails)

Source: Reference [348]

severity. Gastrointestinal complaints are common events [364]. Patients on treatment with this drug may complain of nausea, diarrhea, bloating, dyspepsia, epigastric pain, and other less-frequent gastrointestinal symptoms [365]. More rare effects are skin reactions, cholestatic hepatitis and taste loss [366–369]. The frequency of taste loss ranges between 0.1 and 1% [369]. Risk factors associated with taste loss due to terbinafine are age (65 years and older, OR: 4.4, 95% CI: 1.4–16.1) and body mass index below 21 kg m^2 (OR:4.4, 95% CI: 1.6–14.2). The OR of taste loss in patients 55 years and older with a BMI below 21 kg m^2 is 12.8 when comparing with patients below 35 years old (95% CI: 1.9–88.6) [370].

Cutaneous adverse effects of terbinafine have a wide spectrum of presentation and are infrequent, having been reported in less than 2% of the patients. Recently, a case report of terbinafine-induced lichenoid drug eruption in a patient receiving the antifungal drug for 2 weeks was published. Lesions disappeared totally after 8 weeks of drug withdrawal [371]. Other rare cutaneous side effect induced by terbinafine is pityriasis rosea [372]. Also, the induction of subacute cutaneous lupus erythematosus and exacerbation of systemic lupus erythematosus by terbinafine have been reported [373, 374].

The incidence of serious side effect is less than 1%. Cases of Stevens-Johnson syndrome, neutropenia/agranulocytosis, thrombocytopenia, and aplastic anemia have been rarely reported [375–384]. Hepatotoxicity (including acute hepatitis, cholestasis, acute liver failure and vanishing bile duct syndrome) due to terbinafine has an incidence of 0.5–3/100,000. In most of these events improve after discontinuation of the drug [385–387]. A mixed hepatitis-cholestatic liver injury has also been reported [388].

2.5.7 *Drug Interactions*

Terbinafine does not inhibit or induce CYP 3A4, but it inhibits CYP2D6. Then, interactions with drugs that are metabolized by CYP 2D6 such as tricyclic antidepressants (amitriptyline, amoxapine, clomipramine, doxepin, imipramine, nortriptyline), ß-blockers (atenolol, metoprolol, propranolol, timolol, carvedilol), type B monoamine oxidase inhibitors (rasagiline and selegiline), some antipsychotic (chlorpromazine, haloperidol, risperidone, thioridazine), certain arrhythmics (lidocaine, procainamide), several medications for attention deficit hyperactivity disorder (atomoxetine, methamphetamine, methylphenidate), chloroquine, mirtazapine and dextromethorphan can potentially occur, increasing risk of side effects of these drugs. Concomitant use of tamoxifen with terbinafine should be avoided because it causes a decrease of endoxifen, one of the most important metabolites of tamoxifen, thereby decreasing the effectiveness of tamoxifen [389]. Terbinafine also mildly inhibits the metabolism of cyclosporine with little clinical significance [390].

2.5.8 Use in Special Population

No dosage modification is necessary in elderly patients. Terbinafine is contraindicated for patients with chronic or active hepatic disease [391]. Also, it is not recommended for patients with creatinine clearance lower than 50 mL/min because there are not satisfactory studies to confirm its safety in this population. Terbinafine has not been associated with any teratogenic toxicity in animals. Although it is not known whether terbinafine crosses the human placenta, it is categorized as a Pregnancy Category B drug by the FDA. The product labeling recommends against its use during pregnancy for both, topical and oral formulations. Small amounts of terbinafine are excreted in breast milk contraindicating its use in breastfeeding mothers [392].

References

1. Takemoto K, Yamamoto Y, Ueda Y (2006) Evaluation of antifungal pharmacodynamic characteristics of AmBisome against Candida albicans. Microbiol Immunol 50:579–586
2. Donovick R, Gold WH, Pagano JF, Stout HA (1955–1956) Amphotericins A and B, antifungal antibiotics produced by a streptomycete. I. In vitro studies. Antibiot Annu 3:579–586
3. Mechlinski W, Schaffner CP, Ganis P, Avitabile G (1970) Structure and absolute configuration of the polyene macrolide antibiotic amphotericin B. Tetrahedron Lett 44:3873–3876
4. National Center for Biotechnology Information. PubChem Compound Database; CID=5280965. https://pubchem.ncbi.nlm.nih.gov/compound/amphotericin%20B. Accessed 29 Jan 2017
5. Brajtburg J, Powderly WG, Kobayashi GS, Medoff G (1990) Amphotericin B: current understanding of mechanisms of action. Antimicrob Agents Chemother 34:183–188
6. Nair MP, Schwartz SA (1982) Immunomodulatory effects of amphotericin-B on cellular cytotoxicity of normal human lymphocytes. Cell Immunol 70:287–300
7. Mesa-Arango AC, Scorzoni L, Zaragoza O (2012) It only takes one to do many jobs: amphotericin B as antifungal and immunomodulatory drug. Front Microbiol 3:286
8. Christiansen KJ, Bernard EM, Gold JW, Armstrong D (1985) Distribution and activity of amphotericin B in humans. J Infect Dis 152:1037–1043
9. Collette N, van der Auwera P, Lopez AP, Heymans C, Meunier F (1989) Tissue concentrations and bioactivity of amphotericin B in cancer patients treated with amphotericin B-deoxycholate. Antimicrob Agents Chemother 33:362–368
10. Bekersky I, Fielding RM, Dressler DE, Lee JW, Buell DN, Walsh TJ (2002) Plasma protein binding of amphotericin B and pharmacokinetics of bound versus unbound amphotericin B after administration of intravenous liposomal amphotericin B (AmBisome) and amphotericin B deoxycholate. Antimicrob Agents Chemother 46:834–840
11. Ernst EJ, Klepser ME, Pfaller MA (2000) Postantifungal effects of echinocandin, azole, and polyene antifungal agents against Candida albicans and Cryptococcus neoformans. Antimicrob Agents Chemother 44:1108–1111
12. Atkinson AJ, Bennett JE (1978) Amphotericin B Pharmacokinetics in humans. Antimicrob Agents Chemother 13:271–276
13. Turnidge JD, Gudmundsson S, Vogelman B, Craig WA (1994) The postantibiotic effect of antifungal agents against common pathogenic yeasts. J Antimicrob Chemother 34:83–92
14. Andes D, Stamstad T, Conklin R (2001) Pharmacodynamics of amphotericin B in a neutropenic mouse disseminated-candidiasis model. Antimicrob Agents Chemother 45:922–926

15. Andes D, Safdar N, Marchillo K, Conklin R (2006) Pharmacokinetic-pharmacodynamic comparison of amphotericin B (AMB) and two lipid-associated AMB preparations, liposomal AMB and AMB lipid complex, in murine candidiasis models. Antimicrob Agents Chemother 50:674–684
16. Al-Nakeeb Z, Petraitis V, Goodwin J, Petraitiene R, Walsh TJ, Hope WW (2015) Pharmacodynamics of amphotericin B deoxycholate, amphotericin B lipid complex, and liposomal amphotericin B against Aspergillus fumigatus. Antimicrob Agents Chemother 59:2735–2745
17. Luna B, Drew RH, Perfect JR (2000) Agents for treatment of invasive fungal infections. Otolaryngol Clin N Am 33:277–299
18. Kutty K, Neicheril JC (1987) Treatment of pleural blastomycosis: penetration of amphotericin B into the pleural fluid. J Infect Dis 156:689–690
19. Craven PC, Ludden TM, Drutz DJ, Rogers W, Haegele KA, Skrdlant HB (1979) Excretion pathways of amphotericin B. J Infect Dis 140:329–341
20. Tortorano AM, Prigitano A, Biraghi E, Viviani MA (2005) The European Confederation of Medical Mycology (ECMM) survey of candidaemia in Italy: in vitro susceptibility of 375 *Candida albicans* isolates and biofilm production. J Antimicrob Chemother 56:777–779
21. Blinkhorm RJ, Adelstein D, Spagnuolo PJ (1989) Emergence of a new opportunistic pathogen, *Candida lusitaniae*. J Clin Microbiol 27:236–240
22. Walsh TJ, Melcher GP, Rinaldi MG, Lecciones J, McGough DA, Kelly P, Lee J, Callender D, Rubin M, Pizzo PA (1990) *Trichosporon beigelii*, an emerging pathogen resistant to amphotericin B. J Clin Microbiol 28:1616–1622
23. Messer SA, Jones RN, Fritsche TR (2006) International surveillance of *Candida* spp. and *Aspergillus* spp.: report from the SENTRY Antimicrobial Surveillance Program (2003). J Clin Microbiol 44:1782–1787
24. Martel CM, Parker JE, Bader O et al (2010) A clinical isolate of Candida albicans with mutations in ERG11 (encoding sterol 14alphademethylase) and ERG5 (encoding C22 desaturase) is cross-resistant to azoles and amphotericin B. Antimicrob Agents Chemother 54:3578–3583
25. Sanglard DIscher F, Parkinson T, Falconer D, Bille J (2003) Candida albicans mutations in the ergosterol biosynthetic pathway and resistance to several antifungal agents. Antimicrob Agents Chemother 47:2404–2412
26. Hull CM, Bader O, Parker JE et al (2012) Two clinical isolates of *Candida glabrata* exhibiting reduced sensitivity to amphotericin B both harbor mutations in ERG2. Antimicrob Agents Chemother 56:6417–6421
27. Vandeputte P, Tronchin G, Larcher G et al (2008) A nonsense mutation in the ERG6 gene leads to reduced susceptibility to polyenes in a clinical isolate of *Candida glabrata*. Antimicrob Agents Chemother 52:3701–3709
28. Young LY, Hull CM, Heitman J (2003) Disruption of ergosterol biosynthesis confers resistance to amphotericin B in Candida lusitaniae. Antimicrob Agents Chemother 47:2717–2724
29. Pappas PG, Kauffman CA, Andes DR et al (2016) Clinical practice guideline for the management of candidiasis: 2016 update by the Infectious Diseases Society of America. Clin Infect Dis 62:e1–50
30. Perfect JR, Dismukes WE, Dromer F et al (2010) Clinical practice guidelines for the management of cryptococcal disease: 2010 update by the Infectious Diseases Society of America. Clin Infect Dis 50:291–322
31. Skiada A, Lanternier F, Groll AH et al (2013) Diagnosis and treatment of mucormycosis in patients with hematological malignancies: guidelines from the 3rd European Conference on Infections in Leukemia (ECIL 3). Haematologica 98(4):492–504
32. Chapman SW, Dismukes WE, Proia LA et al, Infectious Diseases Society of America (2008) Clinical practice guidelines for the management of blastomycosis: 2008 update by the Infectious Diseases Society of America. Clin Infect Dis 46:1801–1812
33. Wheat LJ, Freifeld AG, Kleiman MB, et al, Infectious Diseases Society of America (2007) Clinical practice guidelines for the management of patients with histoplasmosis: 2007 update by the Infectious Diseases Society of America. Clin Infect Dis 45:807–825

34. Xia D, Sun WK, Tan MM et al (2015) Aerosolized amphotericin B as prophylaxis for invasive pulmonary aspergillosis: a meta-analysis. Int J Infect Dis 30:78–84
35. Patterson TF, Thompson GR 3rd, Denning DW et al (2016) Practice guidelines for the diagnosis and management of aspergillosis: 2016 update by the Infectious Diseases Society of America. Clin Infect Dis 63:e1–e60
36. Naidoff MA, Green WR (1975) Endogenous Aspergillus endophthalmitis occurring after kidney transplant. Am J Ophthalmol 79:502–509
37. Roney P, Barr CC, Chun CH, Raff MJ (1986) Endogenous Aspergillus endophthalmitis. Rev Infect Dis 8:955–958
38. Weishaar PD, Flynn HW Jr, Murray TG et al (1998) Endogenous Aspergillus endophthalmitis: clinical features and treatment outcomes. Ophthalmology 105:57–65
39. Wilmarth SS, May DR, Roth AM, Cole RJ, Nolan S, Goldstein E (1983) Aspergillus endophthalmitis in an intravenous drug user. Ann Ophthalmol 15:470–472
40. Essman TF, Flynn HW Jr, Smiddy WE et al (1997) Treatment outcomes in a 10-year study of endogenous fungal endophthalmitis. Ophthalmic Surg Lasers 28:185–194
41. Bae JH, Lee SC (2015) Intravitreal liposomal amphotericin B for treatment of endogenous candida endophthalmitis. Jpn J Ophthalmol 59:346–352
42. Drew RH, Arthur RR, Perfect JR (2005) Is it time to abandon the use of amphotericin B bladder irrigation? Clin Infect Dis 40:1465–1470
43. Sau K, Mambula SS, Latz E, Henneke P, Golenbock DT, Levitz SM (2003) The antifungal drug amphotericin B promotes inflammatory cytokine release by a tolllike receptor and CD14-dependent mechanism. J Biol Chem 278:37561–37568
44. Burke D, Lal R, Finkel KW, Samuels J, Foringer JR (2006) Acute amphotericin B overdose. Ann Pharmacother 40:2254–2259
45. Wiwanitkit V (2006) Severe hypertension associated with the use of amphotericin B: an appraisal on the reported cases. J Hypertens 24:1445
46. Rodrigues CA, Yamamoto M, Arantes Ade M, Chauffaille Mde L, Colombo AL, Bordin JO (2006) Amphotericin B-induced severe hypertension in a young patient: case reportand review of the literature. Ren Fail 28:185–187
47. Walker RW, Rosenblum MK (1992) Amphotericin B associated leukoencephalopathy. Neurology 42:2005–2010
48. Barton CH, Palh M, Vaziri ND, Cesario T (1984) Renal magnesium wasting associated with amphotericin B therapy. Am J Med 77:471–474
49. Lucas da Silva PS, Iglesias SB, Waisberg J (2007) Hypokalemic rhabdomyolysis in a child due to amphotericin B therapy. Eur J Pediatr 166:169–171
50. Sutherland SM, Hong DK, Balagtas J, Gutierrez K, Dvorak CC, Sarwal M (2008) Liposomal amphotericin b associated with severe hyperphosphatemia. Pediatr Infect Dis J 27:77–79
51. Olin JL, Spooner LM (2006) Amphotericin B-associated hyperbilirubinemia: case report and review of the literature. Pharmacotherapy 26:1011–1017
52. Bicanic T, Bottomley C, Loyse A et al (2015) Toxicity of Amphotericin B deoxycholate-based induction therapy in patients with HIV-associated cryptococcal meningitis. Antimicrob Agents Chemother 59:7224–7231
53. Day JN, Chau TT, Wolbers M et al (2013) Combination antifungal therapy for cryptococcal meningitis. N Engl J Med 368:1291–1302
54. Hamill RJ, Sobel JD, El-Sadr W et al (2010) Comparison of 2 doses of liposomal amphotericin B and conventional amphotericin B deoxycholate for treatment of AIDS associated acute cryptococcal meningitis: a randomized, double-blind clinical trial of efficacy and safety. Clin Infect Dis 51:225–2329
55. Wingard JR, Kubilis P, Lee L et al (1999) Clinical significance of nephrotoxicity in patients treated with amphotericin B for suspected or proven aspergillosis. Clin Infect Dis 29:1402–1407

56. White MH, Bowden RA, Sandler ES et al (1998) Randomized, double-blind clinical trial of amphotericin B colloidal dispersion vs. amphotericin B in the empirical treatment of fever and neutropenia. Clin Infect Dis 27:296–302
57. Lemke A, Kiderlen AF, Kayser O (2005) Amphotericin B. Appl Microbiol Biotechnol 68:151–162
58. Deray G (2002) Amphotericin B nephrotoxicity. J Antimicrob Chemoth 49(Suppl S1):37–41
59. Bicanic T, Wood R, Meintjes G et al (2008) High-dose amphotericin B with flucytosine for the treatment of cryptococcal meningitis in HIV-infected patients: a randomized trial. Clin Infect Dis 47:123–130
60. Llanos A, Cieza J, Bernardo J et al (1991) Effect of salt supplementation on amphotericin B nephrotoxicity. Kidney Int 40:302–308
61. Stein R, Alexander J (1989) Sodium protects against nephrotoxicity in patients receiving amphotericin B. Am J Med Sci 298:299–304
62. Girmenia C, Cimino G, Di Cristofano F, Micozzi A, Gentile G, Martino P (2005) Effects of hydration with salt repletion on renal toxicity of conventional amphotericin B empirical therapy: a prospective study in patients with hematological malignancies. Support Care Cancer 13:987–992
63. Bahr NC, Rolfes MA, Musubire A et al (2014) Standardized electrolyte supplementation and fluid management improves survival during amphotericin therapy for cryptococcal meningitis in resource-limited settings. Open Forum Infect Dis 1(2):ofu070. doi:10.1093/ofid/ofu070
64. Borro JM, Solé A, de la Torre M et al (2008) Efficiency and safety of inhaled amphotericin B lipid complex (Abelcet) in the prophylaxis of invasive fungal infections following lung transplantation. Transplant Proc 40:3090–3093
65. Rijnders BJ, Cornelissen JJ, Slobbe L et al (2008) Aerosolized liposomal amphotericin B for the prevention of invasive pulmonary aspergillosis during prolonged neutropenia: a randomized, placebo-controlled trial. Clin Infect Dis 46:1401–1408
66. Behre GF, Schwartz S, Lenz K et al (1995) Aerosol amphotericin B inhalations for prevention of invasive pulmonary aspergillosis in neutropenic cancer patients. Ann Hematol 71:287–291
67. Knechtel SA, Klepser ME (2007) Safety of aerosolized amphotericin B. Expert Opin Drug Saf 6:523–532
68. Andreucci M, Solomon R, Tasanarong A (2014) Side effects of radiographic contrast media: pathogenesis, risk factors, and prevention. Biomed Res Int 2014:741018
69. AmBisome (amphotericin B) liposome for injection, package insert. Astellas Pharma US, San Dimas, CA. Revised May 2012. https://www.astellas.us/docs/ambisome.pdf. Accessed 28 Oct 2016
70. Viread (Tenofovir disoproxil fumarate) Package Insert. Gilead Sciences, Inc., Foster City, CA. Revised Feb 2016. http://gilead.com/~/media/files/pdfs/medicines/liver-disease/viread/viread_pi.pdf. Accessed 23 Sept 2016
71. Retrovir (Zidovudine) Package Insert (2008) Glaxo SmithKline, Research Triangle Park, NC. http://www.accessdata.fda.gov/drugsatfda_docs/label/2008/019910s033lbl.pdf. Accessed 23 Sept 2016
72. Baley JE, Meyers C, Kliegman RM, Jacobs MR, Blumer JL (1990) Pharmacokinetics, outcome of treatment, and toxic effects of amphotericin B and 5-fluorocytosine in neonates. J Pediatr 116:791–797
73. Benson JM, Nahata MC (1989) Pharmacokinetics of amphotericin B in children. Antimicrob Agents Chemother 33:1989–1993
74. Koren G, Lau A, Klein J et al (1988) Pharmacokinetics and adverse effects of amphotericin B in infants and children. J Pediatr 113:559–563
75. Nath CE, McLachlan AJ, Shaw PJ, Gunning R, Earl JW (2001) Population pharmacokinetics of amphotericin B in children with malignant diseases. Br J Clin Pharmacol 52:671–680
76. Starke JR, Mason EO Jr, Kramer WG, Kaplan SL (1987) Pharmacokinetics of amphotericin B in infants and children. J Infect Dis 155:766–774

77. Le J, Adler-Shohet FC, Nguyen C, Lieberman JM (2009) Nephrotoxicity associated with amphotericin B deoxycholate in neonates. Pediatr Infect Dis J 28:1061–1063
78. Van den Anker JN, van Popele NM, Sauer PJ (1995) Antifungal agents in neonatal systemic candidiasis. Antimicrob Agents Chemother 39:1391–1397
79. Steinbach WJ (2005) Antifungal agents in children. Pediatr Clin N Am 52:895–915
80. Federal Register (2008) 73:30832. https://www.gpo.gov/fdsys/pkg/FR-2008-05-29/pdf/E8-11806.pdf. Accessed 20 Sept 2016
81. Abelcet (amphotericin B lipid complex injection), package insert. Enzon Pharmaceuticals, Inc., Bridgewater, NJ. Revised October 2010. http://dailymed.nlm.nih.gov/dailymed/archives/fdaDrugInfo.cfm?archiveid=41233. Accessed 28 Oct 2016
82. Ismail MA, Lerner SA (1982) Disseminated blastomycosis in a pregnant woman: review of amphotericin B usage during pregnancy. Am Rev Respir Dis 126:350–353
83. McCoy MJ, Ellenberg JF, Killam AP (1980) Coccidioidomycosis complicating pregnancy. Am J Obstet Gynecol 137:739–740
84. Kuo D (1962) A case of torulosis of the central nervous system during pregnancy. Med J Aust 49:558–560
85. Silberfarb PM, Sarosi GA, Tosh FE (1972) Cryptococcosis and pregnancy. Am J Obstet Gynecol 112:714–720
86. Curole DN (1981) Cryptococcal meningitis in pregnancy. J Reprod Med 26:317–319
87. Harris RE (1966) Coccidioidomycosis complicating pregnancy. Report of 3 cases and review of the literature. Obstet Gynecol 28:401–405
88. Smale LE, Waechter KG (1970) Dissemination of coccidioidomycosis in pregnancy. Am J Obstet Gynecol 107(3):356–361
89. Hadsall FJ, Acquarelli MJ (1973) Disseminated coccidioidomycosis presenting as facial granulomas in pregnancy: a report of two cases and a review of the literature. Laryngoscope 83:51–58
90. Hager H, Welt SI, Cardasis JP, Alvarez S (1988) Disseminated blastomycosis in a pregnant woman successfully treated with amphotericin-B. A case report. J Reprod Med 33:485–488
91. Neiberg AD, Mavromatis F, Dyke J, Fayyad A (1977) Blastomyces dermatitidis treated during pregnancy: report of a case. Am J Obstet Gynecol 128:911–912
92. Philpot CR, Lo D (1972) Cryptococcal meningitis in pregnancy. Med J Aust 2:1005–1007
93. Aitken GW, Symonds EM (1962) Cryptococcal meningitis in pregnancy treated with amphotericin B. A case report. J Obstet Gynaecol Br Emp 69:677–679
94. Youssef D, Raval B, El-Abbassi A, Patel P (2013) Pulmonary blastomycosis during pregnancy: case report and review of the literature. Tenn Med 106:37–39
95. Nayak SU, Talwani R, Gilliam B, Taylor G, Ghosh M (2011) Cryptococcal meningitis in an HIV-positive pregnant woman. J Int Assoc Physicians AIDS Care (Chic) 10:79–82
96. Crum NF, Ballon-Landa G (2006) Coccidioidomycosis in pregnancy: case report and review of the literature. Am J Med 119:993.e11–993.e17
97. Ilett KF, Kristensen JH (2005) Drug use and breastfeeding. Expert Opin Drug Saf 4:745–768
98. Mueller M, Balasegaram M, Koummuki Y, Ritmeijer K, Santana MR, Davidson R (2006) A comparison of liposomal amphotericin B with sodium stibogluconate for the treatment of visceral leishmaniasis in pregnancy in Sudan. J Antimicrob Chemother 58:811–815
99. Pipitone MA, Gloster HM (2005) A case of blastomycosis in pregnancy. J Am Acad Dermatol 53:740–741
100. Dean JL, Wolf JE, Ranzini AC, Laughlin MA (1994) Use of amphotericin B during pregnancy: case report and review. Clin Infect Dis 18:364–368
101. Adler-Moore JP, Gangneux JP, Pappas PG (2016) Comparison between liposomal formulations of amphotericin B. Med Mycol 54:223–231
102. Olson JA, Adler-Moore JP, Jensen GM, Schwartz J, Dignani MC, Proffitt RT (2008) Comparison of the physicochemical, antifungal, and toxic properties of two liposomal Amphotericin B Products. Antimicrob Agents Chemother 52:259–268

103. Wingard JR, White MH, Anaissie E, Raffalli J, Goodman J, Arrieta A, L Amph/ABLC Collaborative Study Group (2000) A randomized, double-blind comparative trial evaluating the safety of liposomal amphotericin B versus amphotericin B lipid complex in the empirical treatment of febrile neutropenia. L Amph/ABLC Collaborative Study Group. Clin Infect Dis 31:1155–1163
104. Torrado JJ, Espada R, Ballesteros MP, Torrado-Santiago S (2008) Amphotericin B formulations and drug targeting. J Pharm Sci 97:2405–2425
105. Coukell AJ, Brogden RN (1998) Liposomal amphotericin B: therapeutic use in the management of fungal infections and visceral leishmaniasis. Drugs 55:585–612
106. Vogelsinger H, Weiler S, Djanani A et al (2006) Amphotericin B tissue distribution in autopsy material after treatment with liposomal amphotericin B and amphotericin B colloidal dispersion. Antimicrob Chemother 57:1153–1160
107. Perkins WR, Minchey SR, Boni LT et al (1992) Amphotericin B phospholipid interactions responsible for reduced mammalian cell toxicity. Biochim Biophys Acta 1107:271–282
108. Adler-Moore J (1994) AmBisome targeting to fungal infections. Bone Marrow Transplant 14(Suppl 5):S3–S7
109. Stone NR, Bicanic T, Salim R, Hope W (2016) Liposomal Amphotericin B (AmBisome(®)): a review of the pharmacokinetics, pharmacodynamics, clinical experience and future directions. Drugs 76:485–500
110. Latour JF, Fuhrmann C, Lagallarde C, Loreuil F (1996) Amphotericin B intralipid formulation: stability and particle size. J Antimicrob Chemother 37:1165–1169
111. Mehta J (1997) Do variations in molecular structure affect the clinical efficacy and safety of lipid-based amphotericin B preparations? Leuk Res 21:183–188
112. Boswell GW, Buell D, Bekersky I (1998) AmBisome (liposomal amphotericin B): a comparative review. J Clin Pharmacol 38:583–592
113. Bekersky I, Fielding RM, Dressler DE, Lee JW, Buell DN, Walsh TJ (2002) Pharmacokinetics, excretion, and mass balance of liposomal amphotericin B (AmBisome) and amphotericin B deoxycholate in humans. Antimicrob Agents Chemother 46:828–833
114. Felton T, Troke PF, Hope WW (2014) Tissue penetration of antifungal agents. Clin Microbiol Rev 27:68–88
115. Wade RL, Chaudhari P, Natoli JL, Taylor RJ, Nathanson BH, Horn DL (2013) Nephrotoxicity and other adverse events among inpatients receiving liposomal amphotericin B or amphotericin B lipid complex. Diagn Microbiol Infect Dis 76:361–367
116. Ashley ESD, Lewis R, Lewis JS, Martin C, Andes D (2006) Pharmacology of systemic antifungal agents. Clin Infect Dis 43:S28–S39
117. Wong-Beringer A, Jacobs RA, Guglielmo BJ (1998) Lipid formulations of amphotericin B: clinical efficacy and toxicities. Clin Infect Dis 27:603–618
118. Groll AH, Piscitelli SC, Walsh TJ (1998) Clinical pharmacology of systemic antifungal agents: a comprehensive review of agents in clinical use, current investigational compounds, and putative targets for antifungal drug development. Adv Pharmacol 44:343–499
119. Groll AH, Giri N, Petraitis V et al (2000) Comparative efficacy and distribution of lipid formulations of amphotericin B in experimental Candida albicans infection of the central nervous system. J Infect Dis 182:274–282
120. Dupont B (2002) Overview of the lipid formulations of amphotericin B. J Antimicrob Chemother 49(Suppl 1):31–36
121. Kethireddy S, Andes D (2007) CNS pharmacokinetics of antifungal agents. Expert Opin Drug Metab Toxicol 3:573–581
122. Strenger V, Meinitzer A, Donnerer J et al (2014) Amphotericin B transfer to CSF following intravenous administration of liposomal amphotericin B. J Antimicrob Chemother 69:2522–2526
123. Ostrosky-Zeichner L, Marr KA, Rex JH, Cohen SH (2003) Amphotericin B: time for a new "gold standard". Clin Infect Dis 37:415–425

124. Richard JH, Amphotericin B (2013) Formulations: a comparative review of efficacy and toxicity. Drugs 73:919–934
125. Montagna MT, Lovero G, Coretti C et al (2014) In vitro activities of amphotericin B deoxycholate and liposomal amphotericin B against 604 clinical yeast isolates. J Med Microbiol 63:1638–1643
126. Johnson EM, Ojwang JO, Szekely A, Wallace TL, Warnock DW (1998) Comparison of in vitro antifungal activities of free and liposome-encapsulated nystatin with those of four amphotericin B formulations. Antimicrob Agents Chemother 42:1412–1416
127. Carrillo-Muñoz AJ, Quindós G, Tur C et al (1999) In-vitro antifungal activity of liposomal nystatin in comparison with nystatin, amphotericin B cholesteryl sulphate, liposomal amphotericin B, amphotericin B lipid complex, amphotericin B desoxycholate, fluconazole and itraconazole. J Antimicrob Chemother 44:397–401
128. Clark JM, Whitney RR, Olsen SJ et al (1991) Amphotericin B lipid complex therapy of experimental fungal infections in mice. Antimicrob Agents Chemother 35:615–621
129. Hostetler JS, Clemons KV, Hanson LH, Stevens DA (1992) Efficacy and safety of amphotericin B colloidal dispersion compared with those of amphotericin B deoxycholate suspension for treatment of disseminated murine cryptococcosis. Antimicrob Agents Chemother 36:2656–2560
130. Pahls S, Schaffner A (1994) Comparison of the activity of free and liposomal amphotericin B in vitro and in a model of systemic and localized murine candidiasis. J Infect Dis 169:1057–1061
131. Clark AD, McKendrick S, Tansey PJ, Franklin IM, Chopra R (1998) A comparative analysis of lipid-complexed and liposomal amphotericin B preparations in haematological oncology. Br J Haematol 103:198–204
132. Enoch DA, Ludlam HA, Brown NM (2006) Invasive fungal infections: a review of epidemiology and management options. J Med Microbiol 55:809–818
133. Hiemenz JW, Walsh TJ (1996) Lipid formulations of amphotericin B: recent progress and future directions. Clin Infect Dis 22(Suppl 2):S133–S144
134. Mehta J, Kelsey S, Chu P et al (1997) Amphotericin B lipid complex (ABLC) for the treatment of confirmed or presumed fungal infections in immunocompromised patients with hematologic malignancies. Bone Marrow Transplant 20:39–43
135. Saliba F, Dupont B (2008) Renal impairment and amphotericin B formulations in patients with invasive fungal infections. Med Mycol 46:97–112
136. Sharkey PK, Graybill JR, Johnson ES et al (1996) Amphotericin B lipid complex compared with amphotericin B in the treatment of cryptococcal meningitis in patients with AIDS. Clin Infect Dis 22:315–321
137. Walsh TJ, Hiemenz JW, Seibel NL et al (1998) Amphotericin B lipid complex for invasive fungal infections: analysis of safety and efficacy in 556 cases. Clin Infect Dis 26:1383–1396
138. Walsh TJ, Finberg RW, Arndt C (1999) at al. Liposomal amphotericin B for empirical therapy in patients with persistent fever and neutropenia. National Institute of Allergy and Infectious Diseases Mycoses Study Group. N Engl J Med 340:764–771
139. Leenders AC, Daenen S, Jansen RL et al (1998) Liposomal amphotericin B compared with amphotericin B deoxycholate in the treatment of documented and suspected neutropenia-associated invasive fungal infections. Br J Haematol 103:205–212
140. Leenders AC, Reiss P, Portegies P et al (1997) Liposomal amphotericin B (AmBisome) compared with amphotericin B both followed by oral fluconazole in the treatment of AIDS-associated cryptococcal meningitis. AIDS 11:1463–1471
141. Johnson PC, Wheat LJ, Cloud GA, et al, U.S. National Institute of Allergy and Infectious Diseases Mycoses Study Group (2002) Safety and efficacy of liposomal amphotericin B compared with conventional amphotericin B for induction therapy of histoplasmosis in patients with AIDS. Ann Intern Med 137:105–109
142. Kauffman CA, Bustamante B, Chapman SW, Pappas PG, Infectious Diseases Society of America (2007) Clinical practice guidelines for the management of sporotrichosis: 2007 update by the Infectious Diseases Society of America. Clin Infect Dis 45:1255–1265

143. Bergman SJ, Tyagi I, Ronald K (2010) Antifungal dosing in critically ill patients. Curr Fungal Infect Rep 4:78–86
144. Würthwein G, Groll AH, Hempel G, Adler-Shohet FC, Lieberman JM, Walsh TJ (2005) Population pharmacokinetics of amphotericin B lipid complex in neonates. Antimicrob Agents Chemother 49:5092–5098
145. Gallis HA, Drew RH, Pickard WW (1990) Amphotericin B: 30 years of clinical experience. Rev Infect Dis 12:308–329
146. Barrett JP, Vardulaki KA, Conlon C, et al., Amphotericin B Systematic Review Study Group (2003) A systematic review of the antifungal effectiveness and tolerability of amphotericin B formulations. Clin Ther 25:1295–1320
147. Martino R (2004) Efficacy, safety and cost-effectiveness of Amphotericin B Lipid Complex (ABLC): a review of the literature. Curr Med Res Opin 20:485–504
148. Ueda S, Miyamoto S, Kaida K et al (2016) Safety and efficacy of treatment with liposomal amphotericin B in elderly patients at least 65 years old with hematological diseases. J Infect Chemother 22:287–291
149. Roden MM, Nelson LD, Knudsen TA et al (2003) Triad of acute infusion-related reactions associated with liposomal amphotericin B: analysis of clinical and epidemiological characteristics. Clin Infect Dis 36:1213–1220
150. Rex JH, Stevens DA (2015) Drugs active against fungi, pneumocystis and microsporidia. In: Mandell, Douglas, Bennett's principles and practice of infectious diseases, 8th edn. Elsevier Saunders, Philadelphia, PA, pp 485–490
151. Peyton LR, Gallagher S, Hashemzadeh M (2015) Triazole antifungals: a review. Drugs Today 51:705–718
152. National Center for Biotechnology Information. PubChem Compound Database; CID:55283. https://pubchem.ncbi.nlm.nih.gov/compound/itraconazole. Accessed 29 Jan 2017
153. National Center for Biotechnology Information. PubChem Compound Database; CID:3365. https://pubchem.ncbi.nlm.nih.gov/compound/fluconazole. Accessed 29 Jan 2017
154. National Center for Biotechnology Information. PubChem Compound Database; CID: 468595. https://pubchem.ncbi.nlm.nih.gov/compound/posaconazole. Accessed 29 Jan 2017
155. National Center for Biotechnology Information. PubChem Compound Database; CID:71616. https://pubchem.ncbi.nlm.nih.gov/compound/voriconazole. Accessed 29 Jan 2017
156. National Center for Biotechnology Information. PubChem Compound Database; CID: 6918485. https://pubchem.ncbi.nlm.nih.gov/compound/isavuconazole. Accessed 29 Jan 2017
157. Mast N, Zheng W, Stout CD, Pikuleva IA (2013) Antifungal azoles: structural insights into undesired tight binding to cholesterol-metabolizing CYP46A1. Mol Pharmacol 84:86–94
158. Shyadehi AZ, Lamb DC, Kelly SL et al (1996) The mechanism of the acyl-carbon bond cleavage reaction catalyzed by recombinant sterol 14 alpha-demethylase of Candida albicans (other names are: lanosterol 14 alpha-demethylase, P-45014DM, and CYP51). J Biol Chem 271:12445–12450
159. Rybak JM, Marx KR, Nishimoto AT, Rogers PD (2015) Isavuconazole: pharmacology pharmacodynamics, and current clinical experience with a new triazole antifungal agent. Pharmacotherapy 35:1037–1051
160. Courtney R, Wexler D, Radwanski E, Lim J, Laughlin M (2004) Effect of food on the relative bioavailability of two oral formulations of posaconazole in healthy adults. Br J Clin Pharmacol 57:218–222
161. Dekkers BG, Bakker M, van der Elst KC et al (2016) Therapeutic drug monitoring of posaconazole: an update. Curr Fungal Infect Rep 10:51–61
162. Moriyama B, Kadri K, Henning SA, Danner RL, Penzak SR, Walsh TJ (2015) Therapeutic drug monitoring and genotypic screening in the clinical use of voriconazole. Curr Fungal Infect Rep 9:74–87
163. Pascual A, Calandra T, Bolay S, Buclin T, Blle J, Marchetti O (2008) Voriconazole therapeutic drug monitoring in patients with invasive mycoses improves efficacy and safety outcomes. Clin Infect Dis 46:201–211

164. Brüggemann RJ, Aarnoutse RE (2015) Fundament and prerequisites for the application of an antifungal TDM service. Curr Fungal Infect Rep 9:122–129
165. Ashbee HR, Barnes RA, Johnson EM, Richardson MD, Gorton R, Hope WW (2014) Therapeutic drug monitoring (TDM) of antifungal agents: guidelines from the British Society for Medical Mycology. J Antimicrob Chemother 69:1162–1176
166. Wiederhold NP, Pennick GJ, Dorsey SA et al (2014) A reference laboratory experience of clinically achievable voriconazole, posaconazole, and itraconazole concentrations within the bloodstream and cerebral spinal fluid. Antimicrob Agents Chemother 58:424–431
167. Nagappan V, Deresinski S (2007) Reviews of anti-infective agents. Posaconazole: a broad-spectrum triazole antifungal agent. Clin Infect Dis 45:1610–1617
168. Thompson GR 3rd, Rendon A, Dos Santos RR et al (2016) Isavuconazole treatment of cryptococcosis and dimorphic mycoses. Clin Infect Dis 63:356–362
169. Parker JE, Warrilow AG, Price CL, Mullins JG, Kelly DE, Kelly SL (2014) Resistance to antifungals that target CYP51. J Chem Biol 7:143–161
170. Cuenca-Estrella M (2014) Antifungal drug resistance mechanisms in pathogenic fungi: from bench to bedside. Clin Microbiol Infect 20(Suppl 6):54–59
171. Alastruey-Izquierdo A, Melhem MS, Bonfietti LX, Rodriguez-Tudela JL (2015) Susceptibility test for fungi: clinical and laboratorial correlations in medical mycology. Rev Inst Med Trop Sao Paulo 57(Suppl 19):57–64
172. Wilson DT, Dimondi VP, Johnson SW, Jones TM, Drew RH (2016) Role of isavuconazole in the treatment of invasive fungal infections. Ther Clin Risk Manag 12:1197–1206
173. Barr VO, Zdyb EG, Postelnick M (2015) The clinical significance of azole antifungals' effects on the liver and transaminase levels. Curr Fungal Infect Rep 9:190–195
174. Feist A, Lee R, Osborne S, Lane J, Yung G (2012) Increased incidence of cutaneous squamous cell carcinoma in lung transplant recipients taking long-term voriconazole. J Heart Lung Transplant 31:1177–1181
175. Zwald FO, Spratt M, Lemos BD et al (2012) Duration of voriconazole exposure: an independent risk factor for skin cancer after lung transplantation. Dermatol Surg 38:1369–1374
176. Nix DE (2014) Cardiotoxicity induced by antifungal drugs. Curr Fungal Infect Rep 8:129–138
177. Panos G, Velissaris D, Karamouzos V, Matzaroglou C, Tyllianakis M (2016) Long QT syndrome leading to multiple cardiac arrests after posaconazole administration in an immune-compromised patient with sepsis: an unusual case report. Am J Case Reports 17:295–300
178. Brüggemann RJM, Alffenaar JC, Blijlevens NMA et al (2009) Clinical relevance of the pharmacokinetic interactions of azole antifungal drugs with other coadministered agents. Clin Infect Dis 48:1441–1458
179. Gubbins PO, Heldenbrand S (2009) Clinically relevant drug interactions of current antifungal agents. Mycoses 53:95–113
180. Miceli MH, Kauffman CA (2015) Isavuconazole: a new broad-spectrum triazole antifungal agent. Clin Infect Dis 61:1558–1565
181. Vadlapatla RK, Patel M, Paturi DK, Pal D, Mitra AK (2014) Clinically relevant drug-drug interactions between antiretrovirals and antifungals. Expert Opin Drug Metab Toxicol 10:561–580
182. Nivoix Y, Levêque D, Herbrecht R, Koffel JC, Beretz L, Ubeaud-Sequier G (2008) The enzymatic basis of drug-drug interactions with systemic triazole antifungals. Clin Pharmacokinet 47:779–792
183. Lempers VJC, Martial LC, Schreuder MF et al (2015) Drug-interactions of azole antifungals with selected immunosuppressants in transplant patients: strategies for optimal management in clinical practice. Curr Op Pharmacol 24:38–44
184. Hohmann C, Kang EM, Jancel T (2010) Rifampin and posaconazole coadministration leads to decreased serum posaconazole concentrations. Clin Infect Dis 50:939–940
185. Autmizguine J, Guptill JT, Cohen-Wolkowiez M, Benjamin DK Jr, Capparelli EV (2014) Pharmacokinetics and pharmacodynamics of antifungals in children: clinical implications. Drugs 74:891–909

186. Ramos-Martín V, O'Connor O, Hope W (2015) Clinical pharmacology of antifungal agents in pediatrics: children are not small adults. Curr Opin Pharmacol 24:128–134
187. Goldman JM, Abdel-Rahman SM (2016) Pharmacokinetic considerations in treating invasive pediatric fungal infections. Exp Op Drug Metabol Toxicol 12:645–655
188. Drogouti E, Pana ZD, Tragiannidis A, Hempel G, Groll A (2015) Clinical pharmacology of itraconazole in children and adolescents. Curr Fungal Infect Rep 9:65–73
189. Dokos C, Pieper S, Lehrnbecher T, Groll AH (2012) Pharmacokinetics, safety and efficacy of voriconazole in pediatric patients: an update. Curr Fungal Infect Rep 6:121–126
190. Pilmis B, Jullien V, Sobel J, Lecuit M, Lortholary O, Charlier C (2015) Antifungal drugs during pregnancy: an updated review. J Antimicrob Chemother 70:14–22
191. Nair AS (2014) Safety of intravenous voriconazole in renal failure. Med J DY Patil Univ 7:105–107
192. Payne KD, Hall RG (2016) Dosing of antifungal agents in obese people. Expert Rev Anti-Infect Ther 14:257–267
193. Kullberg BJ, Arendrup MC (2015) Invasive candidiasis. N Engl J Med 373:1445–1456
194. Wang JF, Xue Y, Zhu XB, Fan H (2015) Efficacy and safety of echinocandins versus triazoles for the prophylaxis and treatment of fungal infections: a meta-analysis of RCTs. Eur J Clin Microbiol Infect Dis 34:651–659
195. Pappas PG, Kauffman CA, Andes D, et al, Infectious Diseases Society of America (2009) Clinical practice guidelines for the management of candidiasis: 2009 update by the Infectious Diseases Society of America. Clin Infect Dis 48:503–535
196. Andes DR, Safdar N, Baddley JW,et al, Mycoses Study Group (2012) Impact of treatment strategy on outcomes in patients with candidemia and other forms of invasive candidiasis: a patient-level quantitative review of randomized trials. Clin Infect Dis 54:1110–1122
197. Denning DW, Ribaud P, Milpied N et al (2002) Efficacy and safety of voriconazole in the treatment of acute invasive aspergillosis. Clin Infect Dis 34:563–571
198. Herbrecht R, Denning DW, Patterson TF, et al, Invasive Fungal Infections Group of the European Organisation for Research and Treatment of Cancer and the Global Aspergillus Study Group (2002) Voriconazole versus amphotericin B for primary therapy of invasive aspergillosis. N Engl J Med 347:408–415
199. Perfect JR, Marr KA, Walsh TJ et al (2003) Voriconazole treatment for less-common, emerging, or refractory fungal infections. Clin Infect Dis 36:1122–1131
200. Patterson TF, Boucher HW, Herbrecht R,et al, European Organization for Research and Treatment of Cancer (EORTC) Invasive Fungal Infections Group (IFIG); Pfizer Global Aspergillus Study Group (2005) Strategy of following voriconazole versus amphotericin B therapy with other licensed antifungal therapy for primary treatment of invasive aspergillosis: impact of other therapies on outcome. Clin Infect Dis 41:1448–1452
201. Singh N, Limaye AP, Forrest G et al (2006) Combination of voriconazole and caspofungin as primary therapy for invasive aspergillosis in solid organ transplant recipients: a prospective, multicenter, observational study. Transplantation 81:320–326
202. Panackal AA, Parisini E, Proschan M (2014) Salvage combination antifungal therapy for acute invasive aspergillosis may improve outcomes: a systematic review and meta-analysis. Int J Infect Dis 28:80–94
203. Marr KA, Schlamm HT, Herbrecht R et al (2015) Combination antifungal therapy for invasive aspergillosis: a randomized trial. Ann Intern Med 162:81–89
204. Panackal AA (2016) Combination antifungal therapy for invasive aspergillosis revisited. Med Mycol Open Access 2(2). pii: 12
205. Ullmann AJ, Cornely OA, Burchardt A et al (2006) Pharmacokinetics, safety, and efficacy of posaconazole in patients with persistent febrile neutropenia or refractory invasive fungal infection. Antimicrob Agents Chemother 50:658–666
206. Krishna G, Moton A, Ma L, Medlock MM, McLeod J (2009) Pharmacokinetics and absorption of posaconazole oral suspension under various gastric conditions in healthy volunteers. Antimicrob Agents Chemother 53:958–966

207. Zoller E, Valente C, Baker K, Klepser ME (2010) Development, clinical utility, and place in therapy of posaconazole for prevention and treatment of invasive fungal infections. Drug Des Devel Ther 4:299–311
208. Percival KM, Bergman SJ (2014) Update on posaconazole pharmacokinetics: comparison of old and new formulations. Curr Fungal Infect Rep 8:139–145
209. Wiederhold NP (2015) Pharmacokinetics and safety of posaconazole delayed-release tablets for invasive fungal infections. Clin Pharmacol 8:1–8
210. Cornely OA, Maertens J, Winston DJ et al (2007) Posaconazole vs. fluconazole or itraconazole prophylaxis in patients with neutropenia. N Engl J Med 356:348–359
211. Bertz H, Drognitz K, Lübbert M (2014) No difference between posaconazole and fluconazole antifungal prophylaxis and mycological diagnostics except costs in patients undergoing AML chemotherapy: a 1-year "real-life" evaluation. Ann Hematol 93:165–167
212. Walsh TJ, Raad I, Patterson TF et al (2007) Treatment of invasive aspergillosis with posaconazole in patients who are refractory to or intolerant of conventional therapy: an externally controlled trial. Clin Infect Dis 44:2–12
213. ClinicalTrials.gov. https://clinicaltrials.gov/ct2/show/NCT01782131?term=posaconazole+aspergillosis&rank=1. Accessed 17 Nov 2016
214. Greenberg RN, Mullane K, van Burik JA et al (2006) Posaconazole as salvage therapy for zygomycosis. Antimicrob Agents Chemother 50:126–133
215. Chitasombat MN, Kontoyiannis DP (2016) Treatment of mucormycosis in transplant patients: role of surgery and of old and new antifungal agents. Curr Opin Infect Dis 29:340–345
216. Durani U, Tosh PK, Barreto JN, Estes LL, Jannetto PJ, Tande AJ (2015) Retrospective comparison of posaconazole levels in patients taking the delayed-release tablet versus the oral suspension. Antimicrob Agents Chemother 59:4914–4918
217. Riley TT, Muzny CA, Swiatlo E, Legendre DP (2016) Breaking the mold: a review of mucormycosis and current pharmacological treatment options. Ann Pharmacother 50:747–757
218. Maertens JA, Raad II, Marr KA et al (2016) Isavuconazole versus voriconazole for primary treatment of invasive mould disease caused by Aspergillus and other filamentous fungi (SECURE): a phase 3, randomised-controlled, non-inferiority trial. Lancet 387:760–769
219. Falci DR, Pasqualotto AC (2013) Profile of isavuconazole and its potential in the treatment of severe invasive fungal infections. Infect Drug Resist 6:163–174
220. Carrillo-Muñoz AJ, Giusiano G, Arechavala A et al (2015) Clinical usefulness of triazole derivatives in the management of fungal infections. Rev Esp Quimioter 28:169–182
221. Gupta AK, Leonardi C, Stolz RR, Pierce PF, Conetta B, Ravuconazole Onychomycosis Group (2005) A phase I/II randomized, double-blind, placebo controlled, dose-ranging study evaluating the efficacy, safety and pharmacokinetics of ravuconazole in the treatment of onychomycosis. J Eur Acad Dermatol Venereal 19:437–443
222. Jo Siu WJ, Tatsumi Y, Senda H et al (2013) Comparison of in vitro antifungal activities of efinaconazole and currently available antifungal agents against a variety of pathogenic fungi associated with onychomycosis. Antimicrob Agents Chemother 57:1610–1616
223. Hector RF (1993) Compounds active against cell walls of medically important fungi. Clin Microbiol Rev 6:1
224. Debono M, Gordee RS (1994) Antibiotics that inhibit fungal cell wall development. Annu Rev Microbiol 48:471
225. Denning DW (2003) Echinocandin antifungal drugs. Lancet 362:1142
226. Lewis RE. Pharmacology of echinocandins. Up to date. Literature review current through: May 2016. This topic last updated: Feb 05, 2016
227. National Center for Biotechnology Information. PubChem Compound Database; CID: 2826718. https://pubchem.ncbi.nlm.nih.gov/compound/caspofungin. Accessed 29 Jan 2017
228. National Center for Biotechnology Information. PubChem Compound Database; CID: 477468. https://pubchem.ncbi.nlm.nih.gov/compound/micafungin. Accessed 29 Jan 2017
229. National Center for Biotechnology Information. PubChem Compound Database; CID: 166548. https://pubchem.ncbi.nlm.nih.gov/compound/anidulafungin. Accessed 29 Jan 2017

230. Fleet GH (1985) Composition and structure of yeast cell walls. Curr Top Med Mycol 1:24
231. Bowman JC, Hicks PS, Kurtz MB et al (2002) The antifungal echinocandin caspofungin acetate kills growing cells of Aspergillus fumigatus in vitro. Antimicrob Agents Chemother 46(9):3001–3012
232. Marr KA, Boeckh M, Carter RA et al (2004) Combination antifungal therapy for invasive aspergillosis. Clin Infect Dis 39(6):797–802
233. Lamaris GA, Lewis RE, Chamilos G et al (2008) Caspofungin-mediated beta-glucan unmasking and enhancement of human polymorphonuclear neutrophil activity against Aspergillus and non-Aspergillus hyphae. J Infect Dis 198:186
234. Eschenauer G, Depestel DD, Carver PL (2007) Comparison of echinocandin antifungals. Ther Clin Risk Manag 3(1):71–97
235. Kauffman CA, Carver PL (2008) Update on echinocandin antifungals. Semin Respir Crit Care Med 29(2):211–219
236. Dodds Ashley ES, Lewis R, Lewis JS, Martin C, Andes D (2006) Pharmacology of systemic antifungal agents. Clin Infect Dis 43(S1):S28–S39
237. Sucher AJ, Chahine EB, Balcer HE (2009) Echinocandins: the newest class of antifungals. Ann Pharmacother 43:1647–1657
238. Azanza Perea JR (2016) Echinocandins: applied pharmacology. Rev Iberoam Micol 33(3):140–144
239. Theuretzbacher U (2004) Pharmacokinetics/pharmacodynamics of echinocandins. Eur J Clin Microbiol Infect Dis 23:805
240. Cappelletty D, Eiselstein-McKitrick K (2007) The echinocandins. Pharmacotherapy 27:369
241. Pfaller MA, Boyken L, Hollis RJ et al (2008) In vitro susceptibility of invasive isolates of Candida spp. to anidulafungin, caspofungin, and micafungin: six years of global surveillance. J Clin Microbiol 46:150
242. Barchiesi F, Spreghini E, Tomassetti S et al (2006) Effects of caspofungin against Candida guilliermondii and Candida parapsilosis. Antimicrob Agents Chemother 50(8):2719–2727
243. Espinel-Ingroff A (1998) Comparison of in vitro activities of the new triazole Sch56592 and the echinocandins mk-0991 (l-743,872) and ly303366 against opportunistic filamentous and dimorphic fungi and yeasts. J Clin Microbiol 36(10):2950–2956
244. Nakai T, Uno J, Otomo K et al (2002) In vitro activity of FK463, a novel lipopeptide antifungal agent, against a variety of clinically important molds. Chemotherapy 48(2):78–81
245. Messer SA, Kirby JT, Sader HS et al (2004) Initial results from a longitudinal international surveillance programme for anidulafungin (2003). J Antimicrob Chemother 54(6):1051–1056
246. Kirkpatrick WR, Perea S, Coco BJ et al (2002) Efficacy of caspofungin alone and in combination with voriconazole in a guinea pig model of invasive aspergillosis. Antimicrob Agents Chemother 46(8):2564–2568
247. Diekema DJ, Messer SA, Hollis RJ et al (2003) Activities of caspofungin, itraconazole, posaconazole, ravuconazole, voriconazole, and amphotericin B against 448 recent clinical isolates of filamentous fungi. J Clin Microbiol 41(8):3623–3626
248. Almyroudis NG, Sutton DA, Fothergill AW et al (2007) In vitro susceptibilities of 217 clinical isolates of zygomycetes to conventional and new antifungal agents. Antimicrob Agents Chemother 51(7):2587–2590
249. Cuenca-Estrella M, Ruiz-Diez B, Martinez-Suarez JV et al (1999) Comparative in-vitro activity of voriconazole (UK-109,496) and six other antifungal agents against clinical isolates of Scedosporium prolificans and Scedosporium apiospermum. J Antimicrob Chemother 43(1):149–151
250. Tawara S, Ikeda F, Maki K et al (2000) In vitro activities of a new lipopeptide antifungal agent, FK463, against a variety of clinically important fungi. Antimicrob Agents Chemother 44:57
251. Kahn JN, Hsu MJ, Racine F et al (2006) Caspofungin susceptibility in Aspergillus and non-Aspergillus molds: inhibition of glucan synthase and reduction of beta-D-1,3 glucan levels in culture. Antimicrob Agents Chemother 50:2214
252. Ito M, Nozu R, Kuramochi T et al (2000) Prophylactic effect of FK463, a novel antifungal lipopeptide, against Pneumocystis carinii infection in mice. Antimicrob Agents Chemother 44:2259

253. Schmatz DM, Powles M, McFadden DC et al (1991) Treatment and prevention of Pneumocystis carinii pneumonia and further elucidation of the P. carinii life cycle with 1,3-beta-glucan synthesis inhibitor L-671,329. J Protozool 38:151S
254. Ramage G, VandeWalle K, Bachmann SP et al (2002) In vitro pharmacodynamic properties of three antifungal agents against preformed Candida albicans biofilms determined by time-kill studies. Antimicrob Agents Chemother 46:3634
255. Kuhn DM, George T, Chandra J et al (2002) Antifungal susceptibility of Candida biofilms: unique efficacy of amphotericin B lipid formulations and echinocandins. Antimicrob Agents Chemother 46:1773
256. Marcos-Zambrano LJ, Escribano P, Bouza E, Guinea J (2016) Comparison of the antifungal activity of micafungin and amphotericin B against Candida tropicalis biofilms. J Antimicrob Chemother 71(9):2498–2501
257. Pham CD, Iqbal N, Bolden CB et al (2014) Role of FKS Mutations in Candida glabrata: MIC values, echinocandin resistance, and multidrug resistance. Antimicrob Agents Chemother 58:4690
258. Perlin DS, Shor E, Zhao Y (2015) Update on antifungal drug resistance. Curr Clin Microbiol Rep 2(2):84–95
259. Zimbeck AJ, Iqbal N, Ahlquist AM et al (2010) FKS mutations and elevated echinocandin MIC values among Candida glabrata isolates from U.S. population-based surveillance. Antimicrob Agents Chemother 54:5042
260. Kofteridis DP, Lewis RE, Kontoyiannis DP (2010) Caspofungin-non-susceptible Candida isolates in cancer patients. J Antimicrob Chemother 65:293
261. Alexander BD, Johnson MD, Pfeiffer CD et al (2013) Increasing echinocandin resistance in Candida glabrata: clinical failure correlates with presence of FKS mutations and elevated minimum inhibitory concentrations. Clin Infect Dis 56:1724
262. Pfaller MA, Castanheira M, Lockhart SR et al (2012) Frequency of decreased susceptibility and resistance to echinocandins among fluconazole-resistant bloodstream isolates of Candida glabrata. J Clin Microbiol 50(4):1199–1203
263. Naicker SD, Magobo RE, Zulu TG et al (2016) Two echinocandin-resistant Candida glabrata FKS mutants from South Africa. Med Mycol Case Rep 11:24–26
264. Tan TY, Hsu LY, Alejandria MM et al (2016) Antifungal susceptibility of invasive Candida bloodstream isolates from the Asia-Pacific region. Med Mycol 54(5):471–477
265. Beyda ND, John J, Kilic A et al (2014) FKS mutant Candida glabrata: risk factors and outcomes in patients with candidemia. Clin Infect Dis 59:819
266. Wang E, Farmakiotis D, Yang D et al (2015) The ever-evolving landscape of candidaemia in patients with acute leukaemia: non-susceptibility to caspofungin and multidrug resistance are associated with increased mortality. J Antimicrob Chemother 70:2362
267. Imbert S, Castain L, Pons A et al (2016) Discontinuation of echinocandin and azole treatments led to the disappearance of an FKS alteration but not azole resistance during clonal Candida glabrata persistent candidemia. Clin Microbiol Infect 22(10):891
268. van Burik JA, Ratanatharathorn V, Stepan DE et al (2004) Micafungin versus fluconazole for prophylaxis against invasive fungal infections during neutropenia in patients undergoing hematopoietic stem cell transplantation. Clin Infect Dis 39(10):1407–1416
269. Walsh TJ, Teppler H, Donowitz GR et al (2004) Caspofungin versus liposomal amphotericin B for empirical antifungal therapy in patients with persistent fever and neutropenia. N Engl J Med 351(14):1391–1402
270. Reboli AC, Rotstein C, Pappas PG et al (2007) Anidulafungin versus fluconazole for invasive candidiasis. N Engl J Med 356(24):2472–2482
271. Mora-Duarte J, Betts R, Rotstein C et al (2002) Comparison of caspofungin and amphotericin B for invasive candidiasis. N Engl J Med 347(25):2020–2029
272. Villanueva A, Gotuzzo E, Arathoon EG et al (2002) A randomized double-blind study of caspofungin versus fluconazole for the treatment of esophageal candidiasis. Am J Med 113(4):294–299

273. de Wet NT, Bester AJ, Viljoen JJ et al (2005) A randomized, double blind, comparative trial of micafungin (FK463) vs. fluconazole for the treatment of oesophageal candidiasis. Aliment Pharmacol Ther 21(7):899–907
274. de Wet N, Llanos-Cuentas A, Suleiman J et al (2004) A randomized, double-blind, parallel-group, dose-response study of micafungin compared with fluconazole for the treatment of esophageal candidiasis in HIV-positive patients. Clin Infect Dis 39(6):842–849
275. Villanueva A, Arathoon EG, Gotuzzo E et al (2001) A randomized double-blind study of caspofungin versus amphotericin for the treatment of Candidal esophagitis. Clin Infect Dis 33(9):1529–1535
276. Krause DS, Simjee AE, van Rensburg C et al (2004) A randomized, double-blind trial of anidulafungin versus fluconazole for the treatment of esophageal candidiasis. Clin Infect Dis 39(6):770–775
277. Arathoon EG, Gotuzzo E, Noriega LM et al (2002) Randomized, double-blind, multicenter study of caspofungin versus amphotericin B for treatment of oropharyngeal and esophageal candidiases. Antimicrob Agents Chemother 46(2):451–457
278. Aliff TB, Maslak PG, Jurcic JG et al (2003) Refractory Aspergillus pneumonia in patients with acute leukemia: successful therapy with combination caspofungin and liposomal amphotericin. Cancer 97(4):1025–1032
279. Maertens J, Raad I, Petrikkos G et al (2004) Efficacy and safety of caspofungin for treatment of invasive aspergillosis in patients refractory to or intolerant of conventional antifungal therapy. Clin Infect Dis 39(11):1563–1571
280. Maertens J, Glasmacher A, Herbrecht R et al (2006) Multicenter, noncomparative study of caspofungin in combination with other antifungals as salvage therapy in adults with invasive aspergillosis. Cancer 107(12):2888–2897
281. Kontoyiannis DP, Hachem R, Lewis RE et al (2003) Efficacy and toxicity of caspofungin in combination with liposomal amphotericin B as primary or salvage treatment of invasive aspergillosis in patients with hematologic malignancies. Cancer 98(2):292–299
282. Grau S, Luque S, Echeverría-Esnal D et al (2016) Urinary micafungin levels are sufficient to treat urinary tract infections caused by Candida spp. Int J Antimicrob Agents 48(2):212–214
283. Cancidas (caspofungin acetate for injection). Highlights of prescribing information, revised April 2016. www.merck.com/product/usa/pi_circulars/c/cancidas/cancidas_pi.pdf. Accessed 15 Jan 2017
284. Mycamine (micafungin sodium for injection). Highlights of prescribing information, revised August 2016. www.astellas.us/docs/mycamine.pdf. Accessed 15 Jan 2017
285. Eraxis (Anidulafungin for injection). Highlights of prescribing information, revised July 2012. www.accessdata.fda.gov/drugsatfda_docs/label/2012/021632s011lbl.pdf. Accessed 15 Jan 2017
286. Lehrnbecher T, Groll AH (2010) Micafungin: a brief review of pharmacology, safety, and antifungal efficacy in pediatric patients. Pediatr Blood Cancer 55:229
287. Ryan DM, Lupinacci RJ, Kartsonis NA (2011) Efficacy and safety of caspofungin in obese patients. Med Mycol 49:748
288. Krishnan BR, James KD, Polowy K et al (2017) CD101, a novel echinocandin with exceptional stability properties and enhanced aqueous solubility. J Antibiot (Tokyo) 70(2):130–135
289. National Center for Biotechnology Information. PubChem Compound Database; CID: 3366. https://pubchem.ncbi.nlm.nih.gov/compound/flucytosine. Accessed 29 Jan 2017
290. Vermes A, Guchelaar HJ, Dankert J (2000) Flucytosine: a review of its pharmacology, clinical indications, pharmacokinetics, toxicity and drug interactions. J Antimicrob Chemother 46:171–179
291. Onishi J, Meinz M, Thompson J et al (2000) Discovery of novel antifungal (1,3)-beta-D-glucan synthase inhibitors. Antimicrob Agents Chemother 44:368–377
292. Cutler RE, Blair AD, Kelly MR (1978) Flucytosine kinetics in subjects with normal and impaired renal function. Clin Pharmacol Ther 24:333–342
293. Wade DN, Sudlow G (1972) The kinetics of 5-fluorocytosine elimination in man. Aust NZ J Med 2:153–158

294. Schönebeck J, Polak A, Fernex M, Scholer HJ (1973) Pharmacokinetic studies on the oral antimycotic agent 5-fluorocytosine in individuals with normal and impaired kidney function. Chemotherapy 18:321–336
295. Drouhet E, Babinet Chapusot JP, Kleinknecht D (1973) 5-fluorocytosine in the treatment of candidiasis with acute renal insufficiency. Biomedicine 19:408–414
296. Peman J, Canton E, Espinel-Ingroff A (2009) Antifungal drug resistance mechanisms. Expert Rev Anti-Infect Ther 7:453–460
297. Espinel-Ingroff A (2008) Mechanisms of resistance to antifungal agents: yeasts and filamentous fungi. Rev Iberoam Micol 25:101–106
298. Chapeland-Leclerc F, Bouchoux J, Goumar A, Chastin C, Villard J, Noel T (2005) Inactivation of the *FCY2* gene encoding purine-cytosine permease promotes cross-resistance to flucytosine and fluconazole in *Candida lusitaniae*. Antimicrob Agents Chemother 49:3101–3108
299. Vandeputte P, Pineau L, Larcher G, Noel T, Brèthes D, Chabasse D, Bouchara JP (2011) Molecular mechanisms of resistance to 5-fluorocytosine in laboratory mutants of Candida glabrata. Mycopathologia 171:11–21
300. Kontoyiannis DP, Lewis RE (2002) Antifungal drug resistance of pathogenic fungi. Lancet 359:1135–1144
301. Costa C, Ponte A, Pais P et al (2015) New mechanisms of flucytosine resistance in C. glabrata unveiled by a chemogenomics analysis in S. cerevisiae. PLoS One 10:e0135110
302. Chowdhary A, Meis JF, Guarro J,et al, European Society of Clinical Microbiology and Infectious Diseases Fungal Infection Study Group; European Confederation of Medical Mycology (2014) ESCMID and ECMM joint clinical guidelines for the diagnosis and management of systemic phaeohyphomycosis: diseases caused by black fungi. Clin Microbiol Infect 20(Suppl 3):47–75
303. Kauffman CA, Frame PT (1977) Bone marrow toxicity associated with 5-fluorocytosine therapy. Antimicrob Agents Chemother 11:244–247
304. Wise GJ, Goldberg P, Kozinn PJ, Nawabi IU (1976) Agranulocytosis associated with flucytosine for urinary candidiasis. Urology 8:490–491
305. Stamm AM, Diasio RB, Dismukes WE et al (1987) Toxicity of amphotericin B plus flucytosine in 194 patients with cryptococcal meningitis. Am J Med 83:236–242
306. Bennett JE, Dismukes WE, Duma RJ et al (1979) A comparison of amphotericin B alone and combined with flucytosine in the treatment of cryptococcal meningitis. N Engl J Med 301:26–131
307. White CA, Traube J (1982) Ulcerating enteritis associated with flucytosine therapy. Gastroenterology 83:1127–1129
308. Harder EJ, Hermans PE (1975) Treatment of fungal infections with flucytosine. Arch Intern Med 135:231–237
309. Vermes A, van der Sijs IH, Guchelaar HJ (2000) Flucytosine: correlation between toxicity and pharmacokinetic parameters. Chemotherapy 46:86–94
310. Vermes A, Mathot RAA, van der Sijs IH, Dankert J, Guchelaar HJ (2000) Population pharmacokinetics of flucytosine: comparison and validation of three models using STS, NPEM, and NONMEM. Ther Drug Monit 22:676–687
311. Fond B, Bentata-Pessayre M, Krivitzky A, Callard P, Dupont B, Delzant G (1983) Iatrogenic colitis during flucytosine treatment for neuromeningeal cryptococcosis. Sem Hop 59:1187
312. Sohail MA, Ikram U (2014) Flucytosine-induced colitis. BMJ Case Rep. pii: bcr2013203381. doi:10.1136/bcr-2013-203381
313. Cappell MS (2004) Colonic toxicity of administered drugs and chemicals. Am J Gastroenterol 99:1175–1190
314. Folk A, Cotoraci C, Balta C et al (2016) Evaluation of hepatotoxicity with treatment doses of flucytosine and amphotericin B for invasive fungal infections. Biomed Res Int 5398730
315. Holt RJ (1978) Clinical problems with 5-fluorocytosine. Mykosen 21:363–369
316. Richardson MD, Warnock DW (2003) Fungal infection: diagnosis and management, 3rd edn. Blackwell, Oxford, pp 66–69

317. Kunka ME, Cady EA, Woo HC, Thompson Bastin ML (2015) Flucytosine pharmacokinetics in a critically ill patient receiving continuous renal replacement therapy. Case Rep Crit Care 2015:927496
318. Richardson MD, Jones BL (2003) Therapeutic guidelines in systemic fungal infection, 3rd edn. Current Medical Literature, London. Publication link: ca9d0625-7a45-49c0-a3e1-41c6fd17e3c3, pp 53–55
319. Kuang D, Ronco C (2007) Adjustment of antimicrobial regimen in critically ill patients undergoing continuous renal replacement therapy. In: J-L Vincent (ed) Yearbook of intensive care and emergency medicine, pp 592–606
320. Stafford CR, Fisher JF, Fadel HE, Espinel-Ingroff AV, Shadomy S, Hamby M (1983) Cryptococcal meningitis in pregnancy. Obstet Gynecol 62:35S–37S
321. Njoku JC, Gumeel D, Hermsen ED (2010) Antifungal therapy in pregnancy and breastfeeding. Curr Fungal Infect Rep 4:62–69
322. National Center for Biotechnology Information. PubChem Compound Database; CID: 1549008. https://pubchem.ncbi.nlm.nih.gov/compound/terbinafine. Accessed 29 Jan 2017
323. Petranyi G, Ryder NS, Stütz A (1984) Allylamine derivatives: new class of synthetic antifungal agents inhibiting fungal squalene epoxidase. Science 224:1239–1419
324. Ryder NS (1992) Terbinafine: mode of action and properties of the squalene epoxidase inhibition. Br J Dermatol 126(Suppl 39):2–7
325. Hosseini-Yeganeh M, McLachlan AJ (2001) Tissue distribution of terbinafine in rats. J Pharm Sci 90:1817–1828
326. Hosseini-Yeganeh M, McLachlan AJ (2002) Physiologically based pharmacokinetic model for terbinafine in rats and humans. Antimicrob Agents Chemother 46:2219–2228
327. Leyden J (1998) Pharmacokinetics and pharmacology of terbinafine and itraconazole. J Am Acad Dermatol 38:S42–S47
328. Faergemann J, Zehender H, Jones T, Maibach HI (1990) Terbinafine levels in serum, stratum corneum, dermis epidermis (without stratum corneum), hair, sebum, and sweat. Acta Derm Venereol (Stockh) 71:322–326
329. Faergemann J, Zehender H, Denouël J, Millerioux L (1993) Levels of terbinafine in plasma, stratum corneum, dermis-epidermis (without stratum corneum), sebum, hair and nails during and after 250 mg terbinafine orally once per day for four weeks. Acta Derm Venereol 73:305–309
330. Kovarik JM, Kirkesseli S, Humbert H et al (1992) Dose-proportional pharmacokinetics of terbinafine and its A-demethylated metabolite in healthy volunteers. Br J Dermatol 126(Suppl 39):8–13
331. De Doncker P (1997) Pharmacokinetics of oral antifungal agents. Dermatol Ther 3:46–57
332. Villars V, Jones TC (1990) Present status of the efficacy and tolerability of terbinafine (Lamisil) used systemically in the treatment of dermatomycoses of skin and nails. J Dermatol Treat 1(Suppl. 2):33–38
333. Jensen JC (1990) Pharmacokinetics of Lamisil in humans. J Dermatol Treat 1(Suppl 2):15–18
334. Zehender H, Cabiac MD, Denouei J et al (1994) Elimination kinetics of terbinafine from human plasma and tissues following multiple-dose administration, and comparison with 3 main metabolites. Drug Invest 8:203–210
335. Debruyne D, Coquerel A (2001) Pharmacokinetics of antifungal agents in onychomycoses. Clin Pharmacokinet 40:441–472
336. Jensen JC (1989) Clinical pharmacokinetics of terbinafine (Lamisil). Clin Exp Dermatol 14:110–113
337. Meletiadis J, Chanock S, Walsh TJ (2006) Human pharmacogenomic variations and their implications for antifungal efficacy. Clin Microbiol Rev 19:763–787
338. Lynch T, Price A (2007) The effect of cytochrome P450 metabolism on drug response, interactions, and adverse effects. Am Fam Physician 76:391–396
339. Vickers AE, Sinclair JR, Zollinger M et al (1999) Multiple cytochrome P-450s involved in the metabolism of terbinafine suggest a limited potential for drug-drug interactions. Drug Metab Dispos 27:1029–1038

340. Nejjam F, Zagula M, Cabiac MD et al (1995) Pilot study of terbinafine in children suffering from tinea capitis: evaluation of efficacy, safety and pharmacokinetics. Br J Dermatol 132:98–105
341. Ghannoum MA, Wraith LA, Cai B, Nyirady J, Isham N (2008) Susceptibility of dermatophyte isolates obtained from a large worldwide terbinafine tinea capitis clinical trial. Br J Dermatol 159:711–713
342. Fernández-Torres B, Carrillo AJ, Martín E et al (2001) In vitro activities of 10 antifungal drugs against 508 dermatophyte strains. Antimicrob Agents Chemother 45:2524–2528
343. Carrillo-Muñoz AJ, Giusiano G, Cárdenes D, Fernández-Molina JM, Eraso E, Quindós G, Guardia C, del Valle O, Tur-Tur C, Guarro J (2008) Terbinafine susceptibility patterns for onychomycosis-causative dermatophytes and Scopulariopsis brevicaulis. Int J Antimicrob Agents 31:540–543
344. Borba-Santos LP, Rodrigues AM, Gagini TB et al (2015) Susceptibility of Sporothrix brasiliensis isolates to amphotericin B, azoles, and terbinafine. Med Mycol 53:178–188
345. Ottonelli Stopiglia CD, Magagnin CM, Castrillón MR et al (2014) Antifungal susceptibilities and identification of species of the Sporothrix schenckii complex isolated in Brazil. Med Mycol 52:56–64
346. Daboit TC, Massotti Magagnin C, Heidrich D et al (2014) In vitro susceptibility of chromoblastomycosis agents to five antifungal drugs and to the combination of terbinafine and amphotericin B. Mycoses 57:116–120
347. van Belkum A, Fahal AH, van de Sande WW (2011) In vitro susceptibility of Madurella mycetomatis to posaconazole and terbinafine. Antimicrob Agents Chemother 55:1771–1773
348. Ameen M, Lear JT, Madan V, Mohd Mustapa MF, Richardson M (2014) British Association of Dermatologists' guidelines for the management of onychomycosis 2014. Br J Dermatol 171:937–958
349. de Sá DC, Lamas AP, Tosti A (2014) Oral therapy for onychomycosis: an evidence-based review. Am J Clin Dermatol 15:17–36
350. Gupta AK, Daigle D, Foley KA (2015) Network meta-analysis of onychomycosis treatments. Skin Appendage Disord 1:74–81
351. Gupta AK, Gregurek-Novak T (2001) Efficacy of itraconazole, terbinafine, fluconazole, griseofulvin and ketoconazole in the treatment of Scopulariopsis brevicaulis causing onychomycosis of the toes. Dermatology 202:235–238
352. Onsberg P (1980) Scopulariopsis brevicaulis in nails. Dermatologica 161:259–264
353. Chen X, Jiang X, Yang M et al (2016) Systemic antifungal therapy for tinea capitis in children. Cochrane Database Syst Rev 5:CD004685
354. Gupta AK, Drummond-Main C (2013) Meta-analysis of randomized, controlled trials comparing particular doses of griseofulvin and terbinafine for the treatment of tinea capitis. Pediatr Dermatol 30:1–6
355. Howden BP, Slavin MA, Schwarer AP, Mijch AM (2003) Successful control of disseminated Scedosporium prolificans infection with a combination of voriconazole and terbinafine. Eur J Clin Microbiol Infect Dis 22:111–113
356. Bhat SV, Paterson DL, Rinaldi MG, Veldkamp PJ (2007) Scedosporium prolificans brain abscess in a patient with chronic granulomatous disease: successful combination therapy with voriconazole and terbinafine. Scand J Infect Dis 39:87–90
357. Gosbell IB, Toumasatos V, Yong J, Kuo RS, Ellis DH, Perrie RC (2003) Cure of orthopaedic infection with Scedosporium prolificans, using voriconazole plus terbinafine, without the need for radical surgery. Mycoses 46:233–236
358. Li JY, Yong TY, Grove DI, Coates PT (2008) Successful control of Scedosporium prolificans septic arthritis and probable osteomyelitis without radical surgery in a long-term renal transplant recipient. Transpl Infect Dis 10:63–65
359. Tortorano AM, Richardson M, Roilides E, et al, European Society of Clinical Microbiology and Infectious Diseases Fungal Infection Study Group (2014) European ESCMID and ECMM joint guidelines on diagnosis and management of hyalohyphomycosis: Fusarium spp., Scedosporium spp. and others. Clin Microbiol Infect 20(Suppl 3):27–46

360. N'diaye B, Dieng MT, Perez A, Stockmeyer M, Bakshi R (2006) Clinical efficacy and safety of oral terbinafine in fungal mycetoma. Int J Dermatol 45:154–157
361. Rothe A, Seibold M, Hoppe T et al (2004) Combination therapy of disseminated Fusarium oxysporum infection with terbinafine and amphotericin B. Ann Hematol 83:394–397
362. Neuburger S, Massenkeil G, Seibold M et al (2008) Successful salvage treatment of disseminated cutaneous fusariosis with liposomal amphotericin B and terbinafine after allogeneic stem cell transplantation. Transpl Infect Dis 10:290–293
363. Tavakkol A, Fellman S, Kianifard F (2006) Safety and efficacy of oral terbinafine in the treatment of onychomycosis: analysis of the elderly subgroup in Improving Results in Onychomycosis-Concomitant Lamisil and Debridement (IRON-CLAD), an open-label, randomized trial. Am J Geriatr Pharmacother 4:1–13
364. Villars VV, Jones TC (1983) Special features of the clinical use of oral terbinafine in the treatment of fungal diseases. Br J Med 308:1275–1279
365. O'Sullivan DP, Needham CA, Bangs A, Atkin K, Kendall FD (1996) Postmarketing surveillance of oral terbinafine in the UK: report of a large cohort study. Br J Clin Pharmacol 42:559–565
366. Van't Wout JW, Herrmann WA, De Vries RA, Stricker BHC (1994) Terbinafine-associated hepatic injury. J Hepatol 21:115–117
367. Juhlin L (1992) Loss of taste and terbinafine. Lancet 339:1483
368. Doty RL, Haxel BR (2005) Objective assessment of terbinafine-induced taste loss. Laryngoscope 115:2035–2037
369. Beutler M, Hartmann K, Kuhn M, Gartmann J (1993) Taste disorders and terbinafine. Br Med J 307:26
370. Stricker BH, Van Riemsdijk MM, Sturkenboom MC, Ottervanger JP (1996) Taste loss to terbinafine: a case-control study of potential risk factors. Br J Clin Pharmacol 42:313–318
371. Zheng Y, Zhang J, Chen H, Lai W, Maibach HI (2016) Terbinafine-induced lichenoid drug eruption. Cutan Ocul Toxicol 30:1–3
372. George A, Bhatia A, Kanish B, Williams A (2015) Terbinafine induced pityriasis rosea-like eruption. Indian J Pharmacol 47:680–681
373. Bonsmann G, Schiller M, Luger TA, Ständer S (2001) Terbinafine-induced subacute cutaneous lupus erythematosus. J Am Acad Dermatol 44:925–931
374. Lorentz K, Booken N, Goerdt S, Goebeler M (2008) Subacute cutaneous lupus erythematosus induced by terbinafine: case report and review of literature. J Dtsch Dermatol Ges 6:823–827
375. Pillans PI, Boyd IW (2007) Toenails and agranulocytosis. Intern Med J 37:572–575
376. Gupta AK, Soori GS, Del Rosso JQ, Bartos PB, Shear NH (1998) Severe neutropenia associated with oral terbinafine therapy. J Am Acad Dermatol 38:765–767
377. Ornstein DL, Ely P (1998) Reversible agranulocytosis associated with oral terbinafine for onychomycosis. J Am Acad Dermatol 39:1023–1024
378. Shapiro M, Li LJ, Miller J (1999) Terbinafine-induced neutropenia. Br J Dermatol 140:1196–1197
379. Conjeevaram G, Vongthavaravat V, Sumner R, Koff RS (2001) Terbinafine-induced hepatitis and pancytopenia. Dig Dis Sci 46:1714–1716
380. Aguilar C, Mueller KK (2001) Reversible agranulocytosis associated with oral terbinafine in a pediatric patient. J Am Acad Dermatol 45:632–634
381. Kovacs MJ, Alshammari S, Guenther L, Bourcier M (1994) Neutropenia and pancytopenia associated with oral terbinafine. J Am Acad Dermatol 31:806
382. Tsai HH, Lee WR, Hu CH (2002) Isolated thrombocytopenia associated with oral terbinafine. Br J Dermatol 147:627–628
383. Grunwald MH (1998) Thrombocytopenia associated with oral terbinafine. Int J Dermatol 37:634
384. Kantarcıoğlu B, Türköz HK, Yılmaz G et al (2014) Aplastic anemia associated with oral terbinafine: a case report and review of the literature. Turk J Haematol 31:411–416

385. Chitturi S, Farrell GC (2007) Drug-induced liver disease. In: Schiff ER, Sorrell MF, Maddrey WC (eds) Schiff's diseases of the liver, 10th edn. Lippincott Williams and Wilkins, Philadelphia, pp 924–1005
386. Gupta AK, del Rosso JQ, Lynde CW, Brown GH, Shear NH (1998) Hepatitis associated with terbinafine therapy: three case reports and a review of the literature. Clin Exp Dermatol 23:64–67
387. Anania FA, Rabin L (2002) Terbinafine hepatotoxicity resulting in chronic biliary ductopenia and portalfibrosis. Am J Med 112:741–742
388. Choudhary NS, Kotecha H, Saraf N, Gautam D, Saigal S (2014) Terbinafine induced liver injury: a case report. J Clin Exp Hepatol 4:264–265
389. Dürrbeck A, Nenoff P (2016) Terbinafine: relevant drug interactions and their management. Hautarzt 67:718–723
390. Jensen P, Lehne G, Fauchald P, Simonsen S (1996) Effect of oral terbinafine treatment on cyclosporin pharmacokinetics in organ transplant recipients with dermatophyte nail infection. Acta Derm Venereol 76:280–281
391. WHO Pharmaceuticals. Newsletter 2001, No. 02&03, p 6
392. (2007) Terbinafine hydrochloride [package insert]. Novartis Pharmaceuticals, East Hanover, NJ
393. Wagner C, Graninger W, Presterl E, Joukhadar C (2006) The echinocandins: comparison of their pharmacokinetics, pharmacodynamics and clinical applications. Pharmacology 78(4):161–177
394. Arendrup MC, Boekhout T, Akova M, Meis JF, Cornely OA, Lortholary O, European Society of Clinical Microbiology and Infectious Diseases Fungal Infection Study Group; European Confederation of Medical Mycology (2014) ESCMID and ECMM joint clinical guidelines for the diagnosis and management of rare invasive yeast infections. Clin Microbiol Infect 20(Suppl 3):76–98

Chapter 3
Candida and Candidiasis

Luis A. Pérez-García, José R. Macías-Pérez, Ángel León-Buitimea, Brenda Alvarado-Sánchez, Mayra D. Ramírez-Quijas, María J. Navarro-Arias, and Saraí C. Rodríguez-Reyes

Abstract *Candida* species comprehend a wide group of opportunistic pathogens that usually form a part of the microbiota on human linings. Under specific conditions such as immune depression, tissue invasion and infection may occur. *Candida* infections are ranked as the sixth most common cause of nosocomial infections according to studies by the Centers for Disease Control and Prevention, thus constituting a serious public health problem. *Candida albicans* has been the most studied member of this species; nevertheless, non-*albicans* species have risen as emergent pathogens in many regions across the globe. This text presents an outlook of *Candida* and its infections, including the historical aspects of the fungus, epidemiology and clinical manifestations of the infection, diagnostic techniques as well as treatment guidelines.

L.A. Pérez-García, Ph.D. (✉) • J.R. Macías-Pérez, Ph.D.
Á. León-Buitimea, Ph.D. • B. Alvarado-Sánchez, Ph.D.
Unidad Académica Multidisciplinaria Zona Huasteca, Universidad Autónoma de San Luis Potosí, Romualdo del Campo 501, Fracc. Rafael Curiel, C.P., 79060 Cd. Valles, San Luis Potosi, Mexico
e-mail: luisantonio.perez@uaslp.mx; roberto.macias@uaslp.mx; angel.leon@uaslp.mx; balvarado@uaslp.mx

M.D. Ramírez-Quijas, M.Sc.
Departamento de Ingeniería Biomédica, Universidad Politécnica del Bicentenario, Carretera Estatal Silao-Romita Km 2, Col. San Juan de los Durán, C.P., 36283 Silao de la Victoria, Guanajuato, Mexico
e-mail: mramirezq@upbicentenario.edu.mx

M.J. Navarro-Arias, M.Sc.
División de Ciencias Naturales y Exactas, Departamento de Biología, Universidad de Guanajuato, Campus Guanajuato, Noria Alta s/n, Col. Noria Alta, C.P., 36050 Guanajuato, Guanajuato, Mexico
e-mail: bt.mnavarro@hotmail.com

S.C. Rodríguez-Reyes, Ph.D.
Centro Universitario de Ciencias de la Salud, Instituto de Investigación en Ciencias Biomédicas, Universidad de Guadalajara, Sierra Mojada 950, Col. Independencia, C.P., 44340 Guadalajara, Jalisco, Mexico
e-mail: scitlalic.rdgz.r@gmail.com

H.M. Mora-Montes, L.M. Lopes-Bezerra (eds.), *Current Progress in Medical Mycology*, DOI 10.1007/978-3-319-64113-3_3

3.1 Introduction

Candidiasis or candidosis is the name of a group of varied infections caused by *Candida* species, ranging from colonization of the mucosa to invasive diseases. When the fungus reaches the bloodstream, the infection is called candidemia, being one of the most important opportunistic mycosis in the world, especially among hospitalized patients [1]. *Candida albicans* and related species are ranked as the sixth cause of most common nosocomial infections according to studies by the Centers for Disease Control and Prevention [2]. In this regard, invasive fungal infections have reached high morbidity and mortality rates, thus constituting a serious public health problem [3].

3.2 History of *C. albicans*

C. albicans remains among the most widespread fungal species implicated in human infections, and the history of its discovery extends from the ancients Greeks to modern-day researchers. Table 3.1 summarizes some of the discoveries about this organism.

From 1940 to the present there are 35,037 papers listed in PUBMED (https://www.ncbi.nlm.nih.gov/pubmed/) that have studied *C. albicans*. Table 3.2 compiles some of the works described from 2000 to 2017.

3.3 Etiology

The genus *Candida* comprises around 150 different species [50], of which approximately 10% are considered human pathogens. *C. albicans* is the most frequently isolated species in the blood cultures of patients with candidiasis, with an incidence of up to 70% [51]. However, in the last three decades, other species of *Candida* have been found to cause candidiasis in human patients such as *C. glabrata*, *C. parapsilosis*, *C. tropicalis* and *C. krusei* [3]. Other species such as *C. dubliniensis*, *C. orthopsilosis*, *C. guilliermondii*, *C. metapsilosis*, *C. inconspicua*, *C. lusitaniae*, *C. norvegensis*, *C. nivariensis*, *C. bracarensis,* or *C. auris* have been isolated as well from some patients with candidiasis [52–54].

The genus *Candida* has a characteristic oval morphology of 2–7 μm in diameter, it reproduces by blastoconidia and is able to assimilate and differentially ferment carbohydrates [55]. In healthy people, it inhabits the skin, mucosa, upper respiratory tract, and the genitourinary and digestive tracts. However, *Candida* can become an opportunistic pathogen, causing different types of candidiasis when the human host immune system is compromised.

Table 3.1 History of *C. albicans*

Event		Refs.
460–370 B. C.	The term "aphthous" was originated by **Hippocrates** in reference to "Mouths affected with aphthous ulcerations" the currently recognized oral candidiasis in the form of "thrush".	[4, 5]
1764	Thrush was first described by **Rosén von Rosenstein**, who mentions a disease of the mouth, more severe if it spreads into the lungs.	[6]
1839	**B. Langenbeck** was the first to describe in detail what is now recognized to be septate hyphae, branched pseudohyphae, and blastoconidia "Under the microscope magnified, the pseudomembranes consisted of an immense number of fungi grown in confusion".	[7]
1849	**J. S. Wilkinson** described thrush in the female genital tract when he found the fungus, which he called epiphytes in the vaginal discharge of a 77 years old woman.	[8]
1853	**C. P. Robin** called the fungus causing thrush as *Oidium albicans. Oidium* reflecting the egg shape of the yeast cell.	[9]
1864	**M. Burchardt** noted the filaments were branched and were periodically blocked by internal barriers called septa. Such septate filaments are now known as hyphae. *Candida* has pseudohyphae which did not have these septa.	[8]
1868	**C. Quinquaud** renamed the fungus as *Syringospora robinin.*	[10]
1869	**J. Parrot**, found that this fungus was not restricted to the mouth and could invade other organs (linings of the stomach, intestines, larynx, trachea, and bronchi). This is now called invasive candidiasis.	[8]
1877	**P. Grawitz**, supported the idea that this fungus was dimorphic. He cultivated the thrush fungus in acidic conditions and grows only as unicellular yeast cells and not as filaments. At the same time, in an independent work **M. Reess** described the hypha was capable of penetrating epithelial cells.	[8, 10]
1887	**C. Audry** showed that thrush was due to a single fungus, which he called *Saccharomyces albicans*, and its form varies according to the medium on which it was grown.	[8]
1889	**E. Laurent** classified the fungus as *Dematium albicans.*	[11]
1890	**W. F. Zopf** classified the fungus as *Monilia albicans.*	[12]
1918	**Mello and L.G. Fern** classified the fungus as *Parasaccharomyces albicans.*	[13]
1920	**Brownlie and Shattock** classified the fungus as *Blastomyces albicans.*	[14]
1923	**C. M. Berkhout** proposed the genus *Candida*, separating this genus from the universal *Monilia* genus that affects fruit and vegetables.	[15]

Table 3.1 (continued)

Event		Refs.
1928	**M. Ota** classified the fungus as *Myceloblastanon albicans.*	[16]
1931	**N.M Dekker** classified the fungus as *Endomycopsis albicans.*	[17, 18]
1931	**R. W. Benham** found that corn-meal agar is an important medium in studying the morphology of the *Mycotoruloideae*, especially for the development of the diagnostic chlamydospores of *Candida albicans.*	[19, 20]
1935	**C. W. Dodge** classified the fungus as *Syringospora albicans.*	[21]
1938	**M. Langeron and P. Guerra**, in France listed 102 synonyms for the thrush fungus *Candida albicans.*	[22]
1953	**W.J. Nickerson**, when inoculated onto two different agar media containing a complex bismuthyl-polyhydroxy-polysulfite (termed Bi-SO_3), *Candida albicans* develops black colonies.	[23, 24]
1954	**Eighth Botanical Congress** officially endorsed the binomial *Candida albicans* as the *nomen conservandum* formally ending the 200-year uncertainty over the etiology and taxonomy of *Candida.*	[8]
1980	**Hans Rieth** started the campaign of anti-*Candida* diet, promoting food with low-sugar content.	[25]
1985	**B. Slutsky, J. Buffo and D. R. Soll** discovered that *C. albicans* switches heritably and at high frequency between at least seven general phenotypes identified by colony morphology on agar.	[26]
1996	**Stanford Genome Technology Center**, supported by grants from the NIDCR, NIH, and the Burroughs Wellcome Fund, began whole genome shotgun sequencing of strain SC5314 of *C. albicans.*	http://www.sequence.stanford.edu/group/candida/index.html
1998	**H. Chibana, B. B. Magee, S. Grindle, Y. Ran, S. Scherer and P. T. Magee**, described the complete physical map of chromosome 7 from *C. albicans.*	[27]
1999	**C.M. Hull and A.D. Johnson** performed the identification of a mating-type locus in the asexual pathogenic yeast *C. albicans.*	[28]

Table 3.2 Discoveries of *C. albicans* (2000–2017)

Date	Report	Refs.
2000	**H. Chibana, J. L. Beckerman, and P. T. Magee** found four types of events to explain the genomic diversity in *C. albicans*: (1) Chromosome length polymorphism results from expansion and contraction of the RPS; (2) reciprocal translocation occurs at the Major Repeat Sequence loci; (3) chromosomal deletion; and (4) trisomy of individual chromosomes.	[29]
2000	**R. Kandasamy, G. Vediyappan, and W. L. Chaffin** demonstrated the existence of β1,3- glucan linked proteins in *C. albicans*, which are related to Pir family proteins of *S. cerevisiae*.	[30]
2000	**P. L. Jr. Fidel, J. Cutright and C. Steele** reported that estrogen but not progesterone, is an important factor in hormone-associated susceptibility to *C. albicans* vaginitis**.**	[31]
2001	**J. B. Anderson, C. Wickens, M. Khan et al.** described the infrequent genetic exchange and recombination in the mitochondrial genome of *C. albicans*.	[32]
2004	**T. Jones, N. A. Federspiel, H. Chibana, J. Dungan et al.** published a whole-genome description of heterozygosity in strain SC5314.	[33]
2004	**P. W. de Groot, A. D. de Boer, J. Cunningham et al**. identified by proteomic analysis several proteins of *C. albicans*: (1) five CWPs are predicted carbohydrate-active enzymes (Cht2p, Crh11p, Pga4p, Phr1p, and Scw1p); (2) Als1p and Als4p are believed to be adhesion proteins; (3) Sod4p/Pga2p is a putative superoxide dismutase and is possibly involved in counteracting host defense reactions.	[34]
2005	**H. Chibana, N. Oka, H. Nakayama et al.** reported the sequence finishing and gene mapping for *C. albicans* chromosome 7 and systemic analysis against the *Saccharomyces cerevisiae* genome.	[35]
2006	**D. P. Thomas, S. P. Bachmann, J. L. Lopez-Ribot** reported the use of proteomics for the comparative analysis of subcellular fractions obtained from *C. albicans* biofilm and planktonic cultures, including cell surface-associated proteins and secreted components present in liquid culture supernatants (for planktonic cultures) and exopolymeric substance (for biofilms).	[36]
2007	**M. van het Hoog, T. J. Rast, M. Martchenko et al.** published the assembly of the *C. albicans* genome into sixteen supercontigs aligned on the eight chromosomes.	[37]
2007	**Oliveira, et al.** showed in their study an association between clinical forms of oral candidiasis and the number of colonies of *C. albicans* in saliva, and that a systemic immune response characterized by the production of TNF-alpha and IFN-gamma is observed in patients with oral candidiasis.	[38]
2007	**K. M. Yeater, J. Chandra, G. Cheng et al.** used microarrays to identify changes in gene expression associated with *C. albicans* biofilm development.	[39]
2009	**G. Butler, M. D. Rasmussen, M. F. Lin et al.** reported the evolution of pathogenicity and sexual reproduction in eight *Candida* genomes.	[40]
2009	**M. Martínez-Gomariz, P. Perumal, S. Mekala et al.** published the proteomic analysis of cytoplasmic and surface proteins from *C. albicans*	[41]
2011	**S. Nicholls, et al.** demonstrated that Hsf1 activation, and thermal adaptation, contribute significantly to the virulence of *C. albicans*.	[42]
2013	**D. Muzzey, K. Schwartz, J. S. Weissman, and G. Sherlock** published the assembly of a phased diploid *C. albicans* genome that facilitates allele-specific measurements and provides a simple model for repeat and indel structure.	[43]

(continued)

Table 3.2 (continued)

Date	Report	Refs.
2013	**T. F. Bartelli, R. C. Ferreira, A. L. Colombo et al.** sequenced and assembled, with eightfold coverage, the mitochondrial genomes of two *C. albicans* clinical isolates (L296 and L757) and compared these sequences with the genome sequence of reference strain SC5314 and revealed 372 polymorphisms (230 coding/142 noncoding).	[44]
2015	**Maiti, et al.** performed the mapping of functional domains and characterize the transcription factor Cph1 that mediates morphogenesis in *C. albicans*.	[45]
2016	**R. Rajendran, A. May, L. Sherry, R. Kean et al.** integrated the *C. albicans* metabolism with biofilm heterogeneity by transcriptome mapping.	[46]
2016	**M. D. Leach, et al.** described a mechanism by which *C. albicans* responds to temperature via Hsf1 and Hsp90, to orchestrate gene expression and chromatin architecture, thereby enabling thermal adaptation and virulence.	[47]
2016	**R. Cevik, R. Tekin, and M. Gem** reported an unusual manifestation of *Candida* arthritis in a patient diagnosed with spondyloarthritis.	[48]
2017	**P. R. Zhou, H. Hua, X. S. Liu** reported that the value of 270 CFU/mL was considered a threshold for distinguishing oral candidiasis from healthy carriage.	[49]

C. albicans is a dimorphic species that can grow as hyphae or yeast (Fig. 3.1), is one of the few species that form true hyphae and chlamydoconidia [56]. Its genome consists of 8 pairs of homologue chromosomes with a total size of 16 Mb [29]; it is the most prevalent and pathogenic species of the genus *Candida*, since it causes between 37% and 70% of the fungal infections caused by this genus [3, 57].

C. parapsilosis sesu lato is responsible for 15.5% of candidiasis in North America, 16.3% in Europe and 23.4% in Latin America [58], mainly affecting neonates and immunocompromised patients [59]. This emerging pathogen is in fact a complex composed of three genetically distinct species: *C. parapsilosis sensu stricto*, *C. metapsilosis,* and *C. orthopsilosis* [60]. The genome of *C. parapsilosis sensu stricto* harbors 5733 genes, with a total size of 13.1 Mb [40].

C. tropicalis is one of the three most commonly isolated non-*albicans* species [61–63]. This is a diploid, dimorphic yeast that exists as either ellipsoidal budding cells or as a pseudomycelium, consisting of long, branched elements bearing conidia, in short chains or clusters [64]. In rare cases, *C. tropicalis* can form true hyphae [65]. Additionally, *C. tropicalis* has the ability to ferment and assimilate sucrose and maltose [66]. The number of chromosomes and the genomic size of *C. tropicalis* are not known precisely, but pulse-field gels reveal approximately 5–6 pairs of homologous chromosomes and a genome size of ~30 Mb [67].

C. glabrata is a nondimorphic species that grows as a yeast, but does not form hypha; and has been reported as the third cause of candidiasis in human patients [68], accounting for 18–26% of all systemic infections caused by *Candida* in the USA [69–71]. This pathogen presents a natural resistance to azole compounds, which are widely used in the treatment against candidiasis; therefore, infections caused by *C. glabrata* are difficult to treat due to inadequate patient management, which can lead to death, causing high mortality rates in immunocompromised and hospitalized patients [72]. The *C. glabrata* genome has 13 chromosomes, with a total size of 12.3 Mb [73].

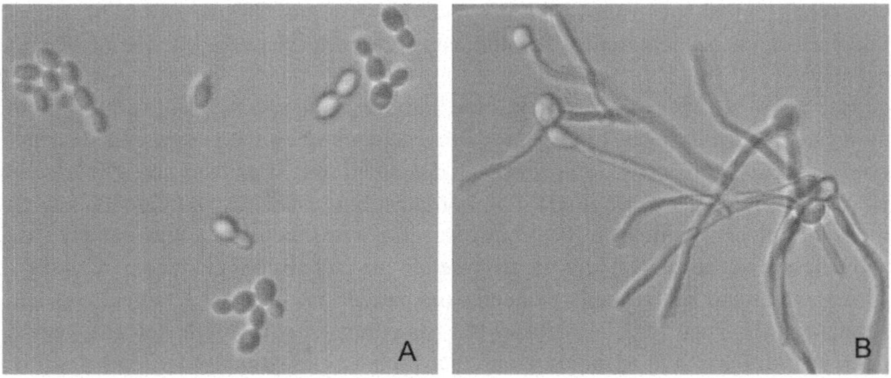

Fig. 3.1 Cellular morphology of *Candida* sp. (**A**) Yeast form and (**B**) hyphal form, cultured in liquid media, observed under 40× objective by means of bright field microscopy (Source: The authors)

3.4 Epidemiology

Candidiasis is a mycosis caused by the *Candida* genus, which has been detected on all continents since the 1980s. It has been classified as one of the most common causes of bloodstream infections [74]: incidences of 6–10 cases in 100,000 people have been reported in most population-based studies in the USA recent study in this country analyzed data from 33 states collected during 2002–2012, and found the average annual rate of invasive candidiasis was 5.3 hospitalizations per 100,000 population ratio [75]. In most European surveys, incidences of 1.4 to 5.7 cases per 100,000 habitants have been reported [76].

Several *Candida* species are commensal and colonize the skin and mucosal surfaces of humans. Critically ill, or otherwise immunocompromised patients, are prone to develop both superficial and life-threatening *Candida* infections [77].They also constitute the most common fungal infections in Acquired Immune Deficiency Syndrome (AIDS) patients, who predominantly develop oropharyngeal candidiasis, which can lead to malnutrition and interfere with the absorption of medication.

C. albicans is the predominant cause of candidiasis [78], with an incidence ranging from 11.5% in Turkey to 32% in Mexico and Taiwan, and over 60% in Austria and Sweden. Currently the incidence of *C. albicans* has decreased, while the incidence of infections due to *Candida* non-*albicans* species (*C. parapsilosis*, *C. tropicalis*, *C. glabrata,* and *C. krusei*) is increasing [79]. A series of studies have shown that the incidence of *C. albicans* has declined from 70% to 50% [80]. This has been reported both in North America and in European countries, which have gradually decreased the number of infections caused by *C. albicans* and have been replaced it by *C. glabrata* infections; whereas the incidence of *C. parapsilosis* or *C. tropicalis* infections has increased in other regions [79]. This change in epidemiology could be associated with severe immunosuppression or illness, prematurity, exposure to broad-spectrum antibiotics and older patients [78], as well as the use of fluconazole in the treatment of candidiasis and the increase in the use of venous catheters [3, 81, 82].

C. parapsilosis is considered by some authors as the second or third most common cause of candidiasis, after *C. albicans*, depending on the geographic distribution. Early reports of *C. parapsilosis* described the organism as a relatively nonpathogenic yeast belonging to the normal biota of healthy individuals, having modest clinical significance [83]; however, important factors such as catheter use and parental nutrition [84] have contributed to the increased incidence of *C. parapsilosis*, which has been mainly isolated from patients with cancer, young adults and neonates [58]. Therefore *C. parapsilosis* has emerged as a significant nosocomial pathogen with clinical manifestations that include endophthalmitis, endocarditis, septic arthritis, peritonitis and fungemia, usually associated with invasive procedures or prosthetic devices [85]. The percentage of *C. parapsilosis* isolates in candidiasis varies considerably depending on the region [86, 87]. This species is predominant in candidemia reported in Australia, Latin America, the Mediterranean area, and in countries of Africa, Asia, and Europe [59, 85]. In studies conducted by Wingar (1995), in a period from 1952–1992, *C. parapsilosis* covers only 7% of cases of candidemia in patients with cancer [88]. In 1999, Kao et al. conducted a prospective study of candidemia in two cities in the United States, Atlanta and San Francisco, from 1992 to 1993. *C. parapsilosis* was found in 21% of the isolates of patients from these populations [89]. A study from 1992 to 2001, recorded the distribution of Bloodstream Infection Isolates (BSI) of different *Candida* species. *C. parapsilosis* accounted for 13% and was the third most common *Candida* species isolated from BSI. The overall incidence in six different studies showed a prevalence between 7 and 21% of *C. parapsilosis* [3]. More recently, a retrospective study from 1991 to 2008 in the Clinic Hospital of Barcelona found *C. parapsilosis* as the second cause of candidemia after *C. albicans*, with an incidence between 14 and 20% [90]. This species presents many interesting biological characteristics that are supposed to be directly related to its virulence, such as its selective adhesion to prosthetic materials, the formation of biofilms on plastic surfaces [91], and the secretion of extracellular proteases [92, 93]. *C. parapsilosis* is usually susceptible to most antifungal agents, but there are reports of clinical isolates with decreased susceptibility to azoles and echinocandins.

C. tropicalis is one of the three most commonly isolated *Candida* non-*albicans* species [61–63, 77]. Its relative importance seems to be even greater in cases reported from Southern Europe and Latin America. *C. tropicalis* has been isolated from patients with solid tumors or hematologic diseases, and reported as the second etiological agent of invasive candidiasis in Asia and some parts of Latin America (Colombia and Brazil). In a recent epidemiological survey conducted in 12 Brazilian medical centers, *C. tropicalis* was the second most frequently recovered *Candida* species, accounting for 33–48% of all candidemia cases [63, 94]; however, a retrospective study from 1991 to 2008 at the Clinic Hospital in Barcelona found *C. tropicalis* as the third cause of candidemia after *C. albicans*, with an incidence between 13 and 14% [90]. Moreover, a review by Hobson in 2003 places *C. tropicalis* as the fourth cause of candidiasis infections worldwide as well [95]. *C. tropicalis* is commonly associated with patients with neutropenia and malignancy [63]. Additionally, it is often found in patients admitted to intensive care units, especially in patients requiring prolonged catheterization, receiving broad-spectrum antibiotics or with cancer [96–98]. Furthermore, *C. tropicalis* appears to display a higher potential to disseminate in neutropenic individuals, when compared with *C. albicans* and other non-*albicans* species [63]. This propensity of *C. tropicalis* for dissemination, and the associated high mortality, may be related to virulence factors exhibited by this species such as biofilm formation, proteinase secretion and dimorphism [99].

Several authors reported *C. glabrata* as the third or fourth common species (18.1%) isolated from candidiasis cases. In a prospective systematic epidemiological study carried out in Intensive Care Unit (ICU) at 27 centers of India, from all cases of acquired candidemia, *C. glabrata* was the fourth common species (7.1%) in adult patients [100]; however, in a retrospective study carried out in Taiwan from July 2009 to July 2012, the proportion of *C. glabrata* increased greatly from 1.1% to 21.6% [101]. Another retrospective study from 1991 to 2008 at the Clinic Hospital in Barcelona found *C. glabrata* as the fourth cause of candidemia, with an incidence between 7 and 14% [90]. Candidiasis caused by *C. glabrata* also demands great care due to its high mortality rate (53.8% in 30 days), and relatively higher resistance to azoles, especially fluconazole. In 2015, Chakrabarti indicated that broad spectrum antibiotics, central venous catheter, mechanical ventilation, diabetes, and elderly patients (>65 years) as risk factors for *C. glabrata* candidemia [100]. It has also been found that *C. glabrata* tends to affect oncology patients with solid tumors [84].

3.5 Virulence Factors of *Candida* spp.

Virulence factors of *Candida* are traits that facilitate colonization and infection of the host cell by the fungus [102]. Many authors define these attributes as "all traits required to establish the disease", "factors that directly interact with mammalian host cells" or as "a component of pathogen that damage the host" [103]. Here, some of them are briefly summarized.

Adhesion The adhesion of *Candida* to the host is essential to colonization and establishment of disease. The adhesion to surfaces of medical devices results in raised candidemia and antifungal resistance related to catheter insertion [103]. Adhesins are proteins that promote the adherence of *C. albicans* to the host or ligands. In *C. albicans*, hyphae are considered the best for adherence; nonetheless, the bulk of initial adhesion is between yeasts and epithelial cells, with hyphae formation being induced after the primary contact with a host cell surface. Once hyphae are formed, adhesins expressed solely on the hyphal cell surface are the main proteins for future adhesion processes [104]. *ALS* (agglutinin-like sequence) gene family encode the most studied adhesins, where the first, *ALS1–4*, encode adhesins specific for hyphae and the products of *ALS5–7* and *ALS9* are found on the surface of yeast cells [105]. *HWP1* encodes an outer surface mannoprotein that is implicated in adherence and virulence, since the *hwp1Δ/hwp1Δ* mutant strain was greatly impaired in the ability to form stable attachments to buccal epithelial cells, and this mutant is less virulent than wild-type strain in the mouse model of systemic candidiasis [104, 106]. There are also other kinds of proteins involved in adherence in an indirect manner, as the Mp65 protein which modifies the cell wall structure, by its glucanase activity, thus facilitating the function of other adhesins [107].

Invasion Secreted aspartyl proteinases (SAPs) are proteins reported to be present in *C. albicans*, *C. parapsilosis and C. guilliermondii* [108]. The function of these proteins is to degrade many human proteins at lesion sites, such as albumin, hemoglobin, keratin, and secretory immunoglobulin, and this proteolytic activity is involved with tissue invasion by the fungus. There are nine known proteins of this type, Sap1–3 are expressed only in yeast, Sap4–6 are expressed in hyphae, Sap7 has never been detected in in vitro assays, Sap8 expressed only in yeast cells grown at 25 °C in defined medium, and Sap9 which is expressed in the later grown phase [103]. Additionally, the phospholipases are hydrolytic enzymes that contributes to invasion by degradation of phospholipids of the cell membrane, resulting in the loss of barrier functions. Four phospholipases genes were identified *PLA*, *PLB*, *PLC*, and *PLD*. Plb1 is a glycoprotein expressed in hyphae that has both, hydrolase and lysophospholipase–transacylase activity and it is probably secreted. In vivo experiments have demonstrated that Plb1 is necessary for virulence and tissue invasion [103, 108].

Phenotypic Switch The transition from yeast to hyphal form is considered the most important virulence factor [109]. Since yeast seems to be essential for the dissemination of the infection and hyphae contribute to host tissue damage, both morphologies are thought to be important for pathogenicity [110]. Phenotypic switch is directed by the expression of activators like Efg1 and Cph1 via cAMP/PKA or MAPK pathway as a consequence of environmental stimuli [111]. At high cell densities, quorum sensing molecule farnesol also promotes hypha formation [112]. This morphological switch is associated with the expression of hypha-specific proteins such as Hwp1, Als3, Sap 4–6, Ece1, and Hyr1 that are not required for hypha formation and maintenance but are important for virulence [111, 112].

3.6 Common Clinical Manifestations

The clinical manifestations range from asymptomatic infections to septic shock syndrome. Additionally, candidemia can be recurrent if the underlying conditions, as retained intravascular catheters, neutropenia, and corticosteroids remain unchanged. Regrettably, the mortality rates associated with candidemia are repeatedly over 40% [113, 114].

Mucocutaneous candidiasis (MC) includes noninvasive symptomatic infections of the skin, nails, and mucous membranes caused by *Candida* species [115]. Presence of some host factors such as autosomal-dominant hyper-IgE syndrome, gain of function STAT1 and IL-17 pathway defects may result in increased susceptibility to mucocutaneous candidiasis and in some cases, may become a chronic infection [116]. Direct signs of MC are *Candida* plaques on the mucosa or thickened skin and nails [117]. The symptoms usually include pain, weight loss or failure to thrive, and may derive in complications such as squamous cell carcinoma, esophageal stricture or cerebral aneurysms [118–120].

Intra-abdominal candidiasis (IAC) is a very common type of deep-seated candidiasis [121], occurring in patients with a set of underlying conditions that specifically involves the gastrointestinal tract and digestive system [122]. In such type of infections, *Candida* may be mixed with bacteria, thus complicating the treatment, since in some cases antibiotic therapy and source control, without antifungal therapy, is enough to the improvement of the patients, but in some others, the presence of *Candida* greatly contributes to bad outcomes [123, 124]. Recent studies described that the most common types of IAC were, in order of occurrence, intra-abdominal abscesses, secondary peritonitis, primary peritonitis, infected pancreatic necrosis and cholecystitis/cholangitis, mainly caused by *C. albicans*, although non-albicans species were also reported. From the cases analyzed, over 60% exhibit mixed bacterial infections and 10% showed presence of two *Candida* species [122].

Septic arthritis caused by *Candida* is commonly developed as a complication of disseminated candidiasis [125–127], and frequently with the presence of underlying conditions such as surgery, solid organ transplantation, trauma, intravenous drug use, hematologic malignancies, among others [128]. An analysis of *Candida* arthritis cases from 1967 through 2014 showed that the apparent mechanism of infection mainly consisted of hematogenous dissemination and, in a minor level, of direct inoculation. The clinical manifestations include local pain, edema, limitation of function and movement, localized erythema and sometimes the disease is accompanied with a febrile process [128].

3.7 Diagnostic Strategies

The diagnosis of any pathological agent requires high sensitivity and specificity levels, being the microbiological culture the test of choice widely used over time. However, there are different tools other than the microbiological culture, which facilitate the diagnosis. These include smears, biopsies, molecular techniques, and serological tests.

3.7.1 Smear

Direct Exam. To make a smear, a sample is taken by scraping the affected area using a swab or spatula, the most commonly used specimens are urine, cerebrospinal fluid, feces vaginal discharge, and expectoration. The sample is spread on a slide, and treated with 10–20% KOH in the case of sputum samples. Finally, the sample is observed by conventional microscopy to search the characteristic structures of

Candida. We can observe individual or budding cells, but the number of cells found is more important [129].

Stains. The smear can be stained with different dyeing methods. The Gram staining is one of the most used, being *Candida* yeasts Gram positive structures. In this test, individual or budding cells (blastoconidia) can be detected with or without presence of pseudomycelium [129].

The smear is one of the most economical and quick tools for identifying a yeast from a bacterium. However, in both cases, direct exam or stain, the final diagnosis should consider the origin of the sample, since the presence of some species of *Candida* are considered normal biota in some parts of the body. Hence, the number of cells detected is of greater clinical importance.

3.7.2 Biopsies

The tissues most used to perform biopsies for the detection of *Candida* are esophagus, lungs, liver, muscle, skin, among others. These samples can be stained with different techniques such as hematoxylin-eosin, Periodic Acid-Schiff (PAS), Gram, Papanicolau, Gomori, Grocott, or Gridley. Through this technique, as with the smear, it is possible to observe single or budding yeasts, with or without the development of pseudohypha. Unlike the fresh examination, the biopsy allows to evaluate the invasion of the tissue by *Candida*, which demonstrates the presence of the infection. Thus, the detection of *Candida* in a biopsy is considered of great utility for the diagnosis of invasive candidiasis [129].

3.7.3 Culture

The microbiological culture has the disadvantage of being a delayed process that in turns delays the diagnosis of the patient and therefore the appropriate treatment. In addition, a sufficient quantity of microorganisms in the inoculum is required to obtain a positive diagnosis [130]. Even with these disadvantages, *Candida* culture is considered for many as the gold standard for the diagnosis of invasive candidiasis [129, 131]. The primary culture of *Candida* is achieved in 2–5 days, using Sabouraud Dextrose Agar, under aerobiosis conditions at room temperature. In this culture, it can be observed the growth of unique or budding colonies, forming pseudohypha in some occasions. From this culture, rapid tests such as germ-tube formation or thermotolerance, formation of chlamydoconidia in bile-agar medium or corn flour agar can be performed [132, 133]. In addition, tests can be performed for isolation and identification using chromogenic culture media, urease test, resistance to cycloheximide, fermentation of sugars, carrot-potatoes agar, among others. Currently, commercial automated systems allow the identification of yeast in a short time (<15 h), with time reduction in the diagnosis [130, 134].

It is important to note that a positive result for *Candida* culture does not always confirm the diagnosis of *Candida* infection, since this microorganism is considered normal biota in some parts of the body. To achieve a proper diagnosis of candidiasis, the patient's signs and symptoms must be taken into account, in addition to the laboratory results. If the culture comes from biopsy or is a blood culture, the positive result confirms the diagnosis of invasive candidiasis. On the other hand, a negative result for the culture of *Candida* is of greater diagnostic value since it allows to discard the infection by this yeast [131].

3.7.4 Molecular Tests

Like the aforementioned techniques, molecular tests will have diagnostic value when invasive candidiasis is suspected. These tools allow a quick diagnosis (<7 h); however, the cost is much higher. In addition, depending on the technique used, it is possible to detect the presence of *Candida* genetic material from nonviable microorganisms or contamination prior to amplification, resulting in a false positive, which represents the greatest problem for its use as a diagnostic tool. One of the advantages of molecular testing is the possibility of identifying a single or multiple species of *Candida* in a sample, which cannot be done by simple observation under a microscope. This is of clinical relevance due to the influence it has on the patient pharmacological treatment and the therapeutic implications.

Some of the most used molecular techniques are PCR-REA, PCR-LiPA, PCRsn, multiplex PCR, nested PCR, PCR-SB, PCR-rt., among others. However, in half the cases the studies have been performed on samples from blood cultures, so the advantage of speed in the emission of the result is lost. Hence, the use of PCR-rt. for the identification of *Candida* from serum and plasma samples has been tested, having better results than from whole blood [135–145].

In 2000, reverse cross blot hybridization assay was used for rapid detection of PCR-amplified DNA from *Candida* species, *Cryptococcus neoformans*, and *Saccharomyces cerevisiae* in clinical samples [146]. Five years later was developed a diagnostic microarray for the rapid and simultaneous identification of the 12 most common pathogenic *Candida* and *Aspergillus* species (*C. albicans*, *C. dubliniensis*, *C. krusei*, *C. glabrata*, *C. tropicalis*, *C. parapsilosis*, *C. guilliermondii*, *C. lusitaniae*, *Aspergillus fumigatus*, *A. flavus*, *A. niger*, and *A. terreus)*. This microarray system was able detecting and clearly identifying the fungal pathogens within 4 h after DNA extraction [147]. L. Basková et al., described a real-time quantitative PCR assay developed for the detection of the most common pathogenic *Candida* and *Aspergillus* species. The single-reaction PCR assay targets a region of the 28S subunit of the fungal rDNA gene. The design of the universal primer/probe, including a pan-*Aspergillus* and pan-*Candida* (Pan-AC) hydrolysis probe, facilitated the detection of numerous *Aspergillus* and *Candida* species (e.g. *C. albicans*, *C. glabrata*, *C. krusei*, *C. tropicalis*, *C. parapsilosis*, *C. kefyr*, *C. guilliermondii*, *C. lusitaniae* and *C. dubliniensis*). The assay permits highly reproducible detection of 10 fg fungal DNA, and facilitates accurate quantification of fungal load. Also, it provides an economic approach to the screening and monitoring of invasive candidiasis as a routine test in clinical diagnosis [148]. Schabereiter-Gurtner and colleagues developed a novel real-time PCR assays, targeting the fungal ITS2 region, for detection and differentiation of *C. albicans*, *C. dubliniensis*, *C. glabrata*, *C. krusei*, *C. parapsilosis*, and *C. tropicalis*) [149]. In 2013, was developed a quantitative PCR assay based on fluorescence resonance energy transfer (FRET) for *C. albicans* in blood. The assay was highly specific and sufficiently sensitive to detect the fungal load for early diagnosis of invasive candidiasis [150]. Molecular phylogenetics of *C. albicans* has been analyzed by multilocus sequence typing (MLST), ABC typing, mating type-like locus (MAT) status, and antifungal susceptibility to a panel of 1391 *C. albicans* isolates [151, 152]. Afsarian et al., performed the first nationwide study on the genotypic relationships of *C. albicans* strains obtained from oropharynx and bronchoalveolar lavage samples from patients, based on MLST, and directly analyze the polymorphism within DNA sequences [153]. All these mentioned examples show the importance of molecular methods for *Candida* identification without disregarding conventional procedures.

3.7.5 Serological Tests

Serum tests represent an important alternative for the diagnosis of candidiasis due to the speed in generating the result, and therefore the timely treatment of the patient. There are tests for either the search for antigens or antibodies against *C. albicans*, although in most cases they exhibit cross reactions with other species. Among the most used are the detection of mannan and anti-mannan antibodies by ELISA; whose combined use yields more reliable results. The detection of *Candida* mannan alone has a sensitivity of 50–100% for the different trademarks used. The species that can be detected through this technique include *C. albicans*, *C. dubliniensis*, *C. famata*, *C. glabrata*, *C. guilliermondii*, *C. lusitaniae*, and *C. tropicalis*. Although some commercial tests show positive results for *Candida* in the presence of *Aspergillus* and *Cryptococcus*. In any case, the clinical signs and symptoms of the patient should be considered, in addition to the positive results for antigens or anti-mannan antibodies. In the case of using the test for antibody detection, special care should be taken in the results obtained from immunosuppressed patients, given their low or null response, which could generate a false negative [154]. Several methods have been used, including serum determination of β-D-glucan, which is a component of the cell wall of *Candida* and *Aspergillus*, with sensitivities ranging from 57 to 100% and 44 to 92% of specificity, with the possibility of obtaining results in approximately 1–2 h [155]. This makes serology a good diagnostic tool in the case of invasive candidiasis, because of the relatively high sensitivity and specificity and the prompt emission of results. This type of test has the characteristic of detecting β-D-glucans at the picogram level and does not react with other polysaccharides. However, false positives can be generated by contamination of β-glucans from antibiotics and other drugs such as immunoglobulins, antitumorals, and sulfamines, dialysis membranes, materials used in surgeries, among others. The detection of β-D-glucans as a test for the diagnosis of *Candida* infection is widely used in the USA and Europe [155].

There are also commercial kits to determine the presence of *C. albicans* anti-mycelium antibodies by immunofluorescence. This test has cross reactivity with *C. tropicalis*, *C. parapsilosis*, *C. krusei*, *C. glabrata*, *C. guilliermondi,i* and *C. dubliniensis*.

Detection by indirect immunofluorescence of *C. albicans* germ-tube antibodies (CAGTA), is a useful diagnostic tool that discriminates between colonization and invasion. However, the standardization of this technique is complicated by its reliance on subjective interpretation [156].

A rapid enzymatic fluorometric assay for measuring D-arabinitol in serum was developed using recombinant D-arabinitol dehydrogenase from *C. albicans* (rArDH). NADH was fluorometrically measure when rArDH and NAD were added to serum extracts, and D-arabinitol concentrations were calculated from standard curves derived from pooled human serum containing known amounts of D-arabinitol. The method was precise and rapid and showed a significant improvement in terms of accuracy, cost, simplicity, specificity, and speed, compared with gas chromatography, mass spectrometry, and earlier enzymatic assays. The rArDH was highly specific for D-arabinitol, cross-reacting only with xylitol (4.9%); however, it has practically fallen into disuse [157].

Recent advances in the serological diagnosis of invasive candidiasis have allowed the development of a Lateral Flow Immunoassay for the detection of antibodies against enolase from *C. albicans*. In this test, a specificity of 98% and a sensitivity of 84% were achieved [134]. The serological diagnosis is relevant because of the speed of the results and the high sensitivity and specificity recorded for the different techniques, which has an impact on the timely therapy of patients with invasive candidiasis. In addition, it requires little investment for its implementation and the technical staff requires minimal training. In the case of patients with oral and vaginal *Candida* infection, laboratory results should be taken with caution and only used to confirm diagnosis when there are signs and symptoms compatible with the infectious process.

3.8 Treatment and Management

Pharmacological treatment used in handling of infections produced by different species of *Candida* is based on the anatomical situation of infection, the specific species of *Candida* responsible for the infection, the sensitivity of *Candida* species to specific antifungal agents, and the patient's immunological condition.

Broadly speaking, the pharmacological treatment comprises three antifungal groups: amphotericin B-based preparations, azoles antifungals, and echinocandins.

In the past years, management of infections caused by different species of *Candida* has undergone various changes. In 2016, the Infectious Disease Society of America (IDSA) reference guides were updated, and the appropriate use of echinocandins (caspofungin, micafungin, and anidulafungin) was included, as well as the incorporation of expanded-spectrum azoles, and lipid formulation of amphotericin B.

Schemes of first-line treatment for infections caused by species of *Candida* under the IDSA guide, the expert panel from the European Society of Clinical Microbiology and Infectious Diseases (ESCMID), and the consensus document of the Spanish Society of Infectious Diseases and Clinical Microbiology (SEIMC) as reference, are as follows:

3.8.1 Candidemia

All patients affected by candidemia, regardless of the presence or not of neutropenia, require the use of pharmacological treatment.

Candidemia in Non-neutropenic Patients Echinocandin is recommended as initial therapy. Echinocandins are broad-spectrum antifungals, which include caspofungin: 70 mg loading dose, followed by 50 mg IV (intravenous) once a day; micafungin: 100 mg IV once a day; anidulafungin: 200 mg loading dose, followed by 100 mg IV once a day. In every case, treatment must follow for at least a 2-week period after observing clinical improvement, and in obtaining a negative blood culture result.

Fluconazole is the drug of choice in most cases of candidemia. Initial treatment is a loading dose, 800 mg or 12 mg/kg PO (latin *per os*; by mouth) or IV respectively, then 400 mg or 6 mg/kg per day at least during 2 weeks after observing clinical improvement, and in obtaining a negative blood culture result. Fluconazole is an alternative to echinocandin as an initial treatment in stable patients without previous intake of azoles. An alternative in the treatment of candidemia with fluconazole is the use of voriconazole at a dose of 400 mg or 6 mg/kg, twice a day per two doses, followed by 200 mg or 3 mg/kg, twice a day.

A treatment with lipid formulation amphotericin B (LFAmB) at a dose of 3–5 mg/kg, once a day, is a good choice in patients with proven intolerance, limited availability, or tolerance (confirmed suspicion) to other antifungal agents.

The implementation of a change in the treatment scheme from echinocandin or LFAmB to fluconazole is recommended after a 5–7-day period in patients whose *Candida* species is sensitive to fluconazole, those who are clinically stable, and in cases in which negative repeated cultures are obtained during antifungal therapy [158, 159].

Candidemia in Neutropenic Patients Echinocandin is recommended in the treatment of candidemia gross patients with neutropenia (caspofungin: 70 mg loading dose, followed by a 50 mg IV once a day; micafungin: 100 mg IV once a day, and anidulafungin: 200 mg loading dose, followed by a 100 mg IV once a day). Provided its broad spectrum and effectiveness, LFAmB is an alternative in the treatment of candidemia (3–5 mg/kg once a day) in neutropenic patients, although its potential toxicity limits the use. On the other hand, fluconazole is an alternative in stable patients, and without prior azole intake (800 mg or 12 mg/kg loading dose, PO or IV respectively, at a dose of 400 mg or 6 mg/kg, once a day). Voriconazol can be used when the antifungal spectrum is desired to be broadened, with an extended coverage to infections caused by mold. Initial treatment scheme is 400 mg or 6 mg/kg twice a day per 2 doses, followed 200–300 mg or 3–4 mg/kg dose, twice a day. It is recommended, in every case, that antifungal treatment must be continued in neutropenic patients until neutropenia is controlled, since it represents a major risk for treatment failure [158, 159].

3.8.2 Candidiasis

Mucocutaneous Candidiasis Patients with oropharyngeal candidiasis can be treated either with antifungal topical agents (e.g.: nystatin, clotrimazole, amphotericin B oral suspension) or with systemic azoles (fluconazole, itrazonazole, or pozaconazole). For mild cases, administration of clotrimazole is recommended in 10 mg, five times a day; or miconazole, 50 mg oral mucoadhesive tablet placed on the surface of the mucosa in the canine fossa, once a day during a 7–14-day period. An alternative to the treatment is the administration of 4–6 mL (100,000 U/mL) nystatin suspension, four times a day, or nystatin (200,000 U/tablet), 1–2 tablets, four times a day during a 7–14-day period. For patients with moderate to severe oropharyngeal candidiasis, the most suited treatment is fluconazole 100–200 mg PO once a day, during a 7–14-day period. In refractory patients to fluconazole, a solution of itraconazole (200 mg, once a day) is indicated, or a posaconazole suspension (start with 400 mg, twice a day during 3 days, then 400 mg, once a day for 28 days) [158, 160]. Other treatment options include voriconazole (200 mg, twice a day), AmB deoxycholate oral suspension (100 mg/mL, four times a day), AmB deoxycholate intravenous (0.3 mg/kg, once a day) or echinocandin intravenous (caspofungin: 70 mg loading dose, then 50 mg on a daily basis; micafungin: 100 mg, once a day; or anidulafungin: 200 mg loading dose, followed by 100 mg on a daily basis) [159–161].

The treatment of esophageal candidiasis always requires systemic therapy with fluconazole (200–300 mg orally, once a day) during a 14–21-day period. Intravenous administration is indicated in those patients with oral therapy intolerance, in which cases the recommendation is fluconazole dose at 6 mg/kg once a day, or caspofungin: 70 mg loading dose, then 50 mg on a daily basis; micafungin: 150 mg once a day; or anidulafungin: 200 mg daily. In refractory patients to fluconazole, itraconazole solution is indicated (200 mg once a day) or voriconazole (200 mg twice a day via oral or IV), during a 14–21-day period. Other treatment options in refractory patients to fluconazole, include: AmB deoxycholate (0.3 mg/kg once a day) during 21 days or the administration of one of the echinocandins in accordance to the previously described treatment Scheme [159, 160].

Genitourinary Tract Candidiasis Vaginal candidiasis can be classified in two groups: noncomplex and complex. 90% of patients report noncomplex vaginitis, and response to treatment is quick to short period, topical, or oral schemes. The indicated treatment is fluconazole 150 mg single dose. In contrast, complex vaginitis requires fluconazole, 150 mg, 2–3 doses, every 72 h. Treatment with azoles for infections caused by non-*albicans* species is not indicated. Infections by *C. glabrata*, *C. krusei*, and other non-*albicans* species frequently respond to topical treatment with boric acid (600 mg per day) or flucytosine at 17%, during a 14-day period. It is recommended that patients with recurring vaginal candidiasis be treated orally or topically under induction with fluconazole during a 2-week period, followed by a 6-month maintenance regimen [161, 162].

Most of the hospitalized patients affected by candiduria are asymptomatic, where the predisposing factor is the presence of a urinary catheter. For these patients, the removal of the urinary catheter represents the first treatment strategy, since removal is equivalent to the administration of antifungal treatment. It is important to determine if presence of candiduria is not a reflection of a disseminated infection by *Candida*, especially in newborns and immunocompromised patients (particularly neutropenic); if so, they must be treated with recommended scheme for candidemia. Therefore, pharmacological treatment with antifungals is only indicated for patients that will undergo urologic manipulation and that have asymptomatic candiduria. Treatment scheme includes administration of 400 mg (6 mg/kg) fluconazole, once a day, or AmB deoxycolate, 0.3–1 mg/kg per day during a 1–7-day period, taking into account the pre- and post-procedure [159].

Symptomatic patients with cystitis and pyelonephritis must be treated with fluconazole during a 2-week period when they have had no prior treatment with azole antifungals, or the isolate is fluconazole susceptible. If there is evidence of fluconazole-resistant isolates, the recommended treatment is AmB deoxycholate with or without flucytosine. The treatment scheme for cystitis in fluconazole-susceptible isolates is fluconazole 200 mg orally, once a day for 14 days. For fluconazole-resistant isolates (e.g., *C. glabrata* infections), AmB deoxycholate 0.3–1 mg/kg must be administered daily during 1–7 days, or flucytosine 25 mg/kg, every 6 h for 7–10 days. For infections caused by *C. krusei*, the previously described AmB deoxycholate treatment is recommended. It is strongly advised that in patients with a urinary catheter, this should be removed as first step in treatment [159, 162].

Treatment for patients with pyelonephritis implies the use of antifungal systemic therapy with fluconazole (200–400 mg daily) PO or IV during a 2-week period minimum, or AmB deoxycholate, 0.5–0.7 mg/kg daily IV can be administered with or without flucytosine at a dose of 25 mg/kg, every 6 h. Infections caused by fluconazole-resistant *C. glabrata* can be treated with flucytosine or AmB deoxycholate monotherapy, or the combination of both in the previously mentioned doses [159, 160].

Endocarditis Treatment for native valve endocarditis is indicated with LFAmB 3–5 mg/kg on a daily basis, and treatment can be combined with 25 mg/kg flucytosine every 6 h, or high doses of echinocandins (caspofungin: 150 mg, micafungin: 150 mg, or anidulafungin: 200 mg); in every case, indicated dosage is on a daily basis. After initial treatment, it is recommended that treatment be changed to fluconazole (400–800 mg, on a daily basis) in patients with fluconazole-susceptible *Candida* strains, who are clinically stable and in who repeat cultures on antifungal therapy are negative. In fluconazole-resistant *Candida* species, voriconazole can be used (200–300 mg, twice a day), or posaconazole (300 mg, on a daily basis) as an alternative. Patients with contraindications for surgery must receive suppressive, prolonged antifungal treatment on fluconazole 400–800 mg during 6 weeks in endocarditis or pacemaker electrode cable infection and a 4-week period in case of bag or generator infection, always after surgery [158, 159].

Central Nervous System Candidiasis (Meningitis) Central nervous system (CNS) candidiasis can result as a consequence of neurosurgical procedures, being an expression of disseminated candidiasis, or an isolated chronic infection. Meningitis treatment requires the use of drugs that penetrate the blood-brain barrier. Treatment is based on monotherapy with liposomal AmB (5 mg/kg/day), or AmB deoxycholate (0.7–1 mg/kg/day), or combined with flucytosine (25 mg/kg, every 6 h). Fluconazole (400–800 mg, once a day) also obtains excellent levels in the cerebrospinal fluid, and is a useful treatment after AmB [159].

Chronic Disseminated Candidiasis (Hepatosplenic Candidiasis) The recommended treatment for this clinical case (in stable patients) is fluconazole (6 mg/kg per day), AmB deoxycholate (0.6–0.7 mg/kg/day), or liposomal AmB (3–5 mg/kg/kg). Patients in critical condition, with a more extended disease, or refractory to fluconazole must be treated with AmB (deoxycholate or lipidic formulation). Treatment must be extended until lesions calcify or heal (especially in patients being treated with continuous chemotherapy or immunosuppressed), which can happen within weeks or months after initial treatment [159].

Neonatal Candidiasis Neonatal candidiasis syndrome is characterized by the presence of an extended dermatitis in newborns infected with *Candida*, as a consequence of contamination of amniotic fluid. Premature and underweight newborns, or infants with prolonged membrane ruptures, and with clinical findings associated to disseminated newborn cutaneous candidiasis must be considered for systemic treatment. AmB deoxycholate (0.5–1 mg/kg/day) is usually the first-line treatment. Fluconazole is considered as a second-line treatment (12 mg/kg PO or IV), mainly in patients with no fluconazole prophylactic treatment. Recommended time of treatment for candidemia, with or without metastasis, is 2 weeks after observing negative results in blood cultures [158, 159, 161].

Note In 2013, the US Food and Drug Administration (FDA) emitted an advisory to not prescribe ketoconazole tablets as an elected treatment for any fungal disease (including infections by *Candida* and dermatophytes) due to extreme risk of liver damage, adrenal failure, and interaction with other drugs. In the same year, the European Union withdrew it from the market.

3.9 Basic Features of the Immune Response to *Candida* Infection

The onset of the host immune response to *Candida* species is the recognition of the conserved fungal pathogen associated molecular patterns (PAMPs) via pattern recognition receptors (PPRs), including C-type lectin receptors (CLRs), Toll-like receptors (TLRs), NOD-like receptors (NLRs) and RIG-I-like receptors (RLRs). These PAMPs mainly belong to cell wall components and its recognition by PPRs is always a cooperative and interactive process which induces the release of pro- and anti-inflammatory cytokines in order to mediate the immune response, thus resulting in opsonization, induction of antimicrobial peptides, inflammasome activation, pyroptosis, and phagocytosis, among others [163]. It also has been demonstrated that innate immune cells may suffer epigenetic reprogramming resulting in a phenomenon referred as "trained immunity" [164]. This comprises an enhancement on cytokine release by monocytes and macrophages in an T-cell independent manner, prior stimulation with attenuated *Candida* strains, broadening the conception of what we used to call "immune memory" [165]. Interestingly, *Candida* species have developed evasion mechanisms that can circumvent the host immune response ranging from the most elaborated, i.e., as inducing macrophages to switch from a more inflammatory M1 phenotype to a less inflammatory M2 phenotype, to a simpler shielding of PAMPs, in detriment of the elicitation of the immune response [166, 167].

3.10 Conclusion

Fungal infections caused by *Candida* species have increased their morbidity and mortality rates during the past decades, especially in nosocomial environments. These infections vary from superficial mucocutaneous candidiasis to systemic candidemias, depending on the existence of underlying conditions and the host immune competence. Mention aside, *Candida* species has also developed a set of virulence factors that enable them to switch from a friendly commensal to an invasive pathogen, along with evasive mechanisms oriented to outwit the protective host immune response. In favor of primarily limit and further eradicate the infection, the accurate identification of *Candida* is of great relevance. This is usually achieved or confirmed by laboratory methods that allow to identify the fungi from different specimen by means going from microscopic observation to biochemical or molecular tests. Finally, all patients affected by infections caused by species of *Candida*, regardless of the presence or not of neutropenia, require the use of pharmacological treatment.

References

1. Sharma Y, Chumber S, Kaur M (2017) Studying the prevalence, species distribution, and detection of in vitro production of phospholipase from Candida isolated from cases of invasive candidiasis. J Global Infect Dis 9(1):8–11
2. Sheevani, Sharma P, Aggarwal A (2013) Nosocomial Candida infection in a rural tertiary care hospital. J Clin Diagn Res 7(2):405–406
3. Pfaller MA, Diekema DJ (2007 Jan) Epidemiology of invasive candidiasis: a persistent public health problem. Clin Microbiol Rev 20(1):133–163
4. Leff S (1956) From witchcraft to world health. Lawrence and Wishart, London
5. Ship JA (1996) Recurrent aphthous stomatitis. An update. Oral Surg Oral Med Oral Pathol Oral Radiol Endod 81(2):141–147
6. Rosén von Rosenstein N (1764) Underrälttelser om barn-sjukdomar och deras botemedel. Kongl. Wet. Acad., Stockholm
7. Langenbeck B (1839) Auffindung von Pilzen auf der Schleimhaut der Speiseröhre einer Typhus-Leiche. Neue Notizen aus dem Gebiete der Natur-und Heilkunde 12(252): columns 145–147
8. Dla G (2014) Tapeworms, lice and prions : a compendium of unpleasant infections. Oxford University Press, Oxford
9. Robin CP (1853) Hlstotre Naturelle des vegdtaux parasites qut crolssent Sur l'homme et sur les animaux vivants, Paris
10. Barnett JA (2008) A history of research on yeasts 12: medical yeasts part 1, Candida albicans. Yeast 25(6):385–417
11. Laurent E (1889) Observations sur le champignon du muguet. Bulletin de la Societé Belge de Microscopie 16:14–18
12. Zopf WF (1890) Die Pilze in morphologischer, physiologischer, biologischer und systematischer beziehung. Encyklopaedie der Naturwissenschaften. The University of Michigan, E. Trewendt
13. Mello FLG (1918) Arq Hig Pat Exot:207–316
14. Brownlie & Shattock (1920) J Laryng Rhin Otol 34:425–431
15. Berkhout CM (1923) De Schimmelgeslachten Monilia, Oidium, Oospora en Torula (Thesis). Rijksuniversiteit te Utrecht
16. Ota M (1928) Champignons parasites de l'homme. VI (Études morphologiques et systématiques) [In Japanese]. Jpn J Dermatol Urol 28(16–23):381–424
17. Dekker NM (1931) Endomycopsis albicans (Vuill.) Verh. K. Akad. Wet., tweede sect 28:231
18. Stelling-Dekker NM (1931) Die sporogenen Hefen. Koninklijke Akademie van Wetenschappen te A, editor, Amsterdam
19. Benham RW (1931) Certain Monilias parasitic on man. Their identification by morphology and by agglutination. J Infect Dis 49:183–215
20. Benham RW (1932) Monilias, yeasts, and cryptococci: their pathogenicity, classification, and identification. Am J Public Health Nations Health 22(5):502–504
21. Dodge CW (1935) Medical mycology. Fungous diseases of men and other mammals. St. Louis, The C.V. Mosby Company
22. Langeron M, Guerra P (1938) Nouvelles recherches de zymologie medicare. Ann Parasitol Hum Comp 16(5):429–476
23. Nickerson WJ (1953) Reduction of inorganic substances by yeasts. I. Extracellular reduction of sulfite by species of Candida. J Infect Dis 93(1):43–56
24. Nickerson WJ (1954) An enzymatic locus participating in cellular division of a yeast. J Gen Physiol 37(4):483–494
25. Meinhof W, Schirren C (2014) Hans Rieth, 1914-1994: to commemorate the occasion of his 100th birthday. Hautarzt 65(11):998–1003. Hans Rieth, 1914-1994: Zum Gedenken anlasslich seines 100. Geburtstages

26. Slutsky B, Buffo J, Soll DR (1985) High-frequency switching of colony morphology in Candida albicans. Science 230(4726):666–669
27. Chibana H, Magee BB, Grindle S, Ran Y, Scherer S, Magee PT (1998) A physical map of chromosome 7 of Candida albicans. Genetics 149(4):1739–1752
28. Hull CM, Johnson AD (1999) Identification of a mating type-like locus in the asexual pathogenic yeast Candida albicans. Science 285(5431):1271–1275
29. Chibana H, Beckerman JL, Magee PT (2000) Fine-resolution physical mapping of genomic diversity in Candida albicans. Genome Res 10(12):1865–1877
30. Kandasamy R, Vediyappan G, Chaffin WL (2000) Evidence for the presence of pir-like proteins in Candida albicans. FEMS Microbiol Lett 186(2):239–243
31. Fidel PL Jr, Cutright J, Steele C (2000) Effects of reproductive hormones on experimental vaginal candidiasis. Infect Immun 68(2):651–657
32. Anderson JB, Wickens C, Khan M, Cowen LE, Federspiel N, Jones T et al (2001) Infrequent genetic exchange and recombination in the mitochondrial genome of Candida albicans. J Bacteriol 183(3):865–872
33. Jones T, Federspiel NA, Chibana H, Dungan J, Kalman S, Magee BB et al (2004) The diploid genome sequence of Candida albicans. Proc Natl Acad Sci U S A 101(19):7329–7334
34. de Groot PW, de Boer AD, Cunningham J, Dekker HL, de Jong L, Hellingwerf KJ et al (2004) Proteomic analysis of Candida albicans cell walls reveals covalently bound carbohydrate-active enzymes and adhesins. Eukaryot Cell 3(4):955–965
35. Chibana H, Oka N, Nakayama H, Aoyama T, Magee BB, Magee PT et al (2005) Sequence finishing and gene mapping for Candida albicans chromosome 7 and syntenic analysis against the Saccharomyces cerevisiae genome. Genetics 170(4):1525–1537
36. Thomas DP, Bachmann SP, Lopez-Ribot JL (2006) Proteomics for the analysis of the Candida albicans biofilm lifestyle. Proteomics 6(21):5795–5804
37. van het Hoog M, Rast TJ, Martchenko M, Grindle S, Dignard D, Hogues H et al (2007) Assembly of the Candida albicans genome into sixteen supercontigs aligned on the eight chromosomes. Genome Biol 8(4):R52
38. Oliveira MA, Carvalho LP, Gomes Mde S, Bacellar O, Barros TF, Carvalho EM (2007) Microbiological and immunological features of oral candidiasis. Microbiol Immunol 51(8):713–719
39. Yeater KM, Chandra J, Cheng G, Mukherjee PK, Zhao X, Rodriguez-Zas SL et al (2007) Temporal analysis of Candida albicans gene expression during biofilm development. Microbiology 153(Pt 8):2373–2385
40. Butler G, Rasmussen MD, Lin MF, Santos MAS, Sakthikumar S, Munro CA et al (2009) Evolution of pathogenicity and sexual reproduction in eight Candida genomes. Nature 459(7247):657–662
41. Martinez-Gomariz M, Perumal P, Mekala S, Nombela C, Chaffin WL, Gil C (2009 Apr) Proteomic analysis of cytoplasmic and surface proteins from yeast cells, hyphae, and biofilms of Candida albicans. Proteomics 9(8):2230–2252
42. Nicholls S, MacCallum DM, Kaffarnik FA, Selway L, Peck SC, Brown AJ (2011) Activation of the heat shock transcription factor Hsf1 is essential for the full virulence of the fungal pathogen Candida albicans. Fungal Genet Biol 48(3):297–305
43. Muzzey D, Schwartz K, Weissman JS, Sherlock G (2013) Assembly of a phased diploid Candida albicans genome facilitates allele-specific measurements and provides a simple model for repeat and indel structure. Genome Biol 14(9):R97
44. Bartelli TF, Ferreira RC, Colombo AL, Briones MR (2013) Intraspecific comparative genomics of Candida albicans mitochondria reveals non-coding regions under neutral evolution. Infect Genet Evol 14:302–312
45. Maiti P, Ghorai P, Ghosh S, Kamthan M, Tyagi RK, Datta A (2015) Mapping of functional domains and characterization of the transcription factor Cph1 that mediate morphogenesis in Candida albicans. Fungal Genet Biol 83:45–57

46. Rajendran R, May A, Sherry L, Kean R, Williams C, Jones BL et al (2016) Integrating Candida albicans metabolism with biofilm heterogeneity by transcriptome mapping. Sci Rep 6:35436
47. Leach MD, Farrer RA, Tan K, Miao Z, Walker LA, Cuomo CA et al (2016) Hsf1 and Hsp90 orchestrate temperature-dependent global transcriptional remodelling and chromatin architecture in Candida albicans. Nat Commun 7:11704
48. Cevik R, Tekin R, Gem M (2016) Candida arthritis in a patient diagnosed with spondyloarthritis. Rev Soc Bras Med Trop 49(6):793–795
49. Zhou PR, Hua H, Liu XS (2017) Quantity of Candida colonies in saliva: a diagnostic evaluation for oral candidiasis. Chin J Dent Res 20(1):27–32
50. Spampinato C, Leonardi D (2013) Candida infections, causes, targets, and resistance mechanisms: traditional and alternative antifungal agents. Biomed Res Int 2013:204237
51. Gow NAR, van de Veerdonk FL, Brown AJP, Netea MG (2011) Candida albicans morphogenesis and host defence: discriminating invasion from colonization. Nat Rev Microbiol 10(2):112–122
52. Odds FC (2008) Evolution of pathogenic Candida species. Evolutionary biology of bacterial and fungal pathogens. American Society of Microbiology
53. Pemán J, Cantón E, Quindós G, Eraso E, Alcoba J, Guinea J et al (2012) Epidemiology, species distribution and in vitro antifungal susceptibility of fungaemia in a Spanish multicentre prospective survey. J Antimicrob Chemother 67(5):1181–1187
54. Ruiz Gaitan AC, Moret A, Lopez Hontangas JL, Molina JM, Aleixandre Lopez AI, Cabezas AH et al (2017) Nosocomial fungemia by Candida auris: first four reported cases in continental Europe. Rev Iberoam Micol 34(1):23–27
55. Hazen KC (1995) New and emerging yeast pathogens. Clin Microbiol Rev 8(4):462–478
56. BA MM, Coleman DC (2014) Molecular epidemiology, phylogeny and evolution of Candida albicans. Infect Genet Evol 21:166–178
57. Fidel PL, Vazquez JA, Sobel JD (1999) Candida glabrata: review of epidemiology, pathogenesis, and clinical disease with comparison to C. albicans. Clin Microbiol Rev 12(1):80–96
58. Trofa D, Gácser A, Nosanchuk JD (2008) Candida parapsilosis, an emerging fungal pathogen. Clin Microbiol Rev 21(4):606–625
59. Almirante B, Rodriguez D, Cuenca-Estrella M, Almela M, Sanchez F, Ayats J et al (2006) Epidemiology, risk factors, and prognosis of Candida parapsilosis bloodstream infections: case-control population-based surveillance study of patients in Barcelona, Spain, from 2002 to 2003. J Clin Microbiol 44(5):1681–1685
60. de Toro M, Torres MJ, Maite R, Aznar J (2011) Characterization of Candida parapsilosis complex isolates. Clin Microbiol Infect 17(3):418–424
61. Álvarez-Lerma F, Nolla-Salas J, León C, Palomar M, Jordá R, Carrasco N et al (2003) Candiduria in critically ill patients admitted to intensive care medical units. Intensive Care Med 29(7):1069–1076
62. Binelli CA, Moretti ML, Assis RS, Sauaia N, Menezes PR, Ribeiro E et al (2006) Investigation of the possible association between nosocomial candiduria and candidaemia. Clin Microbiol Infect 12(6):538–543
63. Colombo AL, Guimarães T, LRBF S, LPDA M, AKB C, Rady P et al (2007) Prospective Observational Study of Candidemia in São Paulo, Brazil: incidence rate, epidemiology, and predictors of mortality. Infect Control Hosp Epidemiol 28(5):570–576
64. Ann Chai LY, Denning DW, Warn P (2010) Candida tropicalis in human disease. Crit Rev Microbiol 36(4):282–298
65. Suzuki T, Miyamae Y, Ishida I (1991) Variation of colony morphology and chromosomal rearrangement in Candida tropicalis pK233. Microbiology 137(1):161–167
66. Martin MV (1979) Germ-tube formation by oral strains of Candida tropicalis. J Med Microbiol 12(2):187–194
67. Doi M, Homma M, Chindamporn A, Tanaka K (1992) Estimation of chromosome number and size by pulsed-field gel electrophoresis (PFGE) in medically important Candida species. Microbiology 138(10):2243–2251

68. Pfaller MA, Andes DR, Diekema DJ, Horn DL, Reboli AC, Rotstein C et al (2014) Epidemiology and outcomes of invasive candidiasis due to non-albicans species of Candida in 2,496 patients: data from the prospective antifungal therapy (PATH) registry 2004–2008. PLoS One 9(7):e101510
69. Presterl E, Daxböck F, Graninger W, Willinger B (2007) Changing pattern of candidaemia 2001–2006 and use of antifungal therapy at the University Hospital of Vienna, Austria. Clin Microbiol Infect 13(11):1072–1076
70. Pfaller MA, Diekema DJ (2010) Epidemiology of invasive mycoses in North America. Crit Rev Microbiol 36(1):1–53
71. Li XS, Reddy MS, Baev D, Edgerton M (2003) Candida albicans Ssa1/2p is the cell envelope binding protein for human salivary histatin 5. J Biol Chem 278(31):28553–28561
72. Baddley JW, Patel M, Bhavnani SM, Moser SA, Andes DR (2008) Association of fluconazole pharmacodynamics with mortality in patients with candidemia. Antimicrob Agents Chemother 52(9):3022–3028
73. Kaufmann CS, Merz WG (1989) Electrophoretic karyotypes of Torulopsis glabrata. J Clin Microbiol 27(10):2165–2168
74. DiazGranados CA, Martinez A, Deaza C, Valderrama S (2008) An outbreak of Candida spp. bloodstream infection in a tertiary care center in Bogotá, Colombia. Braz J Infect Dis 12:390–394
75. Strollo S, Lionakis MS, Adjemian J, Steiner CA, Prevots DR (2017) Epidemiology of hospitalizations associated with invasive candidiasis, United States, 2002–2012. Emerg Infect Dis J 23(1):7
76. Tortorano AM, Peman J, Bernhardt H, Klingspor L, Kibbler CC, Faure O et al (2004) Epidemiology of candidaemia in Europe: results of 28-month European Confederation of Medical Mycology (ECMM) hospital-based surveillance study. Eur J Clin Microbiol Infect Dis 23(4):317–322
77. Hasan F, Xess I, Wang X, Jain N, Fries BC (2009) Biofilm formation in clinical Candida isolates and its association with virulence. Microbes Infect 11(8–9):753–761
78. Horn DL, Neofytos D, Anaissie EJ, Fishman JA, Steinbach WJ, Olyaei AJ et al (2009) Epidemiology and outcomes of candidemia in 2019 patients: data from the prospective antifungal therapy alliance registry. Clin Infect Dis 48(12):1695–1703
79. Sardi JCO, Scorzoni L, Bernardi T, Fusco-Almeida AM, Mendes Giannini MJS (2013) Candida species: current epidemiology, pathogenicity, biofilm formation, natural antifungal products and new therapeutic options. J Med Microbiol 62(1):10–24
80. Richardson M, Lass-Flörl C (2008) Changing epidemiology of systemic fungal infections. Clin Microbiol Infect 14:5–24
81. Lass-Flörl C (2009) The changing face of epidemiology of invasive fungal disease in Europe. Mycoses 52(3):197–205
82. Pappas PG. Invasive candidiasis. Infect Dis Clin North Am 2006;20(3):485–506
83. Weems JJ Jr (1992) Candida parapsilosis: epidemiology, pathogenicity, clinical manifestations, and antimicrobial susceptibility. Clin Infect Dis 14(3):756–766
84. Krcmery V, Barnes AJ (2002) Non-albicans Candida spp. causing fungaemia: pathogenicity and antifungal resistance. J Hosp Infect 50(4):243–260
85. Canton E, Peman J, Quindos G, Eraso E, Miranda-Zapico I, Alvarez M et al (2011) Prospective multicenter study of the epidemiology, molecular identification, and antifungal susceptibility of Candida parapsilosis, Candida orthopsilosis, and Candida metapsilosis isolated from patients with candidemia. Antimicrob Agents Chemother 55(12):5590–5596
86. Pfaller MA, Boyken L, Hollis RJ, Messer SA, Tendolkar S, Diekema DJ (2006) Global surveillance of in vitro activity of micafungin against Candida: a comparison with caspofungin by CLSI-recommended methods. J Clin Microbiol 44(10):3533–3538
87. Pfaller MA, Boyken L, Hollis RJ, Messer SA, Tendolkar S, Diekema DJ. In vitro susceptibilities of Candida spp. to caspofungin: four years of global surveillance. J Clin Microbiol. 2006;44(3):760–763

88. Wingard JR (1995) Importance of Candida species other than C. albicans as pathogens in oncology patients. Clin Infect Dis 20(1):115–125
89. Kao AS, Brandt ME, Pruitt WR, Conn LA, Perkins BA, Stephens DS et al (1999) The epidemiology of candidemia in two United States cities: results of a population-based active surveillance. Clin Infect Dis 29(5):1164–1170
90. Ortega M, Marco F, Soriano A, Almela M, Martínez JA, López J et al (2011) Candida species bloodstream infection: epidemiology and outcome in a single institution from 1991 to 2008. J Hosp Infect 77(2):157–161
91. Branchini ML, Pfaller MA, Rhine-Chalberg J, Frempong T, Isenberg HD (1994) Genotypic variation and slime production among blood and catheter isolates of Candida parapsilosis. J Clin Microbiol 32(2):452–456
92. Fusek M, Smith EA, Monod M, Foundling SI (1993) Candida parapsilosis expresses and secretes two aspartic proteinases. FEBS Lett 327(1):108–112
93. Merkerova M, Dostal J, Hradilek M, Pichova I, Hruskova-Heidingsfeldova O (2006) Cloning and characterization of Sapp2p, the second aspartic proteinase isoenzyme from Candida parapsilosis. FEMS Yeast Res 6(7):1018–1026
94. Miranda LN, van der Heijden IM, Costa SF, Sousa AP, Sienra RA, Gobara S et al (2009) Candida colonisation as a source for candidaemia. J Hosp Infect 72(1):9–16
95. Hobson RP (2003) The global epidemiology of invasive Candida infections – is the tide turning? J Hosp Infect 55(3):159–168. quiz 233
96. Brito LR, Guimaraes T, Nucci M, Rosas RC, Paula Almeida L, Da Matta DA et al (2006) Clinical and microbiological aspects of candidemia due to Candida parapsilosis in Brazilian tertiary care hospitals. Med Mycol 44(3):261–266
97. Kauffman CA, Vazquez JA, Sobel JD, Gallis HA, McKinsey DS, Karchmer AW et al (2000) Prospective multicenter surveillance study of funguria in hospitalized patients. The National Institute for Allergy and Infectious Diseases (NIAID) Mycoses Study Group. Clin Infect Dis 30(1):14–18
98. Rho J, Shin JH, Song JW, Park MR, Kee SJ, Jang SJ et al (2004) Molecular investigation of two consecutive nosocomial clusters of Candida tropicalis candiduria using pulsed-field gel electrophoresis. J Microbiol 42(2):80–86
99. Negri M, Goncalves V, Silva S, Henriques M, Azeredo J, Oliveira R (2010) Crystal violet staining to quantify Candida adhesion to epithelial cells. Br J Biomed Sci 67(3):120–125
100. Chakrabarti A, Sood P, Rudramurthy SM, Chen S, Kaur H, Capoor M et al (2015) Incidence, characteristics and outcome of ICU-acquired candidemia in India. Intensive Care Med 41(2):285–295
101. Hii I-M, Chang H-L, Lin L-C, Lee Y-L, Liu Y-M, Liu C-E et al (2015) Changing epidemiology of candidemia in a medical center in middle Taiwan. J Microbiol Immunol Infect 48(3):306–315
102. Hofs S, Mogavero S, Hube B (2016) Interaction of Candida albicans with host cells: virulence factors, host defense, escape strategies, and the microbiota. J Microbiol 54(3):149–169
103. Yang YL (2003) Virulence factors of Candida species. J Microbiol Immunol Infect 36(4):223–228
104. Moyes DL, Richardson JP, Naglik JR (2015) Candida albicans-epithelial interactions and pathogenicity mechanisms: scratching the surface. Virulence 6(4):338–346
105. Modrzewska B, Kurnatowski P (2015) Adherence of Candida sp. to host tissues and cells as one of its pathogenicity features. Ann Parasitol 61(1):3–9
106. Staab JF, Bahn YS, Tai CH, Cook PF, Sundstrom P (2004) Expression of transglutaminase substrate activity on Candida albicans germ tubes through a coiled, disulfide-bonded N-terminal domain of Hwp1 requires C-terminal glycosylphosphatidylinositol modification. J Biol Chem 279(39):40737–40747
107. Zhu W, Filler SG (2010) Interactions of Candida albicans with epithelial cells. Cell Microbiol 12(3):273–282

108. Calderone RA, Fonzi WA (2001) Virulence factors of Candida albicans. Trends Microbiol 9(7):327–335
109. Biswas S, Van Dijck P, Datta A (2007) Environmental sensing and signal transduction pathways regulating morphopathogenic determinants of Candida albicans. Microbiol Mol Biol Rev 71(2):348–376
110. Jacobsen ID, Wilson D, Wachtler B, Brunke S, Naglik JR, Hube B (2012) Candida albicans dimorphism as a therapeutic target. Expert Rev Anti-Infect Ther 10(1):85–93
111. Sudbery PE (2011) Growth of Candida albicans hyphae. Nat Rev Micro 9(10):737–748
112. Mayer FL, Wilson D, Hube B (2013) Candida albicans pathogenicity mechanisms. Virulence 4(2):119–128
113. Clancy CJ, Barchiesi F, Falconi DiFrancesco L, Morris AJ, Snydman DR, Yu VL et al (2000) Clinical manifestations and molecular epidemiology of late recurrent candidemia, and implications for management. Eur J Clin Microbiol Infect Dis 19(8):585–592
114. Tang HJ, Liu WL, Lin HL, Lai CC (2014) Clinical manifestations and prognostic factors of central line-associated candidemia. Am J Infect Control 42(11):1238–1240
115. Patel SK, Shah DR, Tiwari S (2015) Bioadhesive films containing fluconazole for mucocutaneous candidiasis. Indian J Pharm Sci 77(1):55–61
116. Khullar G, Vignesh P, Lau YL, Rudramurthy SM, Rawat A, De D et al (2017) Chronic mucocutaneous candidiasis. J Allergy Clin Immunol Pract 5(4):1119–1121
117. Puel A, Picard C, Cypowyj S, Lilic D, Abel L, Casanova JL (2010) Inborn errors of mucocutaneous immunity to Candida albicans in humans: a role for IL-17 cytokines? Curr Opin Immunol 22(4):467–474
118. Marazzi MG, Bondi E, Giannattasio A, Strozzi M, Savioli C (2008) Intracranial aneurysm associated with chronic mucocutaneous candidiasis. Eur J Pediatr 167(4):461–463
119. Rautemaa R, Hietanen J, Niissalo S, Pirinen S, Perheentupa J (2007) Oral and oesophageal squamous cell carcinoma – a complication or component of autoimmune polyendocrinopathy-candidiasis-ectodermal dystrophy (APECED, APS-I). Oral Oncol 43(6):607–613
120. Eyerich K, Eyerich S, Hiller J, Behrendt H, Traidl-Hoffmann C (2010) Chronic mucocutaneous candidiasis, from bench to bedside. Eur J Dermatol 20(3):260–265
121. Bassetti M, Marchetti M, Chakrabarti A, Colizza S, Garnacho-Montero J, Kett DH et al (2013) A research agenda on the management of intra-abdominal candidiasis: results from a consensus of multinational experts. Intensive Care Med 39(12):2092–2106
122. Vergidis P, Clancy CJ, Shields RK, Park SY, Wildfeuer BN, Simmons RL et al (2016) Intra-Abdominal Candidiasis: The Importance of Early Source Control and Antifungal Treatment. PLoS One 11(4):e0153247
123. Bassetti M, Righi E, Ansaldi F, Merelli M, Scarparo C, Antonelli M et al (2015) A multicenter multinational study of abdominal candidiasis: epidemiology, outcomes and predictors of mortality. Intensive Care Med 41(9):1601–1610
124. Montravers P, Dupont H, Gauzit R, Veber B, Auboyer C, Blin P et al (2006) Candida as a risk factor for mortality in peritonitis. Crit Care Med 34(3):646–652
125. Murray HW, Fialk MA, Roberts RB (1976) Candida arthritis. A manifestation of disseminated candidiasis. Am J Med 60(4):587–595
126. Fainstein V, Gilmore C, Hopfer RL, Maksymiuk A, Bodey GP (1982) Septic arthritis due to Candida species on patients with cancer: report of five cases and review of the literature. Rev Infect Dis 4(1):78–85
127. Munoz-Fernandez S, Macia MA, Pantoja L, Cardenal A, Pena JM, Martin Mola E et al (1993) Osteoarticular infection in intravenous drug abusers: influence of HIV infection and differences with non drug abusers. Ann Rheum Dis 52(8):570–574
128. Gamaletsou MN, Rammaert B, Bueno MA, Sipsas NV, Moriyama B, Kontoyiannis DP et al (2016) Candida arthritis: analysis of 112 pediatric and adult cases. Open Forum Infect Dis 3(1):ofv207
129. Jones JM (1990) Laboratory diagnosis of invasive candidiasis. Clin Microbiol Rev 3(1):32–45

130. Pfeiffer CD, Samsa GP, Schell WA, Reller LB, Perfect JR, Alexander BD (2011) Quantitation of Candida CFU in initial positive blood cultures. J Clin Microbiol 49(8):2879–2883
131. Clancy CJ, Nguyen ML, Cheng S, Huang H, Fan G, Jaber RA et al (2008) Immunoglobulin G responses to a panel of Candida albicans antigens as accurate and early markers for the presence of systemic candidiasis. J Clin Microbiol 46(5):1647–1654
132. Martinez-Jimenez MC, Munoz P, Valerio M, Vena A, Guinea J, Bouza E (2015) Combination of Candida biomarkers in patients receiving empirical antifungal therapy in a Spanish tertiary hospital: a potential role in reducing the duration of treatment. J Antimicrob Chemother 70(11):3107–3115
133. Martinez-Jimenez MC, Munoz P, Valerio M, Vena A, Guinea J, Bouza E (2016) Combination of Candida biomarkers in patients receiving empirical antifungal therapy in a Spanish tertiary hospital: a potential role in reducing the duration of treatment. J Antimicrob Chemother 71(9):2679
134. He Z-X, Shi L-C, Ran X-Y, Li W, Wang X-L, Wang F-K (2016) Development of a lateral flow immunoassay for the rapid diagnosis of invasive candidiasis. Front Microbiol 7:1451
135. Borst A, Theelen B, Reinders E, Boekhout T, Fluit AC, Savelkoul PHM (2003) Use of amplified fragment length polymorphism analysis to identify medically important Candida spp., including C. dubliniensis. J Clin Microbiol 41(4):1357–1362
136. Coleman D, Sullivan D, Harrington B, Haynes K, Henman M, Shanley D et al (1997) Molecular and phenotypic analysis of Candida dubliniensis: a recently identified species linked with oral candidosis in HIV-infected and AIDS patients. Oral Dis 3(Suppl 1):S96–101
137. Elie CM, Lott TJ, Reiss E, Morrison CJ (1998) Rapid identification of Candida species with species-specific DNA probes. J Clin Microbiol 36(11):3260–3265
138. Ellepola AN, Hurst SF, Elie CM, Morrison CJ (2003) Rapid and unequivocal differentiation of Candida dubliniensis from other Candida species using species-specific DNA probes: comparison with phenotypic identification methods. Oral Microbiol Immunol 18(6):379–388
139. Graf B, Trost A, Eucker J, Gobel UB, Adam T (2004) Rapid and simple differentiation of C. dubliniensis from C. albicans. Diagn Microbiol Infect Dis 48(2):149–151
140. Joly S, Pujol C, Rysz M, Vargas K, Soll DR (1999) Development and characterization of complex DNA fingerprinting probes for the infectious yeast Candida dubliniensis. J Clin Microbiol 37(4):1035–1044
141. McCullough M, Ross B, Reade P (1995) Characterization of genetically distinct subgroup of Candida albicans strains isolated from oral cavities of patients infected with human immunodeficiency virus. J Clin Microbiol 33(3):696–700
142. Fatima A, Bashir G, Wani T, Jan A, Kohli A, Khan MS (2017) Molecular identification of Candida species isolated from cases of neonatal candidemia using polymerase chain reaction-restriction fragment length polymorphism in a tertiary care hospital. Indian J Pathol Microbiol 60(1):61–65
143. Khodadadi H, Karimi L, Jalali-Zand N, Adin H, Mirhendi H (2017) Utilization of size polymorphism in ITS1 and ITS2 regions for identification of pathogenic yeast species. J Med Microbiol 66(2):126–133
144. McMullan R, Metwally L, Coyle PV, Hedderwick S, McCloskey B, O'Neill HJ et al (2008) A prospective clinical trial of a real-time polymerase chain reaction assay for the diagnosis of candidemia in nonneutropenic, critically ill adults. Clin Infect Dis 46(6):890–896
145. Lucignano B, Ranno S, Liesenfeld O, Pizzorno B, Putignani L, Bernaschi P et al (2011) Multiplex PCR allows rapid and accurate diagnosis of bloodstream infections in newborns and children with suspected sepsis. J Clin Microbiol 49(6):2252–2258
146. Posteraro B, Sanguinetti M, Masucci L, Romano L, Morace G, Fadda G (2000) Reverse cross blot hybridization assay for rapid detection of PCR-amplified DNA from Candida species, Cryptococcus neoformans, and Saccharomyces cerevisiae in clinical samples. J Clin Microbiol 38(4):1609–1614
147. Leinberger DM, Schumacher U, Autenrieth IB, Bachmann TT (2005) Development of a DNA microarray for detection and identification of fungal pathogens involved in invasive mycoses. J Clin Microbiol 43(10):4943–4953

148. Baskova L, Landlinger C, Preuner S, Lion T (2007) The Pan-AC assay: a single-reaction real-time PCR test for quantitative detection of a broad range of Aspergillus and Candida species. J Med Microbiol 56(Pt 9):1167–1173
149. Schabereiter-Gurtner C, Selitsch B, Rotter ML, Hirschl AM, Willinger B (2007) Development of novel real-time PCR assays for detection and differentiation of eleven medically important Aspergillus and Candida species in clinical specimens. J Clin Microbiol 45(3):906–914
150. Nabili M, Ashrafi M, Janbabaie G, Hedayati MT, Ali-Moghaddam K, Shokohi T (2013) Quantification and optimization of Candida albicans DNA in blood samples using Real- Time PCR. Rep Biochem Mol Biol 2(1):42–47
151. Odds FC, Bougnoux ME, Shaw DJ, Bain JM, Davidson AD, Diogo D et al (2007) Molecular phylogenetics of Candida albicans. Eukaryot Cell 6(6):1041–1052
152. Odds FC (2010) Molecular phylogenetics and epidemiology of Candida albicans. Future Microbiol 5(1):67–79
153. Afsarian SM, Badali H, Shokohi T, Najafipour S (2015) Molecular diversity of Candida albicans isolated from immunocompromised patients, based on MLST method. Iran J Public Health 44(9):1262–1269
154. Marot-Leblond A, Beucher B, David S, Nail-Billaud S, Robert R (2006) Development and evaluation of a rapid latex agglutination test using a monoclonal antibody to identify Candida dubliniensis colonies. J Clin Microbiol 44(1):138–142
155. Karageorgopoulos DE, Vouloumanou EK, Ntziora F, Michalopoulos A, Rafailidis PI, Falagas ME (2011) β-D-Glucan assay for the diagnosis of invasive fungal infections: a meta-analysis. Clin Infect Dis 52(6):750–770
156. Saez-Roson A, Sevilla MJ, Moragues MD (2014) Identification of superficial Candida albicans germ tube antigens in a rabbit model of disseminated candidiasis. A proteomic approach. Int Microbiol 17(1):21–29
157. Yeo SF, Zhang Y, Schafer D, Campbell S, Wong B (2000) A rapid, automated enzymatic fluorometric assay for determination of D-arabinitol in serum. J Clin Microbiol 38(4):1439–1443
158. Pappas PG, Rex JH, Sobel JD, Filler SG, Dismukes WE, Walsh TJ et al (2004) Guidelines for treatment of candidiasis. Clin Infect Dis 38(2):161–189
159. Pappas PG, Kauffman CA, Andes DR, Clancy CJ, Marr KA, Ostrosky-Zeichner L et al (2016) Clinical practice guideline for the management of candidiasis: 2016 update by the infectious diseases Society of America. Clin Infect Dis 62(4):e1–50
160. Bondaryk M, Kurzątkowski W, Staniszewska M (2013) Antifungal agents commonly used in the superficial and mucosal candidiasis treatment: mode of action and resistance development. Adv Dermatol Allergol/Postępy Dermatol Alergol 30(5):293–301
161. Dias MFRG, Bernardes-Filho F, Quaresma-Santos MVP, AGDF A, Schechtman RC, Azulay DR (2013) Treatment of superficial mycoses: review – part II. An Bras Dermatol 88(6):937–944
162. Achkar JM, Fries BC (2010) Candida infections of the genitourinary tract. Clin Microbiol Rev 23(2):253–273
163. Netea MG, Joosten LA, van der Meer JW, Kullberg BJ, van de Veerdonk FL (2015) Immune defence against Candida fungal infections. Nat Rev Immunol 15(10):630–642
164. Netea MG, Quintin J, van der Meer JW (2011) Trained immunity: a memory for innate host defense. Cell Host Microbe 9(5):355–361
165. Bistoni F, Verducci G, Perito S, Vecchiarelli A, Puccetti P, Marconi P et al (1988) Immunomodulation by a low-virulence, agerminative variant of Candida albicans. Further evidence for macrophage activation as one of the effector mechanisms of nonspecific anti-infectious protection. J Med Vet Mycol 26(5):285–299
166. Reales-Calderon JA, Aguilera-Montilla N, Corbi AL, Molero G, Gil C (2014) Proteomic characterization of human proinflammatory M1 and anti-inflammatory M2 macrophages and their response to Candida albicans. Proteomics 14(12):1503–1518
167. Gantner BN, Simmons RM, Underhill DM (2005) Dectin-1 mediates macrophage recognition of Candida albicans yeast but not filaments. EMBO J 24(6):1277–1286

Chapter 4
When *Aspergillus fumigatus* Meets the Man

Sarah Sze Wah Wong and Jean-Paul Latgé

Abstract *Aspergillus fumigatus* is one of the most ubiquitous opportunistic fungal pathogen, which can cause life-threatening invasive pulmonary infections in immunocompromised populations. Upon the inhalation of the *A. fumigatus* conidia, the encounter between the fungus and the host presents a complex interplay. This chapter will summarize the host innate immunity against *A. fumigatus*, and emphasize on the host immune evasion mechanisms of *A. fumigatus*.

4.1 Introduction

Aspergillus species are saprotrophic thermophilic fungi living in decaying material in the soil. When the fungus is starved, it produces millions of aerial conidia which are responsible for the propagation of the fungus (Fig. 4.1). Some of the *Aspergillus* species cause clinical manifestations ranging from chronic to invasive pulmonary infections, following the inhalation of the conidia (hundreds per day in normal environments) [1, 2]. *Aspergillus fumigatus* is the most prevalent etiologic agent of aspergillosis, followed by *Aspergillus flavus*, *A. niger*, *A. nidulans*, *A. terreus* [3]. In immunocompetent individuals, the inhaled conidia rarely cause infections since the host innate immunity is efficient in the clearance of the fungal pathogen [4–9] (Fig. 4.2). In the populations with pre-existing pulmonary cavities who are otherwise immunocompetent, colonization of *Aspergillus* species will only lead to chronic pulmonary aspergillosis (CPA) [10–12]. Unlike CPA, invasive pulmonary aspergillosis (IPA) predominantly affects immunocompromised individuals [13]. Individuals with primary or secondary immunodeficiency, such as chronic granulomatous disease, hematologic malignancy, hematopoietic stem cells transplantation, solid organ transplantation, and neutropenia consecutive to cancer chemotherapy or immunosuppressive treatment are typically at risk for IPA (Fig. 4.2) [2, 14]. The mortality rates of IPA vary among groups of host with different underlying risk factors, but it is generally high [2, 14]. *A. fumigatus* benefits from immune deficiencies, and it also possesses various well-established and specific strategies which helps the

S.S.W. Wong, Ph.D. • J.-P. Latgé, Ph.D. (✉)
Department of Mycology, Unité des Aspergillus, Institut Pasteur, Paris, France
e-mail: sarah.wong@pasteur.fr; jean-paul.latge@pasteur.fr

H.M. Mora-Montes, L.M. Lopes-Bezerra (eds.), *Current Progress in Medical Mycology*, DOI 10.1007/978-3-319-64113-3_4

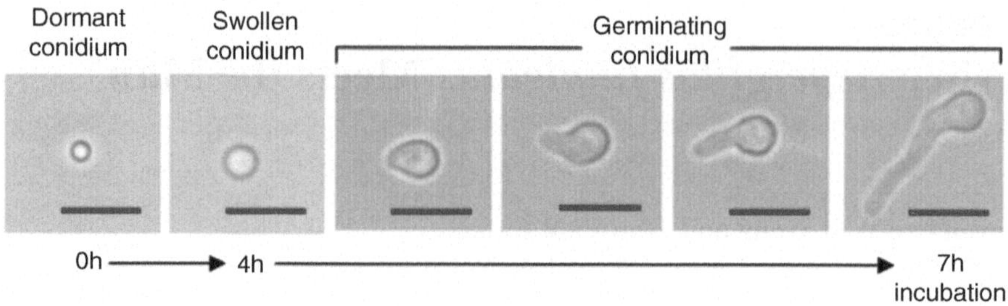

Fig. 4.1 Light microscopy (40×, scale bar = 10 μm) of the morphological change of *A. fumigatus* under incubation at 37 °C in Sabouraud medium

fungus to evade from the host debilitated immune system and colonize the lung parenchyma [15–20].

This chapter will dissect both the host innate immunity and the pathogen anti-immune strategies on the early stages of the fungal infection. Since contact between membrane-bound and soluble pattern recognition receptors (PRRs) of the host and surface pathogen-associated molecular patterns of *A. fumigatus* is the first event leading to the reciprocal recognition of the host and the pathogen (Fig. 4.3). It will especially focus on the host and pathogen molecules favoring their respective recognition, as well as the role of the innate immune cells involved in the anti-*A. fumigatus* defense. A thorough understanding of this complex "tug of war" between the host and the pathogen is fundamental in providing new insights in developing prophylactic strategies of IPA.

4.2 Molecules Responsible for the "Hide-and-Seek" Between Host and *Aspergillus fumigatus*

The fungal cell is surrounded by a cell wall with a specific composition which is very different from the phospholipid bilayer of the host cell plasma membrane. The *A. fumigatus* cell wall which surrounds the fungal cell is mainly composed of polysaccharides which are interlinked alkali insoluble β-1,3-glucans, chitin, and galactomannan and, alkali soluble α-1,3-glucans [21, 22]. These fungal structural polysaccharides are absent in mammalian cells and are therefore pathogen-associated molecular patterns (PAMPs), which are recognized by various membrane-bound and soluble PRRs of the host as foreign objects [21, 23–26].

In resting conidia, this polysaccharide core which is of similar composition both in the conidium and hyphal cell wall, is covered by a bilayered outer layer composed of hydrophobins (the most external) and melanin. The external hydrophobin rodlet layer, which is responsible for the hydrophobicity of the conidia is exclusively composed of the amyloid hydrophobic RodA proteins [27–29]. One way for *A. fumigatus* to become a pathogen is its capacity to "hide" from the host defensive

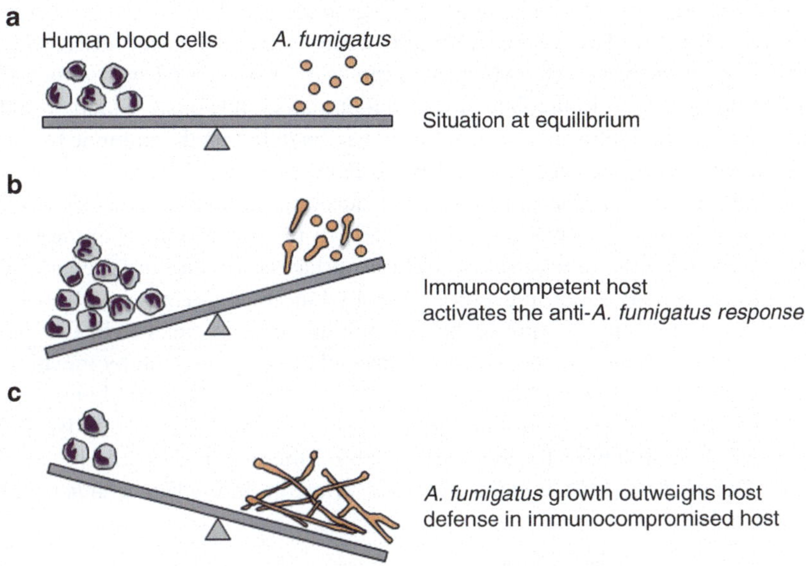

Fig. 4.2 Immune system–*Aspergillus fumigatus* interplay. The fungal disease is only established in the immunodeficient host, where the fungus can resist and escape the immune surveillance and establish the infection

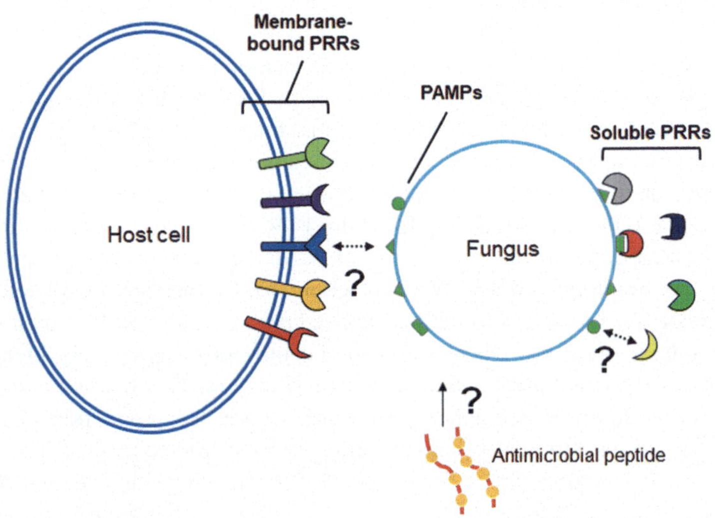

Fig. 4.3 Interaction between various membrane-bound and soluble PRRs which have been shown to bind to the PAMPs on the conidial surface. However, the corresponding PAMPs for each of the PRRs have not been elucidated (as indicated with *question mark*)

response immediately after inhalation and to go undetected after entering the respiratory tract [30–32]. This is due to the presence of the rodlet and melanin outer layers of the dormant conidia which hide the immunogenic cell wall polysaccharides. By doing so, an immediate strong inflammatory response, which would be detrimental for the fungus, is avoided. Moreover, by delaying the immune response, the initial survival of the fungus is prolonged in the host.

The loss of the rodlet and melanin layer during germination leads to an entire modification of the surface layer, which leads to the apparition of immunogenic polysaccharides on the hyphal surface. These surface molecules are recognized by PRRs that are mostly lectin receptors (carbohydrate-recognizing) involved in the initiation of the antifungal response. Indeed, conidial germination can be considered by the fungus as a form suicide. However, instead of being covered by melanin and RodA protein, the cell wall of the mycelium is covered by a specific hyphal galactosaminogalactan, which is immunosuppressive and favors the vegetative fungal growth. All these strategies for the host to seek for the fungal alien and for the fungus to counteract the host immune response after the immediate contact will be discussed below.

4.2.1 Membrane-Bound Pattern Recognition Receptors

4.2.1.1 Toll-Like Receptors

Toll-like receptors (TLRs) are a family of membrane-bound and soluble receptors on sentinel cells involved in the recognition of specific PAMPs [33]. Although the role of TLR2 and TLR4 in the immune defense against fungal pathogens have been extensively studied in the past decades [34]; so far, their precise functions have not yet been clearly elucidated. Various in vitro studies have demonstrated that TLR2 is involved in recognizing *A. fumigatus* [35, 36]. Blocking TLR2 led to a reduced phagocytic rate of *A. fumigatus* conidia, but not of control beads, suggesting that the phagocytic machinery is not impaired by TLR2 blocking [37]. Neutrophil-depleted TLR2-deficient mice had however a lower survival and produced less TNF-α upon stimulation with *A. fumigatus* [38]. The recruitment of neutrophils was severely attenuated in non-immunosuppressed mice deficient in both TLR2 and TLR4, when compared to that in mice with single deficiency [39]. Furthermore, the production of different cytokines in response to *A. fumigatus* is individually mediated by TLR2 and TLR4. For instance, the production of IL-12 and IL-6 from monocytes-derived dendritic cells are mediated by TLR2 and TLR4 respectively [40]. TLR4-mediated pro-inflammatory signals were lost during phenotypic switch from conidia to hyphae, contributing to the escape of the pathogens from the host immune defense [36, 41]. The presence of an optimal innate immune response required both TLR2 and TLR4. Although a direct binding has not been shown, polysaccharides of the cell wall seem to be recognized by TLR2 and TLR4. The IL-6 production via TLR2 and TLR4 stimulation in PBMCs (incubated with TLR2 and TLR4 ligands respectively) was attenuated by pre-incubation with α-glucan and galactomannan, while pre-incubation with β-glucan attenuated the IL-6 production via only TLR4 stimulation [42]. However, TLR2- and TLR4-knockout immunocompetent mice were not more susceptible to challenge of invasive aspergillosis, suggesting that these two receptors are not essential in preventing *Aspergillus* infection in an immunocompetent status [43].

Unlike TLR2 and TLR4, TLR9 is located intracellularly in the endosome compartment of the immune system cells [44]. TLR9 was originally thought to be activated by unmethylated CpG sequences in DNA of bacterial and viral origins only [45]. It was later found that TLR9 can also be activated by unmethylated CpG sequences in fungal DNA [46]. The *A. fumigatus* DNA contains unmethylated CpG sequences, and is therefore capable of activating TLR9 and induce the production of pro-inflammatory cytokines [47]. Since TLR9 is an intracellular receptor, it was suggested that the activation would follow the release of DNA content after fungal lysis in the phagosome. TLR9 is recruited to the phagosomes that contain internalized *A. fumigatus* conidia [48]. However, TLR9 is already recruited to the phagosome merely 1h after phagocytosis of resting conidia [48]. At this stage, the internalized conidia should still be dormant and intact, and the fungal DNA is not exposed. This leaves the underlying recognition mechanism of the fungal DNA and the activation of TLR9 unresolved. Paradoxically, in vivo, TLR9-deficient mice, immunosuppressed by cyclophosphamide, had lower fungal burden than the wild-type mice upon intranasal challenge by *A. fumigatus* [49]. Consistently, in another in vivo study, the neutrophil-depleted TLR9-deficient mice was less susceptible to challenge of dormant or swollen *A. fumigatus* conidia by showing delayed mortality [50]. The expression of dectin-1 was also significantly lowered in the bone marrow-derived dendritic cells from TLR9-deficient mice 14 days post-infection following challenge of swollen conidia [50].

Taken together, the role of TLR2, 4 and 9 in host defense against *A. fumigatus* remains to be further explored.

4.2.1.2 C-Type Lectin Receptors

C-type (Ca^{2+}-dependent) lectin receptors (CLRs), including dectin-1, dectin-2, mannose receptor, Mincle, and DC-SIGN, are important in recognizing the major carbohydrate moieties in the fungal cell wall [51].

Dectin-1 is a major transmembrane receptor for β-1,3-glucan which is found mainly on myeloid cells [52–54]. Since β-1,3-glucan is the major constituent of the fungal cell wall, it was expected that dectin-1 plays a crucial role in the control of fungal infection and especially favors recognition and phagocytosis of fungal particles [37, 55]. In the case of *A. fumigatus*, dectin-1 preferentially recognizes swollen and germinating conidia, as the β-1,3-glucan is exposed on the conidial surface [56–58]. Dectin-1 contains an extracellular lectin-like carbohydrate recognition domain and a cytoplasmic tail, which is phosphorylated at the tyrosine, upon binding with β-1,3-glucan. The phosphorylated cytoplasmic tail then can interact with spleen tyrosine kinase (SYK), which induces cellular responses, such as respiratory burst, phagocytosis and production of pro- and anti-inflammatory cytokines [54, 58, 59]. Dectin-1 induces different cytokine responses independent or in conjunction with TLR: for example, dectin-1 alone induces the production of IL-10, but requires the adaptor MyD88 for the induction of IL-8 and IL-12 [54, 56, 60].

The ligands for the other membrane-bound CLRs remained to be fully determined. Another member of the dectin family, dectin-2, has been shown recently to recognize *A. fumigatus* galactomannan, but its role in the innate response has not been really evaluated [61–63]. Mannose receptor, Mincle and DC-SIGN recognize the N-linked mannan and N-acetylglucosamine (GlcNAc) in the fungal cell wall [51, 64]. Being a polymer of GlcNAc, chitin could also be recognized by mannose receptor [65]. However, it was recently found that chitin is mainly recognized by the Fcγ receptor [66]. Further studies should be warranted to better investigate the respective roles of these polysaccharides tested alone or in association, in the immune response.

4.2.1.3 Crosstalk Between TLRs and CLRs

Some PRRs collaborate to produce a synergetic induction of immune response [67]. For instance, dectin-1, which recognizes β-1,3-glucan, interact with TLR2, which recognizes zymosan, but not β-1,3-glucan [68]. Interestingly, crosstalk between PRRs does not always lead to enhanced immune response. The CLR Mincle suppresses the antifungal immune response mediated by dectin-1 and Mincle [69, 70]. This has also been suggested to be an evasion mechanism of fungal pathogen from the host immune defense [70]. However, this has not been studied in *A. fumigatus*.

4.2.2 Humoral Pattern Recognition Receptors

4.2.2.1 Complement Components

The complement system is often considered an irrelevant immunological weapon against *A. fumigatus* for the following reasons. First, owing to the presence of the thick fungal cell wall, the membrane attack complex (MAC) of the complement system is ineffective on *A. fumigatus* [71, 72]. Second, complement deficiency in human was not found to increase the risk of aspergillosis. However, all three of the complement pathways, classical, alternative and the lectin pathway, are indeed involved in the host defense against *A. fumigatus*, through opsonization and thus, facilitation of phagocytosis [73]. Complement component C3 can directly, or through other components, such as mannose-binding lectin or immunoglobulin, deposit on the surface of the conidia and mycelia [73–78]. The exact chemical nature of the ligand(s) of the complement components on the surface of the conidia or mycelia remain unknown. Furthermore, *A. fumigatus* can evade from the complement attack by scavenging complement regulators, Factor H and C4b binding protein [79, 80], and secreting proteases that degrades complement proteins [81]. Alp1p, a major protease secreted by *A. fumigatus*, can degrade complement proteins C3, C4 and C5 [81].

4.2.2.2 Collectins and Ficolins

Collectins (Collagen + lectin) are a group of soluble pattern recognition receptors characterized by a collagen-domain and carbohydrate recognition domain (CRD), which is responsible in the binding to the carbohydrate moiety in a Ca^{2+}-dependent manner. Collectins in the lung bind to *A. fumigatus* conidial surface as opsonins, and facilitate phagocytosis [51, 82, 83].

Surfactant Proteins

There are four surfactant proteins (SP) in human, SP-A, SP-B, SP-C, and SP-D, that are secreted by alveolar type II cells into the alveolar space [84–86]. Both SP-B and SP-C are small hydrophobic proteins that are mainly responsible in reducing surface tension at the air–liquid interface in the lungs, and their role during *Aspergillus* infection have not been studied [87]. It was only recently found that SP-C binds to bacterial lipopolysaccharides [85, 88]. Apart from that, the immunomodulatory role of SP-C is still uncertain. Meanwhile, SP-A and SP-D, which are hydrophilic proteins, are primarily responsible for the host defense against pulmonary pathogens by facilitating phagocytosis after opsonization [84, 86, 89–91]. SP-A and SP-D are not involved in the lectin complement pathway [92].

Human SP-A and SP-D bind directly to surface of dormant conidia, swelling conidia and hyphae of *A. fumigatus* [83, 93], and the responsible ligand(s) are under investigation by our group. When bound to microorganisms via its CRD region, SP-D binds to the calreticulin/CD91 complex on the surface of macrophages, via its collagenous region, which then mediates the uptake of SP-D opsonized microorganisms by phagocytes and, stimulates inflammatory response by activating NF-κB [94, 95]. Although it was found that the mortality of SP-D-deficient mice was similar to that of the wild-type mice following intranasal challenge of *A. fumigatus* conidia, under immunosuppression of corticosteroids, shorter survival duration, higher hyphal density and more tissue damage were observed in the SP-D deficient mice in comparison to the wild-type [96, 97]. This observation suggested that SP-D plays a role in immune recognition and killing, which have been however insufficiently investigated.

Compared to SP-D, the role of SP-A in the immune defense against *A. fumigatus* is less significant. Although human SP-A bind to conidial surface, the binding is reduced in the presence of extracellular alveolar lipids, while that of SP-D remained unchanged [93]. Immunosuppressed SP-A deficient mice are resistance to lethal

IPA challenge when compared to wild-type mice [97], which could be explained by the increased level of SP-D in SP-A-deficient mice [98].

At present, no evasion mechanism from the opsonization of collectins has yet been observed in *A. fumigatus*.

Mannose-Binding Lectin and Ficolins

Mannose-binding lectin (MBL) binds to the conidial surface of *A. fumigatus* [99]. Mannose-binding lectin by binding to the MBL-associated serine proteases (MASP1, MASP2, and MASP3), activate the lectin complement pathway [100]. Non-immunosuppressed MBL-knockout mice were less susceptible to systemic invasive aspergillosis [101]. It was suggested that the lack of MBL reduced the recruitment of neutrophils, which in turn leads to less tissue damage from inflammation. In contrast, in human, a deficiency in MBL is associated with chronic and invasive aspergillosis [102, 103]. The role of MBL should be revisited.

Ficolins (fibrinogen + collagen + lectin) are a family of lectins that consist of a collagen-like domain and a fibrinogen-like domain described as recognizing specifically N-acetylglucosamine (GlcNAc) [104]. There are three types of ficolins in human, ficolin-1, 2, and 3, which are secreted into the human plasma and binding to ficolin activate the lectin complement pathway [104, 105]. Ficolin-2 binds to *A. fumigatus* [74, 106, 107]. The in vitro binding of ficolin-2 to *A. fumigatus* is inhibited by the presence of GlcNAc and Curdlan (β-1,3-glucan) [107], suggesting that chitin and β-1,3-glucan are the responsible ligands. Previously, ficolin-1 and ficolin-3 were not found to bind to *A. fumigatus*. However, recently, the role of ficolin-3 in lung immune defense was discovered. Ficolin-3 indeed binds to *A. fumigatus* resting or swollen conidia in a calcium-independent manner [74, 82], suggesting that the binding does not involve the carbohydrate recognition domain. Unlike ficolin-2, which is produced in the liver, ficolin-3 is produced by type II alveolar cells and secreted into the alveolar space. Moreover, ficolin-A (the orthologue of ficolin-2 in mouse) increases the release of IL-8 (a proinflammatory cytokine and chemokine for neutrophils) [74]. Upon binding to the carbohydrate ligand on the pathogen surface, ficolins facilitate opsonization, phagocytosis and the activation of the lectin complement pathway. However, complement activation was not impaired in ficolin-knockout mice [74, 108]. This could be explained by the similar role of MBL and ficolins, that both trigger the lectin complement pathway.

Although many elements of the immune system in the anti-*A. fumigatus* response have been discovered or rediscovered in recent years, a comprehensive picture of their roles as well as the ranking of their importance in the immunocompetent, as well as immunocompromised host remains to be elucidated.

4.3 The Cell Actors

4.3.1 Airway Epithelial Cells

Airway epithelial cells (AECs) are the first type of cells to enter in contact with the inhaled conidia [109, 110]. AECs, being nonprofessional phagocytes, are capable of internalizing and killing *A. fumigatus* conidia [111, 112]. Comparing with murine macrophage cell line (J774), the internalized conidia survive longer in the human airway epithelial cell line (A549) than in the alveolar macrophages [113]. Accordingly, a small portion of the internalized conidia are not killed by the AECs, which then germinate and break the epithelial barrier [113]. The inefficient killing of the internalized conidia by the AECs is probably due to inefficient respiratory burst. In addition, the adherence of *A. fumigatus* conidia to the A549 cell line inhibits the release of IL-6, IL-8, and TNF-α, which thus inhibits the recruitment of immune cells and apoptosis of the epithelial cells [114, 115].

AECs produce antimicrobial peptides such as human β-defensins, human lactoferrin (hLF), and histatin 5, which possess some antifungal activity against *A. fumigatus* [4, 6, 109, 110, 116, 117]. The gene expression of human β-defensins hBD2 and hBD9 were found to be induced in A549 cells that are exposed to swollen conidia [116]. hBD2 displayed direct, but low antifungal activity in vitro against *A. fumigatus* [118] and does not enhance the antifungal killing activity of neutrophils [118]. However, hBD could act as chemoattractant for immune cells and activation of professional antigen-presenting cells [119].

4.3.2 Alveolar Macrophages

The resident alveolar macrophages (AMs), which are responsible of the recognition and phagocytosis of the *A. fumigatus* conidia, serve as the major immune cells in the defense of *A. fumigatus*. Upon internalization of *A. fumigatus* conidia, phagolysosomes were not acidified [18]. Dihydroxynaphthalene-melanin (DHN-melanin) on the outer layer of dormant *A. fumigatus* conidia was shown to inhibit the acidification of phagolysosome, and thus, prevent intracellular killing [120–122]. The mutant of polyketide synthase (Δ*pksP*), the enzyme responsible for the initial step in DHN-melanin formation is devoid of DHN-melanin and avirulent [123]. Interestingly, the melanin of *A. niger*, which is different from the melanin in *A. fumigatus* [124, 125], does not inhibit phagolysosome acidification [18]. The intracellular killing of *A. fumigatus* conidia in AMs is triggered by the swelling of conidia. The phagolysosome acidification and the increase in the production of reactive oxygen species (ROS) following phagocytosis is essential for conidial killing [126, 127]. Upon swelling, the outer layer of hydrophobins and melanin is shed. In the absence of inhibition by melanin, the phagolysosome acidifies. On the other hand, the absence of the outer layer of rodlet and melanin unmasks the PAMPs in the inner layer. The exposure of PAMPs recruits PRRs to the phagosome, which triggers production of ROS, cytokines and chemokines and the recruitment of autophagy proteins, LC3 II, Atg5 and Atg7 to the phagolysosome [57, 128, 129]. Pyomelanin, another type of melanin that *A. fumigatus* produces, can also protects the fungus from ROS attack [13, 120, 130]. However, pyomelanin-minus mutant are as pathogenic as their parental strain. In addition, DHN-melanin inhibits LC3-associated phagocytosis (LAP) [123, 131, 132]. Taken together, pyomelanin and DHN-melanin permits the fungus to escape host immune defense by inhibiting phagolysome acidification, quenching ROS and inhibiting LC3-associated phagocytosis.

4.3.3 Neutrophils

Neutrophils act as a strong second line of innate immune defense against *A. fumigatus* [133]. Upon recruitment to the alveolar space, neutrophils constitute the majority of phagocytes to eliminate the conidia [4, 134]. Even though AMs by themselves are able to eradicate resting conidia in a mouse deprived of neutrophils, neutrophils are more important than AMs in the pathogen clearance [127, 135]. Indeed, low number of neutrophils in the host (neutropenia) renders patients at great risk of IPA [2]. Moreover, molecules of *A. fumigatus* which impairs the neutrophil action are true virulence factors. Galactosaminogalactan is one of them. This polysaccharide, which is secreted by *A. fumigatus* hyphae, suppresses the recruitment of neutrophils, which then favors the fungal survival in the host [136, 137]. In addition, it induces anti-inflammatory response by the induction of interleukin-1 receptor antagonist (IL-1Ra) and reduces neutrophil recruitment [136, 138].

An important function unique to neutrophils is its capacity of attacking *A. fumigatus* hyphae, which are too large to be phagocytosed by immune cells. Neutrophils attack the fungal hyphae extracellularly by the secretion of enzymes, which degrade and permeabilize the cell wall and makes the fungus more sensitive to the neutrophil granular toxic molecules. Recently, neutrophil extracellular traps (NETs) which are mainly composed of nuclear DNA and antimicrobial proteins, have been mentioned as playing a major role against microbial pathogens [139]. Although the production of NETs is stimulated by *Aspergillus* hyphae [140], the hyphae are reported to be only slightly susceptible to the killing by NETs [141]. It is suggested that NETs only has a fungistatic effect on the hyphae and prevent further spreading of the disease [141]. This finding is in contrast to the one regarding *Candida albicans*, which is completely susceptible to the killing by NETs [142]. The fungistatic activity of NETs towards *A. fumigatus* hyphae is attributed by calprotectin, a Zn^{2+} chelator, [143]. Interestingly, it was shown recently that *A. fumigatus* hyphae generate hyphal branching upon interaction with neutrophils, which allows lower host immune interference and increases invasive growth [144].

Moreover, *A. fumigatus* seems to have other mechanisms to evade the neutrophil antagonistic action since *A. fumigatus* is less sensitive than *A. niger* for which the germination and hyphal length are more significantly reduced by neutrophils [18].

For example, fumagillin, secreted by the *A. fumigatus* hyphae, inhibits the antifungal function of neutrophils by inhibiting the formation of NADPH oxidase and reduces the degranulation [145]. Neutrophils exposed to fumagillin in vitro also demonstrated reduced rate of phagocytosis of *A. fumigatus* conidia [145]. An analysis of the differences between *A. fumigatus* and other non- or poorly pathogenic *Aspergillus* species, such as melanin structure and cell wall composition, should definitely reveal more information of the success of *A. fumigatus* as pathogen.

4.4 Perspectives

The current arsenal of antifungal agents available for the treatment of IA is mainly limited to azoles, echinocandins, and polyenes [146]. However, there are major drawbacks regarding the drug resistance, efficiency, and toxicity of these current agents [147, 148]. Furthermore, the pipeline of antifungal development has not been introducing any new classes of antifungal agents, especially since there is not a market for drugs for rare diseases such as aspergillosis. Can immunotherapy lead to the development of specific drugs or antibodies that are able to block virulence factors that allow the evasion of *A. fumigatus* from the host immune system? The past decade has witnessed a shift in paradigm of antifungal development, in which the virulence factors are targeted for drug candidates instead of essential genes [149–151]. Analysis of the transcriptomic and proteomic changes occurring after internalization of the pathogen and especially in the phagolysosome, where the fungus is inhibited in the immunocompetent host but not in the immunocompromised host, may lead to the identification of essential virulence factors in vivo. Such approach focused on the inhibition of virulence factors would exert less selective pressure on the pathogens, and thus, the chance of resistance development is lower [152]. However, such immunotherapeutic approach has not been yet undertaken with *A. fumigatus* but the situation may be difficult in the case of pulmonary aspergillosis, which occurs mainly among immunocompromised patients. The identification of monoclonal antibodies able to block hyphal emergence could be also possible. Such strategy has been proposed a few years ago [153, 154], however it has not been pursued yet. Although a lot of progresses have been obtained in recent years, the overall picture to fully understand the innate immune defense against *A. fumigatus* is missing. For example, the respective role of the different PRRs and the lack of definition of all the ligands binding to complement proteins, collectins and other PRRs are essential gaps to fill. The reasons for the ineffectiveness of NETs to kill *A. fumigatus* hyphae should also be investigated. Comparative characterization of major *Aspergillus* species may also provide valuable understanding of how *A. fumigatus* took the "throne" and became the most predominant and successful pathogen among the *Aspergillus* species. Finally, it has been reported that many cytokines and chemokines plays a major role in the defense reactions (not discussed here but see for review [155, 156]. However, the interweaving network between all these chemokines, cytokines and the defense against

A. fumigatus remains obscure. In conclusion, the early events "when *A. fumigatus* meets the man" remain enigmatic and this chapter paves the way for the direction of future exploration.

References

1. Kwon-Chung KJ, Sugui JA (2013) *Aspergillus fumigatus* – what makes the species a ubiquitous human fungal pathogen? PLoS Pathog 9:1–4. doi:10.1371/journal.ppat.1003743
2. Latgé JP (1999) *Aspergillus fumigatus* and aspergillosis. Clin Microbiol Rev 12:310–350
3. Sugui JA, Kwon-Chung KJ, Juvvadi PR et al (2014) *Aspergillus fumigatus* and related species. Cold Spring Harb Perspect Med. doi:10.1101/cshperspect.a019786
4. Balloy V, Chignard M (2009) The innate immune response to *Aspergillus fumigatus*. Microbes Infect 11:919–927. doi:10.1016/j.micinf.2009.07.002
5. Cramer RA, Rivera A, Hohl TM (2011) Immune responses against *Aspergillus fumigatus*: what have we learned? Curr Opin Infect Dis 24:315–322. doi:10.1097/QCO.0b013e328348b159
6. Espinosa V, Rivera A (2016) First line of defense: innate cell-mediated control of pulmonary Aspergillosis. Front Microbiol 7:1–12. doi:10.3389/fmicb.2016.00272
7. Lass-Flörl C, Roilides E, Löffler J et al (2013) Minireview: host defence in invasive aspergillosis. Mycoses 56:403–413. doi:10.1111/myc.12052
8. Margalit A, Kavanagh K (2015) The innate immune response to *Aspergillus fumigatus* at the alveolar surface. FEMS Microbiol Rev 39:670–687. doi:10.1093/femsre/fuv018
9. Romani L (2004) Immunity to fungal infections. Nat Rev Immunol 11:275–288
10. Denning DW, Pleuvry A, Cole DC (2011) Global burden of chronic pulmonary aspergillosis as a sequel to pulmonary tuberculosis. Bull World Health Organ 89:864–872. doi:10.2471/BLT.11.089441
11. Hayes GE, Novak-Frazer L (2016) Chronic pulmonary aspergillosis—where are we? and where are we going? J Fungi 2:18. doi:10.3390/jof2020018
12. Schweer KE, Bangard C, Hekmat K, Cornely OA (2014) Chronic pulmonary aspergillosis. Mycoses 57:257–270. doi:10.1111/myc.12152
13. Dagenais TRT, Keller NP (2009) Pathogenesis of *Aspergillus fumigatus* in invasive aspergillosis. Clin Microbiol Rev 22:447–465
14. Steinbach WJ, Marr KA, Anaissie EJ et al (2012) Clinical epidemiology of 960 patients with invasive aspergillosis from the PATH Alliance registry. J Infect 65:453–464. doi:10.1016/j.jinf.2012.08.003
15. Chai L, Netea MG, Vonk AG, Kullberg B-J (2009a) Fungal strategies for overcoming host innate immune response. Med Mycol 47:227–236. doi:10.1080/13693780802209082
16. Chotirmall SH, Mirkovic B, Lavelle GM, McElvaney NG (2014) Immunoevasive *aspergillus* virulence factors. Mycopathologia 178:363–370. doi:10.1007/s11046-014-9768-y
17. Collette JR, Lorenz MC (2011) Mechanisms of immune evasion in fungal pathogens. Curr Opin Microbiol 14:668–675. doi:10.1016/j.mib.2011.09.007
18. Escobar N, Ordonez SR, Wösten HAB et al (2016) Hide, keep quiet, and keep low: Properties that make *Aspergillus fumigatus* a successful lung pathogen. Front Microbiol 7:1–13. doi:10.3389/fmicb.2016.00438
19. Krappmann S (2016) How to invade a susceptible host: cellular aspects of aspergillosis. Curr Opin Microbiol 34:136–146
20. Marcos CM, de Oliveira HC, de Melo W, da Silva JF et al (2016) Anti-immune strategies of pathogenic fungi. Front Cell Infect Microbiol 6:1–22. doi:10.3389/fcimb.2016.00142
21. Latgé JP (2010) Tasting the fungal cell wall. Cell Microbiol 12:863–872. doi:10.1111/j.1462-5822.2010.01474.x

22. Mouyna I, Fontaine T (2009) Cell wall of *Aspergillus fumigatus*: a dynamic structure. In: Latgé J, Steinbach W (eds) *Aspergillus fumigatus* and Aspergillosis. American Society of Microbiology, Washington, DC, pp 169–183
23. Bidula S, Schelenz S (2016) A sweet response to a sour situation: the role of soluble pattern recognition receptors in the innate immune response to invasive *Aspergillus fumigatus* infections. PLoS Pathog 12:1–6. doi:10.1371/journal.ppat.1005637
24. Latgé JP, Calderone R (2002) Host-microbe interactions: fungi invasive human fungal opportunistic infections. Curr Opin Microbiol 5:355–358
25. Levitz SM (2010) Innate recognition of fungal cell walls. PLoS Pathog 6:e1000758. doi:10.1371/journal.ppat.1000758
26. van de Veerdonk FL, Kullberg BJ, van der Meer JW et al (2008) Host-microbe interactions: innate pattern recognition of fungal pathogens. Curr Opin Microbiol 11:305–312. doi:10.1016/j.mib.2008.06.002
27. Beauvais A, Fontaine T, Aimanianda V, Latgé JP (2014) *Aspergillus* cell wall and biofilm. Mycopathologia 178:371–377. doi:10.1007/s11046-014-9766-0
28. Paris S, Debeaupuis JP, Crameri R et al (2003) Conidial hydrophobins of *Aspergillus fumigatus*. Appl Environ Microbiol 69:1581–1588
29. Pihet M, Vandeputte P, Tronchin G et al (2009) Melanin is an essential component for the integrity of the cell wall of *Aspergillus fumigatus* conidia. BMC Microbiol 11:1–11. doi:10.1186/1471-2180-9-177
30. Aimanianda V, Bayry J, Bozza S et al (2009) Surface hydrophobin prevents immune recognition of airborne fungal spores. Nature 460:1117–1121. doi:10.1038/nature08264
31. Aimanianda V, Latgé JP (2010) Fungal hydrophobins form a sheath preventing immune recognition of airborne conidia. Virulence 1:185–187. doi:10.4161/viru.1.3.11317
32. Rappleye CA, Goldman WE (2008) Fungal stealth technology. Trends Immunol 29:18–24. doi:10.1016/j.it.2007.10.001
33. Akira S, Takeda K (2004) Toll-like receptor signalling. Nature 4:88–88. doi:10.1038/nri1391
34. Underhill DM, Ozinsky A, Hajjar AM et al (1999) The Toll-like receptor 2 is recruited to macrophage phagosomes and discriminates between pathogens. Nature 401:811–815. doi:10.1038/44605
35. Mambula SS, Sau K, Henneke P et al (2002) Toll-like receptor (TLR) signaling in response to *Aspergillus fumigatus*. J Biol Chem 277:39320–39326. doi:10.1074/jbc.M201683200
36. Netea MG, Warris A, Van der Meer JW et al (2003) *Aspergillus fumigatus* evades immune recognition during germination through loss of toll-like receptor-4-mediated signal transduction. J Infect Dis 188:320–326. doi:10.1086/376456
37. Luther K, Torosantucci A, Brakhage AA et al (2007) Phagocytosis of *Aspergillus fumigatus* conidia by murine macrophages involves recognition by the dectin-1 beta-glucan receptor and Toll-like receptor 2. Cell Microbiol 9:368–381. doi:10.1111/j.1462-5822.2006.00796.x
38. Balloy V, Si-Tahar M, Takeuchi O et al (2005) Involvement of toll-like receptor 2 in experimental invasive pulmonary aspergillosis. Infect Immun 73:5420–5425. doi:10.1128/iai.73.9.5420-5425.2005
39. Meier A, Kirschning CJ, Nikolaus T et al (2003) Toll-like receptor (TLR) 2 and TLR4 are essential for *Aspergillus*-induced activation of murine macrophages. Cell Microbiol 5:561–570. doi:10.1046/j.1462-5822.2003.00301.x
40. Braedel S, Radsak M, Einsele H et al (2004) *Aspergillus fumigatus* antigens activate innate immune cells via toll-like receptors 2 and 4. Br J Haematol 125:392–399. doi:10.1111/j.1365-2141.2004.04922.x
41. Chai L, Kullberg BJ, Vonk AG et al (2009b) Modulation of Toll-like receptor 2 (TLR2) and TLR4 responses by *Aspergillus fumigatus*. Infect Immun 77:2184–2192. doi:10.1128/iai.01455-08
42. Chai L, Vonk AG, Kullberg BJ et al (2011) *Aspergillus fumigatus* cell wall components differentially modulate host TLR2 and TLR4 responses. Microbes Infect 13:151–159. doi:10.1016/j.micinf.2010.10.005

43. Dubourdeau M, Athman R, Balloy V et al (2006) *Aspergillus fumigatus* induces innate immune responses in alveolar macrophages through the MAPK pathway independently of TLR2 and TLR4. J Immunol 177:3994–4001
44. Kawai T, Akira S (2011) Toll-like receptors and their crosstalk with other innate receptors in infection and immunity. Immunity 34:637–650. doi:10.1016/j.immuni.2011.05.006
45. Hemmi H, Takeuchi O, Kawai T et al (2000) A Toll-like receptor recognizes bacterial DNA. Nature 408:740–745. doi:10.1038/35047123
46. Kasperkovitz PV, Khan NS, Tam JM et al (2011) Toll-like receptor 9 modulates macrophage antifungal effector function during innate recognition of *Candida albicans* and *Saccharomyces cerevisiae*. Infect Immun 79:4858–4867. doi:10.1128/IAI.05626-11
47. Ramirez-Ortiz ZG, Specht CA, Wang JP et al (2008) Toll-like receptor 9-dependent immune activation by unmethylated CpG motifs in *Aspergillus fumigatus* DNA. Infect Immun 76:2123–2129. doi:10.1128/iai.00047-08
48. Kasperkovitz PV, Cardenas ML, Vyas JM (2010) TLR9 is actively recruited to *Aspergillus fumigatus* phagosomes and requires the N-terminal proteolytic cleavage domain for proper intracellular trafficking. J Immunol 185:7614–7622. doi:10.4049/jimmunol.1002760
49. Bellocchio S, Montagnoli C, Bozza S et al (2004) The contribution of the Toll-like/IL-1 receptor superfamily to innate and adaptive immunity to fungal pathogens in vivo. J Immunol 172:3059–3069. doi:10.4049/jimmunol.172.5.3059
50. Ramaprakash H, Ito T, Standiford TJ et al (2009) Toll-like receptor 9 modulates immune responses to *Aspergillus fumigatus* conidia in immunodeficient and allergic mice. Infect Immun 77:108–119
51. Hardison SE, Brown GD (2012) C-type lectin receptors orchestrate antifungal immunity. Nat Immunol. doi:10.1016/j.tim.2007.10.012
52. Brown GD, Taylor PR, Reid DM et al (2002) Dectin-1 is a major-glucan receptor on macrophages. J Exp Med 196:407–412. doi:10.1084/jem.20020470
53. Brown GD, Herre J, Williams DL et al (2003) Dectin-1 mediates the biological effects of beta-glucans. J Exp Med 197:1119–1124. doi:10.1084/jem.20021890
54. Brown GD (2006) Dectin-1: a signalling non-TLR pattern-recognition receptor. Nat Rev Immunol 6:33–43. doi:10.1038/nri1745
55. Taylor PR, Tsoni SV, Willment JA et al (2007) Dectin-1 is required for beta-glucan recognition and control of fungal infection. Nat Immunol 8:31–38. doi:10.1038/ni1408
56. Gersuk GM, Underhill DM, Zhu L, Marr KA (2006) Dectin-1 and TLRs permit macrophages to distinguish between different *Aspergillus fumigatus* cellular states. J Immunol 176:3717–3724. doi:10.4049/jimmunol.176.6.3717
57. Hohl TM, Van Epps HL, Rivera A et al (2005) *Aspergillus fumigatus* triggers inflammatory responses by stage-specific beta-glucan display. PLoS Pathog 1:e30. doi:10.1371/journal.ppat.0010030
58. Steele C, Rapaka RR, Metz A et al (2005) The beta-glucan receptor dectin-1 recognizes specific morphologies of *Aspergillus fumigatus*. PLoS Pathog 1:e42
59. Underhill DM, Rossnagle E, Lowell CA et al (2005) Dectin-1 activates Syk tyrosine kinase in a dynamic subset of macrophages for reactive oxygen production. Blood 106:2543–2550. doi:10.1182/blood-2005-03-1239
60. Rogers NC, Slack EC, Edwards AD et al (2005) Syk-dependent cytokine induction by dectin-1 reveals a novel pattern recognition pathway for C type lectins. Immunity 22:507–517. doi:10.1016/j.immuni.2005.03.004
61. Reedy J, Wuethrich M, Latgé J, Vyas J (2016) Dectin-2 is a receptor for galactomannan. In: 7th Advances Against Aspergillosis. Manchester, United Kingdom
62. Saijo S, Ikeda S, Yamabe K et al (2010) Dectin-2 recognition of α-mannans and induction of Th17 cell differentiation is essential for host defense against *Candida albicans*. Immunity 32:681–691. doi:10.1016/j.immuni.2010.05.001
63. Sun H, Xu XY, Shao HT et al (2013) Dectin-2 is predominately macrophage restricted and exhibits conspicuous expression during *Aspergillus fumigatus* invasion in human lung. Cell Immunol 284:60–67. doi:10.1016/j.cellimm.2013.06.013

64. Serrano-Gómez D, Domínguez-Soto A, Ancochea J et al (2004) Dendritic cell-specific intercellular adhesion molecule 3-grabbing nonintegrin mediates binding and internalization of *Aspergillus fumigatus* conidia by dendritic cells and macrophages. J Immunol 173:5635–5643. doi:10.4049/jimmunol.173.9.5635
65. Shibata Y, Metzger WJ, Myrvik QN (1997) Chitin particle-induced cell-mediated immunity is inhibited by soluble mannan: mannose receptor-mediated phagocytosis initiates IL-12 production. J Immunol 159:2462–2467
66. Becker K, Aimanianda V, Wang X et al (2016) Cytokines in human PBMCs via the Fc-gamm receptor/Syk/PI3K pathway. MBio 7:1–11. doi:10.1128/mBio.01823-15.Editor
67. Hontelez S, Sanecka A, Netea MG et al (2012) Molecular view on PRR cross-talk in antifungal immunity. Cell Microbiol 14:467–474. doi:10.1111/j.1462-5822.2012.01748.x
68. Gantner BN, Simmons RM, Canavera SJ et al (2003) Collaborative induction of inflammatory responses by dectin-1 and Toll-like receptor 2. J Exp Med 197:1107–1117. doi:10.1084/jem.20021787
69. Rivera A (2014) When PRRs collide: Mincle meddles with dectin and toll. Cell Host Microbe 15:397–399. doi:10.1016/j.chom.2014.03.013
70. Wevers BA, Kaptein TM, Zijlstra-Willems EM et al (2014) Fungal engagement of the C-type lectin mincle suppresses dectin-1-induced antifungal immunity. Cell Host Microbe 15:494–505. doi:10.1016/j.chom.2014.03.008
71. Speth C, Rambach G, Lass-Flörl C et al (2004) The role of complement in invasive fungal infections. Mycoses 47:93–103
72. Speth C, Rambach G (2012) Complement attack against *Aspergillus* and corresponding evasion mechanisms. Interdiscip Perspect Infect Dis. doi:10.1155/2012/463794
73. Kozel TR, Wilson MA, Farrell TP, Levitz SM (1989) Activation of C3 and binding to *Aspergillus fumigatus* conidia and hyphae. Infect Immun 57:3412–3417
74. Bidula S, Kenawy H, Ali YM et al (2013) Role of ficolin-A and lectin complement pathway in the innate defense against pathogenic *aspergillus* species. Infect Immun 81:1730–1740. doi:10.1128/IAI.00032-13
75. Braem SGE, Rooijakkers SHM, van Kessel KPM et al (2015) Effective neutrophil phagocytosis of *Aspergillus fumigatus* is mediated by classical pathway complement activation. J Innate Immun 7:364–374. doi:10.1159/000369493
76. Dumestre-Perard C, Lamy B, Aldebert D et al (2008) *Aspergillus* conidia activate the complement by the mannan-binding lectin C2 bypass mechanism. J Immunol 181:7100–7105. doi:10.4049/jimmunol.181.10.7100
77. Rosbjerg A, Genster N, Pilely K et al (2016) Complementary roles of the classical and lectin complement pathways in the defense against *Aspergillus fumigatus*. Front Immunol 7:1–10. doi:10.3389/fimmu.2016.00473
78. Sturtevant J, Latgé JP (1992) Participation of complement in the phagocytosis of the conidia of *Aspergillus fumigatus* by human polymorphonuclear cells. J Infect Dis 166:580–586
79. Behnsen J, Hartmann A, Schmaler J et al (2008) The opportunistic human pathogenic fungus *Aspergillus fumigatus* evades the host complement system. Infect Immun 76:820–827
80. Vogl G, Lesiak I, Jensen DB et al (2008) Immune evasion by acquisition of complement inhibitors: the mould *Aspergillus* binds both factor H and C4b binding protein. Mol Immunol 45:1485–1493. doi:10.1016/j.molimm.2007.08.011
81. Behnsen J, Lessing F, Schindler S et al (2010) Secreted *Aspergillus fumigatus* protease Alp1 degrades human complement proteins C3, C4, and C5. Infect Immun 78:3585–3594. doi:10.1128/iai.01353-09
82. Bidula S, Sexton DW, Yates M et al (2015) H-ficolin binds *Aspergillus fumigatus* leading to activation of the lectin complement pathway and modulation of lung epithelial immune responses. Immunology 146:281–291. doi:10.1111/imm.12501
83. Madan T, Eggleton P, Kishore U et al (1997) Binding of pulmonary surfactant proteins A and D to *Aspergillus fumigatus* conidia enhances phagocytosis and killing by human neutrophils and alveolar macrophages. Infect Immun 65:3171–3179

84. Kishore U, Greenhough TJ, Waters P et al (2006) Surfactant proteins SP-A and SP-D: structure, function and receptors. Mol Immunol 43:1293–1315. doi:10.1016/j.molimm.2005.08.004
85. Mulugeta S, Beers MF (2006) Surfactant protein C: its unique properties and emerging immunomodulatory role in the lung. Microbes Infect 8:2317–2323. doi:10.1016/j.micinf.2006.04.009
86. Wright JR (2005) Immunoregulatory functions of surfactant proteins. Nat Rev Immunol 5:58–68. doi:10.1038/nri1528
87. Tafel O, Latzin P, Paul K et al (2008) Surfactant proteins SP-B and SP-C and their precursors in bronchoalveolar lavages from children with acute and chronic inflammatory airway disease. BMC Pulm Med 8:6. doi:10.1186/1471-2466-8-6
88. Augusto L, Synguelakis M, Johansson J et al (2003) Interaction of pulmonary surfactant protein C with CD14 and lipopolysaccharide. Infect Immun 71:61–67
89. Carreto-Binaghi LE, Aliouat EM, Taylor ML (2016) Surfactant proteins, SP-A and SP-D, in respiratory fungal infections: their role in the inflammatory response. Respir Res 17:66. doi:10.1186/s12931-016-0385-9
90. Kishore U, Madan T, Sarma PU et al (2002) Protective roles of pulmonary surfactant proteins, SP-A and SP-D, against lung allergy and infection caused by *Aspergillus fumigatus*. Immunobiology 205:610–618. doi:10.1078/0171-2985-00158
91. Sano H, Kuroki Y (2005) The lung collectins, SP-A and SP-D, modulate pulmonary innate immunity. Mol Immunol 42:279–287. doi:10.1016/j.molimm.2004.07.014
92. Willment JA, Brown GD (2008) C-type lectin receptors in antifungal immunity. Trends Microbiol 16:27–32. doi:10.1016/j.tim.2007.10.012
93. Allen MJ, Harbeck R, Smith B et al (1999) Binding of rat and human surfactant proteins A and D to *Aspergillus fumigatus* conidia. Infect Immun 67:4563–4569
94. Mehrad B, Strieter RM, Standiford TJ (1999) Role of TNF-α in pulmonary host defense in murine invasive aspergillosis. J Immunol 162:1633–1640
95. Vandivier RW, Ogden CA, Fadok VA et al (2002) Role of surfactant proteins A, D, and C1q in the clearance of apoptotic cells in vivo and in vitro: calreticulin and CD91 as a common collectin receptor complex. J Immunol 169:3978–3986. doi:10.4049/jimmunol.169.7.3978
96. Madan T, Kishore U, Singh M et al (2001) Protective role of lung surfactant protein D in a murine model of invasive pulmonary aspergillosis. Infect Immun 69:2728–2731. doi:10.1128/IAI.69.4.2728
97. Madan T, Reid KBM, Clark H et al (2010) Susceptibility of mice genetically deficient in SP-A or SP-D gene to invasive pulmonary aspergillosis. Mol Immunol 47:1923–1930. doi:10.1016/j.molimm.2010.02.027
98. LeVine AM, Hartshorn K, Elliott J et al (2002) Absence of SP-A modulates innate and adaptive defense responses to pulmonary influenza infection. Am J Physiol Lung Cell Mol Physiol 282:L563–L572. doi:10.1152/ajplung.00280.2001
99. Neth O, Jack DL, Dodds AW et al (2000) Mannose-binding lectin binds to a range of clinically relevant microorganisms and promotes complement deposition. Infect Immun 68:688–693. doi:10.1128/iai.68.2.688-693.2000
100. Oikonomopoulou K, Edimara S, Lambris JD (2012) Complement system and its role in immune responses. Encycl life Sci. doi:10.1002/9780470015902.a0000508.pub3
101. Clemons KV, Martinez M, Tong AJ, Stevens DA (2010) Resistance of MBL gene-knockout mice to experimental systemic aspergillosis. Immunol Lett 128:105–107. doi:10.1016/j.imlet.2009.12.021
102. Crosdale DJ, Poulton K V, Ollier WE, et al (2001) Mannose-binding lectin gene polymorphisms as a susceptibility factor for chronic necrotizing pulmonary aspergillosis. J Infect Dis 184:653–656. doi:10.1086/322791. JID001096 [pii]
103. Lambourne J, Agranoff D, Herbrecht R et al (2009) Association of mannose-binding lectin deficiency with acute invasive aspergillosis in immunocompromised patients. Clin Infect Dis 49:1486–1491. doi:10.1086/644619

104. Matsushita M (2009) Ficolins: complement-activating lectins involved in innate immunity. J Innate Immun 2:24–32. doi:10.1159/000228160
105. Endo Y, Matsushita M, Fujita T (2011) The role of ficolins in the lectin pathway of innate immunity. Int J Biochem Cell Biol 43:705–712. doi:10.1016/j.biocel.2011.02.003
106. Hummelshøj T, Ma YJ, Munthe-Fog L et al (2012) The interaction pattern of murine serum ficolin-A with microorganisms. PLoS One 7:1–12. doi:10.1371/journal.pone.0038196
107. Ma YJ, Doni A, Hummelshøj T et al (2009) Synergy between ficolin-2 and pentraxin 3 boosts innate immune recognition and complement deposition. J Biol Chem 284:28263–28275. doi:10.1074/jbc.M109.009225
108. Genster N, Præstekjær Cramer E, Rosbjerg A et al (2016) Ficolins promote fungal clearance in vivo and modulate the inflammatory cytokine response in host defense against Aspergillus fumigatus. J Innate Immun. doi:10.1159/000447714
109. Bals R, Hiemstra PS (2004) Innate immunity in the lung: how epithelial cells fight against respiratory pathogens. Eur Respir J 23:327–333. doi:10.1183/09031936.03.00098803
110. Croft CA, Culibrk L, Moore MM, Tebbutt SJ (2016) Interactions of *Aspergillus fumigatus* conidia with airway epithelial cells: a critical review. Front Microbiol 7:1–15. doi:10.3389/fmicb.2016.00472
111. Botterel F, Gross K, Ibrahim-Granet O et al (2008) Phagocytosis of *Aspergillus fumigatus* conidia by primary nasal epithelial cells in vitro. BMC Microbiol 8:97. doi:10.1186/1471-2180-8-97
112. Paris S, Boisvieux-Ulrich E, Crestani B et al (1997) Internalization of *Aspergillus fumigatus* conidia by epithelial and endothelial cells. Infect Immun 65:1510–1514
113. Wasylnka JA, Moore MM (2003) *Aspergillus fumigatus* conidia survive and germinate in acidic organelles of A549 epithelial cells. J Sell Sci 116:1579–1587. doi:10.1242/jcs.00329
114. Berkova N, Lair-Fulleringer S, Femenia F et al (2006) *Aspergillus fumigatus* conidia inhibit tumour necrosis factor- or staurosporine-induced apoptosis in epithelial cells. Int Immunol 18:139–150. doi:10.1093/intimm/dxh356
115. Zhang Z, Liu R, Noordhoek JA, Kauffman HF (2005) Interaction of airway epithelial cells (A549) with spores and mycelium of *Aspergillus fumigatus*. J Infect 51:375–382. doi:10.1016/j.jinf.2004.12.012
116. Alekseeva L, Huet D, Femenia F et al (2009) Inducible expression of beta defensins by human respiratory epithelial cells exposed to *Aspergillus fumigatus* organisms. BMC Microbiol 9:33. doi:10.1186/1471-2180-9-33
117. Lupetti A, van Dissel J, Brouwer C, Nibbering P (2008) Human antimicrobial peptides' antifungal activity against *Aspergillus fumigatus*. Eur J Clin Microbiol Infect Dis 27:1125–1129. doi:10.1007/s10096-008-0553-z
118. Okamoto T, Tanida T, Wei B et al (2004) Regulation of fungal infection by a combination of amphotericin B and peptide 2, a lactoferrin peptide that activates neutrophils. Clin Diagn Lab Immunol 11:1111–1119. doi:10.1128/CDLI.11.6.1111
119. Soruri A, Grigat J, Forssmann U et al (2007) β-Defensins chemoattact macrophages and mast cells but not lymphocytes and dendritic cells: CCR6 is not involved. Eur J Immunol 37:2474–2486. doi:10.1002/eji.200737292
120. Heinekamp T, Thywißen A, Macheleidt J et al (2012) *Aspergillus fumigatus* melanins: Interference with the host endocytosis pathway and impact on virulence. Front Microbiol 3:1–7. doi:10.3389/fmicb.2012.00440
121. Jahn B, Langfelder K, Schneider U et al (2002) PKSP-dependent reduction of phagolysosome fusion and intracellular kill of *Aspergillus fumigatus* conidia by human monocyte-derived macrophages. Cell Microbiol 4:793–803. doi:10.1046/j.1462-5822.2002.00228.x
122. Thywißen A, Heinekamp T, Dahse HM et al (2011) Conidial dihydroxynaphthalene melanin of the human pathogenic fungus *Aspergillus fumigatus* interferes with the host endocytosis pathway. Front Microbiol 2:1–12. doi:10.3389/fmicb.2011.00096

123. Akoumianaki T, Kyrmizi I, Valsecchi I et al (2016) *Aspergillus* cell wall melanin blocks LC3-associated phagocytosis to promote pathogenicity. Cell Host Microbe 19:79–90. doi:10.1016/j.chom.2015.12.002
124. Eisenman HC, Casadevall A (2012) Synthesis and assembly of fungal melanin. Appl Microbiol Biotechnol 93:931–940. doi:10.1007/s00253-011-3777-2
125. Pal AK, Gajjar DU, Vasavada AR (2014) DOPA and DHN pathway orchestrate melanin synthesis in *Aspergillus* species. Med Mycol 52:10–18. doi:10.3109/13693786.2013.826879
126. Ibrahim-Granet O, Philippe B, Boleti H et al (2003) Phagocytosis and intracellular fate of *Aspergillus fumigatus* conidia in alveolar macrophages. Infect Immun 71:891–903
127. Philippe B, Ibrahim-Granet O, Prevost MC et al (2003) Killing of *Aspergillus fumigatus* by alveolar macrophages is mediated by reactive oxidant intermediates. Infect Immun 71:3034–3042
128. Chai L, Netea MG, Sugui J et al (2010) *Aspergillus fumigatus* conidial melanin modulates host cytokine response. Immunobiology 215:915–920. doi:10.1016/j.imbio.2009.10.002
129. Kyrmizi I, Gresnigt MS, Akoumianaki T et al (2013) Corticosteroids block autophagy protein recruitment in *Aspergillus fumigatus* phagosomes via targeting Dectin-1/Syk kinase signaling. J Immunol 191:1287–1299. doi:10.4049/jimmunol.1300132
130. Jacobson ES (2000) Pathogenic roles for fungal melanins. Clin Microbiol Rev 13:708–717. doi:10.1128/CMR.13.4.708-717.2000
131. Chamilos G, Akoumianaki T, Kyrmizi I et al (2016) Melanin targets LC3-associated phagocytosis (LAP): a novel pathogenetic mechanism in fungal disease. Autophagy 12:888–889. doi:10.1080/15548627.2016.1157242
132. Sprenkeler EGG, Gresnigt MS, van de Veerdonk FL (2016) LC3-associated phagocytosis: a crucial mechanism for antifungal host defence against *Aspergillus fumigatus*. Cell Microbiol 18:1208–1216. doi:10.1111/cmi.12616
133. Cunha C, Kurzai O, Löffler J et al (2014) Neutrophil responses to aspergillosis: new roles for old players. Mycopathologia. doi:10.1007/s11046-014-9796-7
134. Levitz SM, Diamond RD (1985) Mechanisms of resistance of *Aspergillus fumigatus* conidia to killing by neutrophils in vitro. J Infect Dis 152:33–42. doi:10.1093/infdis/152.1.33
135. Mircescu MM, Lipuma L, van Rooijen N et al (2009) Essential role for neutrophils but not alveolar macrophages at early time points following *Aspergillus fumigatus* infection. J Infect Dis 200:647–656. doi:10.1086/600380
136. Fontaine T, Delangle A, Simenel C et al (2011) Galactosaminogalactan, a new immunosuppressive polysaccharide of *Aspergillus fumigatus*. PLoS Pathog. doi:10.1371/journal.ppat.1002372
137. Robinet P, Baychelier F, Fontaine T et al (2014) A polysaccharide virulence factor of a human fungal pathogen induces neutrophil apoptosis via NK cells. J Immunol 192:5332–5342. doi:10.4049/jimmunol.1303180
138. Gresnigt MS, Bozza S, Becker KL et al (2014) A polysaccharide virulence factor from *Aspergillus fumigatus* elicits anti-inflammatory effects through induction of Interleukin-1 receptor antagonist. PLoS Pathog 10:e1003936. doi:10.1371/journal.ppat.1003936
139. Gazendam R, van Hamme JL, Tool ATJ et al (2016) Human neutrophils use different mechanisms to kill *Aspergillus fumigatus* conidia and hyphae: evidence from phagocyte defects. J Immunol. doi:10.4049/jimmunol.1501811
140. Brinkmann V, Reichard U, Goosmann C et al (2004) Neutrophil extracellular traps and kill bacteria. Science 303:1532–1535. doi:10.1126/science.1092385
141. Bruns S, Kniemeyer O, Hasenberg M et al (2010) Production of extracellular traps against *Aspergillus fumigatus in vitro* and in infected lung tissue is dependent on invading neutrophils and influenced by hydrophobin RodA. PLoS Pathog 6:e1000873. doi:10.1371/journal.ppat.1000873
142. Urban CF, Reichard U, Brinkmann V, Zychlinsky A (2006) Neutrophil extracellular traps capture and kill *Candida albicans* and hyphal forms. Cell Microbiol 8:668–676. doi:10.1111/j.1462-5822.2005.00659.x

143. McCormick A, Heesemann L, Wagener J et al (2010) NETs formed by human neutrophils inhibit growth of the pathogenic mold *Aspergillus fumigatus*. Microbes Infect 12:928–936. doi:10.1016/j.micinf.2010.06.009
144. Ellett F, Jorgensen J, Frydman GH et al (2017) Neutrophil interactions stimulate evasive hyphal branching by *Aspergillus fumigatus*. PLoS Pathog 13:e1006154. doi:10.1371/journal.ppat.1006154
145. Fallon JP, Reeves EP, Kavanagh K (2010) Inhibition of neutrophil function following exposure to the *Aspergillus fumigatus* toxin fumagillin. J Med Microbiol 59:625–633. doi:10.1099/jmm.0.018192-0
146. Bassetti M, Pecori D, Della Siega P et al (2014) Current and future therapies for invasive aspergillosis. Pulm Pharmacol Ther 32:1–11. doi:10.1016/j.pupt.2014.06.002
147. Arendrup MC (2014) Update on antifungal resistance in *Aspergillus* and *Candida*. Clin Microbiol Infect 8:42–48. doi:10.1111/1469-0691.12513
148. Hadrich I, Makni F, Neji S et al (2012) Invasive aspergillosis: resistance to antifungal drugs. Mycopathologia 174:131–141. doi:10.1007/s11046-012-9526-y
149. Barczak AK, Hung DT (2009) Productive steps toward an antimicrobial targeting virulence. Curr Opin Microbiol 12:490–496. doi:10.1016/j.mib.2009.06.012
150. Clatworthy AE, Pierson E, Hung DT (2007) Targeting virulence: a new paradigm for antimicrobial therapy. Nat Chem Biol 3:541–548. doi:10.1038/nchembio.2007.24
151. Gauwerky K, Borelli C, Korting HC (2009) Targeting virulence: a new paradigm for antifungals. Drug Discov Today 14:214–222. doi:10.1016/j.drudis.2008.11.013
152. Allen RC, Popat R, Diggle SP, Brown SP (2014) Targeting virulence: can we make evolution-proof drugs? Nat Rev Microbiol 12:300–308. doi:10.1038/nrmicro3232
153. Bromuro C, Romano M, Chiani P et al (2010) Beta-glucan-CRM197 conjugates as candidates antifungal vaccines. Vaccine 28:2615–2623. doi:10.1016/j.vaccine.2010.01.012
154. Torosantucci A, Bromuro C, Chiani P et al (2005) A novel glyco-conjugate vaccine against fungal pathogens. J Exp Med 202:597–606. doi:10.1084/jem.20050749
155. Chotirmall SH, Al-Alawi M, Mirkovic B et al (2013) *Aspergillus*-associated airway disease, inflammation, and the innate immune response. Biomed Res Int 2013:723129. doi:10.1155/2013/723129
156. Phadke AP, Mehrad B (2005) Cytokines in host defense against *Aspergillus*: recent advances. Med Mycol 43:173–176. doi:10.1080/13693780500052099

Chapter 5
Histoplasma capsulatum and Histoplasmosis

Chelsea Bueter, George S. Deepe, and Chad A. Rappleye

Abstract The fungal pathogen *Histoplasma capsulatum* infects tens to hundreds of thousands of individuals every year. Ensuing disease severity is largely dependent on the host immune response with activation of cell-mediated immunity essential for eventual pathogen control. During the early stages of infection normally fungicidal phagocytes provide a permissive intracellular host niche for Histoplasma yeasts. These fungal elements avoid destruction by phagocytes by minimizing exposure of immunostimulatory molecules on their surface and by detoxifying antimicrobial reactive compounds produced by phagocytes. Within the phaogcyte, yeasts overcome nutrient limitations including organic sulfur, iron, and zinc. Activation of T-cells and the ensuing production of pro-inflammatory Th1 and Th17 cytokines (especially TNFα, IFNγ, and GM-CSF) stimulate phagocyte reactive oxygen and nitric oxide production, thereby changing the host–pathogen balance to favor phagocyte clearance of intracellular Histoplasma yeasts. Clinical management of histoplasmosis involves identification of susceptible individuals with functional or therapeutic reduction of factors mediating activation of cell mediated immunity and augmentation of fungal control through administration of azole- and polyene-class antifungals.

5.1 Introduction

Histoplasma capsulatum is the causative agent of the respiratory and systemic disease histoplasmosis. Areas of endemicity as defined by clinical case prevalence and skin reactivity to fungal antigens include the Midwest and Eastern parts of the United States (especially around the Mississippi and Ohio River valleys) and many parts of Latin America (particularly Colombia, Venezuela, Brazil, and Argentina)

C. Bueter, Ph.D. • G.S. Deepe, M.D.
Division of Infectious Diseases, College of Medicine, University of Cincinnati, Cincinnati, OH, USA
e-mail: bueterca@ucmail.uc.edu; george.deepe@uc.edu

C.A. Rappleye, Ph.D. (✉)
Department of Microbiology, Ohio State University, Columbus, OH, USA
e-mail: rappleye.1@osu.edu

H.M. Mora-Montes, L.M. Lopes-Bezerra (eds.), *Current Progress in Medical Mycology*, DOI 10.1007/978-3-319-64113-3_5

[1–4]. *Histoplasma* spp. are also found in Africa and reports are emerging of cases in China and India indicating *Histoplasma* species are more globally distributed than originally thought [5–8]. Multiple phylogenetic groups of strains have been defined from both clinical and animal isolates which correlate with these different geographic regions and likely represent distinct subspecies of *Histoplasma* [9, 10]. The major phylogenetic groups defined include two North American clades (NAm1 and NAm2), two Latin American clades (LAmA and LamB) and the African clade. Consequently, we will refer to them collectively using the genus name *Histoplasma* without reference to species or subspecies. Although the annual incidence is likely under reported, estimates of hundreds of thousands of infections occur each year [11, 12]. Fortunately, the majority of infections are self-limited and do not require clinical intervention.

Histoplasma infects both immunocompromised and immunocompetent individuals. *Histoplasma* is initially acquired through inhalation of conidia released from an environmental site. Upon germination, the conidia establish a respiratory infection. Histoplasmosis manifests as a gradient of disease which is ultimately dependent on the balance of *Histoplasma* virulence factors versus the efficiency of immune defense mechanisms. Clinical disease typically manifests in individuals lacking a critical attribute of immune defenses (e.g., CD4 deficiency in AIDS patients or cytokine inhibition therapies) or individuals receiving a large inoculum of conidia/spores [12, 13]. One study of hospitalizations in the United States for histoplasmosis indicated less than half were associated with immunocompromising comorbidities indicating the existence of some, as yet undetermined, additional susceptibility factors [2]. Control and ultimate resolution of *Histoplasma* infection requires cell-mediated immunity and the consequent activation of phagocytes. This indicates that the innate immune system, by itself, is insufficient and that *Histoplasma* cells (yeasts and conidia) can efficiently escape innate immune defenses. This conveniently divides the course of infection into two phases in otherwise normal individuals: an initial phase during which *Histoplasma* conquers host macrophages and a later phase during which activation of macrophages defeats the intracellular yeasts. In this chapter, we discuss factors at the host–pathogen interface that facilitate these two phases of infection: (1) how *Histoplasma* survives and parasitizes host innate immune cells before activation of cell-mediated immunity and (2) how onset of adaptive immunity reverses this situation, leading to control and elimination of *Histoplasma*.

5.1.1 Thermal Dimorphism

The pathogenesis of *Histoplasma* is linked to thermal dimorphism. In the environment, *Histoplasma* grows as a saprobic mold (Fig. 5.1a) in environments with decaying organic matter. The mycelia produce conidia (Fig. 1a, inset) which facilitate dispersal of *Histoplasma* to new environments where they germinate into mycelia again. *N*-acetylglucosamine, a widespread component of microbes, appears to

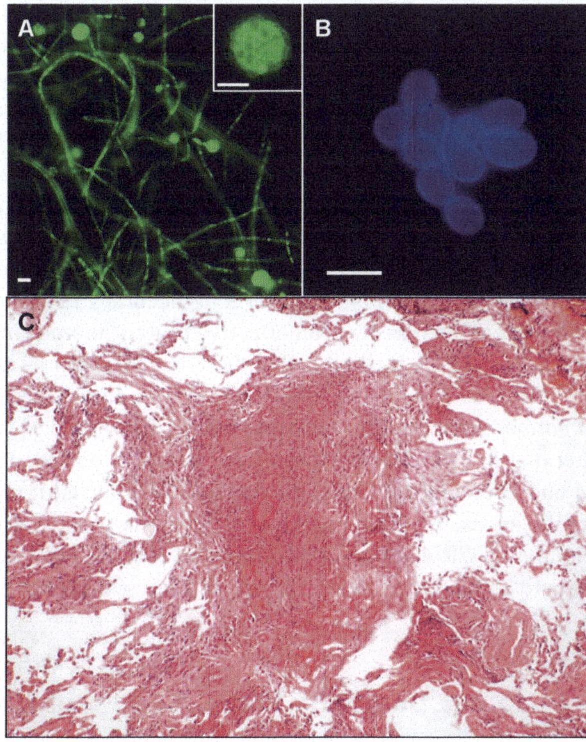

Fig. 5.1 Histoplasma dimorphism. (**a**) At lower temperatures, *Histoplasma* grows as filamentous hyphae that produce microconidia and tuberculated macroconidia (**a**, inset), which function for dispersal. (**b**) At mammalian body temperatures, Histoplasma grows as a budding yeast form. (**c**) Within host tissues, the immune response organizes into granulomas in an attempt to contain the yeasts which have invaded phagocytes. *Scale bars* in each panel indicate 5 μm

provide one cue signaling the soil environment since this monosaccharide enhances the differentiation into filamentous growth [14]. However, upon inhalation into a mammal, the conidia germinate into a yeast form (Fig. 5.1b) and express factors that enable pathogenesis. The principal cue inducing differentiation into the yeast form is the elevated temperature of the mammalian host (i.e., 37 °C). Genetic studies have identified an orphan sensor-kinase that is required for the temperature-dependent switch to the yeast phase as well as a set of transcription factors (Ryp = required for yeast phase) that control expression of the yeast-phase regulon [15–18]. In the absence of these factors, or through pharmacological inhibition of the transition, *Histoplasma* cells remain as mycelia at 37 °C and are unable to establish infections in mammals [16, 19], demonstrating the differentiation into yeasts is necessary for *Histoplasma* pathogenesis.

5.1.2 Symptoms and Diagnosis

Patients with acute *Histoplasma* infection present with fever, chills, headache, respiratory symptoms, cough, and chest discomfort. Histoplasmosis is not often suspected until patients fail to recover after a standard antibiotic therapy. Diagnosis can be confirmed by the presence of yeast cells in tissue samples stained with methenamine silver, or by culturing blood or tissue [20]. Histoplasmosis is a granulomatous disease (Fig. 5.1c). As histoplasmosis can result in respiratory failure or death, rapid diagnosis is vital. Diagnostic tests including: enzyme immunoassay (EIA), complement fixation (CF), immunodiffusion (ID) and PCR assays, all of which deliver quick determinations. EIA detects antigen in urine or serum with a sensitivity of over 80–90% [21–23]. However, relying on just one sample may result in failure to detect histoplasmosis, and consequently testing both blood and serum is recommended [23]. Antigen has also been discovered in the BAL of pulmonary histoplasmosis patients with over 93% accuracy [24]. Aside from its utility in diagnosis, the antigen assay is an effective tool for monitoring treatment responses. One caveat of the EIA method is that there is a 90% cross-reactivity with blastomycosis that may obscure results. CF identifies host antibodies to *Histoplasma*. Persistence of antibodies years after exposure makes this assay less useful; however, a single CF titer of 1:32 may indicate a more recent infection [20, 25]. The ID assay detects the M and H precipitin bands; the M band will persist for years, making it less useful for diagnosis, but the H band is typically only found during active infection [20, 24]. Finally, a PCR assay is in development and has been successful in culture isolates and clinical isolates, though as of yet there is no commercially available test [26].

5.2 Innate Immunity Phase: Success of *Histoplasma*

5.2.1 Infection of Macrophages

Within the lung, *Histoplasma* cells overcome challenges to survival by successfully invading phagocytic cells. The small size of *Histoplasma* microconidia enable them to penetrate into the alveoli of the lung. Within alveolar spaces, *Histoplasma* cells encounter surfactant, a lipid-rich fluid that also contains the collectins SP-A and SP-D [27]. At physiological concentrations, these two surfactant proteins can permeabilize yeast cells [28], and this antifungal activity represents a significant barrier to *Histoplasma* colonization within the lumen of the alveoli. To overcome this, *Histoplasma* cells find an intracellular refuge by infecting alveolar macrophages, thereby sheltering itself from the activities of SP-A and SP-D.

Resident alveolar macrophages encounter *Histoplasma* first and are the primary host cell. By day 3 post-infection, yeasts are found inside these cells and are also within inflammatory macrophages. By day 7, yeasts reside in inflammatory macrophages and neutrophils [29]. To become intracellular, *Histoplasma* conidia and yeast rely on non-opsonic phagocytosis by macrophages [30, 31]. To accomplish this, *Histoplasma* conidia and yeast bind to β-integrin family complement receptors [e.g., complement receptor 3; CR3 (CD18/CD11b)] on phagocytes, which triggers phagocytic uptake [31, 32]. On the fungal side, this adhesion is mediated by heat shock protein 60 (Hsp60) molecules present on the yeast cell surface [33, 34]. Yeast binding to CR3 is temperature sensitive and requires divalent cations (e.g., Ca2+ and Mg2+) indicating internalization occurs through structural dynamics of the integrins [32]. Vaccination with Hsp60 provides protection to the host, possibly by blocking the critical Hsp60-CR3 ligand-receptor interaction and subsequent phagocytosis or by redirecting phagocytosis through FcγR [35–37].

5.2.2 Avoidance of Detection

Although intracellular residence protects *Histoplasma* from extracellular immune defenses, infection of phagocytic cells carries risks of its own. Upon interaction with fungal cells, phagocytes are normally antifungal. However, CR3-mediated phagocytosis, in the absence of other activating signals, is generally noninflammatory. Thus, to promote uptake into a permissive intracellular environment, *Histoplasma* yeast must bind to phagocytic receptors but avoid stimulating signaling receptors that would activate phagocyte defenses. Signaling pattern recognition receptors (PRRs) that detect fungal pathogen-associated molecular patterns (PAMPs) commonly have a carbohydrate recognition domain that binds to molecules of the glycan/glycoprotein-rich fungal cell wall [38].

5.2.2.1 β-Glucan Concealment

Chief among these PRRs for detecting fungal pathogens is Dectin-1, a C-type lectin receptor that recognizes β-glucans that are ubiquitous constituents of fungal cell walls [39]. Recognition of fungal β-glucans by Dectin-1 receptors on macrophages and neutrophils triggers inflammatory responses that include production of reactive oxygen species (ROS) and the secretion of pro-inflammatory cytokines (e.g., TNFα and IL-6) [40–44]. Mouse studies show that Dectin-1 contributes relatively little to the control of *Histoplasma* yeasts [45, 46]. The cell walls of *Histoplasma* yeasts contain abundant β-glucans and therefore the potential to be recognized by Dectin-1. However, *Histoplasma* yeast minimize this detection through two separate mechanisms. The yeasts cells of most strains of *Histoplasma* also produce an α-linked glucan polysaccharide and this α-glucan constitutes the outermost layer of the cell wall, effectively concealing the underlying immunostimulatory β-glucans from detection [47, 48]. The α-glucan polysaccharide is synthesized by α-glucan synthase (Ags1) which is thought to extend short glucan oligomers formed by an α-amylase-like protein (Amy1) [48, 49]. The second mechanism for minimization of β-glucan detection is through enzymatic removal of β-glucan fragments. *Histoplasma* yeasts secrete an endo-β-glucanase (Eng1) which acts on the fungal cell wall and hydrozlyzes surface-exposed β-glucans [50]. These two mechanisms, shielding of β-glucans and pruning away exposed β-glucans effectively prevent detection of *Histoplasma* yeasts by Dectin-1, thereby preventing macrophage responses. Consistent with their pathogenesis-enabling role, avirulent mycelia do not produce α-glucan or Eng1 [51, 52].

5.2.2.2 Toll-Like Receptors

Histoplasma yeasts can be detected by Toll-like receptors (TLRs) as well as Dectin-2. Immune cells expressing Dectin-2, but not Mincle, respond to *Histoplasma* yeasts suggesting a mannan-type PAMP is present on yeasts [46]. The involvement of TLRs in yeast detection comes from studies in mice lacking the MyD88 adaptor protein which is required for TLR2 and TLR4 signaling. In the absence of MyD88, mice show a modest defect in the control of *Histoplasma* infections beginning at 7 days post-infection which coincides with the activation of cell mediated immunity [45]. Consistent with this, mice lacking MyD88 have lower pro-inflammatory cytokines, notably IL-12 which is required for generation of the Th1 response necessary for efficient control of *Histoplasma* (see section below on the adaptive immunity phase). A yeast cell wall extract can stimulate TLR2 but cytokine production by macrophages interacting with whole yeast cells does not depend on TLR2 [30] suggesting, like β-glucan exposure discussed above, the spatial organization of the yeast cells minimizes detection of PAMPs by TLR2.

Additional fungal molecules have been localized to the yeast cell wall. Surface exposed proteins include the Hsp60 adhesin mentioned above, an extracellular catalase (CatB/M-antigen), histone 2B, and Hsp70 [53–56]. In addition, yeast cells of some, but not all, *Histoplasma* lineages secrete Yps3, a protein that associates with the yeast cell surface via interactions with cell wall chitin [57–59]. Yps3 contributes to extrapulmonary dissemination through an unknown mechanism but is dispensable for pulmonary infection [60].

Minimizing detection by macrophages sets *Histoplasma* apart from opportunistic fungal pathogens which are readily detected by multiple PRRs. The ability to conceal itself from macrophages provides at least a partial explanation for why innate immune cells are unable to fully control *Histoplasma* yeasts.

5.2.3 Neutralization of Macrophage Defenses

5.2.3.1 Reactive Oxygen

Phagocytic cells are armed with multiple factors to either kill or restrict the growth of microbes. Production of antifungal reactive oxygen species (ROS) is initiated by formation of superoxide by the phagocyte NADPH oxidase complex. Superoxide and other reactive oxygen molecules derived from it (i.e., peroxides and hydroxyl radicals) directly damage microbes. Superoxide dismutases are antioxidant enzymes produced by aerobic organisms; however, in most microbes, superoxide dismutases are intracellular to elminate damage from superoxide produced as a by-product of respiration. However, for intramacrophage pathogens, antimicrobial superoxide produced by the phagocyte is extracellular to the microbial cell. Thus, for cells interacting with macrophages or neutrophils, intracellular superoxide dismutases provide little protection.

Histoplasma yeasts secrete two factors that combat phagocyte-derived reactive oxygen and enable its survival in phagocytic cells. *Histoplasma* yeasts produce an extracellular superoxide dismutase (Sod3) as well as an extracellular catalase (CatB) which dismute superoxide and eliminate peroxide ROS, respectively [51, 61, 62]. Both enzymes can associate with the yeast cell wall in addition to being soluble. Importantly, the extracellular localization positions Sod3 and CatB to act on ROS produced by the phagocyte against *Histoplasma*. Without Sod3 or CatB, *Histoplasma* yeasts are readily killed by macrophages and neutrophils [61, 62]. Consistent with its secretion, Sod3 does not act on, nor does it protect against, metabolically-derived ROS within the yeast cells [62]. Thus, *Histoplasma* yeasts release an efficient collaboration of antioxidant enzymes that negate phagocyte-derived ROS. Consequently, *Histoplasma* yeasts survive levels of ROS which are lethal to the opportunistic pathogen *Candida albicans*. The rapid destruction of ROS mediated by these enzymes also explains why earlier studies failed to detect ROS production by phagocytes in response to *Histoplasma* (i.e., the ROS was destroyed by Sod3 and CatB before it could be measured) [63–65]. Interestingly, the expression of extracellular superoxide dismutases is common for bacterial pathogens that infect macrophages suggesting that elimination of superoxide is a necessary defense for survival in phagocytes. As with α-glucan, expression of Sod3 and CatB are restricted to the pathogenic yeast consistent with their function specifically in enabling the pathogenic lifestyle of yeasts [51, 66].

5.2.3.2 Reactive Nitrogen

In addition to ROS, *Histoplasma* yeast are confronted with reactive nitrogen species (RNS) which is fungistatic to *Histoplasma*. Yeast cells respond to RNS by upregulating multiple stress-response genes including a nitric oxide reductase, although its functional role has not been defined [67]. Inhibition of RNS production decreases the ability of macrophages to control *Histoplasma* [68, 69]. RNS production against *Histoplasma* is regulated by pro-inflammatory cytokines. This underscores the necessity of *Histoplasma* yeasts to avoid detection and minimize inflammatory cytokine production.

5.2.3.3 Phagosome Acidification

Within the macrophage, *Histoplasma*-containing phagosomes can fuse with lysosomes exposing *Histoplasma* yeast to hydrolytic enzymes. Regardless of phagolysosomal fusion, *Histoplasma* yeasts prevent complete acidification of the *Histoplasma*-containing phagosome, maintaining a pH around 6 [70–73]. This partial acidification is thought to permit release of one of the iron molecules from transferrin (see below) but prevent activity of acidic hydrolases of the lysosome. Yeast modulation of pH in the phagosome is an active process as killed yeast cells are found in acidified compartments [70]. At least part of this mechanism is exclusion of proton ATPases from the phagosomal membrane that mediate transport of H^+ ions into the phagosome [73]. Yeasts that artificially acidify the phagosome environment are killed, indicating the pathogenesis requirement to maintain a more neutral pH [74].

5.2.4 Overcoming Nutrient Limitations

5.2.4.1 Pyrimidines and Essential Cofactors

Once established in the phagosome, *Histoplasma* yeast scavenge available nutrients in order to proliferate. The *Histoplasma*-containing phagosome lacks available nucleic acids and essential vitamins for *Histoplasma* utilization. Consequently, *Histoplasma* yeast synthesize these compounds from simpler organic molecules derived from central metabolism. For example, growth in macrophages requires the pyrimidine biosynthetic pathway to overcome the paucity of uracil in the phagosome [75]. Similarly, biosynthesis of riboflavin and pantothenate cofactors is necessary for intramacrophage growth of *Histoplasma* yeasts [76]. Without these pathways, *Histoplasma* yeasts can survive macrophage defenses, but are unable to proliferate within macrophages and have severely attenuated virulence in murine models of histoplasmosis. Biotin synthesis is not required, indicating either the availability of biotin within the phagosome or effective scavenging mechanisms [76].

5.2.4.2 Organic Sulfur

Histoplasma yeast, unlike mycelia, are auxotrophic for organic sulfur [77]. This is due to the lack of expression of sulfite reductase by yeast cells, preventing the incorporation of inorganic sulfate into cysteine. Interestingly, the lack of sulfite reductase activity is dependent on temperature with the elevated temperature of mammals correlating with lack of activity [77–79]. Although growth of yeasts in culture requires an organic sulfur source, they proliferate well in macrophages [80]. Together these findings suggest that yeasts have adapted to an abundant supply of organic sulfur within the macrophage phagosome.

5.2.4.3 Iron and Zinc

Sequestration within the phagosome imposes challenges for acquisition of trace metals such as iron and zinc. Indeed, reducing the availability of metals is a defense strategy used by macrophages to restrict microbial pathogens [81]. *Histoplasma* yeast have multiple strategies to acquire iron. Iron in the host is bound to transport molecules such as transferrin and these must be brought to the phagosome for iron to become available to yeasts within the macrophage. Although *Histoplasma* yeast prevent acidification of the phagosome, a drop in pH is necessary to liberate iron from transferrin [82]. Thus, *Histoplasma* maintains a lumenal pH around 6.0 at which it is acidic enough to release one of the two iron atoms bound to transferrin but sufficiently neutral to prevent acidic hydrolase activity. Addition of the weak base chloroquine to infected macrophages prevents *Histoplasma* proliferation by raising the phagosomal pH and reducing available iron [82].

In addition to phagosome pH modulation, *Histoplasma* yeast steal iron by secretion of iron chelating hydroxamate siderophores and iron reduction systems. Restriction of iron causes upregulation of siderophore biosynthesis genes through an iron-responsive GATA family tanscription factor (Sre1) [83, 84]. Prevention of siderophore synthesis reduces proliferation of *Histoplasma* yeasts in macrophages [85, 86]. In addition to siderophore synthesis, some strains of *Histoplasma* express genes encoding a high-affinity iron reduction/transport system (Fet3 and Ftr1) as well as a glutathione-dependent iron reduction system catalyzed by a secreted γ-glutamyl transferase (Ggt1) [87, 88]. Loss of Ggt1 function blocks the ability of yeast to proliferate inside macrophages confirming yeasts must overcome iron limitations [88]. In murine models of disease, the operation of various iron acquisition mechanisms causes differences in *Histoplasma* pathogenesis. For example, strains that lack siderophore production are attenuated early in infection if they also lack the Fet3/Ftr1 iron reduction system whereas strains that express Fet3/Ftr1 do not exhibit decreased virulence until 2 weeks post infection, a time point at which adaptive immunity and the production of IFNγ has activated macrophages to further restrict iron availability [85, 86].

As with iron, zinc availability can also be modulated by macrophages imposing an additional nutritional limitation on intramacrophage yeasts. *Histoplasma* yeasts express a high affinity zinc transporter (Zrt2) when faced with low zinc concentrations [89]. The Zrt2 transporter facilitates zinc import into *Histoplasma* yeasts and facilitates growth in macrophages. However in a murine disease model, Zrt2-dependent zinc transport was not required until after 5 days of infection suggesting that zinc levels in resting macrophages are sufficient for *Histoplasma* yeast proliferation but that activation of macrophages restricts zinc availability. Consistent with this, treatment of *Histoplasma*-infected macrophages with GM-CSF reduces levels of intramacrophage iron and zinc impairs growth of yeasts [90]. This cytokine upregulates zinc-binding metallothioneins and macrophage zinc exporters, the end result of which is to reduce the zinc available to *Histoplasma* yeasts within the phagosome [91]. Together these findings show that *Histoplasma* expresses a zinc transporter to overcome the cytokine-induced zinc limitation in macrophages. Preliminary work indicates a third essential metal, copper, is similarly restricted after activation of macrophages and that *Histoplasma* yeasts express a high-affinity copper transporter to overcome copper limitation (Shen Q and Rappleye CA, personal communication).

5.2.4.4 Cbp1

Within the phagosome, *Histoplasma* yeasts secrete a small calcium-binding protein [92, 93]. Although the protein has nanomolar affinity for calcium, whether Cbp1 influences calcium homeostasis of yeasts is unknown. As with the α-glucan, Eng1, Sod3, and CatB virulence factors, Cbp1 production is restricted to yeasts and absent from mycelia arguing for a role in pathogenesis [94, 95]. Cbp1 has multiple intramolecular disulfide bonds which makes Cbp1 resistant to proteases, a feature consistent with Cbp1 functioning in the macrophage phagosome [96]. Without Cbp1, yeast proliferation in macrophages is impaired confirming Cbp1's importance to the intramacrophage growth of yeasts [97, 98].

Unlike nonpathogenic and opportunistic fungi which are readily controlled by innate immune cells, *Histoplasma* yeast survive their encounter with phagocytes. The mechanisms that distinguish *Histoplasma* from other fungi include the ability to promote their phagocytosis while minimizing detection by host phagocytes and the production of factors to defend the yeasts against phagocyte-produced ROS (Fig. 5.2). *Histoplasma* yeasts not only survive, but they also use the macrophage as their primary host cell. Within the *Histoplasma*-containing phagosome, yeasts successfully combat nutrient limitations or synthesize essential compounds de novo that are unavailable. In this way, yeasts convert the macrophage phagosome into a host cell permissive for yeast proliferation. However, upon activation of cell-mediated immunity, these mechanisms prove insufficient and the macrophage gains the advantage.

5.3 Cell-Mediated Immunity: Host Prevails

5.3.1 Kinetics

Macrophages and DCs produce cytokines to stimulate the type 1 helper T cell (Th1) response and present antigens to T cells, and by day 7 of infection, the adaptive response has been activated. In mice, T cells produce interferon γ (IFNγ) that induces macrophages to kill or inhibit the growth of yeast cells. Human macrophages induce fungistasis independent of IFNγ, at least in vitro, but are dependent on interleukin-3 (IL-3), granulocyte macrophage colony-stimulating factor (GM-CSF) and macrophage colony-stimulating factor (M-CSF) and collagen adherence [99, 100].

Histoplasma infection results in the formation of granulomas, a spatial organization of immune cells formed to wall off the pathogen it cannot eliminate (Fig. 5.1c). In hosts with intact cell-mediated immune responses, these granulomas consist of macrophages and lymphocytes with *Histoplasma* contained in the center [101]. T cell-depleted mice are still able to form granulomas, but they are much less capable of containing the infection and have elevated fungal burdens [102, 103]. Both humans and mice form granulomas, though the latter tend to not caseate [104]. Granulomas can be seen as early as day 7 post intraperitoneal infection in mice [101]. The timing of granuloma

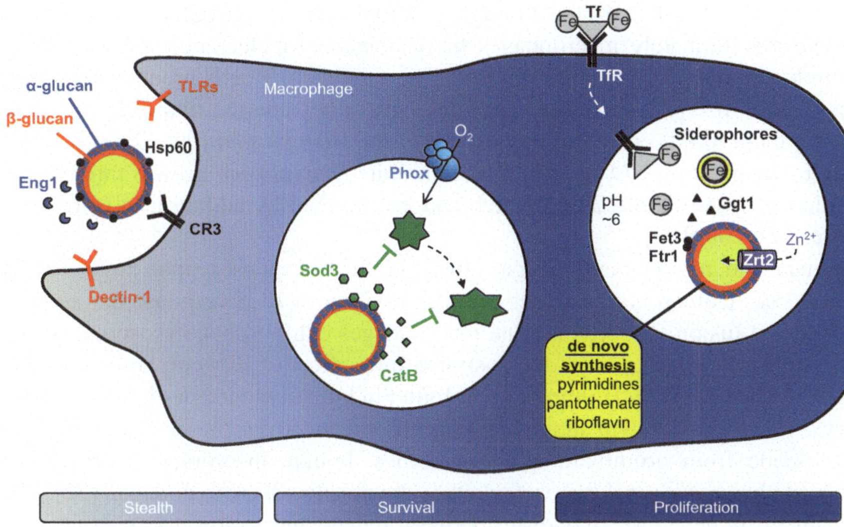

Fig. 5.2 Mechanisms of *Histoplasma* pathogenesis of macrophages. Schematic showing three aspects of macrophage infection: stealth, survival of defenses, and intramacrophage proliferation. *Stealth interaction*: *Histoplasma* yeasts utilize Hsp60 binding to phagocytic receptors (*black*; CR3) to induce uptake while simultaneously reducing detection by signaling receptors (*red*; Toll-like receptors (TLRs) and Dectin-1). *Histoplasma* minimizes β-glucan detection by Dectin-1 by production of α-glucan to conceal cell wall β-glucans and secretion of the endo-β-glucanase (Eng1) which prunes away any remaining exposed β-glucans. *Survival of defenses*: *Histoplasma* produces extracellular superoxide dismutase (Sod3) and extracellular catalase (CatB) to destroy reactive oxygen species generated by the phagocyte NADPH-oxidase (Phox) and ensure survival of yeasts. *Proliferation*: *Histoplasma* prevents complete phagosome acidification and acquires iron from host transferrin (Tf) by production of siderophores (*yellow*), a gamma-glutamyl transferase (Ggt1), and the Fet3/Ftr1 iron reductase. Within the phagosome, a high affinity zinc transporter (Zrt2) aids in zinc acquisition and de novo biosynthesis provides intramacrophage yeasts with pyrimidines and essential vitamin cofactors (riboflavin and pantothenate) to meet metabolic needs

formation in humans is a bit more difficult to determine; specific immune responses are assumed to be initiated between 7 and 18 days post exposure and once activated, formed granulomas begin to resolve often resulting in the lesions being encapsulated and then calcifying [105]. In an immunosuppressed host undergoing granuloma dissolution, previously trapped fungi may be released and start replicating again.

5.3.2 *Innate-Adaptive Immunity Connections*

5.3.2.1 Monocytes/Macrophages

Monocytes/macrophages, along with DCs, assist in the initiation of the adaptive response. Macrophages may promote a Th1 response through cytokine release and presentation of antigen to CD4+ T cells, to act in concert with the stimuli provided by the DCs [99, 106, 107]. Although macrophages are initially permissive to *Histoplasma*, ultimately, macrophages are responsible for clearance of *Histoplasma* infection. Once the adaptive response has been initiated, T cells produce Th1 cytokines such as IFNγ, which impel macrophages to become the final effector cells, enabling them to kill or inhibit the growth of the *Histoplasma* [99]. IFNγ is likely acting by stimulating NO release, which inhibits *Histoplasma* growth through the chelation of iron, the effects of which can be reversed by adding iron to in vitro cultures [99, 108, 109].

Human and mouse macrophages differ in their responses upon ingestion of *Histoplasma*. Human macrophages produce a vigorous ROS response but no phagolysosomal fusion, whereas murine macrophages exhibit phagolysosomal fusion [99]. While most studies have demonstrated little effect of IFNγ on human macrophages in vitro [72, 99], Brummer et al. found in human monocyte-derived macrophages, short-term growth inhibition may be enhanced by the addition of IFNγ [110]. Aside from proinflammatory cytokines, human macrophages do exhibit enhanced phagolysosomal fusion when they are able to adhere to collagen [72]. The mechanisms of activation may differ between human and mouse macrophages, but ultimately both are able to induce fungistasis.

5.3.2.2 Dendritic Cells

DCs recognize *Histoplasma* surface-exposed cyclophilin A via very late antigen 5 (VLA-5) [111, 112]. Unlike macrophages, human DCs are able to directly kill *Histoplasma* without activation by the adaptive immune response. Upon phagocytosis, human DCs kill *Histoplasma* yeasts via lysosomal hydrolases after phagolysosomal fusion occurs [113]. The killed yeasts are degraded and antigenic peptides are generated that can activate T cells while the DCs are also making IL-12 prompting Th1 differentiation. Both MyD88 and TLR 7/9 are essential for infected resident DCs to facilitate proper T cell activation and IFNγ production [45, 114]. Additionally, DCs have been reported to ingest *Histoplasma* infected macrophages and produce antigenic epitopes that are cross-presented to $CD8^{+}$ T cells [115]. DCs stimulate both $CD4^{+}$ and $CD8^{+}$ T cells, making them central in the defense against *Histoplasma*.

There are many described DC subsets. Two of these have important roles during *Histoplasma* infection. $CD8^{+}$ DCs are potent initiators of a proinflammatory Th1 response in response to *Histoplasma*. They are able to prime both $CD4^{+}$ and $CD8^{+}$ T cells and produce IL-12 thereby activating a strong Th1 adaptive response [116]. Additionally, these DCs are able to cross-present antigen: phagocytosing extracellular particles, that would normally be processed through the MHCII pathway, and utilizing the MHCI pathway to present the antigens on their surface to CD8+ T cells [116].

The CD103+ DC subset has also been shown to have a significant role in *Histoplasma* infection. They are the major producer of type I IFN in response to *Histoplasma* infection of the lungs, and Van Prooyen et al. demonstrated type I IFN is required for successful restriction of fungal growth by DCs. Similar to the facilitation of the IFNγ response by DCs, production of type I IFN TLR 7/9 [114].

Contradictory data have been published that type I interferons may actually be deleterious to host resistance since type I IFN receptor deleted animals restricted growth [117].

5.3.3 Important Cytokines and Their Effects

Cytokines produced by innate and adaptive cells that encounter *Histoplasma* are crucial for controlling fungal infection. The proinflammatory cytokines TNFα, IFNγ, GM-CSF and IL-17/23 are all vital for host defense, enabling clearance of the infection. IL-4, IL-10, and IL-33 on the other hand impede an effective host response resulting in enhanced fungal growth. Deficiencies and the interplay of these cytokines determine the outcome of infection (Fig. 5.3).

5.3.3.1 TNFα

TNFα is one of the earliest cytokines induced by *Histoplasma* and it is fundamental in primary *Histoplasma* infection [118]. TNFα is required in primary and secondary infection in mice for successful clearance and neutralization results in accelerated mortality in mice [119, 120]. TNFα signaling is necessary for successful defense as $TNFR1^{-/-}$ and $TNFR2^{-/-}$ mice manifest a high mortality rate, with the former much more susceptible to this fungal pathogen than the latter [121]. The importance of TNFα in human *Histoplasma* infections has been highlighted with the advent of TNFα blocking drugs and the resulting rise in disseminated *Histoplasma* infections (discussed further below).

TNFα combats *Histoplasma* infection in two ways. Neutralization of TNFα is associated with a decrement in CD8+ T cells in the lungs, suggesting it is involved in the recruitment and/or expansion of CD8+ T cells [122, 123]. An absence of TNFα also diminishes NO production in primary infection [119, 120]. In secondary infection, TNFα deficiency impairs fungal clearance by skewing the response towards Th2, and the negative effects can be rescued by also neutralizing IL-4 and IL-10 [119].

5.3.3.2 IFNγ

IFNγ is the necessary defensive cytokine against *Histoplasma* infection. It is produced by T cells and acts on macrophages to induce fungistasis. IFNγ production depends principally on IL-12, which is upregulated within 3 days of infection. After initiating the IFNγ response during this time period, IL-12 is dispensable for host defense [65, 108, 124]. Neutralization of IL-12 at the start of infection leads to accelerated mortality, as its absence results in no induction of IFNγ and therefore no protective effect by this cytokine [125]. While IFNγ is vital for survival of a primary

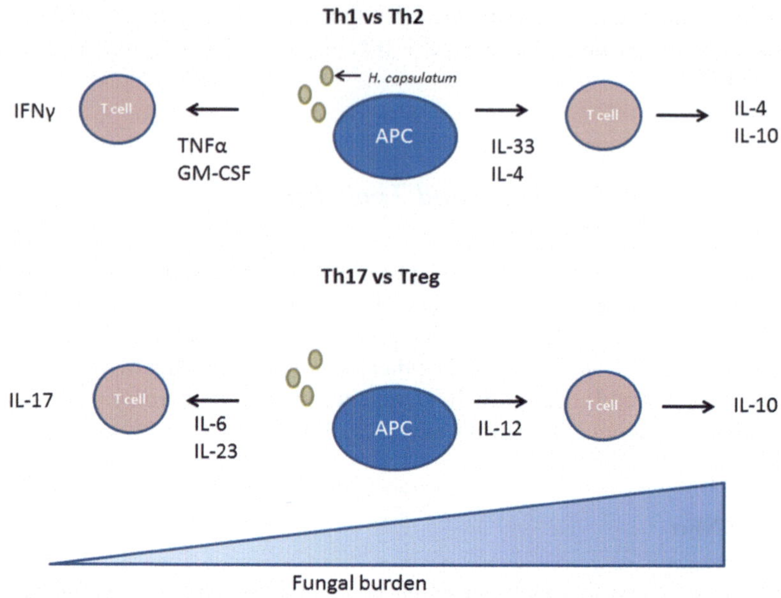

Fig. 5.3 Balancing the T cell response: the good vs. the bad. Different T cell subsets have varied effects on a *Histoplasma* infection. The antigen-presenting cells (APCs) that encounter the yeasts produce cytokines that polarize naïve T cells. Immunocompetent hosts will generate Th1 and Th17 T cells, producing IFNγ and IL-17 respectively, that enable the host to clear the infection. Conversely, hosts skewed towards a Th2 or Treg response have elevated levels of IL-4 and/or IL-10 that impairs their ability to clear the infection

infection in a mouse model, the animal is able to survive a secondary infection without IFNγ, at least in part due to increased TNFα levels [120], suggesting there is some overlap in efficacy with other proinflammatory cytokines able to substitute for the lack of IFNγ.

IFNγ is crucial for defense against *Histoplasma* in both the mouse model and human infections. Most in vitro studies have shown IFNγ does not activate human monocytes/macrophages to inhibit intracellular growth of *Histoplasma*, yet there is evidence that IFNγ is still involved in clearing human *Histoplasma* infections. Human deficiency in the IFNγ receptor is a risk factor for refractory and recurrent disseminated histoplasmosis [126]. Furthermore, Sampaio et al. described patients with dominant STAT1 gain-of-function mutations and disseminated histoplasmosis. Stimulation of IFNγ receptors leads to phosphorylation of STAT1, which homodimerizes and heterodimerizes before translocating to the nucleus and activating interferon-induced genes. These patients with delayed dephosphorylation of STAT1 likely have decreased ability to respond to IFNγ [127]. While IFNγ may not be required for the final macrophage activation step in humans, it is involved in the initiation of the Th1 response and a lack of IFNγ may allow for a stronger Th2 response and therefore impaired fungal clearance.

5.3.3.3 GM-CSF

GM-CSF is another critical cytokine for control of *Histoplasma* infections. Neutralization of it during primary infection increases mortality in association with fewer CD4+ and CD8+ T cells in the lungs, a decrement in TNFα, IFNγ and NO, and an elevation in IL-4 and IL-10 [128]. Similar to IFNγ, neutralization during secondary infection exacerbates infection but does not increase mortality [128]. This result indicates TNFα is the primary cytokine during secondary infection. As discussed earlier, GM-CSF induces nutritional immunity against *Histoplasma* by lowering the levels of intracellular iron and zinc in macrophages and thereby inducing fungistasis [90]. GM-CSF sequesters zinc by upregulating zinc exporters in macrophages and enhancing binding of free zinc to metallothioneins thereby preventing its use by *Histoplasma* [91]. GM-CSF neutralization, and the resulting skew towards a Th2 response, results in more free zinc and an increase in fungal growth.

5.3.3.4 IL-17/23

IL-17 has a role in optimal clearance of *Histoplasma*. Neutralization of IL-17A impedes efficient fungal clearance, however does not lead to progressive infection [129]. In the absence of IL-12, and therefore IFNγ, IL-23 extends survival through an IL-17-dependent mechanism [129]. Furthermore, Galectin-3 negatively regulates IL-17A production; when mice are deficient in Galectin-3, they have enhanced fungal clearance as increased IL-23 production by DCs skews the response towards a Th17 response and away from a Th1 response [130]. There is an optimal balance between Th1 and Th17 responses for clearance of infection; augmenting the Th17 response enhances clearance and determining how to boost IL-17 may aid in the development of treatments for this infection.

5.3.4 Impaired Immunity

IL-4, IL-10, and IL-33 are all detrimental to resolving *Histoplasma* infection. Most individuals with an intact immune system will respond with the appropriate cytokines to infection; however, perturbations to the immune response may skew towards a detrimental Th2 response.

5.3.4.1 IL-4 and IL-33

Infection by *Histoplasma* induces a Th1 response; yet when a Th2 response develops concurrently there is poor control of fungal burden. IL-4 is a Th2 inducing cytokine produced by leukocytes such as eosinophils, basophils and group 2 innate lymphoid cells (ILC2) [131]. CCR2−/− mice exhibit defective inflammatory monocyte recruitment and an influx of eosinophils producing IL-4, skewing their response towards Th2 and resulting in elevated fungal burdens and increased mortality [132]. The CCR2 ligands, CCL2 and CCL7, diminish the Th2 response by constraining IL-4 generation, likely through reducing eosinophil influx [116]. One mechanism of action exerted by IL-4 to negatively affect the outcome of *Histoplasma* infection is by enhancing zinc transport into macrophages [90]. IL-4 also acts to suppress GM-CSF and IFNγ signaling, thereby subduing the antifungal response of the host [128, 133, 134].

Traditionally IL-33 is a precursor to IL-4 [135]. In *Histoplasma* infection the opposite occurs, with IL-4 appearing first and driving IL-33 production in the CCR2$^{-/-}$ model mouse [136]. These mice have elevated fungal burdens that are resolved by neutralizing IL-33 [136]. This suggests that while IL-4 has multiple known roles for subverting an effective Th1 response to *Histoplasma*, in the CCR2−/− model, IL-33 is responsible for the exacerbated infection.

5.3.4.2 IL-10

IL-10 inhibits the antimicrobial effect of phagocytes and dampens T cell activation [137]. GM-CSF in primary *Histoplasma* infection and TNFα in secondary infection suppress IL-10, enabling the Th1 response to effectively control the infection [119, 128]. An absence of IL-10 aids clearance as IL-10−/− mice resolve both primary and secondary *Histoplasma* infections more quickly than wild-type mice [138]. This effect is independent of increases in protective cytokines during the adaptive phase; nonetheless, endogenous IFNγ or TNFα is required for clearance in these animals in a primary infection whereas TNFα is sufficient in a secondary infection [138].

It has recently been demonstrated that HIF-1α is a negative regulator of IL-10 expression in macrophages. A myeloid-specific HIF-1α$^{-/-}$ mouse model exhibits increased IL-10 levels and elevated fungal burdens leading to increased mortality [139]. IL-10, like IL-4, impedes successful resolution of *Histoplasma* infection, but suppression by Th1 cytokines and transcriptional regulation prevent this cytokine from being produced in a normal response.

5.3.5 Primary Immunity

5.3.5.1 Th1 cells

CD4^{+} T cells are required for defense against *Histoplasma*. Without them mice die within 2 weeks from a sublethal challenge [123]. These cells are the main IFNγ producers and a lack of CD4^{+} T cells results in a dramatic decrease in IFNγ levels and therefore no activation of macrophages and no control of *Histoplasma* growth [122, 140]. Antigens are recognized through the T cell receptor (TCR) and the variable element of the β chain (Vβ) allows for many different epitopes to be recognized.

In a primary infection $CD4^+$ $V\beta4^+$ T cells are required to defend against *Histoplasma*, they expand during the peak and early resolution phases of infection and depletion of this subset hinders optimal clearance [141–143]. This suggests a dominant *Histoplasma* antigen is responsible for driving expansion of a single T cell subset.

$CD8^+$ T cells are necessary for optimal clearance of *Histoplasma*, but unlike $CD4^+$ T cells, they are not required. While loss of $CD8^+$ T cells is associated with increased CFUs and depletion of this subset delays clearance, mice survive and eventually recover [123]. Like $CD4^+$ T cells, $CD8^+$ T cells produce IFNγ, but they additionally generate granzyme to kill infected cells as an additional mode of defense [116]. $CD8^+$ T cells recognize antigens presented by DCs on their cell surface from ingested macrophages infected with *Histoplasma* and also act directly against infected macrophages through perforin-mediated killing, thereby destroying a site of replication for the fungus [115].

5.3.5.2 Th17 Cells vs. Regulatory T Cells

The Th17–Treg balance is tightly regulated. While Th17 cells facilitate elimination of *Histoplasma*, an excess of Tregs is likely to dampen the inflammatory response, preventing the host from recruiting cells to properly fight the infection. Nevertheless, Tregs are essential for preventing uncontrolled inflammation and are a key cell type for dampening the inflammatory response once the infection has been cleared. Th17 cells and Tregs exist in a balance; while both are promoted by TGF-β, the Th17 inducing cytokine IL-6 restrains Treg development and Tregs themselves suppress Th17 cells [144]. Reduced numbers of Tregs, such as found in $CCR5^{-/-}$, results in an increase in Th17 cells and enhanced *Histoplasma* clearance [145]. Although Th17 cells are dispensable when it comes to defense against *Histoplasma* in a normal mouse, they augment clearance in mice where the balance of Tregs and Th17 cells is skewed towards Th17.

While fewer Tregs have a positive outcome on infection resolution, more Tregs do not necessarily mean impaired resolution. Animals supplemented with zinc exhibit significant skewing towards Tregs and away from Th17 cells yet have no change in fungal burden when compared to animals that were not supplemented with zinc [146]. These animals have fewer Th17 cells, but their IFNγ producing Th1 cells were unaffected and still able to combat infection efficiently [146]. Tregs induced by zinc supplementation do not impair immunity to *Histoplasma* infection, yet in animals where TNFα has been neutralized there is a population of Tregs that dampens antigen-specific immunity to *Histoplasma* and results in elevated fungal burden [147]. Different subsets of Tregs have different roles during *Histoplasma* infection and dysregulation of these subsets leads to very different outcomes.

5.3.6 Secondary Immunity

5.3.6.1 Memory T Cells

As in the primary response, secondary infection resistance is dominated by T cells. In the latter, protective immunity is hindered only when both $CD4^+$ and $CD8^+$ T cells are depleted, elimination of a single subset does not enhance mortality [122, 148]. The $V\beta4^+$ T cell subset remains important, but there is also a significant role for $V\beta6^+$ T cells. Simultaneous depletion of $V\beta4^+$ and $V\beta6^+$ T cells impairs fungal clearance in secondary infection [142].

Vaccine based immunity to *Histoplasma* requires the generation of memory T cells. The potent *Histoplasma* antigen, Hsp60, has been successfully used to vaccinate mice; with recombinant Hsp60 (rHsp60) conferring protection against lethal challenge [36, 149]. There are variations in the subset of T cells that expand depending on if mice are immunized with a fragment or the whole Hsp60 protein. $CD4^+$ $V\beta6^+$ T cells make up approximately 90% of the T cell clones in mice vaccinated with the immunoactive fragment 3 (F3) of Hsp60, their presence is accompanied with elevated levels of IFNγ, TNFα and GM-CSF, and depletion of these cells abrogated protection [148]. Alternatively, immunization with rHSP60 results in $V\beta8.1/8.2^+$ T cells preferentially expanding and protects from a subsequent challenge [143] In the absence of $CD4^+$ T cells, $CD8^+$ T cells are able to confer vaccine protection, through MHC I and TNFα production [150]. Alternatively the vaccine-based immune response to *Histoplasma* is dependent upon Th17 while Th1 immunity was dispensable [151]. Being able to induce a vaccine response from $CD8^+$ T cells and/or Th17 T cells is imperative as some of the most vulnerable to this fungus are immunocompromised and lack a $CD4^+$ response.

5.3.6.2 Role of B Cells and Antibodies

While DCs, T cells, and IFNγ-activated macrophages are fundamental in protection against *Histoplasma*, the importance of B cells during an in vivo infection is more difficult to determine. Since humoral immunity works in conjunction with cell-mediated immunity, experimental models to determine the efficacy of serum or a specific antibody independently are convoluted [152]. B cell deficient animals do not manifest a difference in fungal burdens compared to wild-type animals [122], but this does not necessarily mean that there are no protective antibodies conferring protection. Specific monoclonal antibodies can be used to augment protective immunity; phagocytosis and phagosomal maturation are enhanced when yeasts are opsonized by antibody and they interfere with the ability of *Histoplasma* to control the phagosomal pH, thereby facilitating killing of the fungus and prolonging survival in a mouse model [37, 56, 153]. The antibodies produced in vivo will not be monoclonal, they will be responsive to a wide variety of epitopes, some may be protective but in small numbers while others may be nonprotective and able to interfere with the protective antibodies [152]. Furthermore, the isotype is also very important, with IgG1 and IgG2a being protective against *Histoplasma*, but not IgG2b [37]. Cell-mediated immunity is undoubtedly vital, and while the role of B cells and antibodies is still undetermined, there is evidence suggesting a protective role.

5.3.7 Perturbations to the Immune Response

Most individuals infected with *Histoplasma* resolve the infection. Less than 1% of those infected are even aware they have been infected [12]. Immunocompromised patients, which are unable to mount an effective cell-mediated immune response to *Histoplasma*, have a difficult time resolving an infection upon exposure and frequently develop disseminated disease. Patients with AIDS, solid organ transplant, Hodgkin's disease, chronic and acute lymphocytic leukemia, and those receiving TNFα blockers or corticosteroids are all at increased risk for developing disease [12]. Immunocompromised hosts develop disease in multiple organs through a primary infection or through reactivation of a previous infection [154]. Older adults are also at risk for chronic progressive disseminated histoplasmosis. This is generally slowly progressive and an often fatal infection that occurs in older adults who are not overtly immunocompromised [155–157]. Understanding why these conditions increases the risk for contracting histoplasmosis allows insight into the immune response to this fungus.

5.3.7.1 HIV/AIDS

The hallmark of AIDS is a low $CD4^+$ T cell count. As these cells are required for successful clearance of *Histoplasma*, AIDS patients infected with *Histoplasma* are more likely to develop disease. In fact, the diagnosis of progressive disseminated histoplasmosis was one of the early definitions of AIDS. With the advent of successful antiretroviral therapy, there has been a significant decrease in the number of disseminated histoplasmosis cases in the United States, nevertheless HIV-associated histoplasmosis hospitalizations remain significant [12, 158]. Areas with poor access to antiretroviral drugs, such as Central America, have a considerable problem with disseminated histoplasmosis in AIDS patients [159]. Those patients with $CD4^+$ T cell counts of less than 150 cells/μL are most at risk [12]. When correctly diagnosed, treatment with antifungals resolves infection in these patients [160].

5.3.7.2 Solid Organ Transplants/Immunosuppression/Corticosteroids

Patients on immunosuppressive drugs and/or corticosteroids after receiving a solid organ transplant (SOT) are at increased risk of developing disseminated histoplasmosis. The first year after transplant is the period of highest risk, with the median time from transplant to diagnosis around 27 months [161, 162]. A year of antifungal treatment successfully resolves the infection, with rare relapse [161]. SOT patients with human leukocyte antigen HLA-B37 may be at slightly higher risk for disseminated histoplasmosis, though [163] did not find statistical significance in their small sample size [163]. SOT patients on immunosuppressant drugs that live in endemic areas may develop a new infection or if they had previously infection that was contained in granulomas, the fungi may escape due to the drugs causing reactivation.

5.3.7.3 TNFα Blockers

While the development of TNFα blockers has been salutary for many with conditions like rheumatoid arthritis and plaque psoriasis, they elevate the risk of developing histoplasmosis. *Histoplasma* complications have been found to be more common in patients taking anti-TNFα monoclonal antibodies than in those taking a soluble TNFα receptor [164]. The monoclonal antibody infliximab neutralizes TNFα, but may also disrupt granulomas, resulting in previously sequestered *Histoplasma* being released and preventing the necessary TNFα response to resist infection [165]. Antifungal therapy for 12 months while stopping TNFα blocker therapy resolves infection in most cases. Resumption of TNFα blocker therapy is safe following successful antifungal treatment, as long as patients are closely clinically monitored [166]. An alternate method of action is enhanced activity and increased number of regulatory T cells upon treatment with TNFα blockers, thereby suppressing inflammation and enabling the *Histoplasma* to escape confinement [167]. Blocking TNFα increases susceptibility for developing histoplasmosis and yet only a small percentage of patients on these drugs in endemic regions develop infections, suggesting some patients are less able to compensate for the loss of TNFα than others. They may be unable to increase IFNγ or GM-CSF levels to counteract the lack of TNFα or have an underlying additional immunodeficiency.

5.4 Clinical Management of Histoplasmosis

5.4.1 Treatment

Most *Histoplasma* infections resolve without treatment. For the more vulnerable immunocompromised patients and the rare immunocompetent patients that develop histoplasmosis, the current Infectious Diseases Society of America (IDSA) guidelines are to treat with a combination of liposomal amphotericin B and itraconazole [168]. Other azoles, including: fluconazole, posaconazole, isavuconazole, and voriconazole, exhibit activity against *Histoplasma* and provide additional options for patients intolerant of itraconazole [168]. However, fluconazole is much less effective than other azoles, and clinical studies with isavuconazole are limited. The echinocandins class of antifungals, which includes caspofungin, are not recommended for treatment as they have minimal effect on Histoplasma yeasts [169, 170].

References

1. Ajello L (1971) Distribution of Histoplasma capsulatum in the United States. In: Ajello L, Chick EW, Furcolow MF (eds) Histoplasmosis. Charles C. Thomas, Springfield, IL, pp 103–122
2. Chu JH, Feudtner C, Heydon K et al (2006) Hospitalizations for endemic mycoses: a population-based national study. Clin Infect Dis 42:822–825. doi:10.1086/500405
3. Edwards LB, Acquaviva FA, Livesay VT et al (1969) An atlas of sensitivity to tuberculin, PPD-B, and histoplasmin in the United States. Am Rev Respir Dis 99(Suppl):1–132
4. Mochi A, Edwards PQ (1952) Geographical distribution of histoplasmosis and histoplasmin sensitivity. Bull World Health Organ 5:259–291
5. Bahr NC, Antinori S, Wheat LJ, Sarosi GA (2015) Histoplasmosis infections worldwide: thinking outside of the Ohio River valley. Curr Trop Med Rep 2:70–80. doi:10.1007/s40475-015-0044-0
6. Gugnani HC (2000) Histoplasmosis in Africa: a review. Indian J Chest Dis Allied Sci 42: 271–277
7. Pan B, Chen M, Pan W, Liao W (2013) Histoplasmosis: a new endemic fungal infection in China? Review and analysis of cases. Mycoses 56:212–221. doi:10.1111/myc.12029
8. Zhao B, Xia X, Yin J et al (2001) Epidemiological investigation of Histoplasma capsulatum infection in China. Chin Med J 114:743–746
9. Kasuga T, Taylor JW, White TJ (1999) Phylogenetic relationships of varieties and geographical groups of the human pathogenic fungus Histoplasma capsulatum darling. J Clin Microbiol 37:653–663
10. Kasuga T, White TJ, Koenig G et al (2003) Phylogeography of the fungal pathogen Histoplasma capsulatum. Mol Ecol 12:3383–3401
11. Hammerman KJ, Powell KE, Tosh FE (1974) The incidence of hospitalized cases of systemic mycotic infections. Sabouraudia 12:33–45
12. Kauffman CA (2007) Histoplasmosis: a clinical and laboratory update. Clin Microbiol Rev 20:115–132. doi:10.1128/CMR.00027-06
13. Kauffman CA (2009) Histoplasmosis. Clin Chest Med 30:217–225. doi:10.1016/j.ccm.2009.02.002
14. Gilmore SA, Naseem S, Konopka JB, Sil A (2013) N-acetylglucosamine (GlcNAc) triggers a rapid, temperature-responsive morphogenetic program in thermally dimorphic fungi. PLoS Genet 9:e1003799. doi:10.1371/journal.pgen.1003799
15. Beyhan S, Gutierrez M, Voorhies M, Sil A (2013) A temperature-responsive network links cell shape and virulence traits in a primary fungal pathogen. PLoS Biol 11:e1001614. doi:10.1371/journal.pbio.1001614
16. Nemecek JC, Wüthrich M, Klein BS (2006) Global control of dimorphism and virulence in fungi. Science 312:583–588. doi:10.1126/science.1124105
17. Nguyen VQ, Sil A (2008) Temperature-induced switch to the pathogenic yeast form of Histoplasma capsulatum requires Ryp1, a conserved transcriptional regulator. Proc Natl Acad Sci 105:4880–4885. doi:10.1073/pnas.0710448105
18. Webster RH, Sil A (2008) Conserved factors Ryp2 and Ryp3 control cell morphology and infectious spore formation in the fungal pathogen Histoplasma capsulatum. Proc Natl Acad Sci 105:14573–14578. doi:10.1073/pnas.0806221105
19. Medoff G, Sacco M, Maresca B et al (1986) Irreversible block of the mycelial-to-yeast phase transition of Histoplasma capsulatum. Science 231:476–479
20. Hage CA, Azar MM, Bahr N et al (2015) Histoplasmosis: up-to-date evidence-based approach to diagnosis and management. Semin Respir Crit Care Med 36:729–745. doi:10.1055/s-0035-1562899
21. Connolly PA, Durkin MM, Lemonte AM et al (2007) Detection of histoplasma antigen by a quantitative enzyme immunoassay. Clin Vaccine Immunol CVI 14:1587–1591. doi:10.1128/CVI.00071-07
22. Hage CA, Ribes JA, Wengenack NL et al (2011b) A multicenter evaluation of tests for diagnosis of histoplasmosis. Clin Infect Dis Off Publ Infect Dis Soc Am 53:448–454. doi:10.1093/cid/cir435

23. Swartzentruber S, Rhodes L, Kurkjian K et al (2009) Diagnosis of acute pulmonary histoplasmosis by antigen detection. Clin Infect Dis Off Publ Infect Dis Soc Am 49:1878–1882. doi:10.1086/648421
24. Hage CA, Davis TE, Fuller D et al (2010) Diagnosis of histoplasmosis by antigen detection in BAL fluid. Chest 137:623–628. doi:10.1378/chest.09-1702
25. Wheat J, French ML, Kohler RB et al (1982) The diagnostic laboratory tests for histoplasmosis: analysis of experience in a large urban outbreak. Ann Intern Med 97:680–685
26. Babady NE, Buckwalter SP, Hall L et al (2011) Detection of Blastomyces dermatitidis and Histoplasma capsulatum from culture isolates and clinical specimens by use of real-time PCR. J Clin Microbiol 49:3204–3208. doi:10.1128/JCM.00673-11
27. Brummer E, Stevens DA (2010) Collectins and fungal pathogens: roles of surfactant proteins and mannose binding lectin in host resistance. Med Mycol Off Publ Int Soc Hum Anim Mycol 48:16–28. doi:10.3109/13693780903117473
28. McCormack FX, Gibbons R, Ward SR et al (2003) Macrophage-independent fungicidal action of the pulmonary collections. J Biol Chem 278:36250–36256. doi:10.1074/jbc.M303086200
29. Deepe GS Jr, Gibbons RS, Smulian AG (2008) Histoplasma capsulatum manifests preferential invasion of phagocytic subpopulations in murine lungs. J Leukoc Biol 84:669–678. doi:10.1189/jlb.0308154
30. Lin J-S, Huang J-H, Hung L-Y et al (2010) Distinct roles of complement receptor 3, Dectin-1, and sialic acids in murine macrophage interaction with Histoplasma yeast. J Leukoc Biol 88:95–106. doi:10.1189/jlb.1109717
31. Newman SL, Bucher C, Rhodes J, Bullock WE (1990) Phagocytosis of Histoplasma capsulatum yeasts and microconidia by human cultured macrophages and alveolar macrophages. Cellular cytoskeleton requirement for attachment and ingestion. J Clin Invest 85:223–230
32. Bullock WE, Wright SD (1987) Role of the adherence-promoting receptors, CR3, LFA-1, and p150,95, in binding of Histoplasma capsulatum by human macrophages. J Exp Med 165:195–210
33. Habich C, Kempe K, Gomez FJ et al (2006) Heat shock protein 60: identification of specific epitopes for binding to primary macrophages. FEBS Lett 580:115–120. doi:10.1016/j.febslet.2005.11.060
34. Long KH, Gomez FJ, Morris RE, Newman SL (2003) Identification of heat shock protein 60 as the ligand on Histoplasma capsulatum that mediates binding to CD18 receptors on human macrophages. J Immunol 170:487–494
35. Deepe GS, Gibbons R, Brunner GD, Gomez FJ (1996) A protective domain of heat-shock protein 60 from Histoplasma capsulatum. J Infect Dis 174:828–834. doi:10.1093/infdis/174.4.828
36. Gomez FJ, Allendoerfer R, Deepe GS (1995) Vaccination with recombinant heat shock protein 60 from Histoplasma capsulatum protects mice against pulmonary histoplasmosis. Infect Immun 63:2587–2595
37. Guimarães AJ, Frases S, Gomez FJ et al (2009) Monoclonal antibodies to heat shock protein 60 alter the pathogenesis of Histoplasma capsulatum. Infect Immun 77:1357–1367. doi:10.1128/IAI.01443-08
38. Willment JA, Brown GD (2008) C-type lectin receptors in antifungal immunity. Trends Microbiol 16:27–32. doi:10.1016/j.tim.2007.10.012
39. Herre J, Willment JA, Gordon S, Brown GD (2004) The role of Dectin-1 in antifungal immunity. Crit Rev Immunol 24:193–203
40. Brown GD, Herre J, Williams DL et al (2003) Dectin-1 mediates the biological effects of beta-glucans. J Exp Med 197:1119–1124. doi:10.1084/jem.20021890
41. Brown GD, Taylor PR, Reid DM et al (2002) Dectin-1 is a major beta-glucan receptor on macrophages. J Exp Med 196:407–412
42. Kennedy AD, Willment JA, Dorward DW et al (2007) Dectin-1 promotes fungicidal activity of human neutrophils. Eur J Immunol 37:467–478. doi:10.1002/eji.200636653

43. Taylor PR, Brown GD, Reid DM et al (2002) The beta-glucan receptor, dectin-1, is predominantly expressed on the surface of cells of the monocyte/macrophage and neutrophil lineages. J Immunol Baltim Md 1950 169:3876–3882
44. Taylor PR, Tsoni SV, Willment JA et al (2007) Dectin-1 is required for beta-glucan recognition and control of fungal infection. Nat Immunol 8:31–38. doi:10.1038/ni1408
45. Coady A, Sil A (2015) MyD88-dependent signaling drives host survival and early cytokine production during Histoplasma capsulatum infection. Infect Immun 83:1265–1275. doi:10.1128/IAI.02619-14
46. Wang H, LeBert V, Hung CY et al (2014) C-type lectin receptors differentially induce th17 cells and vaccine immunity to the endemic mycosis of North America. J Immunol Baltim Md 1950 192:1107–1119. doi:10.4049/jimmunol.1302314
47. Rappleye CA, Eissenberg LG, Goldman WE (2007) Histoplasma capsulatum alpha-(1,3)-glucan blocks innate immune recognition by the beta-glucan receptor. Proc Natl Acad Sci U S A 104:1366–1370. doi:10.1073/pnas.0609848104
48. Rappleye CA, Engle JT, Goldman WE (2004) RNA interference in Histoplasma capsulatum demonstrates a role for alpha-(1,3)-glucan in virulence. Mol Microbiol 53:153–165. doi:10.1111/j.1365-2958.2004.04131.x
49. Marion CL, Rappleye CA, Engle JT, Goldman WE (2006) An alpha-(1,4)-amylase is essential for alpha-(1,3)-glucan production and virulence in Histoplasma capsulatum. Mol Microbiol 62:970–983. doi:10.1111/j.1365-2958.2006.05436.x
50. Garfoot AL, Shen Q, Wüthrich M et al (2016) The Eng1 β-glucanase enhances histoplasma virulence by reducing β-glucan exposure. MBio 7:e01388–e01315. doi:10.1128/mBio.01388-15
51. Holbrook ED, Edwards JA, Youseff BH, Rappleye CA (2011) Definition of the extracellular proteome of pathogenic-phase Histoplasma capsulatum. J Proteome Res 10:1929–1943. doi:10.1021/pr1011697
52. Kanetsuna F, Carbonell LM, Gil F, Azuma I (1974) Chemical and ultrastructural studies on the cell walls of the yeastlike and mycelial forms of Histoplasma capsulatum. Mycopathol Mycol Appl 54:1–13
53. Gomez FJ, Gomez AM, Deepe GS (1991) Protective efficacy of a 62-kilodalton antigen, HIS-62, from the cell wall and cell membrane of Histoplasma capsulatum yeast cells. Infect Immun 59:4459–4464
54. Gomez FJ, Gomez AM, Deepe GS (1992) An 80-kilodalton antigen from Histoplasma capsulatum that has homology to heat shock protein 70 induces cell-mediated immune responses and protection in mice. Infect Immun 60:2565–2571
55. Guimarães AJ, Hamilton AJ, de Matos Guedes HL et al (2008) Biological function and molecular mapping of M antigen in yeast phase of Histoplasma capsulatum. PLoS One 3:e3449. doi:10.1371/journal.pone.0003449
56. Nosanchuk JD, Steenbergen JN, Shi L et al (2003) Antibodies to a cell surface histone-like protein protect against Histoplasma capsulatum. J Clin Invest 112:1164–1175. doi:10.1172/JCI19361
57. Bohse ML, Woods JP (2005) Surface localization of the Yps3p protein of Histoplasma capsulatum. Eukaryot Cell 4:685–693. doi:10.1128/EC.4.4.685-693.2005
58. Bohse ML, Woods JP (2007a) Expression and interstrain variability of the YPS3 gene of Histoplasma capsulatum. Eukaryot Cell 6:609–615. doi:10.1128/EC.00010-07
59. Keath EJ, Painter AA, Kobayashi GS, Medoff G (1989) Variable expression of a yeast-phase-specific gene in Histoplasma capsulatum strains differing in thermotolerance and virulence. Infect Immun 57:1384–1390
60. Bohse ML, Woods JP (2007b) RNA interference-mediated silencing of the YPS3 gene of Histoplasma capsulatum reveals virulence defects. Infect Immun 75:2811–2817. doi:10.1128/IAI.00304-07
61. Holbrook ED, Smolnycki KA, Youseff BH, Rappleye CA (2013) Redundant catalases detoxify phagocyte reactive oxygen and facilitate Histoplasma capsulatum pathogenesis. Infect Immun 81:2334–2346. doi:10.1128/IAI.00173-13

62. Youseff BH, Holbrook ED, Smolnycki KA, Rappleye CA (2012) Extracellular superoxide dismutase protects Histoplasma yeast cells from host-derived oxidative stress. PLoS Pathog 8:e1002713. doi:10.1371/journal.ppat.1002713
63. Eissenberg LG, Goldman WE (1987) Histoplasma capsulatum fails to trigger release of superoxide from macrophages. Infect Immun 55:29–34
64. Fleischmann J, Wu-Hsieh B, Howard DH (1990) The intracellular fate of Histoplasma capsulatum in human macrophages is unaffected by recombinant human interferon-gamma. J Infect Dis 161:143–145
65. Wolf JE, Kerchberger V, Kobayashi GS, Little JR (1987) Modulation of the macrophage oxidative burst by Histoplasma capsulatum. J Immunol 138:582–586
66. Schaffner A, Davis CE, Schaffner T et al (1986) In vitro susceptibility of fungi to killing by neutrophil granulocytes discriminates between primary pathogenicity and opportunism. J Clin Invest 78:511–524
67. Nittler MP, Hocking-Murray D, Foo CK, Sil A (2005) Identification of Histoplasma capsulatum transcripts induced in response to reactive nitrogen species. Mol Biol Cell 16:4792–4813. doi:10.1091/mbc.E05-05-0434
68. Lane TE, Wu-Hsieh BA, Howard DH (1994) Antihistoplasma effect of activated mouse splenic macrophages involves production of reactive nitrogen intermediates. Infect Immun 62:1940–1945
69. Nakamura LT, Wu-Hsieh BA, Howard DH (1994) Recombinant murine gamma interferon stimulates macrophages of the RAW cell line to inhibit intracellular growth of Histoplasma capsulatum. Infect Immun 62:680–684
70. Eissenberg LG, Goldman WE, Schlesinger PH (1993) Histoplasma capsulatum modulates the acidification of phagolysosomes. J Exp Med 177:1605–1611
71. Eissenberg LG, Schlesinger PH, Goldman WE (1988) Phagosome-lysosome fusion in P388D1 macrophages infected with Histoplasma capsulatum. J Leukoc Biol 43:483–491
72. Newman SL, Gootee L, Kidd C et al (1997) Activation of human macrophage fungistatic activity against Histoplasma capsulatum upon adherence to type 1 collagen matrices. J Immunol 158:1779–1786
73. Strasser JE, Newman SL, Ciraolo GM et al (1999) Regulation of the macrophage vacuolar ATPase and phagosome-lysosome fusion by Histoplasma capsulatum. J Immunol Baltim Md 1950 162:6148–6154
74. Isaac DT, Coady A, Van Prooyen N, Sil A (2013) The 3-hydroxy-methylglutaryl coenzyme A lyase HCL1 is required for macrophage colonization by human fungal pathogen Histoplasma capsulatum. Infect Immun 81:411–420. doi:10.1128/IAI.00833-12
75. Retallack DM, Heinecke EL, Gibbons R et al (1999) The URA5 gene is necessary for histoplasma capsulatum growth during infection of mouse and human cells. Infect Immun 67:624–629
76. Garfoot AL, Zemska O, Rappleye CA (2014) Histoplasma capsulatum depends on de novo vitamin biosynthesis for intraphagosomal proliferation. Infect Immun 82:393–404. doi:10.1128/IAI.00824-13
77. Stetler DA, Boguslawski G (1979) Cysteine biosynthesis in a fungus, Histoplasma capsulatum. Sabouraudia 17:23–34
78. Boguslawski G, Akagi JM, Ward LG (1976) Possible role for cysteine biosynthesis in conversion from mycelial to yeast form of Histoplasma capsulatum. Nature 261:336–338
79. Howard DH, Dabrowa N, Otto V, Rhodes J (1980) Cysteine transport and sulfite reductase activity in a germination-defective mutant of Histoplasma capsulatum. J Bacteriol 141:417–421
80. Wu-Hsieh BA, Howard DH (1992) Intracellular growth inhibition of Histoplasma capsulatum induced in murine macrophages by recombinant gamma interferon is not due to a limitation of the supply of methionine or cysteine to the fungus. Infect Immun 60:698–700
81. Becker KW, Skaar EP (2014) Metal limitation and toxicity at the interface between host and pathogen. FEMS Microbiol Rev 38:1235–1249. doi:10.1111/1574-6976.12087
82. Newman SL, Gootee L, Brunner G, Deepe GS (1994) Chloroquine induces human macrophage killing of Histoplasma capsulatum by limiting the availability of intracellular iron and

is therapeutic in a murine model of histoplasmosis. J Clin Invest 93:1422–1429. doi:10.1172/JCI117119
83. Hwang LH, Seth E, Gilmore SA, Sil A (2012) SRE1 regulates iron-dependent and -independent pathways in the fungal pathogen Histoplasma capsulatum. Eukaryot Cell 11:16–25. doi:10.1128/EC.05274-11
84. Winters MS, Spellman DS, Chan Q et al (2008) Histoplasma capsulatum proteome response to decreased iron availability. Proteome Sci 6:36. doi:10.1186/1477-5956-6-36
85. Hilty J, George Smulian A, Newman SL (2011) Histoplasma capsulatum utilizes siderophores for intracellular iron acquisition in macrophages. Med Mycol 49:633–642. doi:10.3109/13693786.2011.558930
86. Hwang LH, Mayfield JA, Rine J, Sil A (2008) Histoplasma requires SID1, a member of an iron-regulated siderophore gene cluster, for host colonization. PLoS Pathog 4:e1000044. doi:10.1371/journal.ppat.1000044
87. Timmerman MM, Woods JP (2001) Potential role for extracellular glutathione-dependent ferric reductase in utilization of environmental and host ferric compounds by Histoplasma capsulatum. Infect Immun 69:7671–7678. doi:10.1128/IAI.69.12.7671-7678.2001
88. Zarnowski R, Cooper KG, Brunold LS et al (2008) Histoplasma capsulatum secreted γ-glutamyltransferase reduces iron by generating an efficient ferric reductant. Mol Microbiol 70:352–368. doi:10.1111/j.1365-2958.2008.06410.x
89. Dade J, DuBois JC, Pasula R et al (2016) HcZrt2, a zinc responsive gene, is indispensable for the survival of Histoplasma capsulatum in vivo. Med Mycol 54:865–875. doi:10.1093/mmy/myw045
90. Winters MS, Chan Q, Caruso JA, Deepe GS (2010) Metallomic analysis of macrophages infected with Histoplasma capsulatum reveals a fundamental role for zinc in host defenses. J Infect Dis 202:1136–1145. doi:10.1086/656191
91. Subramanian Vignesh K, Landero Figueroa JA, Porollo A et al (2013) Granulocyte macrophage-colony stimulating factor induced Zn sequestration enhances macrophage superoxide and limits intracellular pathogen survival. Immunity 39:697–710. doi:10.1016/j.immuni.2013.09.006
92. Batanghari JW, Deepe GS Jr, Di Cera E, Goldman WE (1998) Histoplasma acquisition of calcium and expression of CBP1 during intracellular parasitism. Mol Microbiol 27:531–539
93. Kügler S, Young B, Miller VL, Goldman WE (2000) Monitoring phase-specific gene expression in Histoplasma capsulatum with telomeric GFP fusion plasmids. Cell Microbiol 2:537–547. doi:10.1046/j.1462-5822.2000.00078.x
94. Edwards JA, Chen C, Kemski MM et al (2013) Histoplasma yeast and mycelial transcriptomes reveal pathogenic-phase and lineage-specific gene expression profiles. BMC Genomics 14:695. doi:10.1186/1471-2164-14-695
95. Patel JB, Batanghari JW, Goldman WE (1998) Probing the yeast phase-specific expression of the CBP1 gene in Histoplasma capsulatum. J Bacteriol 180:1786–1792
96. Beck MR, DeKoster GT, Hambly DM et al (2008) Structural features responsible for the biological stability of histoplasma's virulence factor CBP. Biochemistry (Mosc) 47:4427–4438. doi:10.1021/bi701495v
97. Edwards JA, Alore EA, Rappleye CA (2011) The yeast-phase virulence requirement for α-glucan synthase differs among Histoplasma capsulatum chemotypes. Eukaryot Cell 10:87–97. doi:10.1128/EC.00214-10
98. Sebghati TS, Engle JT, Goldman WE (2000) Intracellular parasitism by Histoplasma capsulatum: fungal virulence and calcium dependence. Science 290:1368–1372
99. Newman SL (1999) Macrophages in host defense against Histoplasma capsulatum. Trends Microbiol 7:67–71
100. Newman SL, Gootee L (1992) Colony-stimulating factors activate human macrophages to inhibit intracellular growth of Histoplasma capsulatum yeasts. Infect Immun 60:4593–4597
101. Heninger E, Hogan LH, Karman J et al (2006) Characterization of the Histoplasma capsulatum-induced granuloma. J Immunol Baltim Md 1950 177:3303–3313

102. Gomez AM, Bullock WE, Taylor CL, Deepe GS (1988) Role of L3T4+ T cells in host defense against Histoplasma capsulatum. Infect Immun 56:1685–1691
103. Pincelli C, Fujioka A, Hashimoto A et al (1988) T-cell depletion by monoclonal antibodies does not prevent granuloma formation in mice. Exp Cell Biol 56:229–235
104. Baughman RP, Kim CK, Vinegar A et al (1986) The pathogenesis of experimental pulmonary histoplasmosis. Correlative studies of histopathology, bronchoalveolar lavage, and respiratory function. Am Rev Respir Dis 134:771–776. doi:10.1164/arrd.1986.134.4.771
105. Deepe GS, Bullock WE (1992) Histoplasmosis: a granulomatous inflammatory response. In: Gallin JI, Goldstein IM, Snyderman R (eds) Inflammation: basic principles and clinical correlates. Raven Press, New York, pp 943–958
106. Serbina NV, Jia T, Hohl TM, Pamer EG (2008) Monocyte-mediated defense against microbial pathogens. Annu Rev Immunol 26:421–452. doi:10.1146/annurev.immunol.26.021607.090326
107. Szymczak WA, Deepe GS (2010) Antigen-presenting dendritic cells rescue CD4-depleted CCR2–/– mice from lethal Histoplasma capsulatum infection. Infect Immun 78:2125–2137. doi:10.1128/IAI.00065-10
108. Deepe GS (2000) Immune response to early and late Histoplasma capsulatum infections. Curr Opin Microbiol 3:359–362
109. Lane TE, Wu-Hsieh BA, Howard DH (1991) Iron limitation and the gamma interferon-mediated antihistoplasma state of murine macrophages. Infect Immun 59:2274–2278
110. Brummer E, Kurita N, Yoshida S et al (1991) Killing of Histoplasma capsulatum by gamma-interferon-activated human monocyte-derived macrophages: evidence for a superoxide anion-dependent mechanism. J Med Microbiol 35:29–34. doi:10.1099/00222615-35-1-29
111. Gildea LA, Morris RE, Newman SL (2001) Histoplasma capsulatum yeasts are phagocytosed via very late antigen-5, killed, and processed for antigen presentation by human dendritic cells. J Immunol 166:1049–1056
112. Gomez FJ, Pilcher-Roberts R, Alborzi A, Newman SL (2008) Histoplasma capsulatum cyclophilin A mediates attachment to dendritic cell VLA-5. J Immunol 181:7106–7114
113. Gildea LA, Ciraolo GM, Morris RE, Newman SL (2005) Human dendritic cell activity against Histoplasma capsulatum is mediated via phagolysosomal fusion. Infect Immun 73:6803–6811. doi:10.1128/IAI.73.10.6803-6811.2005
114. Van Prooyen N, Henderson CA, Hocking Murray D, Sil A (2016) CD103+ conventional dendritic cells are critical for TLR7/9-dependent host defense against Histoplasma capsulatum, an endemic fungal pathogen of humans. PLoS Pathog 12:e1005749. doi:10.1371/journal.ppat.1005749
115. Lin J-S, Yang C-W, Wang D-W, Wu-Hsieh BA (2005) Dendritic cells cross-present exogenous fungal antigens to stimulate a protective CD8 T cell response in infection by Histoplasma capsulatum. J Immunol Baltim Md 1950 174:6282–6291
116. Wüthrich M, Deepe GS, Klein B (2012) Adaptive immunity to fungi. Annu Rev Immunol 30:115–148. doi:10.1146/annurev-immunol-020711-074958
117. Inglis DO, Berkes CA, Murray DRH, Sil A (2010) Conidia but not yeast cells of the fungal pathogen Histoplasma capsulatum trigger a type I interferon innate immune response in murine macrophages. Infect Immun 78:3871–3882. doi:10.1128/IAI.00204-10
118. Smith JG, Magee DM, Williams DM, Graybill JR (1990) Tumor necrosis factor-alpha plays a role in host defense against Histoplasma capsulatum. J Infect Dis 162:1349–1353
119. Allendoerfer R, Deepe GS (1998) Blockade of endogenous TNF-α exacerbates primary and secondary pulmonary histoplasmosis by differential mechanisms. J Immunol 160:6072–6082
120. Zhou P, Miller G, Seder RA (1998) Factors involved in regulating primary and secondary immunity to infection with Histoplasma capsulatum: TNF-alpha plays a critical role in maintaining secondary immunity in the absence of IFN-gamma. J Immunol Baltim Md 1950 160:1359–1368
121. Allendoerfer R, Deepe GS Jr (2000) Regulation of infection with Histoplasma capsulatum by TNFR1 and –2. J Immunol Baltim Md 1950 165:2657–2664
122. Allendörfer R, Brunner GD, Deepe GS (1999) Complex requirements for nascent and memory immunity in pulmonary histoplasmosis. J Immunol Baltim Md 1950 162:7389–7396

123. Deepe GS (1994) Role of CD8+ T cells in host resistance to systemic infection with Histoplasma capsulatum in mice. J Immunol Baltim Md 1950 152:3491–3500
124. Cain JA, Deepe GS Jr (1998) Evolution of the primary immune response to Histoplasma capsulatum in murine lung. Infect Immun 66:1473–1481
125. Zhou P, Sieve MC, Bennett J et al (1995) IL-12 prevents mortality in mice infected with Histoplasma capsulatum through induction of IFN-gamma. J Immunol 155:785–795
126. Zerbe CS, Holland SM (2005) Disseminated histoplasmosis in persons with interferon-gamma receptor 1 deficiency. Clin Infect Dis Off Publ Infect Dis Soc Am 41:e38–e41. doi:10.1086/432120
127. Sampaio EP, Hsu AP, Pechacek J et al (2013) Signal transducer and activator of transcription 1 (STAT1) gain-of-function mutations and disseminated coccidioidomycosis and histoplasmosis. J Allergy Clin Immunol 131:1624–1634. doi:10.1016/j.jaci.2013.01.052
128. Deepe GS, Gibbons R, Woodward E (1999) Neutralization of endogenous granulocyte-macrophage colony-stimulating factor subverts the protective immune response to Histoplasma capsulatum. J Immunol 163:4985–4993
129. Deepe GS Jr, Gibbons RS (2009) Interleukins 17 and 23 influence the host response to Histoplasma capsulatum. J Infect Dis 200:142–151. doi:10.1086/599333
130. Wu S-Y, Yu J-S, Liu F-T et al (2013) Galectin-3 negatively regulates dendritic cell production of IL-23/IL-17-axis cytokines in infection by Histoplasma capsulatum. J Immunol Baltim Md 1950 190:3427–3437. doi:10.4049/jimmunol.1202122
131. Verma AH, Bueter CL, Rothenberg ME, Deepe GS (2016) Eosinophils subvert host resistance to an intracellular pathogen by instigating non-protective IL-4 in CCR2(−/−) mice. Mucosal Immunol. doi:10.1038/mi.2016.26
132. Szymczak WA, Deepe GS (2009) The CCL7-CCL2-CCR2 axis regulates IL-4 production in lungs and fungal immunity. J Immunol Baltim Md 1950 183:1964–1974. doi:10.4049/jimmunol.0901316
133. Hamilton JA, Whitty GA, Royston AK et al (1992) Interleukin-4 suppresses granulocyte colony-stimulating factor and granulocyte-macrophage colony-stimulating factor levels in stimulated human monocytes. Immunology 76:566–571
134. Liew FY, Li Y, Severn A et al (1991) A possible novel pathway of regulation by murine T helper type-2 (Th2) cells of a Th1 cell activity via the modulation of the induction of nitric oxide synthase on macrophages. Eur J Immunol 21:2489–2494. doi:10.1002/eji.1830211027
135. Sattler S, Smits HH, Xu D, Huang F-P (2013) The evolutionary role of the IL-33/ST2 system in host immune defence. Arch Immunol Ther Exp 61:107–117. doi:10.1007/s00005-012-0208-8
136. Verma A, Kroetz DN, Tweedle JL, Deepe GS (2015) Type II cytokines impair host defense against an intracellular fungal pathogen by amplifying macrophage generation of IL-33. Mucosal Immunol 8:380–389. doi:10.1038/mi.2014.75
137. Couper KN, Blount DG, Riley EM (2008) IL-10: the master regulator of immunity to infection. J Immunol Baltim Md 1950 180:5771–5777
138. Deepe GS Jr, Gibbons RS (2003) Protective and memory immunity to Histoplasma capsulatum in the absence of IL-10. J Immunol Baltim Md 1950 171:5353–5362
139. Fecher RA, Horwath MC, Friedrich D et al (2016) Inverse correlation between IL-10 and HIF-1α in macrophages infected with Histoplasma capsulatum. J Immunol Baltim Md 1950 197:565–579. doi:10.4049/jimmunol.1600342
140. Wu-Hsieh B (1989) Relative susceptibilities of inbred mouse strains C57BL/6 and A/J to infection with Histoplasma capsulatum. Infect Immun 57:3788–3792
141. Gomez FJ, Cain JA, Gibbons R et al (1998) Vbeta4(+) T cells promote clearance of infection in murine pulmonary histoplasmosis. J Clin Invest 102:984–995. doi:10.1172/JCI2963
142. Gomez FJ, Woodward EO, Pilcher-Roberts R et al (2001) V beta 6+ and V beta 4+ T cells exert cooperative activity in clearance of secondary infection with Histoplasma capsulatum. J Immunol Baltim Md 1950 166:2855–2862

143. Scheckelhoff M, Deepe GS (2002) The protective immune response to heat shock protein 60 of Histoplasma capsulatum is mediated by a subset of V beta 8.1/8.2+ T cells. J Immunol Baltim Md 1950 169:5818–5826
144. Zheng SG (2013) Regulatory T cells vs Th17: differentiation of Th17 versus Treg, are the mutually exclusive? Am J Clin Exp Immunol 2:94–106
145. Kroetz DN, Deepe GS (2010) CCR5 dictates the equilibrium of proinflammatory IL-17+ and regulatory Foxp3+ T cells in fungal infection. J Immunol Baltim Md 1950 184:5224–5231. doi:10.4049/jimmunol.1000032
146. George MM, Subramanian Vignesh K, Landero Figueroa JA et al (2016) Zinc induces dendritic cell tolerogenic phenotype and skews regulatory T cell-Th17 balance. J Immunol Baltim Md 1950 197:1864–1876. doi:10.4049/jimmunol.1600410
147. Deepe GS Jr, Gibbons RS (2008) TNF-alpha antagonism generates a population of antigen-specific CD4+CD25+ T cells that inhibit protective immunity in murine histoplasmosis. J Immunol Baltim Md 1950 180:1088–1097
148. Deepe GS, Gibbons R (2001) V beta 6+ T cells are obligatory for vaccine-induced immunity to Histoplasma capsulatum. J Immunol Baltim Md 1950 167:2219–2226
149. Deepe GS, Gibbons RS (2002) Cellular and molecular regulation of vaccination with heat shock protein 60 from Histoplasma capsulatum. Infect Immun 70:3759–3767
150. Wuthrich M, Filutowicz HI, Warner T et al (2003) Vaccine immunity to pathogenic fungi overcomes the requirement for CD4 help in exogenous antigen presentation to CD8+ T cells: implications for vaccine development in immune-deficient hosts. J Exp Med 197:1405–1416. doi:10.1084/jem.20030109
151. Wüthrich M, Gern B, Hung CY et al (2016) Vaccine-induced protection against 3 systemic mycoses endemic to North America requires Th17 cells in mice. J Clin Invest 126:795. doi:10.1172/JCI85788
152. Casadevall A (1998) Antibody-mediated protection against intracellular pathogens. Trends Microbiol 6:102–107
153. Shi L, Albuquerque PC, Lazar-Molnar E et al (2008) A monoclonal antibody to Histoplasma capsulatum alters the intracellular fate of the fungus in murine macrophages. Eukaryot Cell 7:1109–1117. doi:10.1128/EC.00036-08
154. Johnson PC, Khardori N, Najjar AF et al (1988) Progressive disseminated histoplasmosis in patients with acquired immunodeficiency syndrome. Am J Med 85:152–158
155. Kauffman CA (2001) Fungal infections in older adults. Clin Infect Dis Off Publ Infect Dis Soc Am 33:550–555. doi:10.1086/322685
156. Reddy P, Gorelick DF, Brasher CA, Larsh H (1970) Progressive disseminated histoplasmosis as seen in adults. Am J Med 48:629–636
157. Smith JW, Utz JP (1972) Progressive disseminated histoplasmosis. A prospective study of 26 patients. Ann Intern Med 76:557–565
158. Benedict K, Derado G, Mody RK (2016) Histoplasmosis-associated hospitalizations in the United States, 2001-2012. Open Forum Infect Dis 3:ofv219. doi:10.1093/ofid/ofv219
159. Gutierrez ME, Canton A, Sosa N et al (2005) Disseminated histoplasmosis in patients with AIDS in Panama: a review of 104 cases. Clin Infect Dis Off Publ Infect Dis Soc Am 40:1199–1202. doi:10.1086/428842
160. Myint T, Anderson AM, Sanchez A et al (2014) Histoplasmosis in patients with human immunodeficiency virus/acquired immunodeficiency syndrome (HIV/AIDS): multicenter study of outcomes and factors associated with relapse. Medicine (Baltimore) 93:11–18. doi:10.1097/MD.0000000000000016
161. Assi M, Martin S, Wheat LJ et al (2013) Histoplasmosis after solid organ transplant. Clin Infect Dis Off Publ Infect Dis Soc Am 57:1542–1549. doi:10.1093/cid/cit593
162. Kauffman CA, Israel KS, Smith JW et al (1978) Histoplasmosis in immunosuppressed patients. Am J Med 64:923–932
163. Zimmer A, Miller G, Mallal S, Thomas L (2016) Human leukocyte antigen and risk of disseminated histoplasmosis in solid organ transplant recipients. Transpl Infect Dis Off J Transplant Soc 18:160–161. doi:10.1111/tid.12470

164. Toh S, Li L, Harrold LR et al (2012) Comparative safety of infliximab and etanercept on the risk of serious infections: does the association vary by patient characteristics? Pharmacoepidemiol Drug Saf 21:524–534. doi:10.1002/pds.3238
165. Wallis RS, Broder M, Wong J et al (2005) Reactivation of latent granulomatous infections by infliximab. Clin Infect Dis Off Publ Infect Dis Soc Am 41(Suppl 3):S194–S198. doi:10.1086/429996
166. Vergidis P, Avery RK, Wheat LJ et al (2015) Histoplasmosis complicating tumor necrosis factor-α blocker therapy: a retrospective analysis of 98 cases. Clin Infect Dis Off Publ Infect Dis Soc Am 61:409–417. doi:10.1093/cid/civ299
167. Ehrenstein MR, Evans JG, Singh A et al (2004) Compromised function of regulatory T cells in rheumatoid arthritis and reversal by anti-TNFalpha therapy. J Exp Med 200:277–285. doi:10.1084/jem.20040165
168. Wheat LJ, Freifeld AG, Kleiman MB et al (2007) Clinical practice guidelines for the management of patients with histoplasmosis: 2007 update by the Infectious Diseases Society of America. Clin Infect Dis Off Publ Infect Dis Soc Am 45:807–825. doi:10.1086/521259
169. Hage CA, Connolly P, Horan D et al (2011a) Investigation of the efficacy of micafungin in the treatment of histoplasmosis using two North American strains of Histoplasma capsulatum. Antimicrob Agents Chemother 55:4447–4450. doi:10.1128/AAC.01681-10
170. Kohler S, Wheat LJ, Connolly P et al (2000) Comparison of the echinocandin caspofungin with amphotericin B for treatment of histoplasmosis following pulmonary challenge in a murine model. Antimicrob Agents Chemother 44:1850–1854

Chapter 6
Cryptococcus and Cryptococcosis

Rodrigo Maciel da C. Godinho, Débora L. Oliveira, Priscila C. Albuquerque, Fabianno F. Dutra, Rodrigo de Almeida-Paes, Marcio L. Rodrigues, and Fernanda L. Fonseca

Abstract Cryptococcosis is a fungal disease caused by *Cryptococcus neoformans* and *C. gattii*. Lung disease and meningocephalitis are the most common clinical outcomes of cryptococcosis, especially in immunosuppressed patients. Current estimates suggest 278,000 cases of human cryptococcal meningitis annually. This syndrome is fatal if untreated. Treatment of cryptococcosis is expensive and poorly effective. This complex scenario makes clear the need for innovation in the field of *Cryptococcus* and cryptococcosis.

R.M. da C. Godinho, D.Sc.
Instituto de Microbiologia Paulo de Góes, Universidade Federal do Rio de Janeiro (UFRJ), Rio de Janeiro, Brazil
e-mail: godinho.r@gmail.com

D.L. Oliveira, D.Sc. • P.C. Albuquerque, D.Sc. • F.L. Fonseca, D.Sc. (✉)
Centro de Desenvolvimento Tecnológico em Saúde (CDTS), Fundação Oswaldo Cruz (Fiocruz), Avenida Brasil, 4036, sala 814, Rio de Janeiro, RJ 21040-361, Brazil
e-mail: Debora_leite@yahoo.com.br; priscila.costa@cdts.fiocruz.br; ffonseca@cdts.fiocruz.br

F.F. Dutra, D.Sc.
Laboratório de Inflamação e Imunidade, Departamento de Imunologia, Instituto de Microbiologia, Universidade Federal do Rio de Janeiro (UFRJ), Rio de Janeiro, Brazil
e-mail: fabianno44@yahoo.com.br

R. de Almeida-Paes, D.Sc.
Laboratório de Micologia, Instituto Nacional de Infectologia Evandro Chagas, Fundação Oswaldo Cruz (Fiocruz), Rio de Janeiro, Brazil
e-mail: rodrigo.paes@ini.fiocruz.br

M.L. Rodrigues, D.Sc.
Instituto de Microbiologia Paulo de Góes, Universidade Federal do Rio de Janeiro (UFRJ), Rio de Janeiro, Brazil

Centro de Desenvolvimento Tecnológico em Saúde (CDTS), Fundação Oswaldo Cruz (Fiocruz), Avenida Brasil, 4036, sala 814, Rio de Janeiro, RJ 21040-361, Brazil
e-mail: marciolr@cdts.fiocruz.br

H.M. Mora-Montes, L.M. Lopes-Bezerra (eds.), *Current Progress in Medical Mycology*, DOI 10.1007/978-3-319-64113-3_6

Cryptococcosis is an infectious disease with a worldwide distribution that is commonly caused by two pathogenic encapsulated fungal species: *Cryptococcus neoformans* and *Cryptococcus gattii*. Historically, the *Cryptococcus* genus has been classified into three varieties and five serotypes based on the antigenic determinants of the capsular polysaccharide: *C. neoformans* var *grubii* (serotype A), *C. neoformans* var *neoformans* (serotype D), *C. neoformans* (hybrid AD), and *C. gattii* (serotypes B and C) [1]. To date, molecular methodologies have established at least eight different molecular types for the *C. neoformans*/*C. gattii* complex, and these molecular types correlate with specific serotypes. For *C. neoformans,* these classifications include VNI/AFLP1, serotype A; VNII/AFLP1A, serotype A; VNIII/AFLP3, serotype AD; and VNIV/AFLP2, serotype D. The molecular types of *C. gattii* (VG1/AFLP4; VGII/AFLP6; VGIII/AFLP5; and VGIV/AFLP7) are all associated with both serotype B and serotype C [2].

The first clinical case describing a *Cryptococcus* species as the etiological agent of a fatal disease dates from the late nineteenth century [3]. Nearly one million cases of cryptococcal meningitis occurred in 2006, resulting in approximately 625,000 deaths within 3 months after infection [4]. Estimates of the global incidence of AIDS-related cryptococcal meningitis have been decreasing based on early diagnosis and improvements in antifungal treatment in some countries [5]. However, cryptococcosis remains the most fatal mycosis among AIDS patients worldwide [4].

Cryptococcal infection occurs primarily via the inhalation of fungal particles, which reach the lungs and can disseminate to the brain in immunocompromised hosts. Although an immunosuppressive condition is often associated with cryptococcosis, healthy hosts may also develop the disease, primarily through *C. gattii* infection [6].

Both *C. neoformans* and *C. gattii* display several virulence factors that modulate their interactions with their hosts. The most studied among these virulence factors are the ability to survive at mammalian body temperature, enzyme secretion, the addition of melanin to the fungal cell wall, and the abundant secretion of polysaccharides that may surround the cell to form a capsular structure [1]. All of these virulence factors appear to interact with the human host in a deleterious manner, subverting the immune response or protecting the fungus from immune effector molecules.

Clinical presentations may vary from self-limiting cutaneous infections to fatal meningoencephalitis. Amphotericin B, flucytosine, and fluconazole are recommended for treatment, although the high costs, wide array of side effects, and drug resistance that are associated with these drugs represent problems in health policy implementation [7]. Efforts to identify new fungal therapeutic targets and improve existing antifungal treatments are necessary.

Central aspects related to *Cryptococcus* spp. and cryptococcal infection will be discussed in more detail below.

6.1 Recent Data Related to the Epidemiology of Cryptococcosis

6.1.1 Clinical Epidemiology

In the early twentieth century, the occurrence of cryptococcosis was sporadic, with less than 300 cases reported before the 1950s [8]. In the 1970s, the number of cases increased due to the association of the infection with organ transplantation, immunosuppressive treatments, and malignancy [9]. The emergence of cryptococcosis as an important systemic mycosis worldwide came with the emergence of acquired immunodeficiency syndrome (AIDS) in the 1980s. In fact, the major risk factors for cryptococcosis are HIV infection, immunosuppressive therapies, solid organ transplantation, hematologic malignancy, sarcoidosis, treatment with some monoclonal antibodies or anti-granulocyte-macrophage colony stimulating factor (GM-CSF) antibodies, systemic lupus, rheumatoid arthritis, CD4+ lymphopenia, decompensated chronic liver disease, renal failure, peritoneal dialysis, hyper IgM or IgE syndrome, and tuberculosis [10].

Epidemiologic differences are observed between the agents of cryptococcosis. Most patients infected with *C. neoformans* are immunocompromised, with a few cases reported in individuals without an immunodeficiency-related condition. In contrast, *C. gattii* affects primarily immunocompetent individuals. Another important difference between these two species is that *C. neoformans* infections occur worldwide, while *C. gattii* shows some endemic areas of occurrence [8]. Epidemiological studies have demonstrated a lower incidence for *C. gattii* infections than *C. neoformans* infections. In 1996, only 3 of 248 cases of cryptococcosis were attributed to *C. gattii* (serotype B) in HIV-negative individuals in the US [1]. *C. gattii* isolates have a restricted global distribution, with infections occurring in tropical and subtropical areas. However, some studies have attributed some *C. gattii* outbreaks in uncommon areas, such as Canada and areas in the US that are outside of California, to climate changes [3]. To illustrate this point, in 2006, British Columbia reported an average annual incidence of 5.8 per million persons, while endemic areas in Australia confirmed 0.61 *C. gattii* cases per million. This figure represented the highest incidence of *C. gattii* infection in the world [11].

Infections caused by *C. neoformans* are largely associated with HIV infection [9]. In 2006, Park and colleagues estimated that nearly one million cases of cryptococcal meningitis occur annually in this group of patients, with a mortality rate higher than 60%. Thus, cryptococcal meningitis is one of the leading causes of death in HIV/AIDS patients in Africa, accounting for more deaths than tuberculosis [4]. Current estimates cite 278,000 cases of cryptococcal meningitis annually, and this infection is associated with 100% mortality in untreated AIDS patients [5]. The total death counts were recently 233,000. This reduction can be explained by improved treatment for established cases, with the WHO recommending therapy with amphotericin and flucytosine [3], as well as the detection of asymptomatic antigenemia and oral treatment with fluconazole after excluding meningitis [5].

The majority of these affected patients are from sub-Saharan Africa, which is a region with several difficulties in terms of the diagnosis and treatment of this disease [12]. Primary cutaneous cryptococcosis can occur in both immunocompromised and immunocompetent individuals living in poverty [13]. Pulmonary cryptococcosis is usually seen in immunocompetent individuals but can also occur in immunosuppressed patients, with further dissemination to the central nervous system [14]. The real burden of cryptococcosis caused by *C. neoformans* in at-risk patients without HIV is not well-recognized [12].

The lack of *C. neoformans/C. gattii* differentiation by most clinical laboratories worldwide makes the knowledge of cryptococcosis caused by *C. gattii* inaccurate [15]. In general, most affected individuals are immunocompetent, including a significant number of children [16]. The major endemic areas of this pathogen are Australia, Botswana, Brazil, Canada, Papua New Guinea, South Africa, the United States of America, and Vietnam [15]. The major manifestations of cryptococcosis caused by *C. gattii* in humans are meningoencephalitis and pulmonary disease [17]. Most authors report the central nervous system as the major site of infection by *C. gattii*. However, in North American outbreaks, pulmonary disease predominates [15].

6.1.2 Ecoepidemiology

Current knowledge of *C. neoformans* and *C. gattii* ecology shows that these species do not have specific habitats and instead inhabit diverse niches and substrates in the regions where they can be found [18]. Although both species are ubiquitous yeasts, their main ecological niches have some differences. For instance, *C. neoformans* is usually isolated from pigeon droppings, while *C. gattii* is more frequently isolated from decaying wood and soil [19]. However, both species can be isolated from household dust [20].

The occurrence of *C. neoformans* in soil and pigeon droppings has been widely studied. Poultry manure is considered to be a natural substrate for *C. neoformans*. Several birds, especially *Columba livia* pigeons, can bear this fungus on their beaks, legs, and feathers, as well as present colonization by *Cryptococcus* on the crop. These birds act as fungal dispersers in the environment, generating a source of infection for humans and other mammals [21]. Other animals, such as ants (*Odontomachus bauri*), can also harbor *C. neoformans* and contribute to its dispersal in the environment [22].

One or both species of *Cryptococcus* have been found in different trees from around the world. Some species from which *C. neoformans* has been isolated include *Tipuana tipu*, *Tabebuia avellanedae*, *Thuja* sp., *Peltophorum dubium*, *Cassia grandis* and *Eucalyptus* trees [23]. *C. gattii* was first isolated from *Eucalyptus* trees in Australia, and for several years, these trees were thought to be the habitat of this species. However, at the end of the last century, *C. gattii* began to be isolated from several other trees around the world, and attempts to isolate this species from *Eucalyptus* trees in other countries were unsuccessful. Some species that have been reported to harbor *C. gattii* include *Acacia visco*, *Cassia grandis*, *Guettarda acreana*, *Ficus soatensis*, and *Pinus* spp. [24].

6.1.3 Molecular Epidemiology

With the advance of molecular methodologies and their use in medical mycology, isolates of *C. neoformans* and *C. gattii* were separated into eight different molecular types: for *C. neoformans,* VNI-VNIV, and for *C. gattii,* VGI-VGIV. This improvement in *Cryptococcus* classification clarified several epidemiological aspects of these species, since they exhibit differences in habitats, antifungal susceptibility, clinical manifestations, virulence, antifungal susceptibility and epidemiology [18]. Despite the importance of the molecular typing of *Cryptococcus* species, molecular data are unavailable in several countries in Africa, Asia, and Eastern Europe, as well as the United Kingdom, Ireland, Norway, and Finland [25].

C. neoformans VNI is the most frequent molecular type worldwide, except in Australia and Papua, New Guinea, where *C. gattii* VGI predominates. VNII is less frequent, but has been isolated on all continents. The molecular type VNB was thought to be exclusive to Africa, but this type has been also found in Brazil and Colombia. The frequencies of VNIII and VNIV are very low on all continents, except for Europe, where they account for 18.5% and 18.3% of the *Cryptococcus* strains [25].

As reviewed by Chen [15], *C. gattii* VGII is the most frequently isolated molecular type from environmental samples, followed by the molecular types VGI, VGIII, and VGIV, which is rarely isolated. In clinical samples obtained from humans or animals, however, the frequencies of the molecular types VGI and VGII are very similar; together, these types account for 77% of all *C. gattii* isolates. The frequencies of the molecular types VGIII and VGIV in clinical samples are lower but very similar to each other (13% and 10%, respectively). The geographic distribution of these molecular types is also heterogeneous; the VGII molecular type predominates in the Americas, VGI is most prevalent in Europe, Asia, and Oceania, and VGIV predominates in Africa.

The relevance of *C. gattii* VGII as an emerging pathogen was demonstrated during the Vancouver Island outbreak [2]. The initial molecular analysis of this highly virulent molecular type revealed that these fungi originated in Australia or South America [26]. Further studies confirmed South America as the origin of *C. gattii* VGII dispersal. It was initially proposed that the Amazon rainforest was the original source of *C. gattii* VGII [27], but recent studies changed the origin of this molecular type to the semi-arid desert region in Northeast Brazil [28].

A novel study of *C. gattii* VGIII that used multi-locus sequence typing and whole-genome sequencing revealed that this molecular type is highly diverse genetically, with minor differences between geographic origins, sources of isolation, serotypes, and mating types. This molecular type has two subpopulations, which comprise serotypes B and C of *C. gattii*. The origin of the serotype B VGIII population, which is more virulent, is very likely to be Mexico and the USA, while serotype C VGIII originated in Colombia [29].

6.2 Clinical Aspects of Cryptococcosis

Cryptococcal infection occurs via the inhalation of environmental infectious propagules, resulting in initial lung colonization. In general, the presence of yeast cells in the lung elicits an immune response that results in a lymph node complex [30]. An effective immune response is able to eradicate the fungi, leading to the clearance of lung infection [9]. Asymptomatic cryptococcal infection is prevalent in certain geographic regions, as indicated by serologic analysis [31]. However, depending on the inoculum, the immune status of the host, or the virulence of the strain, the infection may progress to acute disease or become latent [32]. In latent infections, the individual is asymptomatic, although *Cryptococcus* remains alive inside macrophages in the thoracic lymph node, where the fungus can persist for years. When local immunity is suppressed, the dormant yeasts begin to grow, multiply, and disseminate outside of the pulmonary-lymph node complex, leading to disease. Immunosuppressive conditions, such as HIV coinfection and organ transplantation, represent potential risks for the development of acute cryptococcosis from latent infection [9]. The reactivation of latent infections upon immunosuppression is observed more frequently in *C. neoformans*/HIV coinfection, while *C. gattii* is primarily responsible for acute infection in immunocompetent individuals [1].

After cryptococcal cells have reached the alveolar space and come into contact with the immune system, the fungi reach the bloodstream, cross the blood-brain barrier and disseminate into the central nervous system (CNS), causing a meningoencephalitis that is life-threatening if left untreated. Some clinical features of cryptococcosis are dependent on the host immune status and the fungal species. Pulmonary cryptococcal disease is more frequent in immunocompetent hosts (60%) and associated with *C. gattii* infection, while meningitis is common in AIDS patients (86%) [33]. In *C. gattii*-associated pulmonary disease, a progressive inflammatory response and cryptococcoma formation are typical, as is concurrent lung and brain infection [11]. The presence of single mass lesions in the lungs and brain has been mistaken for neoplastic disease or pyogenic abscesses [34], and these lesions commonly produce neurologic sequelae that require surgery or prolonged antifungal therapy [35]. In contrast, HIV-associated cryptococcosis presents with outstanding CNS and extrapulmonary involvement, poor inflammatory reactions and a high burden of *C. neoformans,* which is frequently cultured from the blood, urine and skin lesions [11]. Although *Cryptococcus* species have diverse infection profiles, some clinical features, such as fever and pulmonary or neurological symptoms, are similar [11].

Although cryptococcosis most commonly affects the lung and brain, many sites in the body may be involved, presenting a wide array of clinical presentations. The skin, prostate, eyes, urinary tract and bone are some organs that are infected by these fungi, primarily in HIV-associated cryptococcosis [36].

6.2.1 Pulmonary Cryptococcosis

Pulmonary cryptococcosis ranges from asymptomatic infection of the airways to severe pneumonia with respiratory failure. Typically, symptoms include coughing, pleuritic chest pain, fever, dyspnea, weight loss, malaise, and more rarely, hemopstysis [37]. As these symptoms are similar to those of other pulmonary infections, cryptococcosis may have a delayed diagnosis, which can lead to rapid progression and disease dissemination in some cases [38]. The host immune status influences the clinical presentation of pulmonary cryptococcosis. In immunocompetent individuals, infection may be asymptomatic and can occur in approximately one-third of healthy hosts [39]. Occasionally, these hosts can progress from isolated pulmonary infection to disseminated disease, as occurred in the Vancouver Island *C. gattii* outbreak [40]. In contrast, immunocompromised hosts develop symptomatic cryptococcal pneumonia with the presence of interstitial infiltrates or pleural effusions. These conditions can progress rapidly to acute respiratory distress syndrome (ARDS) [41]. Primarily in AIDS patients, CNS dissemination is a natural consequence after pulmonary involvement and is fatal if untreated. Between 10 and 55% of patients with cryptococcal meningoencephalitis develop pulmonary infection [42]. Although immunosuppressive hosts are considered to be at greatest risk of developing cryptococcosis complications, fatal cases and CNS dissemination have been reported in non-immunocompromised groups [43].

In HIV-positive patients, chest radiograph abnormalities include the presence of interstitial infiltrates or pleural effusions, which are often associated with disseminated disease [38]. In immunocompetent individuals, the presence of single or multiple pulmonary nodules or cryptococcomas in the lung is typical of *C. gattii* cryptococcosis, occurring in 77% of cases [11] (Fig. 6.1).

6.2.2 Meningeal Cryptococcosis

Although both *Cryptococcus* species are able to invade the brain and cause life-threatening disease, *C. neoformans* is more frequently associated with CNS cryptococcosis [44]. *Cryptococcus* infects the meninges and the brain parenchyma, leading to an inflammatory process that is responsible for meningoencephalitis. In addition to the typical meningoencephalitis, many other syndromes have been reported in *C. neoformans*-infected patients, including meningitis, encephalitis, ventriculitis, increased intracranial pressure (ICP) and space-occupying lesions (such as cerebral abscesses, cysts and cryptococcomas) [45]. Headache accompanied by neck stiffness and photophobia, fever, memory loss, lethargy, altered consciousness and personality changes are common symptoms in patients with neurological cryptococcosis [45]. Secondary infections of the lungs, skin, eyes and prostate are less frequent [45]. Depending on the host immune status, the signs and symptoms of disease may be chronic and last for months or occur within days. HIV patients present a high

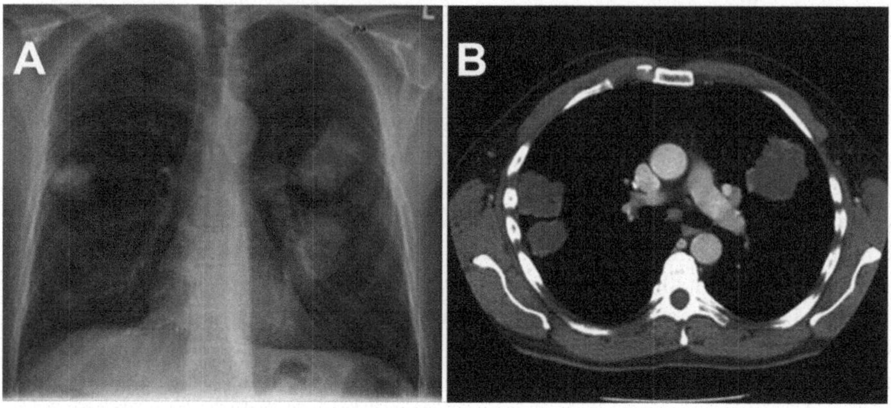

Fig. 6.1 *C. gattii* lung infection. Chest radiograph (**a**) and computed tomographic scan of chest and abdomen (**b**) showing nodular infiltrates near the pleura. Extracted from: Lindberg, J, Hagen, F, Laursen, A, Stenderup, J, Boekhout, T. (2007). *Cryptococcus gattii* Risk for Tourists Visiting Vancouver Island, Canada. Emerging Infectious Disease 13(1): 178–179 DOI: 10.3201/eid1301.060945. Available from https://wwwnc.cdc.gov/eid/article/13/1/06-0945_article

fungal burden, high polysaccharide titers in the CSF (cerebrospinal fluid), elevated ICP, and slow CSF sterilization, even after antifungal treatment initiation. Consequently, these immunosuppressed individuals have a shorter onset of signs and symptoms that may extend for months, leading to death [46]. In contrast, immunocompetent individuals may present subacute or chronic complaints of headaches or altered mental status. Lower rates of fungemia (10%) and mortality (25%) and an intense inflammatory response are common in these hosts [47].

Elevated ICP (>25 cm water) is typical in meningitis caused by both *C. gattii* (50–60% of cases) and *C. neoformans* (30–75% of cases) [44]. This increased ICP is related to the accumulation of cryptococcal polysaccharides, mannitol and/or encapsulated fungal cells, which impairs CSF resorption [48]. The common signs and symptoms of altered ICP include severe headache, hearing loss, seizures, vomiting, altered mental consciousness and lethargy [48]. HIV patients may present with ophthalmologic complications derived from ICP due to underlying meningitis or optical nerve infiltration by yeast cells [49]. In both cases, visual loss occurs, which may be either rapid with few effective treatments during widespread disseminated disease or gradual and reversible with medical management to decrease ICP [9]. In addition, photophobia, diplopia and ocular motility deficits have been reported as common ophthalmologic complaints [49].

Ophthalmoplegia, papilledema, neuroretinitis with or without vitritis and optic atrophy are the ophthalmic complications associated with this infection, and most of these complications are caused by CNS infection [50]. ICP diagnosis is associated with a poor prognosis and neurological sequelae [51].

In general, *C. gattii* has a predisposition to infect the brain parenchyma rather than the meninges and to develop cryptococcomas, which must be considered in the differential diagnostic (Fig. 6.2). These large lesions in the brain may cause symp-

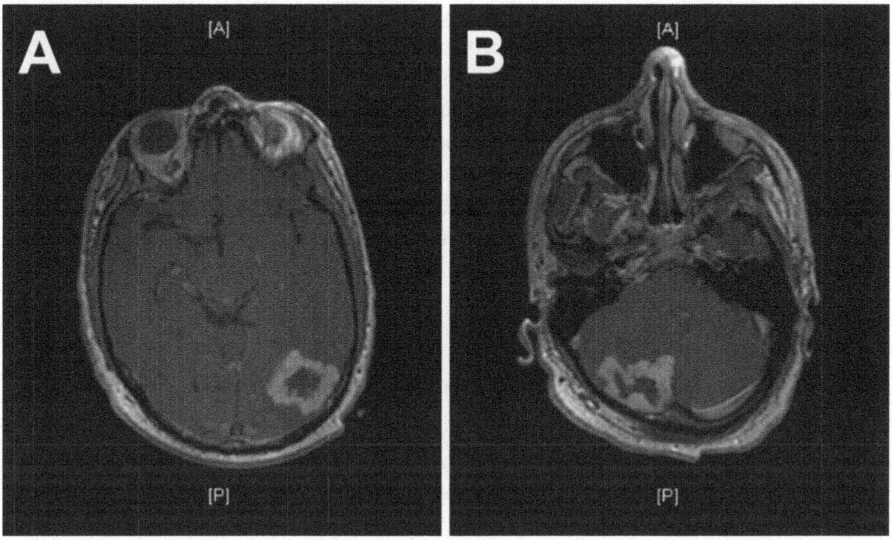

Fig. 6.2 *C. gattii* CNS infection. Magnetic resonance imaging of brain cryptococcomas from a patient with a *C. gattii* infection showing a large mass in left parietal lobe (**a**) and in the right cerebellar hemisphere (**b**). Extracted from: Byrnes EJ III, Li W, Lewit Y, Perfect JR, Carter DA, et al. (2009) First Reported Case of *Cryptococcus gattii* in the Southeastern USA: Implications for Travel Associated Acquisition of an Emerging Pathogen. PLoS One 4(6): e5851. doi:10.1371/journal.pone.0005851

toms that range from headaches to increased ICP or acute obstructive hydrocephalus [44]. Patients who present with this condition typically respond poorly to antifungal therapy, requiring neurosurgical intervention to control pressure or improve treatment efficacy [34].

6.2.3 Cutaneous Cryptococcosis

Cutaneous cryptococcal infection is the third most common clinical manifestation of cryptococcosis, accounting for approximately 10–15% of cases [52]. Skin lesions are found in 5% of patients with cryptococcal meningitis [53]. Local inoculation and dissemination from distant sites of infection represent two ways to establish the disease. Primary cutaneous cryptococcosis is rare and is associated with direct inoculation through a skin injury, which is the main portal of fungal entry and the site at which the lesion will develop. Most cases are observed in older individuals and rural workers who have no underlying disease [54]. A single skin lesion, such as a whitlow or phlegmon, is typically observed and occurs in a sun-exposed area. The absence of fungal dissemination is common in this type of infection. In most cases, the skin lesions are described as ulcers, cellulitis and abscesses that may be located at a site of trauma anywhere on the body [55]. Primary cutaneous infection is caused by *C. neoformans* serotype D [56], although *C. gattii* has been isolated in cutaneous infections, primarily in tropical and subtropical areas with eucalypt reservoirs [57]. Secondary cutaneous cryptococcosis occurs consecutive to hematogenous dissemination, and immunosuppressive conditions are considered to be risk factors. Skin disease presents with multiple skin lesions, which are distributed in clothed and unclothed body areas and frequently present as papules or macropapules with a central ulceration [58]. Other cutaneous lesions include purpura, vesicles, nodules, cellulitis, abscesses, ulcers, pyoderma gangrenosum-like lesions, molluscum-like lesions and herpes-like plaques, among others [59].

6.2.4 Nonmeningeal, Nonpulmonary Cryptococcosis

Cryptococcus species can cause disease in any organ, primarily in immunocompromised individuals, particularly in those with severely impaired cellular immunity, such as patients with AIDS, lymphoreticular malignancy or immunosuppression after corticosteroid therapy or organ transplantation.

Cryptococcal eye infection is rare and a consequence of disseminated disease. However, conjunctivitis, iritis, chorodititis, chorioretinitis, vitritis, endophthalmitis and intraocular cryptococcomas have been reported [60]. HIV patients are the patients most frequently affected by ophthalmologic complications, although some cases involving immunocompetent individuals have been reported [61].

Prostatic infection by *C. neoformans* is also rare and is associated with systemic cryptococcosis in immunocompromised individuals [62]. This infection is usually asymptomatic and may result from fungal dissemination from the CNS and/or the lungs or occur due to descending urinary spread [63]. Although primary prostatitis has been reported infrequently, a latent infection of the prostate may develop into disseminated cryptococcosis after urologic surgery on the prostate or other genitourinary manipulation, as reported previously [64]. Thus, the prostate is considered to be a reservoir of infection from which systemic relapse may occur [65]. Even while undergoing appropriate antifungal therapy, AIDS patients have positive cultures from urine and seminal fluid, suggesting that the fungus persists at this site [66]. Patients with prostatic cryptococcosis present high levels of prostate-specific antigen (PAS), which represents an important inflammatory process and may help track therapeutic responses [63]. In addition to the prostate, other organs that belong to the genitourinary tract, such as the kidney, epididymis and testis, have been involved in cryptococcal infections in individuals with severe immunosuppression [67]. Patients typically present with symptoms that include urgency, dysuria, suprapubic pain, hematuria, and bladder outlet obstruction [68].

Active cryptococcal infection in unusual body sites is rare, although some case reports have been described, such as cryptococcal peritonitis [69], arthritis [70], myositis [71], olecranon bursitis [72] and hepatitis [73]. Infections of the parathyroid [74] and parotid glands [75], infections of the biliary tract and adrenal involvement [76] are infrequent but have been observed. Bone infection is associated with osteolytic lesions in any bone of the body, and cryptococcal osteomyelitis has been associated with underlying sarcoidosis [77]. Rare cases of fungal colonization in the colon have been reported, although peritonitis has been described in patients with cirrhosis [69]. Cryptococcuria and cryptococcemia were reported in several case series [36]. In some conditions, even after initiating antifungal therapy, cryptococcal infection may persist in these uncommon sites, suggesting that these sites may be a source for future recurrence and dissemination [74].

6.2.5 Cryptococcal Immune Reconstitution Inflammatory Syndrome

Immune reconstitution inflammatory syndrome (IRIS) is considered to be a clinical deterioration that is associated with new or worsening clinical manifestations of cryptococcal disease caused by the recovery of *Cryptococcus*-specific immune responses. In this syndrome, a dysregulated reversal of the anti-inflammatory response into a strong proinflammatory immune response occurs [78]. Typically, IRIS affects HIV-infected patients with CNS infection and solid organ transplant (SOT) recipients [36]. However, this syndrome has been observed in healthy hosts, such as pregnant women [46]. In immunosuppressed patients, IRIS development is associated with the withdrawal or reduction of immunosuppression and/or the restoration of immune functions [79].

When observed upon the initiation of highly active antiretroviral therapy (HAART), cryptococcal IRIS can be defined as paradoxical or unmasking. Paradoxical IRIS occurs in up to one-third of patients with pre-HAART cryptococcosis and presents as a worsening or recurrence of previously treated cryptococcal disease in the same or new anatomical sites. In the unmasking form of IRIS, patients with or without a prior diagnosis may develop acute symptoms of cryptococcosis, such as meningitis or necrotizing lymphadenopathy [80].

Epidemiologic data demonstrated that 5–11% of SOT recipients with cryptococcosis develop cryptococcal IRIS 4–6 weeks after the initiation of antifungal therapy [81]. For HIV patients, the incidence of IRIS ranges from 8% to 50% after HAART initiation [82].

IRIS is associated with significant morbidity and mortality and is responsible for diverse sequelae, including neurological deficits, which may be associated with inflammatory mass lesions, hydrocephalus, meningeal enhancement and increased CSF inflammation [44]. Patients are rarely asymptomatic, and most IRIS cases present with CNS involvement and clinical features that are similar to an active cryptococcal infection, although pulmonary manifestations are uncommon [9]. The signs and symptoms in these cases may range from headaches and drowsiness to new or recurrent neurological manifestations, such as ataxia, deafness, blindness, and seizures [83].

6.3 Therapy and Management of Cryptococcosis

The gold standard treatment for cryptococcosis that is recommended by the World Health Organization (WHO) and the Infectious Diseases Society of America (IDSA) guidelines includes three off-patent drugs: amphotericin B (AmB), flucytosine, and fluconazole [84].

AmB belongs to the polyenes class and binds ergosterol, which is present in fungal cell membranes. AmB can form pores in the plasma membrane, disrupting cell integrity, and is also known to induce oxidative stress in the cytoplasmic environment [85]. AmB is the fastest drug for *Cryptococcus* spp. clearance based on different clinical trials [86]. Because of its low bioavailability, AmB cannot be administrated orally using the existing formulations and thus requires hospital admission for intravenous administration [1]. Furthermore, prolonged treatment with AmB is inadvisable due to substantial side effects, such as anemia, nephrotoxicity and local infusion reactions. Thus, AmB usage is limited in regions without a satisfactory medical infrastructure [87].

AmB displays a very low solubility in water and is therefore administrated as a deoxycholate salt (AmBd) [88]. Newer formulations of AmB, such as liposomal AmB, an AmB-lipid complex and an AmB colloidal dispersion, have fewer toxic side effects and are better tolerated, allowing the administration of higher doses [3]. However, the high cost of lipid AmB is a barrier to the implementation of this drug, as its use is economically unfeasible in low- and middle-income countries [89].

Flucytosine or 5-fluorocytosine (5-FC) is a pyrimidine analogue that is converted into 5-fluorouracil (5-FU) inside fungal cells by the enzyme cytosine deaminase [90], which is not present in mammalian cells [91]. The accumulation of 5-FU inside fungal cells inhibits RNA and DNA synthesis, causing an antifungal effect, although the mechanism of action is not fully elucidated [92]. Mammals are unable to produce 5-FU from 5-FC; however, bacteria in the human microflora can perform this conversion, causing some side effects, such as hepatotoxicity, hematological toxicity and gastrointestinal toxicity [93].

Fluconazole inhibits the final step of ergosterol synthesis, leading to the accumulation of toxic steroidal components in cellular membranes [94]. This drug is largely used in the initial phase of treatment and is also used during the maintenance phase to avoid relapses. Fluconazole has excellent bioavailability and easily reaches the central nervous system (CNS) [95]. However, the development of resistance to fluconazole is a critical problem in cryptococcosis therapy, with resistance reported in more than 60% of culture-positive individuals in South Africa [51].

Despite the guideline recommendations and the inclusion of these three drugs in the 19th WHO Model List of Essential Medicines [81], AmBd and 5-FC are still unavailable in most low- and middle-income countries, where cryptococcosis is a leading cause of death among HIV-infected patients, with an alarming mortality rate of 70% for cryptococcosis patients in sub-Saharan Africa [4].

This section will discuss cryptococcosis management and therapy in three groups of patients: HIV-infected individuals, solid organ transplant (SOT) patients and non-HIV/non-SOT individuals. A summary of the first line of therapy proposed by the IDSA for these three groups is presented in Table 6.1.

Table 6.1 First-line treatment for cryptococcal meningitis according to the IDSA guidelines

	HIV-positive patients	SOT recipients	Non-HIV/Non-SOT patients
Induction phase	At least 2 weeks: AmBd (0.7–1.0 mg/kg per day iv) 5-FC (100 mg/kg per day orally in 4 doses)	At least 2 weeks: ABLC (5 mg/kg per day orally in 4 doses)	At least 4 weeks: AmBd (0.7–1.0 mg/kg per day iv) 5-FC (100 mg/kg per day orally in 4 doses)
Consolidation phase	At least 8 weeks: Fluconazole 400 mg (6 mg/kg per day orally)	At least 8 weeks: Fluconazole 400 mg (6–12 mg/kg per day orally)	At least 8 weeks: Fluconazole 400 mg (6–12 mg/kg per day orally)
CSF sterilization			
Maintenance or suppressive phase	At least 12 months: Fluconazole 200 mg (3 mg/kg per day orally)	6–12 months: Fluconazole 200 mg (3–6 mg/kg per day orally)	6–12 months: Fluconazole 200 mg (3–6 mg/kg per day orally)

The treatment is divided into the induction, consolidation and maintenance phases. Complete fungal clearance in the CSF should be reached at the end of the consolidation phase in order to continue to the maintenance phase. Otherwise, the patients should repeat the induction phase. The three groups of patients affected by cryptococcal meningitis (HIV-positive, SOT, and non-HIV/non-SOT) are treated differently

6.3.1 *Cryptococcosis Treatment in the HIV-Infected Population*

HIV infection is considered to be the leading predisposing condition to the development of cryptococcal meningitis [96], with 80–90% of cases in HIV-positive patients [97]. Therefore, the extensive use of highly active anti-retroviral therapy (HAART) starting from the mid-1990s resulted in a remarkable decrease in the cryptococcal meningitis incidence in developed countries [98]. However, in low- and middle-income countries, access to HAART is limited and is initiated during late phases of HIV infection [89]. In fact, the failure to continue monitoring patients after hospitalization is the first barrier to positive outcomes in HIV-associated cryptococcal meningitis treatment in South Africa [99]. Indeed, according to IDSA guidelines [100], patients with disseminated cryptococcosis or cryptococcal meningoencephalitis should be immediately tested for HIV infection.

The first line of therapy recommended by the IDSA guidelines includes an induction phase that is characterized by at least 2 weeks of treatment with intravenous AmBd (0.7–1.0 mg/kg per day) plus 5-FC (100 mg/kg per day orally). This strategy is associated with rapid *Cryptococcus* clearance and is highly effective based on data from randomized clinical trials [86]. Ideally, sterilization of the CSF should be achieved after 14 days of primary therapy, which is associated with a positive outcome [101]. In places where 5-FC is not registered or not available at the supplier level, such as several countries in Africa and Asia [89], alternative regimens should be considered as primary therapies [102]. For instance, AmBd (1 mg/kg per day, intravenously) plus fluconazole (800 mg/day, orally) for at least 2 weeks is the standard primary therapy recommended by the Southern African HIV Clinicians Society guidelines [103]. However, the absence of 5-FC in induction therapy is associated with treatment failure in 3 months [104]. In situations where AmBd is not available or cannot be given safely, fluconazole monotherapy at higher doses (1200 mg/day orally) is largely used, although this regimen is associated with poor outcomes, such as higher mortality rates, a prolonged CSF sterilization time and a risk of relapse [105]. Bicanc and colleagues observed 32 relapse episodes among 27 patients who received fluconazole monotherapy during the induction phase [106].

After 2 or more weeks of induction therapy, the consolidation phase is established, with at least 8 weeks of fluconazole 400 mg (6 mg/Kg) per day orally. In a clinical trial with 306 patients, 72% of patients presented negative CSF cultures after fluconazole treatment, while 60% of patients had negative CSF cultures after itraconazole treatment, supporting the preferential use of fluconazole during the consolidation phase [107].

After fungal clearance, the maintenance or suppressive phase starts. During this phase, patients should take 200 mg fluconazole per day, orally, for a period of 12 months or more. A clinical trial demonstrated that fluconazole maintenance therapy reduced the chance of relapse to zero, while the placebo group displayed a relapse rate of 15% in patients who completed successful induction and consolidation therapies [108]. As an alternative maintenance therapy, the use of itraconazole (400 mg/day orally) or AmBd (1 mg/kg per week intravenously) is suggested for azole-intolerant individuals or patients with fluconazole therapy failure [100]. Both alternative drugs require patients to be monitored due to toxicity-related side effects. The results of two different trials highlighted fluconazole as the most effective drug in the maintenance phase. Patients undergoing fluconazole treatment showed significantly lower relapse rates than patients undergoing itraconazole or AmBd regimens [109]. In the HIV coinfected population, the suppression phase should be maintained for at least 1 year and should be discontinued only after the CD4 cell count is restored and higher than 100 cells/μL and the HIV RNA serum level is undetectable for at least 3 consecutive months [110]. If the CD4 cell count decreases to less than 100 cells/μL or if cryptococcal antigenemia increases, the guidelines recommend the restitution of maintenance therapy [111]. In the case of asymptomatic antigenemia, lumbar puncture and blood cultures are highly recommended [100]. If the culture results are positive, the patients should repeat the induction phase [100].

To date, the most effective strategy for preventing and reducing cryptococcosis mortality and morbidity remains early HIV diagnosis and adhesion to HAART, especially before the CD4 count falls below 350 cells/μL [84]. On the other hand, in the presence of a high fungal burden, HAART initiation can be harmful due to the risk of immune reconstitution inflammatory syndrome (IRIS) development [112]. In a randomized clinical trial, which was designed to address the optimal HAART timing, a cohort was initiated 72 h after antifungal therapy, and this strategy resulted in a high mortality rate [113]. The restoration of the immune status upon HAART initiation may induce an exacerbated inflammatory response, which contributes to tissue damage and increases the severity of cryptococcosis, particularly in the CNS

[114]. However, later HAART initiation and prolonged immunosuppression may pose risks of developing complications associated with HIV infection and the worsening of cryptococcal disease [115].

The optimal starting time for HAART remains to be elucidated, and new trials with larger cohorts are necessary [116]. The Southern Africa HIV Clinicians Society guideline currently recommends a HAART initiation window between 4 and 6 weeks [103]. Management of major IRIS complications has not yet been studied extensively in clinical trials. However, the IDSA guideline considers the use of anti-inflammatory drugs, such as corticosteroids, in patients with CNS inflammation and increased intracranial pressure. Corticosteroid therapy should be monitored carefully and administered concomitant with antifungal therapy [117].

6.3.2 Cryptococcosis in Solid Organ Transplant (SOT) Recipients

Opportunistic *Cryptococcus* infection affects approximately 2.8% of the SOT population [118]. The mortality rate reaches 15% 90 days after cryptococcosis diagnosis and 27% 12 months after cryptococcosis diagnosis [119].

As in other patient groups, cryptococcosis in SOT individuals manifests primarily as meningoencephalitis or pulmonary disease. Between 48 and 89% of the *Cryptococcus*-infected SOT population has CNS involvement [120], and between 25 and 64% of SOT recipients have pulmonary disease, with the disease limited to the lungs in 6–33% of these patients [121]. It is noteworthy that in patients with disseminated disease, several sites can be infected by *Cryptococcus* simultaneously, such as the soft tissues, joints, bones, eyes, and the thyroid and adrenal glands [74].

After a SOT procedure, the use of immunosuppressive drugs is necessary to avoid organ rejection. However, this therapy increases not only the risk of acquiring cryptococcosis but also the severity of this disease [122]. Immunosuppressive regimens include calcineurin inhibitors (such as tacrolimus or cyclosporine) [123], mycophenolate mofetil, which is an inhibitor of lymphocyte de novo purine synthesis [124], and corticosteroids [118]. Cryptococcosis management in the SOT population includes the gradual reduction of immunosuppression prior to the initiation of antifungal therapy [100]. The abrupt removal of immunosuppressive therapy may result in a higher risk of organ rejection or IRIS development [125].

Approximately 5% of SOT recipients with cryptococcosis present renal dysfunction (creatinine level > 2.0 mg/dL) [100]. The use of calcineurin inhibitors and AmBd is known to be highly nephrotoxic [118]. Therefore, AmBd is not recommended as a first-line therapy in SOT patients [100]. Liposomal AmB formulations are used preferentially in these patients according to the IDSA guideline [100] because these formulations are associated with a lower incidence of nephrotoxicity [126].

The induction therapy established by the IDSA for SOT patients consists of liposomal AmB (3–4 mg/kg per day) or an AmB lipid complex (ABLC) (5 mg/kg per day) plus 5-FC (100 mg/kg per day divided into 4 doses) for at least 2 weeks. For the consolidation phase, the guidelines recommend 6–12 mg/kg per day of oral fluconazole for 8 weeks. Finally, the maintenance phase consists of 6–12 months of orally administered fluconazole at a dose of 3–6 mg/kg per day. This regimen is recommended for CNS disease and severe pulmonary or disseminated non-CNS cryptococcosis [100]. For patients with mild to moderate symptoms with no CNS involvement, no evidence of disseminated infection and nondiffused lung infiltrates, the guidelines recommend 6 mg/kg per day of fluconazole for 6–12 months [127]. Patients who received maintenance therapy for 6 months or more had a relapse rate of 1.3% [128]. In the absence of AmB lipid formulations, AmB deoxycholate should be used with extra caution. The guidelines suggest a maximum dosage of 0.7 mg/kg per day of AmBd, with frequent renal function monitoring. In areas with no access to 5-FC, induction therapy guidelines may include liposomal formulations of AmB (6 mg/kg per day) for 4–6 weeks. The absence of 5-FC during this phase is considered to be a risk factor for poor treatment outcomes [104]; however, this drug remains unavailable in most countries in Africa, Asia, and Latin America [89].

The mean CSF sterilization time is 16 days after treatment initiation [121]. After this period, another lumbar puncture is recommended, and if the CSF presents a positive fungal culture, the induction therapy must be repeated. Therapeutic lumbar punctures are recommended to relieve intracranial pressure during cryptococcal meningitis, and this management strategy is associated with decreased cryptococcal meningitis mortality in HIV-infected patients [129].

The gradual reduction of immunosuppressive therapy is highly recommended to restore cellular immunity against *Cryptococcus* in SOT patients [79]. However, the abrupt reduction of immunosuppressive therapy may induce a potent shift in the immune status to a Th1 proinflammatory response, which is related to IRIS development in SOT patients [130]. This strong proinflammatory response may result in tissue lesions in the CNS and the lungs, worsening the clinical condition of the patient [125]. Therefore, IRIS can be misdiagnosed as therapy failure [131]. This syndrome has a prevalence of 4.8% in SOT-related cryptococcosis patients and is associated with a high risk of organ rejection [125].

6.3.3 Cryptococcosis in Non-HIV and Non-SOT Patients

Concerning the treatment of cryptococcosis in virtually immunocompetent hosts, there is no consensus among specialists that is clearly expressed in the IDSA guidelines, primarily because the major controlled clinical trials performed in the non-HIV non-SOT population included individuals with an immunosuppressed status, such as cancer patients and patients undergoing steroidal drug treatment [132]. Patients with systemic lupus erythematosus are at risk of developing cryptococcosis, which reinforces the importance of the host immune status [133]. In a Taiwan study, *C. neoformans* was responsible for 6 deaths among 15 patients with lupus erythematosus [134].

The first line of induction therapy that is recommended for cryptococcal meningitis in this group of patients is AmBd (0.7–1.0 mg/kg per day, intravenously) plus 5-FC (100 mg/kg per day, orally in 4 doses) for 4–6 weeks. The subsequent consolidation phase is characterized by 8 weeks of treatment with fluconazole 400 mg/day, orally. Maintenance therapy is continued for 6–12 months, with fluconazole 200 mg/day, orally. For pulmonary cryptococcosis in non-immunosuppressed patients, the treatment is fluconazole 400 mg per day, orally, for 6–12 months [100]. Severe pulmonary cryptococcosis should be treated as CNS disease [135]. Immunocompetent hosts are particularly susceptible to infections caused by *C. gattii* species [136]. These infections tend to form more cryptococcomas in the lungs and in the CNS than *C. neoformans* infections; therefore, the therapeutic response tends to be slower [137]. The presence of cerebral cryptococcomas may require surgical removal if the lesions are larger than 3 cm [138]. Surgery is also recommended if the cryptococcoma is compressing the optic nerve to avoid the development of blindness [139]. The management and therapy principles used for *C. gattii* infection are the same as those used for *C. neoformans* infection [44].

6.4 Laboratory Diagnosis: Classic and New Diagnosis Trends

The definitive laboratory diagnosis of cryptococcosis is based on the direct observation of the yeast through India ink staining of bodily fluids, the direct examination and culture of *Cryptococcus* from clinical specimens and immunoassays for cryptococcal antigens [9]. Other methods are also used for the diagnosis of this mycosis, including histopathology of infected tissues and molecular diagnosis methods, which are applied extensively for research but not widely used in the clinic [140].

6.4.1 Direct Examination

Direct microscopic observation is the most rapid and low-cost method used for the diagnosis of cryptococcosis, with the detection of encapsulated yeasts in a variety of samples, such as cerebrospinal fluid (CSF), sputum, bronchial lavage, pus from mucosal cutaneous lesions, urine, and macerated tissues obtained through biopsy. In direct examinations, *C. neoformans* is characterized as globular or oval-shaped cells, which may present a single bud, surrounded by a polysaccharide capsule that ranges from 5 to 20 μm in diameter. The visualization of the capsule is evident when the samples, primarily CSF and urine, are stained with India ink. India ink is a negative stain that creates a dark background, with the yeast appearing to be surrounded

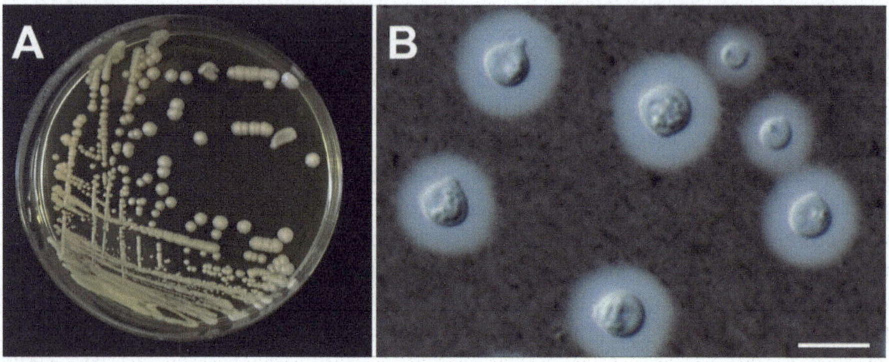

Fig. 6.3 *Cryptococcus* yeasts in culture. (**a**) Isolated colonies in agar Sabouraud. (**b**) India ink negative staining showing the capsule as an exclusion zone surrounding the cell. *Scale bar*: 10 μm

by a clear halo generated by the polysaccharide capsule, which repels the India ink staining (Fig. 6.3). Sputum and other purulent samples can be treated with a 10% potassium hydroxide solution. The purpose of this sample preparation is to disrupt cellular proteinaceous debris, cleaning the specimen for easier fungal detection. The sensitivity of microscopic observation in samples from AIDS patients reaches 80% of positive cases of cryptococcal meningitis, but in immunocompetent patients, this percentage drops to 50% [141].

Histological preparations of tissues, such as lung, skin, bone marrow, brain, and other organs, can reveal the presence of *Cryptococcus* by using special stains that label the polysaccharide capsule, such as mucicarmine, periodic acid-Schiff, and Alcian blue stains. Another reagent, the Fontana–Masson staining agent, identifies melanin deposited in the yeast cell wall. Other staining procedures are able demonstrate the presence of fungal chitin (Calcofluor White) and of the fungal cell wall (Grocott's methenamine silver) in clinical specimens [142].

6.4.2 Culture

Cryptococcus species are easily isolated and maintained in culture on Sabouraud agar, with or without antibiotics. The production of melanin is a feature that is widely used for the identification of *C. neoformans* and *C. gattii*. These yeasts produce the enzyme phenoloxidase, which oxidizes phenolic and diphenolic compounds found in some media, such as Sunflower Seed agar (*Helianthus annus*), Niger seed agar (*Phalaris canariensis*), Bird Seed agar (*Guizotia abyssinica*) and L-3,4-dihydroxyphenylalanine (L-Dopa), which is a chemically defined medium. Under these conditions, the colonies grow with a dark brown color [143]. To differentiate *C. neoformans* from *C. gattii*, L-canavanine glycine bromothymol blue agar (CGB) can be used because *C. gattii* grows in the presence of canavanine and uses glycine as a source of carbon and nitrogen; thus, the pH of the medium increases and the color of the indicator, bromothymol blue, changes to a vivid cobalt blue color after approximately 48 h of incubation at room temperature [144]. In contrast, *C. neoformans* is sensitive to the presence of canavanine and does not use glycine as a sole source of carbon, leaving the medium's pH unaltered and its coloration unchanged [144].

6.4.3 Immunological Methods

Immunological tests that are sensitive to the cryptococcal polysaccharide capsular antigen (CrAg), which is shed during infection, are used to aid in the diagnosis of invasive cryptococcosis. Latex agglutination and immunoenzymatic techniques using serum, CSF, and urine are widely available.

The detection of the capsular antigen using the latex agglutination test has proved to be an important tool in the diagnosis of cryptococcosis, as this test is more sensitive than microscopy [145] and faster than the isolation of the fungus in culture. In addition to being highly specific, this test can be performed as a qualitative and semi-quantitative assay [145].

Although rare, false-positive results can be obtained in the latex agglutination test due to cross-reactions with polysaccharides from other microorganisms. False-negative results are also rare but may occur due to a low antigen concentration, the presence of immunocomplexes, a prozone effect, or thinly encapsulated strains [141].

The lateral flow immunochromatographic assay for the detection of cryptococcal antigen (CrAG LFA) was recently developed for the diagnosis of cryptococcal meningitis as a true point-of-care test, which is accessible in any setting without the need for a laboratory structure (Fig. 6.4). The CrAg LFA uses a combination of two monoclonal antibodies against the most abundant capsule polysaccharide, which is termed glucuronoxylomannan. The first monoclonal antibody reacts with the polysaccharide of serotypes A, B, and C; the second monoclonal antibody is highly reactive with the polysaccharide of serotypes A and D. Combined, the monoclonal antibodies are highly reactive with polysaccharides from *C. neoformans* and *C. gattii.* This test is more accurate than the CrAg-latex test or the enzyme immunoassay (EIA) and has advantages with respect to the detection range. The test can be performed for screening and preemptive treatment of subclinical cryptococcosis and to monitor serum/CSF polysaccharide titers for treatment control. This test is less time consuming than other tests ($\leq$10 min), requires little or no equipment, can be performed without electricity, is stable at room temperature and is relatively inexpensive [142].

6.4.4 Molecular Methods

Molecular methods, although not applied in routine diagnosis, are widely applied in research for several purposes; for example, these methods have high sensitivity and specificity and are important alternative tools for overcoming the limitations of conventional diagnostic tests. These methods can be employed for identification and

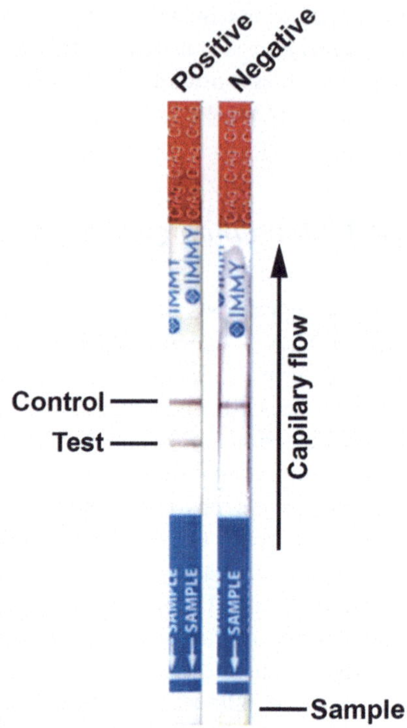

Fig. 6.4 Positive and negative strips from the lateral flow immunochromatographic assay for the detection of cryptococcal antigen (CrAG LFA)

typing and have been applied primarily for molecular epidemiology studies. Although numerous molecular techniques have been applied for typing *C. neoformans* and *C. gattii* strains, only three methods were proved to produce comparable results: PCR fingerprinting, AFLP, and MLST [25]. At present, eight major molecular types are widely recognized: VNI, VNII, VNIII, and VNIV among *C. neoformans* isolates and VGI VGII, VGIII, and VGIV among *C. gattii* isolates [25].

6.5 Polymerase Chain Reaction

The polymerase chain reaction (PCR) is widely used in research laboratories and has high sensitivity and specificity in detecting minimal amounts of DNA from species of the *Cryptococcus* genus from clinical and environmental samples. Nested PCR, multiplex PCR, and real-time PCR are the most common variant PCR methodologies used to identify *C. neoformans* and *C. gattii*. Several target sequences are available, including URA5 (the orotidine monophosphate pyrophosphorylase gene), CAP59 (the capsule synthesis gene), M13 (minisatellite-specific core sequence), and ITS (internal transcribed spacer regions 18S, 5.8S, and 28S) [146].

6.6 Polymerase Chain Reaction: Restriction Fragment Length Polymorphism

Polymerase chain reaction—restriction fragment length polymorphism (PCR–RFLP) is similar to restriction fragment length polymorphism (RFLP), but in this case, the probe hybridization step is not needed because the genomic DNA fragments are previously subjected to PCR with specific primers and then cleaved with restriction enzymes [147]. PCR-RFLP that targets the URA5 gene during the amplification reaction has been employed in molecular epidemiology studies to determine the potential relationships among molecular types in clinical and environmental isolates of *Cryptococcus*. This approach is also recommended when it is necessary to obtain more information about a specific strain [148].

6.7 Amplified Fragment Length Polymorphism

Amplified fragment length polymorphism (AFLP) is another alternative for typing *C. neoformans* complex individuals. AFLP is very useful in detecting genetic variability, since this technique reveals the presence or absence of polymorphisms at restriction sites. The AFLP genotyping technique is both more sensitive and more expensive than PCR fingerprinting and is also performed in several steps. First, the genomic DNA is cleaved with restriction enzymes, followed by the ligation of specific adapters to the sticky ends of the cleaved genomic fragments. The amplification of the generated segments is carried out via PCR with specific primers that recognize sequences in the adapter, and the separation of the subset of amplified fragments is achieved through high-resolution gel electrophoresis. This process enables the grouping of the *C. neoformans* species complex into six molecular types: AFLP1 to AFLP6 [149].

6.8 Multi-locus Sequencing Typing

The multi-locus sequencing typing (MLST) technique is highly reproducible and trustworthy. MLST is based on variations in the nucleotide sequences of multiple housekeeping genes, and the following set of genetic loci that encode capsule, urease, phospholipase, and laccase production was proposed as a consensus international standard for multi-locus sequence typing for *C. neoformans* and *C. gattii*: CAP59, GPD1, LAC1, PLB1, SOD1, URA5, and IGS1 [150]. MLST is of great value for determining the circulating *Cryptococcus* complex genotypes in different parts of the world [29]. After amplification and sequencing of the desired regions, MLST analysis is fully automated. A work-group created and supported by the International Society of Human and Animal Mycoses (ISHAM) proposed MLST as a consensus methodology for genotyping *C. neoformans* and *C. gattii.* This recommendation was supported by all major research groups that are involved in molecular strain typing of the *C. neoformans* species complex [150].

6.9 *Cryptococcus* Virulence Factors

6.9.1 Survival at High Temperatures

The ability to survive at high temperatures is a key characteristic of fungi that are able to become invasive in mammalian hosts due to the mammalian body temperature [151]. Only two species of *Cryptococcus* can cause invasive systemic disease in mammalian hosts and exhibit optimal growth at temperatures higher than 30 °C, despite the presence of classical virulence factors, such as melanin and capsule production, in other members of this clade [152].

Mammalian hosts with increased body temperature, such as rabbits, are naturally resistant to experimental cryptococcosis [153]. However, upon immunosuppression with corticosteroidal drugs, *Cryptococcus* is able to colonize the rabbit CNS [153]. Moreover, an increase in the environmental temperature prolonged the lifespan of infected mice [154]. These findings eventually led to early attempts to treat cryptococcosis patients using hyperthermia; however, such treatments had no success [155].

Cryptococcal thermo-tolerance has been studied extensively, and the application of several molecular techniques revealed that a huge shift in gene expression occurs during yeast exposure to mammalian body temperatures. The temperature-regulated genes involve a long list of processes, including signaling pathways (*RAS1*, *CNA1*, *CNB1*, *MPK1,* and *CTS1*) [156], energetic metabolism (*VPH1*) [157], structural components of the polysaccharide capsule (*UGD1*) [158] and even amino acid metabolism (*ILV2* and *SPE3*) [159] and mitochondrial detoxification functions, such as the manganese-superoxide dismutase (*SOD2*) [160]. Conserved processes, such as sugar metabolism and ribonucleotide synthesis, are also involved in thermo-tolerance, as evidenced by the avirulent temperature-sensitive phenotype of mutants in the trehalose synthesis machinery [161] and de novo pyrimidine ribonucleotide formation [162].

6.9.2 Melanin

Melanin is a widespread pigment that is expressed in a broad range of fungi [163] and was first described in the *Cryptococcus* genus by Staib in 1962 (reviewed in [164]). Two metabolic pathways can lead to melanin synthesis in fungi: the 1,8-dihydroxynaphthalene (DHN) pathway and the L-3,4-dihydroxyphenylalanine (L-DOPA) substrate pathway, which shows closer similarity to the mammalian pathway [165]. The latter pathway is present in *C. neoformans* [166]. Melanin produced is deposited in the cell wall and can be observed commonly in a culture medium containing with D-DOPA (Fig. 6.5).

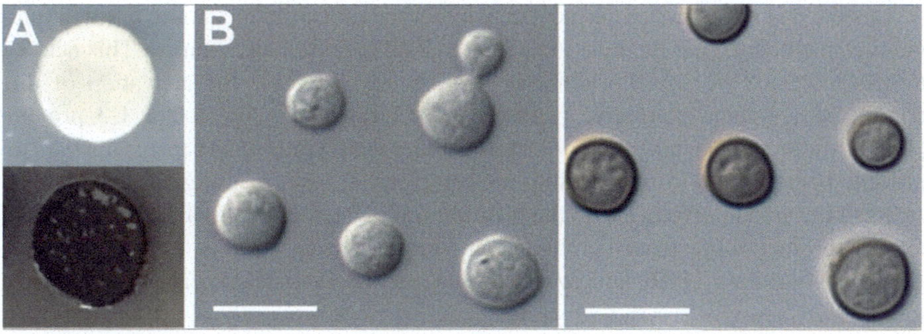

Fig. 6.5 Melanin production. *Cryptococcus* grown on medium containing 0.1% L-DOPA to demonstrate melanization (*brown pigment*). Colony morphology (**a**) and optical microscopy (**b**) in presence and absence of L-DOPA. *Scale bar*: 10 μm

Throughout this process, the enzyme laccase converts L-DOPA molecules into dopaquinone, which is subjected to the addition of amino groups to form cyclodopa. Oxidized cyclodopa is converted to dopachrome, which undergoes tautomerization to generate dihydroxyindoles that ultimately polymerize spontaneously into melanin, which is a negatively charged pigment [166]. The cryptococcal enzyme laccase is encoded by two genes: *LAC1* and *LAC2* [167]. The *LAC1* product is the main component responsible for melanin production. The association of this gene with virulence in *Cryptococcus* was observed using deletion mutants, and those studies showed that deletion mutants for this gene are less virulent [167].

The *LAC1* laccase is targeted to the cell wall in a Sec6p-dependent manner [168]. The *LAC2* laccase was demonstrated to be localized primarily throughout the cytoplasm [169]. It is possible that this enzyme reaches the cell wall and the supernatant inside the lumen of secreted vesicles, as Sec6 molecules are also related to extracellular vesicle secretion according to data that demonstrated that laccase and its activity are present in purified extracellular vesicles produced by *C. neoformans* [170].

During infection, melanin appears to confer resistance against host defenses [171]. Several studies demonstrated protection against phagocytosis in melanized fungi [172]. It is possible that the increased negative charge generated by melanin deposition is one of the factors responsible for this antiphagocytic effect [173].

Inside macrophages, the melanization process increases the ability of *Cryptococcus* to resist oxidative stress produced by host cells, such as oxygen- and nitrogen-derived oxidants [174]. This effect is achieved through melanin's natural ability to combat free radicals by transferring its electrons, as shown by electron spin resonance experiments [173].

Another characteristic of fungal melanin is to interfere with the host immune response [172]. Melanized *Cryptococcus* can reduce TNF-α cytokine secretion by alveolar macrophages and decrease lymphocyte expansion [175]. Melanization also increases the dispersion of this fungus, as it becomes more likely to survive in the spleen, liver, and brain of infected animals [171]. The fungus causes severe tissue damage in the murine central nervous system, with decreased cytokine responses in comparison with nonmelanized strains [175].

Despite its importance in cryptococcal infection, upon immunization, purified fungal melanin can trigger the host innate and adaptive immune responses. This activation is indicated by specific antibody production [176] and the activation of the complement system alternative pathway, with ultimate binding to the C3 protein [177]. However, the localization of melanin under the cryptococcal capsule appears to prevent this molecule from activating the complement system during infection [177].

In addition to interfering with the host immune system, melanin is able to sequester and protect fungi from synthetic anti-fungal agents, such as amphotericin B [178], caspofungin [179] and azoles [180], as observed in experimental murine infection assays.

6.9.3 Secreted Enzymes

Although *Cryptococcus* secretes several classes of enzymes, two of these classes have been directly implicated in fungal virulence: phospholipase B (PLB) [181] and urease [182].

Phospholipases are a heterogeneous class of enzymes [183]. Specifically, fungal PLB is expressed by several fungi (including *Cryptococcus*) and presents three distinct activities: hydrolase activity (the ability to remove *sn*-1 or *sn*-2 fatty acids from a glycerophospholipid), lysophospholipase activity (LPL), and transacylase activity (fatty acid addition) [184]. Genetic studies in *C. neoformans* demonstrated that a single gene (*PLB1*) is responsible for phospholipase activity in this yeast [184].

Known as an "invasin," PLB is important for cryptococcal tissue dissemination during traversal of the alveolar membrane, probably because the primary surfactant present in this environment, namely dipalmitoyl phosphatidylcholine, is a substrate for cryptococcal PLB [185]. This finding is in accordance with early observations by Chen [186] that the amount of PLB secreted is directly related to the fungal burden in a murine model of infection.

Another role for PLB is related to brain colonization during cryptococcosis. Cox [184] demonstrated that a *C. neoformans PLB1* knockout was less able to cross the blood-brain barrier (BBB), and an in vitro model of the BBB was less permeable to the *plb1*Δ strain than to the wild-type strain [187].

Although the absence of PLB does not appear to influence *Cryptococcus* uptake by macrophages, this mutation appears to delay fungal replication inside host phagocytic cells [184]. Once phagocytosed, the fungi inside phagolysosomes come into contact with lysosomal enzymes that are capable of inhibiting fungal growth, such as cathepsins B and L, as demonstrated in vitro [188]. It has been proposed that PLB activity against the phagolysosome membrane increases the permeability of the membrane, reducing the concentrations of these enzymes inside this compartment [184]. This finding explains the observed delay in fungal replication in the absence of PLB and the observation that cathepsin L activity decreases progressively inside the phagolysosome during the phagocytosis of live but not heat-killed *C. neoformans* cells [189].

Another characteristic of PLB is the capacity to generate bioactive eicosanoids using host arachidonic acid [190] to interfere with cellular levels of these molecules. This imbalance in eicosanoid homeostasis can shift the immune response toward a TH2 pattern, which is associated with fungal proliferation and tissue dissemination, as observed in mice infected with the wild-type strain in comparison to mice infected with *plb1*Δ, which exhibited an intense TH1 response and fungal clearance [190].

Another important enzyme for cryptococcal virulence is urease, which is a secreted nickel-dependent metalloenzyme that drives the hydrolysis of urea into ammonia and carbamate. The urease produced by *Cryptococcus* is encoded by a single gene named *URE1* [191]; however, similar to bacteria, accessory proteins are required for urease activity, namely Ure4, Ure6 and Ure7 (homologs of the UreD, UreF, and UreG proteins in bacteria, respectively) [182] plus a nickel/cobalt transporter protein named Nic1. A major difference between cryptococcal and bacterial urease activity is that the former lacks the nickel chaperone UreEp, which is present in bacteria [192]. In *Cryptococcus,* Ure7p contains histidine-rich domains that are capable of nickel binding and may overcome the absence of a UreE homolog [182].

Long studied as a virulence factor in some bacteria [193] and produced by many pathogenic fungi [194], urease activity and its role in cryptococcal virulence have only begun to be clarified.

Although this enzyme does not appear to be essential for the early stages of cryptococcal infection, published data showed a strong induction of IL-4 and IL-13 cytokines, high IgE antibody production and eosinophil infiltration into the lungs of mice infected with the wild-type strain, in contrast to mice infected with the *ure1*Δ strain [195]. Moreover, the same study reported a larger number of immature dendritic cells in the lung-associated lymph nodes of wild-type-infected animals in comparison to those infected with the urease-deficient mutant [195]. Taken together, these results point to a TH2 activation pattern that is implicated in persistent infection by *Cryptococcus* to the detriment of the host. This pattern is absent in the *URE1* knockout [195].

Although urease does not appear to be essential for cryptococcal survival in the brain [196], animals infected with the *ure1*Δ strain showed no apparent neurological damage and a decreased CFU load in organs with small capillary circulation, suggesting that the role of urease in brain colonization could be related to the crossing of the BBB [196]. To elucidate this issue, Shi and collaborators [197] used intravital microscopy to observe the initial steps of brain colonization *in vivo*. Using this approach, those authors demonstrated that *Cryptococcus* is rapidly trapped in brain capillary beds in a microembolic event [197], probably due to its size rather than a receptor-associated mechanism. The lack of urease activity does not influence this process; however, urease activity is essential to the transmigration of the BBB, as it increases transmigration sites through a mechanism that remains unknown [197].

It has also been proposed that the importance of urease in brain cryptococcosis could be related to tissue damage in the BBB caused by ammonia production and/or pH increases caused by the activity of this enzyme. In vitro tracking of a tight junction marker (ZO-1) in HBMEC cells exposed to wild-type and *ure1*Δ fungi showed that the deficient fungi are less efficient in loosening these junctions [182].

6.9.4 Polysaccharide Capsule

Since the early isolation of this fungus, the existence of an extracellular capsule surrounding cryptococcal cells was noted (Fig. 6.6) [198]. It was years before Benham [199] proposed that this structure could play an important role in pathogenesis.

From a poorly structurally and functionally described feature, the cryptococcal capsule became the most important and well-studied virulence factor related to this genus. Nevertheless, many aspects of the synthesis, structure and functions of the cryptococcal capsule remain to be elucidated. The first attempts to investigate the polysaccharide capsule were based on paper chromatography and identified traces of mannose, xylose, glucuronic acid, and galactose [200]. It was then proposed that this structure could be used to separate *Cryptococcus* species into serotypes defined by reactions with rabbit hyperimmune sera [201].

Rebers [202] shed the first light onto the structural problem by demonstrating that the polysaccharides from the cryptococcal capsule could be fractionated into two groups. Later, Batacharjee and colleagues [203] demonstrated that the largest fraction of purified polysaccharides did not contain galactose. Those authors termed this polysaccharide glucuronoxylomannan (GXM). It was only in 1982 that Cherniak [204] published the first detailed study of the minor, galactose-containing capsular polysaccharide, which was then termed galactoxylomannan (GalXM) [later renamed glucuronoxylomannogalactan (GXMGal)].

GXM is by far the most abundant of the two capsular polysaccharide components [205]. The molecular weight of GXM is approximately 1700 to 7000 kDa, comprising 90% of the overall polysaccharide composition of the capsule [206]. The current proposed structure of this molecule consists of the repetition of one or more motifs composed by a core triad of α-1,3-linked mannose with side substitutions of β-1,2-linked glucuronic acid and β-1,2- or β-1,4-linked xylose [207]. In addition, these triads undergo O-acetylation on their mannose residues due to the action of an enzyme called Cas1 [208].

The other polysaccharide component of the capsule, GXMGal, presents a molecular weight range of 101–275 kDa [209]. Recent data concerning the structure of GXMGal purified from an acapsular mutant (which does not produce GXM) showed this polymer to be organized as an α-1,6-linked galactose core with alternating β-1,3 galactose side chains α-1,4-linked to a dimer of mannose-α-1,3-mannose. These alternate side chains can receive substitutions of α-1,3- or β-1,3-linked xylose molecules. In addition, molecules of β-1,3-linked glucuronic acid may also be present [210].

The basic components of these sugars are UDP-glucuronic acid, which is synthesized by the Ugd1 UDP-glucose dehydrogenase [157], UDP-xylose, which is generated via the conversion of UDP-glucuronic acid by Uxs1 decarboxylase [211], UDP-galactose, which is generated via the conversion of UDP-glucose by Uge1 epimerase [212], and GDP-mannose, which is synthesized by a phosphomannose isomerase named Man1 and possibly other enzymes [212].

It was proposed that these sugar molecules are synthesized inside the Golgi cisternae [213], and transporters for GDP-mannose and UDP-galactose that are pres-

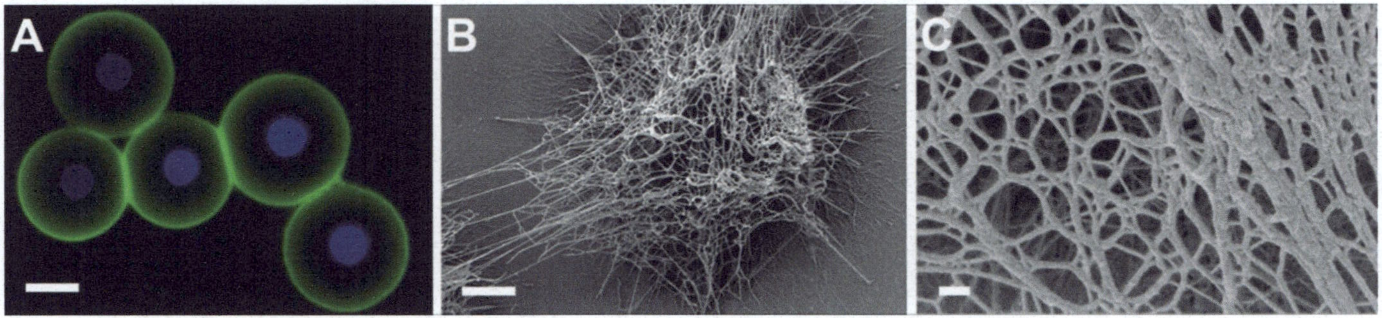

Fig. 6.6 Different views of *C. neoformans* capsular polysaccharide. Immunofluorescence microscopy using monoclonal antibody against GXM and the cell wall stained with Calcofluor White. *Scale bar*: 10 μm (**a**). Scanning electronic microscopy showing fibrillar morphology of capsule (**b** and **c**). *Scale bars*: 1 μm and 200 nm

ent in the membrane of this structure have been described to play a role in capsular polysaccharide production [214].

Once outside of the cell, the anchoring of the polysaccharide capsule appears to be dependent on several components and structures. Both α-1,3 and β-1,6 glucan influence capsule attachment [215]. The first molecule influences the attachment of the whole capsular structure [216], and the second molecule influences the capsule density [215].

Chitin and chitosan, which are also components of the fungal cell wall, are determinants in capsule formation. Mutants containing decreased chitosan levels and increased chitin levels exhibited an augmented capsule size in comparison to wild-type individuals [217]. Reinforcing that finding, chitin and chito-oligomers were shown to bind GXM in vitro [218], which may indicate a role for these molecules in capsule attachment.

Finally, among the proteins that insert into cell wall structures, two GPI-anchored proteins are likely to be related to capsule structure: MP98 and phospholipase. MP98 affects chitosan levels [217], and phospholipase binds to β1,6 glucan, affecting cell wall integrity and capsule density [219].

Once attached to the cell wall, GXM fibers are believed to aggregate with each other via interactions with divalent cations [220]. Newly secreted fibers attach to the outer portion of the preexisting capsular structure, as indicated by irreversible staining of GXM fibers with complement component C3 during capsule enlargement [221]. This finding is consistent with the observation that antigenic determinants appear first at the distal portion of the capsule in vivo [222].

The most established function of the cryptococcal capsule is the avoidance of phagocytosis, which occurs very early during lung infection [223]. This impairment of innate immunity is probably a consequence of the capsule acting as a barrier between the immunogenic epitopes in the cell wall and their receptors on phagocytes [224]. This function is profoundly important, as without internalizing and killing the fungus, these cells cannot correctly present fungal antigens to T cells. Thus, this process negatively impacts the host's ability to mount an adaptive immune response, leading to fungal persistence and dissemination [225].

Among the receptors implicated in the recognition of the fungal capsule is the mannose receptor, which was shown to influence the uptake of *Cryptococcus* by dendritic cells [226]. Fc receptors that are present in these cells also engage capsular polysaccharide, leading to opposite outcomes. When bound to FcRIIB or FcRγII, GXM can trigger phagocytosis [227] or induce inhibitory signals, respectively [228].

After being internalized, the cryptococcal capsule plays a decisive role in survival and replication inside phagocytes. Acapsular mutants are unable to replicate inside the phagolysosome and cannot escape the cell by extrusion [229]. This phenotype can be rescued by coating the acapsular yeast with exogenous GXM [229].

The observed increased survival inside phagocytes may be related to protection against reactive oxygen species, which can either be trapped by the abundant negative charge of the capsular polysaccharide's glucuronic acid residues [230] or diluted by the expansion of the phagolysosome. Phagolysosome expansion may be driven by capsular enlargement and is believed to make this compartment leaky and less concentrated [231].

Encapsulated strains disseminate from the lungs better than acapsular strains [232], and this difference appears to be linked to the engagement of receptors in epithelial cells. GXM was demonstrated to bind and enter the intracellular environment of the pulmonary alveolar epithelium via interactions with CD14 [233].

The role of the polysaccharide capsule in *Cryptococcus* dissemination among host tissues appears to be related not only to capsule enlargement but also to capsule organization. Fungal cells modulate capsule design during spreading, as demonstrated by differential binding to specific antibodies while infecting different organs, which represents a clear modification of the exposed epitopes [222].

Capsular GXM is also shed constitutively into the host environment and can be detected in large quantities in several tissues [234]. This secreted exopolysaccharide produces several deleterious effects on the host. For example, large quantities of polysaccharide can decrease antibody production [235] and lymphocyte proliferation [236]. Purified GXM was also shown to impair T cell activation, either via dendritic cells exposed to this polysaccharide or by directly rendering T cells unresponsive to polyclonal stimuli upon contact with this polysaccharide [237]. In addition, brain microglial cells appear to be insensitive to stimuli, such as LPS, after polysaccharide contact [238].

GXM can interact with TLR4 in macrophages, inducing the expression of FasL, which can promote T cell apoptosis through the activation of caspases 8 and 9 [239]. Macrophages can also undergo apoptosis themselves through the expression of both Fas and FasL receptors.

Neutrophil migration is also impacted by GXM, as this molecule decreases surface levels of the TNFα receptor and L-selectin and engages CD18, which is part of the integrin LFA1, to impact the tight binding of those cells to the activated endothelium [240].

Cryptococcus virulence relies upon the externalization of several molecules, such as melanin, enzymes and polysaccharides. However, several virulence-related molecules do not contain any secretion targeting sequences. Nevertheless, these molecules are commonly observed in the extracellular environment. Rodrigues and colleagues [241] described the presence of extracellular vesicles in *C. neoformans* culture supernatants, and these compartments revealed a complex composition of extracellular vesicle cargo, including polysaccharides, active enzymes, nucleic acids, pigments, toxins, and lipoproteins [242].

The secretion of vesicles during infection has proven difficult to assess. However, indirect observations, such as the recognition of EV proteins by antibodies present in serum from cryptococcosis patients [243], reinforces the idea that these structures play an important role in cryptococcal infection, possibly by acting as virulence delivery bags, as proposed previously [243].

Virulence factors described above and other cryptococcocal components impacting disease progress are directly linked to immunological mechanism of host responses. The immune response against *Cryptococcus* is out of the scope of this chapter and will not be discussed here. Major mechanisms, however, are summarized in Table 6.2.

Table 6.2 Summary of major immune mechanisms against *Cryptococcus*

Cell type	Cell phenotype	Effector molecules	Protection mechanisms	Experimental host phenotype	Fungal evasion mechanisms
Macrophage	M1	TNF, IL-12, NO, ROS	Oxidative killing by ROS and NO [244–246]; cytokine-induced immune response modulation [230, 247–249]	Associated with increased survival and decreased fungal burden [244]	Apoptosis induction [250]; phagocytosis inhibition [223, 224, 251, 252]; protection against ROS and NO [230, 253–255]; M2 polarization induction [255–260]; cytokine production inhibition [227, 261]
Macrophage	M2	IL-10, TGFβ, arginase-1	–	Associated with decreased survival and increased fungal burden [256, 258, 260]	Efficient intracellular replication [262–265]; benefits from phagolysosome low pH [266]; nonlytic exocytosis [267–269]; CNS dissemination ("trojan horse") [222, 270, 271]
Neutrophil	–	ROS, sphingomyelin synthase	Intracellular killing by ROS [272]; extracellular killing [273]; yeast removal from brain microvasculature [274, 275]	No influence in fungal burden; may increase immunopathology [195, 276, 277]	Phagocytosis inhibition; protection against ROS [230]; NETs inhibition [274]; migration inhibition [240, 278]
Innate Lymphoid cells	ILC1 (NK)	IFNγ, perforin	Extracellular killing by perforin-rich granules [279–281]; growth inhibition [282, 283]; early production of IFNγ and macrophage activation [284–286]	–	–
Innate Lymphoid cells	ILC2	IL-5, IL-13	–	Associated with M2 macrophages, increased fungal burden and increased immunopathology [287]	–

Cell type	Cell phenotype	Effector molecules	Protection mechanisms	Experimental host phenotype	Fungal evasion mechanisms
Dendritic cells	–	TNF, IL-6, IL-12, IL-23, ROS, CD40, D80, CD86	Intracellular oxidative killing [188, 288, 289]; growth inhibition [188, 290]; antigen presentation [291]	Immunization with pulsed DCs increases survival, decreases fungal burden and reduces immunopathology [292, 293]	Phagocytosis inhibition [225, 294]; inhibition of cytokine and chemokine production, antigen presentation and costimulatory molecules expression [288, 294–296]
T CD4+ Lymphocytes	Th1	TNF, IFNγ	Stimulates oxidative killing (ROS and NO) by M1 macrophages [244, 256]	Associated with M1 macrophages, increased survival, and decreased fungal burden [244, 247]	Inhibition of proliferation [236, 237]; direct induction of apoptosis [297]; indirect induction of apoptosis [359, 360]
T CD4+ Lymphocytes	Th2	IL-4, IL-5, IL-13, IL-10	–	Associated with M2 macrophages [298–301]; Associated with decreased survival, increased fungal burden and increased immunopathology [300–302]	Inhibition of proliferation [236, 237]; direct induction of apoptosis [297]; indirect induction of apoptosis [239, 250]
T CD4+ Lymphocytes	Th17	IL-17, IL-22, IL-23	Leucocyte migration; Th2 polarization inhibition [303, 304]	No influence in survival, no influence in fungal burden, limits Th2-driven inflammation [300, 303]	Inhibition of proliferation [236, 237]; direct induction of apoptosis [297]; indirect induction of apoptosis [239, 250]
T CD8+ Lymphocytes	–	Granulysin	Killing by granulysin release [305, 306]	Associated with increased survival and decreased fungal burden [307]	–
B Lymphocytes	–	Antibodies	Opsonization [308, 309]; increases killing by phagocytes [282, 283]; growth inhibition [310, 311]; inhibits fungal budding [312, 313]; inhibits capsule formation [310]; inhibits GXM release [314]; inhibits biolfim formation [315]	Associated with increased survival, decreased fungal burden and restriction of detrimental inflammation [316–321]	–

6.10 Final Considerations

Cryptococcal meningitis is considered to be a global problem, based on the high incidence and mortality estimated by the US Centers of Disease and Prevention (CDC). Although AIDS patients in poor countries are the most affected by cryptococcosis, healthy individuals were victims of important outbreaks in the US and Canada [322]. The annual deaths caused by this infectious disease could be decreased to 70,000 if rapid and accurate diagnostics and appropriate fungal treatments were available for all people, especially in high-burden countries [5]. To reach this goal, investments in research, development, and innovation are required to better understand important aspects related to *Cryptococcus* and cryptococcosis and to improve antifungal therapies and diagnostic tools.

References

1. Kwon-Chung KJ et al (2014) Cryptococcus neoformans and Cryptococcus gattii, the etiologic agents of cryptococcosis. Cold Spring Harb Perspect Med 4(7):a019760
2. Kidd SE et al (2004) A rare genotype of Cryptococcus gattii caused the cryptococcosis outbreak on Vancouver Island (British Columbia, Canada). Proc Natl Acad Sci U S A 101(49):17258–17263
3. Dromer F, Casadevall A, Perfect JR, Sorrell T (2011) Cryptococcus neoformans: latency and disease. In: Heitman J, Kozel TR, Kwon-Chung KJ, Perfect JR, Casadevall A (eds) Cryptococcus: from human pathogen to model yeast. American Society for Microbiology, USA, pp 431–439
4. Park BJ et al (2009) Estimation of the current global burden of cryptococcal meningitis among persons living with HIV/AIDS. AIDS 23(4):525–530
5. Denning DW (2016) Minimizing fungal disease deaths will allow the UNAIDS target of reducing annual AIDS deaths below 500 000 by 2020 to be realized. Philos Trans R Soc Lond Ser B Biol Sci 371(1709):1–10
6. Harris J, Lockhart S, Chiller T (2012) Cryptococcus gattii: where do we go from here? Med Mycol 50(2):113–129
7. Rodrigues ML (2016) Funding and innovation in diseases of neglected populations: the paradox of cryptococcal meningitis. PLoS Negl Trop Dis 10(3):e0004429
8. Srikanta D, Santiago-Tirado FH, Doering TL (2014) Cryptococcus neoformans: historical curiosity to modern pathogen. Yeast 31(2):47–60
9. Maziarz EK, Perfect JR (2016) Cryptococcosis. Infect Dis Clin N Am 30(1):179–206
10. Chen M et al (2016) Cryptococcosis and tuberculosis co-infection in mainland China. Emerg Microbes Infect 5(9):e98
11. Sorrell TC, Chen SCA, Phillips P, Marr KA (2011) Clinical perspectives on cryptococcus neoformans and Cryptococcus gattii: implications for diagnosis and management. In: Heitman J, Kozel TR, Kwon-Chung KJ, Perfect JR, Casadevall A (eds) Cryptococcus: from human pathogen to model yeast. American Society for Microbiology, USA, pp 595–606
12. Vallabhaneni S et al (2016) The global burden of fungal diseases. Infect Dis Clin N Am 30(1):1–11
13. Du L et al (2015) Systemic review of published reports on primary cutaneous cryptococcosis in immunocompetent patients. Mycopathologia 180(1–2):19–25
14. Haddad N et al (2015) Pulmonary cryptococcoma: a rare and challenging diagnosis in immunocompetent patients. Autops Case Rep 5(2):35–40

15. Chen SC, Meyer W, Sorrell TC (2014) Cryptococcus gattii infections. Clin Microbiol Rev 27(4):980–1024
16. Santos WRAD et al (2008) Primary endemic Cryptococcosis gattii by molecular type VGII in the state of Pará, Brazil. Mem Inst Oswaldo Cruz 103:813–818
17. Galanis E et al (2009) Clinical presentation, diagnosis and management of Cryptococcus gattii cases: Lessons learned from British Columbia. Can J Infect Dis Med Microbiol 20(1):23–28
18. Gullo FP et al (2013) Cryptococcosis: epidemiology, fungal resistance, and new alternatives for treatment. Eur J Clin Microbiol Infect Dis 32(11):1377–1391
19. Lin X, Heitman J (2006) The biology of the Cryptococcus neoformans species complex. Annu Rev Microbiol 60:69–105
20. Brito-Santos F et al (2015) Environmental isolation of Cryptococcus gattii VGII from indoor dust from typical wooden houses in the deep Amazonas of the Rio Negro basin. PLoS One 10(2):e0115866
21. Takahara DT et al (2013) First report on Cryptococcus neoformans in pigeon excreta from public and residential locations in the metropolitan area of Cuiaba, State of Mato Grosso, Brazil. Rev Inst Med Trop Sao Paulo 55(6):371–376
22. Jesus MSD et al (2012) Cryptococcus neoformans carried by Odontomachus bauri ants. Mem Inst Oswaldo Cruz 107:466–469
23. Cattana ME et al (2014) Native trees of the Northeast Argentine: natural hosts of the Cryptococcus neoformans-Cryptococcus gattii species complex. Rev Iberoam Micol 31(3):188–192
24. Springer DJ, Chaturvedi V (2010) Projecting global occurrence of Cryptococcus gattii. Emerg Infect Dis 16(1):14–20
25. Cogliati M (2013) Global molecular epidemiology of Cryptococcus neoformans and Cryptococcus gattii: an atlas of the molecular types. Scientifica 2013:23
26. Fraser JA et al (2005) Same-sex mating and the origin of the Vancouver Island Cryptococcus gattii outbreak. Nature 437(7063):1360–1364
27. Hagen F et al (2013) Ancient dispersal of the human fungal pathogen Cryptococcus gattii from the Amazon rainforest. PLoS One 8(8):e71148
28. Souto AC et al (2016) Population genetic analysis reveals a high genetic diversity in the Brazilian Cryptococcus gattii VGII population and shifts the global origin from the Amazon rainforest to the semi-arid desert in the northeast of Brazil. PLoS Negl Trop Dis 10(8):e0004885
29. Firacative C et al (2016) MLST and whole-genome-based population analysis of Cryptococcus gattii VGIII links clinical, veterinary and environmental strains, and reveals divergent serotype specific sub-populations and distant ancestors. PLoS Negl Trop Dis 10(8):e0004861
30. Baker RD (1976) The primary pulmonary lymph node complex of crytptococcosis. Am J Clin Pathol 65(1):83–92
31. Goldman DL et al (2001) Serologic evidence for Cryptococcus neoformans infection in early childhood. Pediatrics 107(5):E66
32. Garcia-Hermoso D, Janbon G, Dromer F (1999) Epidemiological evidence for dormant Cryptococcus neoformans infection. J Clin Microbiol 37(10):3204–3209
33. Chen S et al (2000) Epidemiology and host- and variety-dependent characteristics of infection due to Cryptococcus neoformans in Australia and New Zealand. Australasian Cryptococcal Study Group. Clin Infect Dis 31(2):499–508
34. Mitchell DH et al (1995) Cryptococcal disease of the CNS in immunocompetent hosts: influence of cryptococcal variety on clinical manifestations and outcome. Clin Infect Dis 20(3):611–616
35. Speed B, Dunt D (1995) Clinical and host differences between infections with the two varieties of *Cryptococcus neoformans*. Clin Infect Dis 21(1):28–34, discussion 35–36
36. Chayakulkeeree M, Perfect JR (2006) Cryptococcosis. Infect Dis Clin N Am 20(3):507–544, v–vi

37. Chang WC et al (2006) Pulmonary cryptococcosis: comparison of clinical and radiographic characteristics in immunocompetent and immunocompromised patients. Chest 129(2):333–340
38. Shirley RM, Baddley JW (2009) Cryptococcal lung disease. Curr Opin Pulm Med 15(3):254–260
39. Nadrous HF et al (2003) Pulmonary cryptococcosis in nonimmunocompromised patients. Chest 124(6):2143–2147
40. Phillips P et al (2015) Longitudinal clinical findings and outcome among patients with Cryptococcus gattii infection in British Columbia. Clin Infect Dis 60(9):1368–1376
41. Murray RJ et al (1988) Recovery from cryptococcemia and the adult respiratory distress syndrome in the acquired immunodeficiency syndrome. Chest 93(6):1304–1306
42. Brizendine KD, Baddley JW, Pappas PG (2011) Pulmonary cryptococcosis. Semin Respir Crit Care Med 32(6):727–734
43. Nunez M, Peacock JE Jr, Chin R Jr (2000) Pulmonary cryptococcosis in the immunocompetent host. Therapy with oral fluconazole: a report of four cases and a review of the literature. Chest 118(2):527–534
44. McMullan BJ, Sorrell TC, Chen SC (2013) Cryptococcus gattii infections: contemporary aspects of epidemiology, clinical manifestations and management of infection. Future Microbiol 8(12):1613–1631
45. Colombo AC, Rodrigues ML (2015) Fungal colonization of the brain: anatomopathological aspects of neurological cryptococcosis. An Acad Bras Cienc 87(2 Suppl):1293–1309
46. Ecevit IZ et al (2006) The poor prognosis of central nervous system cryptococcosis among nonimmunosuppressed patients: a call for better disease recognition and evaluation of adjuncts to antifungal therapy. Clin Infect Dis 42(10):1443–1447
47. Lui G et al (2006) Cryptococcosis in apparently immunocompetent patients. QJM 99(3):143–151
48. Fries BC et al (2005) Phenotypic switching of Cryptococcus neoformans can produce variants that elicit increased intracranial pressure in a rat model of cryptococcal meningoencephalitis. Infect Immun 73(3):1779–1787
49. Okun E, Butler WT (1964) Ophthalmologic complications of Cryptococcal meningitis. Arch Ophthalmol 71:52–57
50. Custis PH, Haller JA, de Juan E Jr (1995) An unusual case of cryptococcal endophthalmitis. Retina 15(4):300–304
51. Perfect JR, Bicanic T (2015) Cryptococcosis diagnosis and treatment: what do we know now. Fungal Genet Biol 78:49–54
52. Chiriac A et al (2017) Primary cutaneous cryptococcosis during infliximab therapy. Dermatol Ther 30(1):1–3
53. Kwon-Chung KJ et al (2000) Cryptococcosis: clinical and biological aspects. Med Mycol 38(Suppl 1):205–213
54. Wilson ML, Sewell LD, Mowad CM (2008) Primary cutaneous Cryptococcosis during therapy with methotrexate and adalimumab. J Drugs Dermatol 7(1):53–54
55. Revenga F et al (2002) Primary cutaneous cryptococcosis in an immunocompetent host: case report and review of the literature. Dermatology 204(2):145–149
56. Neuville S et al (2003) Primary cutaneous cryptococcosis: a distinct clinical entity. Clin Infect Dis 36(3):337–347
57. Dora JM et al (2006) Cutaneous cryptococccosis due to Cryptococcus gattii in immunocompetent hosts: case report and review. Mycopathologia 161(4):235–238
58. Probst C et al (2010) Cryptococcosis mimicking cutaneous cellulitis in a patient suffering from rheumatoid arthritis: a case report. BMC Infect Dis 10:239
59. Manrique P et al (1992) Polymorphous cutaneous cryptococcosis: nodular, herpes-like, and molluscum-like lesions in a patient with the acquired immunodeficiency syndrome. J Am Acad Dermatol 26(1):122–124
60. Wykoff CC et al (2009) Intraocular cryptococcoma. Arch Ophthalmol 127(5):700–702
61. Alzahrani YA et al (2016) Cryptococcal iridociliary granuloma. Surv Ophthalmol 61(4):498–501

62. Wada R et al (2008) Granulomatous prostatitis due to Cryptococcus neoformans: diagnostic usefulness of special stains and molecular analysis of 18S rDNA. Prostate Cancer Prostatic Dis 11(2):203–206
63. Siddiqui TJ, Zamani T, Parada JP (2005) Primary cryptococcal prostatitis and correlation with serum prostate specific antigen in a renal transplant recipient. J Infect 51(3):e153–e157
64. Allen R et al (1982) Disseminated cryptococcosis after transurethral resection of the prostate. Aust NZ J Med 12(4):296–299
65. Larsen RA et al (1989) Persistent Cryptococcus neoformans infection of the prostate after successful treatment of meningitis. California Collaborative Treatment Group. Ann Intern Med 111(2):125–128
66. Staib F, Seibold M, L'Age M (1990) Persistence of Cryptococcus neoformans in seminal fluid and urine under itraconazole treatment. The urogenital tract (prostate) as a niche for Cryptococcus neoformans. Mycoses 33(7–8):369–373
67. Seo IY et al (2006) Granulomatous cryptococcal prostatitis diagnosed by transrectal biopsy. Int J Urol 13(5):638–639
68. Lief M, Sarfarazi F (1986) Prostatic cryptococcosis in acquired immune deficiency syndrome. Urology 28(4):318–319
69. Bal CK et al (2014) Spontaneous cryptococcal peritonitis with fungemia in patients with decompensated cirrhosis: Report of two cases. Indian J Crit Care Med 18(8):536–539
70. Stead KJ et al (1988) Septic arthritis due to Cryptococcus neoformans. J Infect 17(2):139–145
71. Flagg SD et al (2001) Myositis resulting from disseminated cryptococcosis in a patient with hepatitis C cirrhosis. Clin Infect Dis 32(7):1104–1107
72. Farr RW, Wright RA (1992) Cryptococcal olecranon bursitis in cirrhosis. J Rheumatol 19(1):172–173
73. Daly JS et al (1990) Disseminated, nonmeningeal gastrointestinal cryptococcal infection in an HIV-negative patient. Am J Gastroenterol 85(10):1421–1424
74. Thalla R et al (2009) Sequestration of active Cryptococcus neoformans infection in the parathyroid gland despite prolonged therapy in a renal transplant recipient. Transpl Infect Dis 11(4):349–352
75. Kantarcioglu AS et al (2006) Cryptococcal parotid involvement: an uncommon localization of Cryptococcus neoformans. Med Mycol 44(3):279–283
76. Salyer WR et al (1973) Adrenal involvement in cryptococcosis. Am J Clin Pathol 60(4):559–561
77. Liu PY (1998) Cryptococcal osteomyelitis: case report and review. Diagn Microbiol Infect Dis 30(1):33–35
78. Tan DB et al (2008) Immunological profiles of immune restoration disease presenting as mycobacterial lymphadenitis and cryptococcal meningitis. HIV Med 9(5):307–316
79. Singh N, Perfect JR (2007) Immune reconstitution syndrome associated with opportunistic mycoses. Lancet Infect Dis 7(6):395–401
80. Nunnari G et al (2013) Cryptococcal meningitis in an HIV-1-infected person: relapses or IRIS? Case report and review of the literature. Eur Rev Med Pharmacol Sci 17(11):1555–1559
81. Crypto Collab Tx Study Gp, Alexander BDSC, Forrest G, Johnson L, Lortholary O, Singh N (2008) Cryptococcus-associated Immune Reconstitution Syndrome (IRS) in Solid Organ Transplant (SOT) recipients: results from a prospective, multicenter study. In 48th annual ICAAC/Infectious Diseases Society of America. Washington, DC
82. Shelburne SA III et al (2005) The role of immune reconstitution inflammatory syndrome in AIDS-related Cryptococcus neoformans disease in the era of highly active antiretroviral therapy. Clin Infect Dis 40(7):1049–1052
83. Chen SC et al (2013) Antifungal therapy and management of complications of cryptococcosis due to Cryptococcus gattii. Clin Infect Dis 57(4):543–551
84. World Health Organization (2011) Rapid advice: diagnosis, prevention and management of cryptococcal disease in HIV-infected adults, adolescents and children. Geneva
85. Mesa-Arango AC, Scorzoni L, Zaragoza O (2012) It only takes one to do many jobs: Amphotericin B as antifungal and immunomodulatory drug. Front Microbiol 3:286

86. Day JN et al (2013) Combination antifungal therapy for cryptococcal meningitis. N Engl J Med 368(14):1291–1302
87. Bicanic T et al (2008) High-dose amphotericin B with flucytosine for the treatment of cryptococcal meningitis in HIV-infected patients: a randomized trial. Clin Infect Dis 47(1):123–130
88. Barratt G, Bretagne S (2007) Optimizing efficacy of Amphotericin B through nanomodification. Int J Nanomedicine 2(3):301–313
89. Loyse A et al (2013) Cryptococcal meningitis: improving access to essential antifungal medicines in resource-poor countries. Lancet Infect Dis 13(7):629–637
90. Groll AH et al (2003) Clinical pharmacology of antifungal compounds. Infect Dis Clin N Am 17(1):159–191, ix
91. Brouwer AE et al (2007) Oral versus intravenous flucytosine in patients with human immunodeficiency virus-associated cryptococcal meningitis. Antimicrob Agents Chemother 51(3):1038–1042
92. Loyse A et al (2013) Flucytosine and cryptococcosis: time to urgently address the worldwide accessibility of a 50-year-old antifungal. J Antimicrob Chemother 68(11):2435–2444
93. Harris BE et al (1986) Conversion of 5-fluorocytosine to 5-fluorouracil by human intestinal microflora. Antimicrob Agents Chemother 29(1):44–48
94. Gray KC et al (2012) Amphotericin primarily kills yeast by simply binding ergosterol. Proc Natl Acad Sci U S A 109(7):2234–2239
95. Saag MS, Dismukes WE (1988) Azole antifungal agents: emphasis on new triazoles. Antimicrob Agents Chemother 32(1):1–8
96. Hajjeh RA, Brandt ME, Pinner RW (1995) Emergence of cryptococcal disease: epidemiologic perspectives 100 years after its discovery. Epidemiol Rev 17(2):303–320
97. Antinori S (2013) New Insights into HIV/AIDS-Associated Cryptococcosis. ISRN AIDS 2013:471363
98. Antinori S et al (2009) AIDS-associated cryptococcosis: a comparison of epidemiology, clinical features and outcome in the pre- and post-HAART eras. Experience of a single centre in Italy. HIV Med 10(1):6–11
99. Kendi C et al (2013) Predictors of outcome in routine care for Cryptococcal meningitis in Western Kenya: lessons for HIV outpatient care in resource-limited settings. Postgrad Med J 89(1048):73–77
100. Perfect JR et al (2010) Clinical practice guidelines for the management of cryptococcal disease: 2010 update by the infectious diseases society of america. Clin Infect Dis 50(3):291–322
101. Brouwer AE et al (2004) Combination antifungal therapies for HIV-associated cryptococcal meningitis: a randomised trial. Lancet 363(9423):1764–1767
102. Rajasingham R et al (2012) Cryptococcal meningitis treatment strategies in resource-limited settings: a cost-effectiveness analysis. PLoS Med 9(9):e1001316
103. Southern African HIV Clinicians Society T (2013) Guideline for the prevention, diagnosis and management of cryptococcal meningitis among HIV-infected persons: 2013 update. South Afr J HIV Med 14(2):76
104. Dromer F et al (2008) Major role for amphotericin B-flucytosine combination in severe cryptococcosis. PLoS One 3(8):e2870
105. Day JN et al (2011) Most cases of cryptococcal meningitis in HIV-uninfected patients in Vietnam are due to a distinct amplified fragment length polymorphism-defined cluster of Cryptococcus neoformans var. grubii VN1. J Clin Microbiol 49(2):658–664
106. Bicanic T et al (2006) Symptomatic relapse of HIV-associated cryptococcal meningitis after initial fluconazole monotherapy: the role of fluconazole resistance and immune reconstitution. Clin Infect Dis 43(8):1069–1073
107. van der Horst CM et al (1997) Treatment of cryptococcal meningitis associated with the acquired immunodeficiency syndrome. National Institute of Allergy and Infectious Diseases Mycoses Study Group and AIDS Clinical Trials Group. N Engl J Med 337(1):15–21

108. Bozzette SA et al (1991) A placebo-controlled trial of maintenance therapy with fluconazole after treatment of cryptococcal meningitis in the acquired immunodeficiency syndrome. California Collaborative Treatment Group. N Engl J Med 324(9):580–584
109. Saag MS et al (1999) A comparison of itraconazole versus fluconazole as maintenance therapy for AIDS-associated cryptococcal meningitis. National Institute of Allergy and Infectious Diseases Mycoses Study Group. Clin Infect Dis 28(2):291–296
110. Mussini C et al (2004) Discontinuation of maintenance therapy for cryptococcal meningitis in patients with AIDS treated with highly active antiretroviral therapy: an international observational study. Clin Infect Dis 38(4):565–571
111. Zolopa A et al (2009) Early antiretroviral therapy reduces AIDS progression/death in individuals with acute opportunistic infections: a multicenter randomized strategy trial. PLoS One 4(5):e5575
112. Chang CC et al (2013) Clinical and mycological predictors of cryptococcosis-associated immune reconstitution inflammatory syndrome. AIDS 27(13):2089–2099
113. Makadzange AT et al (2010) Early versus delayed initiation of antiretroviral therapy for concurrent HIV infection and cryptococcal meningitis in sub-saharan Africa. Clin Infect Dis 50(11):1532–1538
114. Musubire AK et al (2012) Challenges in diagnosis and management of Cryptococcal immune reconstitution inflammatory syndrome (IRIS) in resource limited settings. Afr Health Sci 12(2):226–230
115. Haddow LJ et al (2010) Cryptococcal immune reconstitution inflammatory syndrome in HIV-1-infected individuals: proposed clinical case definitions. Lancet Infect Dis 10(11):791–802
116. Bicanic TA et al (2013) Starting ART following cryptococcal meningitis. the optimal time has yet to be defined. South Afr J HIV Med 14(3):105
117. Lesho E (2006) Evidence base for using corticosteroids to treat HIV-associated immune reconstitution syndrome. Expert Rev Anti-Infect Ther 4(3):469–478
118. Husain S, Wagener MM, Singh N (2001) Cryptococcus neoformans infection in organ transplant recipients: variables influencing clinical characteristics and outcome. Emerg Infect Dis 7(3):375–381
119. La Hoz RM, Pappas PG (2013) Cryptococcal infections: changing epidemiology and implications for therapy. Drugs 73(6):495–504
120. Davis JA et al (2009) Central nervous system involvement in cryptococcal infection in individuals after solid organ transplantation or with AIDS. Transpl Infect Dis 11(5):432–437
121. Singh N et al (2005) Antifungal management practices and evolution of infection in organ transplant recipients with Cryptococcus neoformans infection. Transplantation 80(8):1033–1039
122. Singh N et al (2007) Cryptococcus neoformans in organ transplant recipients: impact of calcineurin-inhibitor agents on mortality. J Infect Dis 195(5):756–764
123. Karie-Guigues S et al (2009) Long-term renal function in liver transplant recipients and impact of immunosuppressive regimens (calcineurin inhibitors alone or in combination with mycophenolate mofetil): the TRY study. Liver Transpl 15(9):1083–1091
124. Allison AC, Eugui EM (2005) Mechanisms of action of mycophenolate mofetil in preventing acute and chronic allograft rejection. Transplantation 80(2 Suppl):S181–S190
125. Singh N et al (2005) An immune reconstitution syndrome-like illness associated with Cryptococcus neoformans infection in organ transplant recipients. Clin Infect Dis 40(12):1756–1761
126. Moen MD, Lyseng-Williamson KA, Scott LJ (2009) Liposomal amphotericin B: a review of its use as empirical therapy in febrile neutropenia and in the treatment of invasive fungal infections. Drugs 69(3):361–392
127. Dromer F et al (1996) *Epidemiology of cryptococcosis in France: a 9-year survey (*1985-1993). *French Cryptococcosis Study Group.* Clin Infect Dis 23(1):82–90

128. Sun HY, Wagener MM, Singh N (2009) Cryptococcosis in solid-organ, hematopoietic stem cell, and tissue transplant recipients: evidence-based evolving trends. Clin Infect Dis 48(11):1566–1576
129. Rolfes MA et al (2014) The effect of therapeutic lumbar punctures on acute mortality from cryptococcal meningitis. Clin Infect Dis 59(11):1607–1614
130. Singh N et al (2008) Cryptococcosis in solid organ transplant recipients: current state of the science. Clin Infect Dis 47(10):1321–1327
131. Henao-Martinez AF, Beckham JD (2015) Cryptococcosis in solid organ transplant recipients. Curr Opin Infect Dis 28(4):300–307
132. Dismukes WE et al (1987) Treatment of cryptococcal meningitis with combination amphotericin B and flucytosine for four as compared with six weeks. N Engl J Med 317(6):334–341
133. Tristano AG (2010) Cryptococcal meningitis and systemic lupus erythematosus: a case report and review. Rev Chil Infectol 27(2):155–159
134. Chen HS et al (2007) Invasive fungal infection in systemic lupus erythematosus: an analysis of 15 cases and a literature review. Rheumatology (Oxford) 46(3):539–544
135. Yao Z, Liao W, Chen R (2005) Management of cryptococcosis in non-HIV-related patients. Med Mycol 43(3):245–251
136. Datta K et al (2009) Spread of Cryptococcus gattii into Pacific Northwest region of the United States. Emerg Infect Dis 15(8):1185–1191
137. Arayawichanont A et al (1999) Successful medical treatment of multiple cryptococcomas: report of a case and literature review. J Med Assoc Thail 82(10):991–999
138. Hospenthal DR, Bennett JE (2000) Persistence of cryptococcomas on neuroimaging. Clin Infect Dis 31(5):1303–1306
139. Blackie JD et al (1985) Ophthalmological complications of cryptococcal meningitis. Clin Exp Neurol 21:263–270
140. Kozel TR, Wickes B (2014) Fungal diagnostics. Cold Spring Harb Perspect Med 4(4):a019299
141. Perfect JR, Casadevall A (2002) Cryptococcosis. Infect Dis Clin N Am 16(4):837–874, v–vi
142. Vidal JE, Boulware DR (2015) Lateral flow assay for cryptococcal antigen: an important advance to improve the continuum of HIV care and reduce cryptococcal meningitis-related mortality. Rev Inst Med Trop Sao Paulo 57(Suppl 19):38–45
143. Menezes Rde P, Penatti MP, Pedroso Rdos S (2011) Different culture media containing methyldopa for melanin production by Cryptococcus species. Rev Soc Bras Med Trop 44(5):591–594
144. Klein KR et al (2009) Identification of Cryptococcus gattii by use of L-canavanine glycine bromothymol blue medium and DNA sequencing. J Clin Microbiol 47(11):3669–3672
145. Dominic RS et al (2009) Diagnostic value of latex agglutination in cryptococcal meningitis. J Lab Physicians 1(2):67–68
146. Rivera V et al (2015) Validation and clinical application of a molecular method for the identification of Cryptococcus neoformans/Cryptococcus gattii complex DNA in human clinical specimens. Braz J Infect Dis 19(6):563–570
147. Trilles L et al (2008) Regional pattern of the molecular types of Cryptococcus neoformans and Cryptococcus gattii in Brazil. Mem Inst Oswaldo Cruz 103(5):455–462
148. Trilles L et al (2014) Identification of the major molecular types of Cryptococcus neoformans and C. gattii by Hyperbranched rolling circle amplification. PLoS One 9(4):e94648
149. Sidrim JJ et al (2010) Molecular methods for the diagnosis and characterization of Cryptococcus: a review. Can J Microbiol 56(6):445–458
150. Meyer W et al (2009) Consensus multi-locus sequence typing scheme for Cryptococcus neoformans and Cryptococcus gattii. Med Mycol 47(6):561–570
151. Casadevall A (2012) Fungi and the rise of mammals. PLoS Pathog 8(8):e1002808
152. Petter R et al (2001) A survey of heterobasidiomycetous yeasts for the presence of the genes homologous to virulence factors of Filobasidiella neoformans, CNLAC1 and CAP59. Microbiology 147(Pt 8):2029–2036
153. Perfect JR, Lang SD, Durack DT (1980) Chronic cryptococcal meningitis: a new experimental model in rabbits. Am J Pathol 101(1):177–194

154. Kuhn LR (1949) Effect of elevated body temperatures on cryptococcosis in mice. Proc Soc Exp Biol Med 71(3):341–343
155. Littman ML, Zimmerman LE (1956) Cryptococcosis. Grune & Stratton, New York
156. Fox DS, Cox GM, Heitman J (2003) Phospholipid-binding protein Cts1 controls septation and functions coordinately with calcineurin in Cryptococcus neoformans. Eukaryot Cell 2(5):1025–1035
157. Griffith CL et al (2004) UDP-glucose dehydrogenase plays multiple roles in the biology of the pathogenic fungus Cryptococcus neoformans. J Biol Chem 279(49):51669–51676
158. Moyrand F, Janbon G (2004) UGD1, encoding the Cryptococcus neoformans UDP-glucose dehydrogenase, is essential for growth at 37 degrees C and for capsule biosynthesis. Eukaryot Cell 3(6):1601–1608
159. Kingsbury JM et al (2004) Cryptococcus neoformans Ilv2p confers resistance to sulfometuron methyl and is required for survival at 37 degrees C and in vivo. Microbiology 150(Pt 5):1547–1558
160. Giles SS et al (2005) Cryptococcus neoformans mitochondrial superoxide dismutase: an essential link between antioxidant function and high-temperature growth. Eukaryot Cell 4(1):46–54
161. Ngamskulrungroj P et al (2009) The trehalose synthesis pathway is an integral part of the virulence composite for Cryptococcus gattii. Infect Immun 77(10):4584–4596
162. de Gontijo FA et al (2014) The role of the de novo pyrimidine biosynthetic pathway in Cryptococcus neoformans high temperature growth and virulence. Fungal Genet Biol 70:12–23
163. Nosanchuk JD, Stark RE, Casadevall A (2015) Fungal melanin: what do we know about structure? Front Microbiol 6:1463
164. Williamson PR, Wakamatsu K, Ito S (1998) Melanin biosynthesis in Cryptococcus neoformans. J Bacteriol 180(6):1570–1572
165. Land EJ, Ramsden CA, Riley PA (2004) Quinone chemistry and melanogenesis. Methods Enzymol 378:88–109
166. Eisenman HC, Casadevall A (2012) Synthesis and assembly of fungal melanin. Appl Microbiol Biotechnol 93(3):931–940
167. Pukkila-Worley R et al (2005) Transcriptional network of multiple capsule and melanin genes governed by the Cryptococcus neoformans cyclic AMP cascade. Eukaryot Cell 4(1):190–201
168. Panepinto J et al (2009) Sec6-dependent sorting of fungal extracellular exosomes and laccase of Cryptococcus neoformans. Mol Microbiol 71(5):1165–1176
169. Missall TA et al (2005) Distinct stress responses of two functional laccases in Cryptococcus neoformans are revealed in the absence of the thiol-specific antioxidant Tsa1. Eukaryot Cell 4(1):202–208
170. Eisenman HC et al (2009) Vesicle-associated melanization in Cryptococcus neoformans. Microbiology 155(Pt 12):3860–3867
171. Salas SD et al (1996) Effect of the laccase gene CNLAC1, on virulence of Cryptococcus neoformans. J Exp Med 184(2):377–386
172. Mednick AJ, Nosanchuk JD, Casadevall A (2005) Melanization of Cryptococcus neoformans affects lung inflammatory responses during cryptococcal infection. Infect Immun 73(4):2012–2019
173. Wang Y, Aisen P, Casadevall A (1995) Cryptococcus neoformans melanin and virulence: mechanism of action. Infect Immun 63(8):3131–3136
174. Jacobson ES, Tinnell SB (1993) Antioxidant function of fungal melanin. J Bacteriol 175(21):7102–7104
175. Huffnagle GB et al (1995) Down-regulation of the afferent phase of T cell-mediated pulmonary inflammation and immunity by a high melanin-producing strain of Cryptococcus neoformans. J Immunol 155(7):3507–3516
176. Rosas AL et al (2000) Synthesis of polymerized melanin by Cryptococcus neoformans in infected rodents. Infect Immun 68(5):2845–2853

177. Rosas AL et al (2002) Activation of the alternative complement pathway by fungal melanins. Clin Diagn Lab Immunol 9(1):144–148
178. Ikeda R et al (2003) Effects of melanin upon susceptibility of Cryptococcus to antifungals. Microbiol Immunol 47(4):271–277
179. Martinez LR, Casadevall A (2006) Susceptibility of Cryptococcus neoformans biofilms to antifungal agents in vitro. Antimicrob Agents Chemother 50(3):1021–1033
180. van Duin D et al (2004) Effects of voriconazole on Cryptococcus neoformans. Antimicrob Agents Chemother 48(6):2014–2020
181. Chayakulkeeree M et al (2011) SEC14 is a specific requirement for secretion of phospholipase B1 and pathogenicity of Cryptococcus neoformans. Mol Microbiol 80(4):1088–1101
182. Singh A et al (2013) Factors required for activation of urease as a virulence determinant in Cryptococcus neoformans. MBio 4(3):e00220–e00213
183. Ghannoum MA (2003) Potential role of phospholipases in virulence and fungal pathogenesis. Clin Microbiol Rev 13(1):122–143, table of contents
184. Cox GM et al (2001) Extracellular phospholipase activity is a virulence factor for Cryptococcus neoformans. Mol Microbiol 39(1):166–175
185. Santangelo RT et al (1999) Biochemical and functional characterisation of secreted phospholipase activities from Cryptococcus neoformans in their naturally occurring state. J Med Microbiol 48(8):731–740
186. Chen SC et al (1997) Identification of extracellular phospholipase B, lysophospholipase, and acyltransferase produced by Cryptococcus neoformans. Infect Immun 65(2):405–411
187. Maruvada R et al (2012) Cryptococcus neoformans phospholipase B1 activates host cell Rac1 for traversal across the blood-brain barrier. Cell Microbiol 14(10):1544–1553
188. Hole CR et al (2012) Mechanisms of dendritic cell lysosomal killing of Cryptococcus. Sci Rep 2:739
189. Smith LM, Dixon EF, May RC (2015) The fungal pathogen Cryptococcus neoformans manipulates macrophage phagosome maturation. Cell Microbiol 17(5):702–713
190. Noverr MC et al (2003) Role of PLB1 in pulmonary inflammation and cryptococcal eicosanoid production. Infect Immun 71(3):1538–1547
191. Cox GM et al (2000) Urease as a virulence factor in experimental cryptococcosis. Infect Immun 68(2):443–448
192. Lee MH et al (1992) Klebsiella aerogenes urease gene cluster: sequence of ureD and demonstration that four accessory genes (ureD, ureE, ureF, and ureG) are involved in nickel metallocenter biosynthesis. J Bacteriol 174(13):4324–4330
193. Mobley HL (1996) The role of Helicobacter pylori urease in the pathogenesis of gastritis and peptic ulceration. Aliment Pharmacol Ther 10(Suppl 1):57–64
194. Rappleye CA, Goldman WE (2006) Defining virulence genes in the dimorphic fungi. Annu Rev Microbiol 60:281–303
195. Osterholzer JJ et al (2009) Cryptococcal urease promotes the accumulation of immature dendritic cells and a non-protective T2 immune response within the lung. Am J Pathol 174(3):932–943
196. Olszewski MA et al (2004) Urease expression by Cryptococcus neoformans promotes microvascular sequestration, thereby enhancing central nervous system invasion. Am J Pathol 164(5):1761–1771
197. Shi M et al (2010) Real-time imaging of trapping and urease-dependent transmigration of Cryptococcus neoformans in mouse brain. J Clin Invest 120(5):1683–1693
198. Barnett JA (2010) A history of research on yeasts 14: medical yeasts part 2, Cryptococcus neoformans. Yeast 27(11):875–904
199. Benham RW (1935) Cryptococci: their identification by morphology and by serology. J Infect Dis 57(3):255–274
200. Evans EE, Mehl JW (1951) A qualitative analysis of capsular poly-saccharides from Cryptococcus neoformans by filter paper chromatography. Science (Washington) 114(2949):10–11

201. Evans EE (1950) The antigenic composition of Cryptococcus neoformans. I. A serologic classification by means of the capsular and agglutination reactions. J Immunol 64(5):423–430
202. Rebers PA et al (1958) Precipitation of the specific polysaccharide of Cryptococcus neoformans A by types II and XIV antipneumococcal sera1. J Am Chem Soc 80(5):1135–1137
203. Bhattacharjee AK, Kwon-Chung KJ, Glaudemans CP (1980) Structural studies on the major, capsular polysaccharide from Cryptococcus bacillisporus serotype B. Carbohydr Res 82(1):103–111
204. Cherniak R, Reiss E, Turner SH (1982) A galactoxylomannan antigen of Cryptococcus neoformans serotype A. Carbohydr Res 103(2):239–250
205. O'Meara TR, Alspaugh JA (2012) The Cryptococcus neoformans capsule: a sword and a shield. Clin Microbiol Rev 25(3):387–408
206. Zaragoza O et al (2009) The capsule of the fungal pathogen Cryptococcus neoformans. Adv Appl Microbiol 68:133–216
207. McFadden DC et al (2007) Capsule structural heterogeneity and antigenic variation in Cryptococcus neoformans. Eukaryot Cell 6(8):1464–1473
208. Janbon G et al (2001) Cas1p is a membrane protein necessary for the O-acetylation of the Cryptococcus neoformans capsular polysaccharide. Mol Microbiol 42(2):453–467
209. McFadden DC, De Jesus M, Casadevall A (2006) The physical properties of the capsular polysaccharides from Cryptococcus neoformans suggest features for capsule construction. J Biol Chem 281(4):1868–1875
210. Heiss C et al (2009) The structure of Cryptococcus neoformans galactoxylomannan contains beta-D-glucuronic acid. Carbohydr Res 344(7):915–920
211. Bar-Peled M, Griffith CL, Doering TL (2001) Functional cloning and characterization of a UDP- glucuronic acid decarboxylase: the pathogenic fungus Cryptococcus neoformans elucidates UDP-xylose synthesis. Proc Natl Acad Sci U S A 98(21):12003–12008
212. Moyrand F et al (2008) UGE1 and UGE2 regulate the UDP-glucose/UDP-galactose equilibrium in Cryptococcus neoformans. Eukaryot Cell 7(12):2069–2077
213. Yoneda A, Doering TL (2009) An unusual organelle in Cryptococcus neoformans links luminal pH and capsule biosynthesis. Fungal Genet Biol 46(9):682–687
214. Moyrand F, Fontaine T, Janbon G (2007) Systematic capsule gene disruption reveals the central role of galactose metabolism on Cryptococcus neoformans virulence. Mol Microbiol 64(3):771–781
215. Gilbert NM et al (2010) KRE genes are required for beta-1,6-glucan synthesis, maintenance of capsule architecture and cell wall protein anchoring in Cryptococcus neoformans. Mol Microbiol 76(2):517–534
216. Reese AJ, Doering TL (2003) Cell wall alpha-1,3-glucan is required to anchor the Cryptococcus neoformans capsule. Mol Microbiol 50(4):1401–1409
217. Baker LG et al (2007) Chitosan, the deacetylated form of chitin, is necessary for cell wall integrity in Cryptococcus neoformans. Eukaryot Cell 6(5):855–867
218. Fonseca FL et al (2009) Role for chitin and chitooligomers in the capsular architecture of Cryptococcus neoformans. Eukaryot Cell 8(10):1543–1553
219. Siafakas AR et al (2007) Cell wall-linked cryptococcal phospholipase B1 is a source of secreted enzyme and a determinant of cell wall integrity. J Biol Chem 282(52):37508–37514
220. Frases S et al (2009) Capsule of Cryptococcus neoformans grows by enlargement of polysaccharide molecules. Proc Natl Acad Sci U S A 106(4):1228–1233
221. Zaragoza O et al (2006) The polysaccharide capsule of the pathogenic fungus Cryptococcus neoformans enlarges by distal growth and is rearranged during budding. Mol Microbiol 59(1):67–83
222. Charlier C et al (2005) Capsule structure changes associated with Cryptococcus neoformans crossing of the blood-brain barrier. Am J Pathol 166(2):421–432
223. Mansour MK, Levitz SM (2002) Interactions of fungi with phagocytes. Curr Opin Microbiol 5(4):359–365

224. Kozel TR, Gotschlich EC (1982) The capsule of Cryptococcus neoformans passively inhibits phagocytosis of the yeast by macrophages. J Immunol 129(4):1675–1680
225. Syme RM et al (1999) The capsule of Cryptococcus neoformans reduces T-lymphocyte proliferation by reducing phagocytosis, which can be restored with anticapsular antibody. Infect Immun 67(9):4620–4627
226. Syme RM et al (2002) Primary dendritic cells phagocytose Cryptococcus neoformans via mannose receptors and Fcgamma receptor II for presentation to T lymphocytes. Infect Immun 70(11):5972–5981
227. Monari C et al (2005) Glucuronoxylomannan, a microbial compound, regulates expression of costimulatory molecules and production of cytokines in macrophages. J Infect Dis 191(1):127–137
228. Monari C et al (2006) Microbial immune suppression mediated by direct engagement of inhibitory Fc receptor. J Immunol 177(10):6842–6851
229. Alvarez M, Casadevall A (2006) Phagosome extrusion and host-cell survival after Cryptococcus neoformans phagocytosis by macrophages. Curr Biol 16(21):2161–2165
230. Zaragoza O et al (2008) Capsule enlargement in Cryptococcus neoformans confers resistance to oxidative stress suggesting a mechanism for intracellular survival. Cell Microbiol 10(10):2043–2057
231. Tucker SC, Casadevall A (2002) Replication of Cryptococcus neoformans in macrophages is accompanied by phagosomal permeabilization and accumulation of vesicles containing polysaccharide in the cytoplasm. Proc Natl Acad Sci U S A 99(5):3165–3170
232. Wilder JA et al (2002) Complementation of a capsule deficient Cryptococcus neoformans with CAP64 restores virulence in a murine lung infection. Am J Respir Cell Mol Biol 26(3):306–314
233. Barbosa FM et al (2007) Binding of glucuronoxylomannan to the CD14 receptor in human A549 alveolar cells induces interleukin-8 production. Clin Vaccine Immunol 14(1):94–98
234. De Jesus M et al (2008) Spleen deposition of Cryptococcus neoformans capsular glucuronoxylomannan in rodents occurs in red pulp macrophages and not marginal zone macrophages expressing the C-type lectin SIGN-R1. Med Mycol 46(2):153–162
235. Murphy JW, Cozad GC (1972) Immunological unresponsiveness induced by cryptococcal capsular polysaccharide assayed by the hemolytic plaque technique. Infect Immun 5(6):896–901
236. Mody CH, Syme RM (1993) Effect of polysaccharide capsule and methods of preparation on human lymphocyte proliferation in response to Cryptococcus neoformans. Infect Immun 61(2):464–469
237. Yauch LE, Lam JS, Levitz SM (2006) Direct inhibition of T-cell responses by the Cryptococcus capsular polysaccharide glucuronoxylomannan. PLoS Pathog 2(11):e120
238. Barluzzi R et al (1998) Role of the capsule in microglial cell-Cryptococcus neoformans interaction: impairment of antifungal activity but not of secretory functions. Med Mycol 36(4):189–197
239. Monari C et al (2008) Capsular polysaccharide induction of apoptosis by intrinsic and extrinsic mechanisms. Cell Microbiol 10(10):2129–2137
240. Dong ZM, Murphy JW (1997) Cryptococcal polysaccharides bind to CD18 on human neutrophils. Infect Immun 65(2):557–563
241. Rodrigues ML et al (2007) Vesicular polysaccharide export in Cryptococcus neoformans is a eukaryotic solution to the problem of fungal trans-cell wall transport. Eukaryot Cell 6(1):48–59
242. Peres da Silva R et al (2015) Extracellular vesicles from Paracoccidioides pathogenic species transport polysaccharide and expose ligands for DC-SIGN receptors. Sci Rep 5:14213
243. Rodrigues ML et al (2008) Extracellular vesicles produced by Cryptococcus neoformans contain protein components associated with virulence. Eukaryot Cell 7(1):58–67
244. Zhu J, Yamane H, Paul WE (2010) Differentiation of effector CD4 T cell populations (*). Annu Rev Immunol 28:445–489

245. Rivera J et al (2002) Antibody efficacy in murine pulmonary Cryptococcus neoformans infection: a role for nitric oxide. J Immunol 168(7):3419–3427
246. Hardison SE et al (2010) Pulmonary infection with an interferon-γ-producing Cryptococcus neoformans strain results in classical macrophage activation and protection. Am J Pathol 176(2):774–785
247. Huffnagle GB et al (1996) Afferent phase production of TNF-alpha is required for the development of protective T cell immunity to Cryptococcus neoformans. J Immunol 157(10):4529–4536
248. Xu J et al (2016) Disruption of early tumor necrosis factor alpha signaling prevents classical activation of dendritic cells in lung-associated lymph nodes and development of protective immunity against cryptococcal infection. mBio 7(4):e00510–e00516
249. Herring AC et al (2002) Induction of interleukin-12 and gamma interferon requires tumor necrosis factor alpha for protective T1-cell-mediated immunity to pulmonary Cryptococcus neoformans infection. Infect Immun 70(6):2959–2964
250. Monari C et al (2005) Cryptococcus neoformans capsular glucuronoxylomannan induces expression of fas ligand in macrophages. J Immunol 174(6):3461–3468
251. Small JM, Mitchell TG (1989) Strain variation in antiphagocytic activity of capsular polysaccharides from Cryptococcus neoformans serotype A. Infect Immun 57(12):3751–3756
252. Levitz SM (1994) Macrophage-Cryptococcus interactions. Immunol Ser 60:533–543
253. Aguirre KM, Gibson GW (2000) Differing requirement for inducible nitric oxide synthase activity in clearance of primary and secondary Cryptococcus neoformans infection. Med Mycol 38(5):343–353
254. Naslund PK, Miller WC, Granger DL (1995) Cryptococcus neoformans fails to induce nitric oxide synthase in primed murine macrophage-like cells. Infect Immun 63(4):1298–1304
255. Facchetti F et al (1999) Expression of inducible nitric oxide synthase in human granulomas and histiocytic reactions. Am J Pathol 154(1):145–152
256. Xiao G et al (2008) Cryptococcus neoformans inhibits nitric oxide synthesis caused by CpG-oligodeoxynucleotide-stimulated macrophages in a fashion independent of capsular polysaccharides. Microbiol Immunol 52(3):171–179
257. Arora S et al (2005) Role of IFN-γ in regulating T2 immunity and the development of alternatively activated macrophages during allergic bronchopulmonary mycosis. J Immunol 174(10):6346–6356
258. Qiu Y et al (2012) Immune modulation mediated by cryptococcal laccase promotes pulmonary growth and brain dissemination of virulent Cryptococcus neoformans in mice. PLoS One 7(10):e47853–e47853
259. Eastman AJ et al (2015) Cryptococcal heat shock protein 70 homolog Ssa1 contributes to pulmonary expansion of Cryptococcus neoformans during the afferent phase of the immune response by promoting macrophage M2 polarization. J Immunol 194(12):5999–6010
260. Arora S et al (2011) Effect of cytokine interplay on macrophage polarization during chronic pulmonary infection with Cryptococcus neoformans. Infect Immun 79(5):1915–1926
261. Kawakami K et al (1999) Differential effect of Cryptococcus neoformans on the production of IL-12p40 and IL-10 by murine macrophages stimulated with lipopolysaccharide and gamma interferon. FEMS Microbiol Lett 175(1):87–94
262. Deshaw M, Pirofski LA (1995) Antibodies to the Cryptococcus neoformans capsular glucuronoxylomannan are ubiquitous in serum from HIV+ and HIV− individuals. Clin Exp Immunol 99(3):425–432
263. Abe K et al (2000) Th1-Th2 cytokine kinetics in the bronchoalveolar lavage fluid of mice infected with Cryptococcus neoformans of different virulences. Microbiol Immunol 44(10):849–855
264. Kechichian TB, Shea J, Poeta MD (2007) Depletion of alveolar macrophages decreases the dissemination of a glucosylceramide-deficient mutant of Cryptococcus neoformans in immunodeficient mice. Infect Immun 75(10):4792–4798
265. Levitz SM et al (1999) Cryptococcus neoformans resides in an acidic phagolysosome of human macrophages. Infect Immun 67(2):885–890

266. Voelz K et al (2014) 'Division of labour' in response to host oxidative burst drives a fatal Cryptococcus gattii outbreak. Nat Commun 5:5194–5194
267. Johnston SA, May RC (2010) The human fungal pathogen Cryptococcus neoformans escapes macrophages by a phagosome emptying mechanism that is inhibited by Arp2/3 complex-mediated actin polymerisation. PLoS Pathog 6(8):e1001041–e1001041
268. Nicola AM et al (2011) Nonlytic exocytosis of Cryptococcus neoformans from macrophages occurs in vivo and is influenced by phagosomal pH. MBio 2(4):e00167–e00111
269. Liu T-B, Perlin DS, Xue C (2012) Molecular mechanisms of cryptococcal meningitis. Virulence 3(2):173–181
270. Diamond RD, Bennett JE (1973) Growth of Cryptococcus neoformans within human macrophages in vitro. Infect Immun 7(2):231–236
271. Ngamskulrungroj P et al (2012) The primary target organ of Cryptococcus gattii is different from that of Cryptococcus neoformans in a murine model. MBio 3(3):e00103–e00112
272. Qureshi A et al (2010) Role of sphingomyelin synthase in controlling the antimicrobial activity of neutrophils against Cryptococcus neoformans. PLoS One 5(12):e15587–e15587
273. Rocha JDB et al (2015) Capsular polysaccharides from Cryptococcus neoformans modulate production of neutrophil extracellular traps (NETs) by human neutrophils. Sci Rep 5:8008–8008
274. Qureshi A et al (2011) Cryptococcus neoformans modulates extracellular killing by neutrophils. Front Microbiol 2:193
275. Mednick AJ et al (2003) Neutropenia alters lung cytokine production in mice and reduces their susceptibility to pulmonary cryptococcosis. Eur J Immunol 33(6):1744–1753
276. Wozniak KL, Vyas JM, Levitz SM (2006) In vivo role of dendritic cells in a murine model of pulmonary cryptococcosis. Infect Immun 74(7):3817–3824
277. Wozniak KL, Kolls JK, Wormley FL (2012) Depletion of neutrophils in a protective model of pulmonary cryptococcosis results in increased IL-17A production by gamma/delta T cells. BMC Immunol 13(1):65–65
278. Dong ZM, Murphy JW (1996) Cryptococcal polysaccharides induce L-selectin shedding and tumor necrosis factor receptor loss from the surface of human neutrophils. J Clin Invest 97(3):689–698
279. Wiseman JCD et al (2007) Perforin-dependent cryptococcal microbicidal activity in NK cells requires PI3K-dependent ERK1/2 signaling. J Immunol 178(10):6456–6464
280. Marr KJ et al (2009) Cryptococcus neoformans directly stimulates perforin production and rearms NK cells for enhanced anticryptococcal microbicidal activity. Infect Immun 77(6):2436–2446
281. Islam A et al (2013) An acidic microenvironment increases NK cell killing of Cryptococcus neoformans and Cryptococcus gattii by enhancing perforin degranulation. PLoS Pathog 9(7):e1003439–e1003439
282. Miller MF et al (1990) Human natural killer cells do not inhibit growth of Cryptococcus neoformans in the absence of antibody. Infect Immun 58(3):639–645
283. Cordero RJB et al (2013) Antibody binding to Cryptococcus neoformans impairs budding by altering capsular mechanical properties. J Immunol 190(1):317–323
284. Kawakami K et al (2000) IL-18 contributes to host resistance against infection with Cryptococcus neoformans in mice with defective IL-12 synthesis through induction of IFN-gamma production by NK cells. J Immunol 165(2):941–947
285. Kawakami K et al (2000) NK cells eliminate Cryptococcus neoformans by potentiating the fungicidal activity of macrophages rather than by directly killing them upon stimulation with IL-12 and IL-18. Microbiol Immunol 44(12):1043–1050
286. Zhang T et al (1997) Interleukin-12 (IL-12) and IL-18 synergistically induce the fungicidal activity of murine peritoneal exudate cells against Cryptococcus neoformans through production of gamma interferon by natural killer cells. Infect Immun 65(9):3594–3599

287. Müller U et al (2013) Abrogation of IL-4 receptor-α-dependent alternatively activated macrophages is sufficient to confer resistance against pulmonary cryptococcosis despite an ongoing Th2 response. Int Immunol 25(8):459–470
288. Kelly RM et al (2005) Opsonic requirements for dendritic cell-mediated responses to Cryptococcus neoformans. Infect Immun 73(1):592–598
289. Hole CR et al (2016) Antifungal activity of plasmacytoid dendritic cells against Cryptococcus neoformans in vitro requires expression of dectin-3 (CLEC4D) and reactive oxygen species. Infect Immun 84(9):2493–2504
290. Huston SM et al (2013) Cryptococcus gattii is killed by dendritic cells, but evades adaptive immunity by failing to induce dendritic cell maturation. J Immunol 191(1):249–261
291. Wozniak KL, Levitz SM (2008) Cryptococcus neoformans enters the endolysosomal pathway of dendritic cells and is killed by lysosomal components. Infect Immun 76(10):4764–4771
292. Ueno K et al (2015) Dendritic cell-based immunization ameliorates pulmonary infection with highly virulent Cryptococcus gattii. Infect Immun 83(4):1577–1586
293. Hage CA et al (2003) Pulmonary cryptococcosis after initiation of anti-tumor necrosis factor-α therapy. Chest 124(6):2395–2397
294. Lupo P et al (2008) The presence of capsule in Cryptococcus neoformans influences the gene expression profile in dendritic cells during interaction with the fungus. Infect Immun 76(4):1581–1589
295. Grijpstra J et al (2009) The Cryptococcus neoformans cap10 and cap59 mutant strains, affected in glucuronoxylomannan synthesis, differentially activate human dendritic cells. FEMS Immunol Med Microbiol 57(2):142–150
296. Hardison SE, Brown GD (2012) C-type lectin receptors orchestrate antifungal immunity. Nat Immunol 13(9):817–822
297. Pericolini E et al (2006) Cryptococcus neoformans capsular polysaccharide component galactoxylomannan induces apoptosis of human T-cells through activation of caspase-8. Cell Microbiol 8(2):267–275
298. Stenzel W et al (2009) IL-4/IL-13-dependent alternative activation of macrophages but not microglial cells is associated with uncontrolled cerebral cryptococcosis. Am J Pathol 174(2):486–496
299. Dyken SJV, Locksley RM (2013) Interleukin-4- and interleukin-13-mediated alternatively activated macrophages: roles in homeostasis and disease. Annu Rev Immunol 31(1):317–343
300. Müller U et al (2007) IL-13 induces disease-promoting type 2 cytokines, alternatively activated macrophages and allergic inflammation during pulmonary infection of mice with Cryptococcus neoformans. J Immunol 179(8):5367–5377
301. Zhang Y et al (2009) Robust Th1 and Th17 immunity supports pulmonary clearance but cannot prevent systemic dissemination of highly virulent Cryptococcus neoformans H99. Am J Pathol 175(6):2489–2500
302. Wager CML et al (2015) STAT1 signaling within macrophages is required for antifungal activity against Cryptococcus neoformans. Infect Immun 83(12):4513–4527
303. Szymczak WA, Sellers RS, Pirofski L-a (2012) IL-23 dampens the allergic response to Cryptococcus neoformans through IL-17–independent and –dependent mechanisms. Am J Pathol 180(4):1547–1559
304. Müller U et al (2012) Lack of IL-4 receptor expression on T helper cells reduces T helper 2 cell polyfunctionality and confers resistance in allergic bronchopulmonary mycosis. Mucosal Immunol 5(3):299–310
305. Ma LL et al (2002) CD8 T cell-mediated killing of Cryptococcus neoformans requires granulysin and is dependent on CD4 T cells and IL-15. J Immunol 169(10):5787–5795
306. Mody CH et al (1994) Un vivo depletion of murine CD8 positive T cells impairs survival during infection with a highly virulent strain ofCryptococcus neoformans. Mycopathologia 125(1):7–17

307. Jarvis JN et al (2013) The phenotype of the Cryptococcus-specific cd4+ memory t-cell response is associated with disease severity and outcome in hiv-associated cryptococcal meningitis. J Infect Dis 207(12):1817–1828
308. Zaragoza O, Casadevall A (2006) Monoclonal antibodies can affect complement deposition on the capsule of the pathogenic fungus Cryptococcus neoformans by both classical pathway activation and steric hindrance. Cell Microbiol 8(12):1862–1876
309. Nabavi N, Murphy JW (1986) Antibody-dependent natural killer cell-mediated growth inhibition of Cryptococcus neoformans. Infect Immun 51(2):556–562
310. Rosas ÁL, Nosanchuk JD, Casadevall A (2001) Passive immunization with melanin-binding monoclonal antibodies prolongs survival of mice with lethal Cryptococcus neoformans infection. Infect Immun 69(5):3410–3412
311. McClelland EE et al (2010) Ab binding alters gene expression in Cryptococcus neoformans and directly modulates fungal metabolism. J Clin Invest 120(4):1355–1361
312. Rodrigues ML et al (2000) Human antibodies against a purified glucosylceramide from Cryptococcus neoformans inhibit cell budding and fungal growth. Infect Immun 68(12):7049–7060
313. Rachini A et al (2007) An anti-β-glucan monoclonal antibody inhibits growth and capsule formation of Cryptococcus neoformans in vitro and exerts therapeutic, anticryptococcal activity in vivo. Infect Immun 75(11):5085–5094
314. McClelland EE, Casadevall A (2012) Strain-related differences in antibody-mediated changes in gene expression are associated with differences in capsule and location of binding. Fungal Genet Biol 49(3):227–234
315. Dromer F et al (1987) Protection of mice against experimental cryptococcosis by anti-Cryptococcus neoformans monoclonal antibody. Infect Immun 55(3):749–752
316. Mukherjee S et al (1994) Monoclonal antibodies to Cryptococcus neoformans capsular polysaccharide modify the course of intravenous infection in mice. Infect Immun 62(3):1079–1088
317. Fleuridor R, Zhong Z, Pirofski L-a (1998) A human IgM monoclonal antibody prolongs survival of mice with lethal cryptococcosis. J Infect Dis 178(4):1213–1216
318. Rivera J, Zaragoza O, Casadevall A (2005) Antibody-mediated protection against Cryptococcus neoformans pulmonary infection is dependent on B cells. Infect Immun 73(2):1141–1150
319. Subramaniam KS et al (2010) Improved survival of mice deficient in secretory immunoglobulin M following systemic infection with Cryptococcus neoformans. Infect Immun 78(1):441–452
320. Aguirre KM, Johnson LL (1997) A role for B cells in resistance to Cryptococcus neoformans in mice. Infect Immun 65(2):525–530
321. Baumgarth N (2011) The double life of a B-1 cell: self-reactivity selects for protective effector functions. Nat Rev Immunol 11(1):34–46
322. Espinel-Ingroff A, Kidd SE (2015) Current trends in the prevalence of Cryptococcus gattii in the United States and Canada. Infect Drug Resist 8:89–97

Chapter 7
Pneumocystis jirovecii and Pneumocystosis

Olga Matos, Ana Luisa Tomás, and Francisco Antunes

Abstract *Pneumocystis jirovecii* pneumonia (PcP or pneumocystosis) remains a major cause of respiratory illness among immunocompromised patients, especially those infected with the human immunodeficiency virus (HIV). *P. jirovecii* infection has also been detected in healthy individuals and people with chronic lung diseases, raising the possibility that the spectrum of illness caused by this pathogenic fungus may extend well beyond the immunocompromised host. In industrialized countries, while the number of acquired immunodeficiency syndrome (AIDS) patients suffering from PcP decreases, there is a rising number of non-HIV immunocompromised patients at risk for life-threatening PcP. In contrast, in developing countries a limited number of epidemiological studies have evaluated PcP prevalence; however, recent reports have described an increased rate of the disease in AIDS patients in Africa, Asia and South America.

After nearly three decades of intensive research, a significant progress has been made in the understanding of PcP; however, this disease continues to be a challenge for clinicians and microbiologists; in particular, the processes of detection and interpretation of the PcP testing, because even the most current diagnostic methods cannot adequately differentiate between colonized (asymptomatic) and symptomatic patients.

This chapter reviews the basic biology of *P. jirovecii* and the epidemiology of the caused infection, including genetic diversity. It also discusses pathogenesis and host defences against *P. jirovecii*, and the clinical presentation in HIV-infected and non-HIV-infected immunocompromised patients. Attention will be devoted to classic and modern approaches for diagnosis, as well as to treatment and prophylaxis of

O. Matos, M.D., M.Sc., Ph.D. (✉) • A.L. Tomás, M.Sc., Ph.D. student
Medical Parasitology Unit, Group of Opportunistic Protozoa/HIV and Other Protozoa, Global Health and Tropical Medicine, Instituto de Higiene e Medicina Tropical, Universidade NOVA de Lisboa, Lisboa, Portugal
e-mail: omatos@ihmt.unl.pt; ana.tomas@ihmt.unl.pt

F. Antunes, M.D., Ph.D.
Instituto de Saúde Ambiental, Faculdade de Medicina de Lisboa, Universidade de Lisboa, Lisboa, Portugal
e-mail: fantunes@medicina.ulisboa.pt

H.M. Mora-Montes, L.M. Lopes-Bezerra (eds.), *Current Progress in Medical Mycology*, DOI 10.1007/978-3-319-64113-3_7

PcP. This chapter is designed to present a balanced, comprehensive, and concise examination of the latest advances in the field of *P. jirovecii* pneumonia, and to contain the most up-to-date references.

7.1 Introduction

Pneumocystis jirovecii (formerly *Pneumocystis carinii* f. sp. *hominis*) is an atypical fungus causing pneumonia in humans. This microorganism was first reported in Brazil, in 1909 and 1910, by Carlos Chagas and Antonio Carini, respectively, who misidentified the organism found in murine lungs with stages in the life cycle of *Trypanosoma cruzi* [1, 2]. Antonio Carini, doubting that the cysts were part of the life cycle of *T. cruzi*, sent his specimens to the Pasteur Institute in Paris, where in 1912, the researcher duo of Pierre and Marie Delanoë described the organism identified as a new biological entity, and suggested the name *Pneumocystis carinii* because cysts, their characteristic morphological form, were found only in the lungs of hosts and as a tribute to Carini [3].

For several decades this organism was considered an enigmatic lung pathogen, and in the beginning of the 1980s of the twentieth century it became one of the leading causes of death in the pandemic of acquired immunodeficiency syndrome (AIDS) [4].

Pneumocystis was classified for many years as a protozoan, because of the histological characteristics of its life cycle forms, trophozoite and cyst, inability to maintain in vitro cultures of the organism, successful response to treatment with anti-parasitic drugs and by contrast not responding to a broad spectrum of antifungal agents, such as amphotericin B, and the number of copies of ribosomal ribonucleic acid (rRNA) genes, having only one nuclear gene unlike other fungi studied, which have hundreds of copies for this gene [5]. Later, based on DNA studies it was reclassified as an atypical fungus of the *phylum* Ascomycota [6–9]. *Pneumocystis* infects a variety of mammalian hosts, and the human form because of its host specificity was renamed *Pneumocystis jirovecii* in honor of Otto Jirovec, who linked it to epidemics of interstitial plasma cell pneumonia in neonates in Europe [7], while *P. carinii* is now reserved for the species affecting rats.

The life cycle of *Pneumocystis*, which develops extracellularly in the alveolar cavities of mammals, is not completely defined, containing an asexual mode of replication via binary fission of the trophic form (formerly called 'trophozoite') and a sexual mode of replication that gives rise to a cystic form or ascus containing eight spores [10]. The asexual mode of replication was observed in lung tissues of infected rats and humans by ultrastructural transmission electron microscopy studies [11, 12]. Nevertheless, the sexual replication is still putative, based on the presence of a pheromone receptor protein on the surface of some trophic forms, suggesting mating of two haploid trophic forms, giving rise to a diploid zygote that progresses through meiosis to produce four haploid nuclei [10, 13]. Meiosis is followed by an

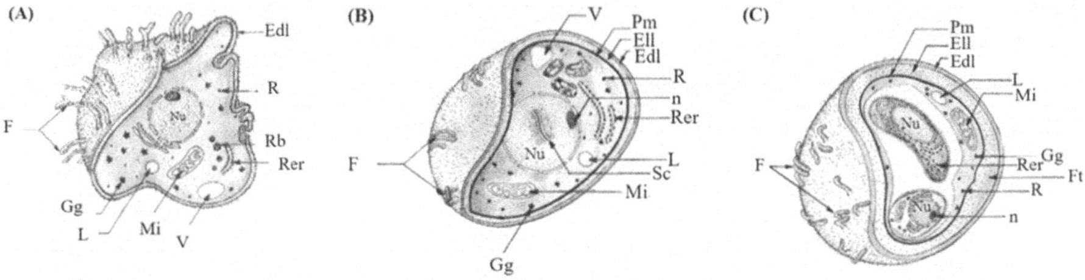

Fig. 7.1 Structure of the developmental stages of *Pneumocystis*. (**a**) Structure of the trophic form of *Pneumocystis*; (**b**) Structure of the sporocyte of *Pneumocystis*; (**c**) Structure of the cystic form of *Pneumocystis*. *F* filopodia, *Gg* glycogen granules, *L* lipid bodies, *Mi* mitochondria, *V* vacuoles, *Rer* rough endoplasmic reticulum, *Rb* round bodies, *R* ribosome, *Nu* nucleus, *Sc* synaptonemal complex, *n* nucleolus, *Pm* plasma membrane, *Ell* electron-lucent layer, *Edl* electron-dense layer, *Ft* focal thickening of the electron-lucent layer (adapted from Matos [14])

additional mitosis to produce eight nuclei that will turn into spores by invagination of the cyst cell membranes to produce eight double-membrane spores, and finally excystment occurs with the release of the spores to the alveolar cavity.

Figure 7.1 presents the structure of the developmental stages of *Pneumocystis*. The trophic forms appear as mononuclear, 2–8 μm in diameter, mostly haploid eukaryotic cells, amoeboid in shape, presenting a thin cell wall consisting of an electron-dense single layer (Fig. 7.1a). They attach specifically to type-1 epithelial alveolar cells, emitting cytoplasmic expansions or filopodia that may penetrate deeply into the cytoplasm of the host cell. However, no disruption of host cell membrane results from either attachment or filopodial activity [15, 16]. The intermediate sporocyte, probably diploid, measures 4–5 μm in diameter and has initially an oval form (Fig. 7.1b). As the cell matures it becomes spherical and an electron-lucent layer appears between the electron-dense layer and the plasma membrane [12]. It further thickens at the late sporocyte stage to finally produce a thick-walled mature cyst that measures 4–7 μm in diameter (Fig. 7.1c) [12, 15]. Trophic forms are the most abundant of all *Pneumocystis* lifecycle stages, representing 90–95% of the total population in the lungs of hosts with PcP [15].

Pneumocystis jirovecii is usually found to the lungs, although extrapulmonary sites have been reported [17].

7.2 Epidemiology of *Pneumocystis jirovecii* Infection

7.2.1 Preponderance of PcP

P. jirovecii is often found in the lungs of healthy people; however, PcP is rare in persons with an intact immune system [18]. As such, prior to the human immunodeficiency virus (HIV)/AIDS pandemic, PcP was an uncommon disease. In the 1980s, with the onset of that pandemic, there were significant changes in the incidence of

Table 7.1 Summary of risk factors for *Pneumocystis jirovecii* pneumonia in different underlying conditions

Underlying condition	Risk factor	References
HIV infection/AIDS	CD4+ T count <200 cells/µl	[22, 23]
	CD4+ T cell percentage < 14%	[23, 24]
	History of oral thrush	[22–24]
	History of previous PcP	[23, 24]
	Viral replication not controlled	[22–24]
	Recurrent bacterial Pneumonia	[23, 24]
	Unintentional weight loss	[23, 24]
	Tobacco	[25]
Haematological malignancies	CD4+ T count <200 cells/µl	[26]
	Lymphocytopenia	[26, 27]
	Steroid and cytotoxic therapy	[26–28]
	Administration of purine analogues	[26, 29]
	Chemotherapy with R-CHOP14 or high-dose methotrexate	[27, 30]
	Administration of Monoclonal antibodies	[26, 30, 31]
Solid organ transplantation	CD4+ T count <200 cells/µl	[32, 33]
	Immunosuppressive therapies	[33–35]
	CMV disease	[33, 36, 37]
	Co-morbidity (tuberculosis, bacterial pneumonia, hepatitis C)	[38]
	Allograft rejection	[33, 35, 38]
	Neutropenia	[33]
	Exposure to cases of PcP	[33, 39]
Solid tumors	Corticosteroid therapy	[40, 41]
	Chemotherapy	[42]
	Radiotherapy	[43]
Inflammatory and collagen-vascular disease	Steroid and cytotoxic therapy	[44]
	Administration of monoclonal antibodies	[45, 46]

PcP, as it turned from a rare infection into a common pneumonia [19–21]. In this period, before the widespread use of PcP prophylaxis and combination antiretroviral therapy (cART), PcP occurred in 70–80% of patients with AIDS [22] and the disease was associated with a mortality rate of 20–40% of these patients. Approximately, 90% of PcP cases occurred in patients with CD4+ T cells count <200 cells/mm^3, or with a percentage < 14%. Other risk factors associated with a higher risk of PcP included, oral thrush, previous episodes of PcP, recurrent bacterial pneumonia, unintentional weight loss, higher plasma HIV RNA levels and tobacco use. Table 7.1 summarizes the risk factors for PcP in HIV-infected and in non-HIV-infected persons. However, with the introduction of PcP prophylaxis since late 1980s, and the development and availability of cART since mid-1990s, there was a decline in PcP cases in the United States and Western Europe [21]. The recent incidence among patients with AIDS in the United States and Western Europe is <1 case per 100 person-years [47]. Most of the recent cases occur in patients who are unaware of their HIV-status or are not receiving ongoing care for HIV due to non-adherence to treatment, those with active substance abuse or psychiatric illness, those who are not receiving or responding to antiretroviral therapy or prophylaxis because of factors related to pharmacokinetics, or unexplained biologic factors, and in those with advanced immunodeficiency (CD4+ T cells <100 cells/mm^3) [48–50]. Data from the European Centre for Disease Prevention and Control (ECDC) and the World Health Organization (WHO) Europe point to PcP as the most common AIDS-defining illness in Europe in 2015, recorded in 20.0% of cases, followed by esophageal candidiasis (11.2%) and wasting syndrome due to HIV (10.8%) [51].

In developed countries, the marked decrease in incidence of PcP in HIV-infected patients is being counterbalanced by the emergence of the disease in non-HIV-infected populations, namely haematologic patients; solid organ transplant recipients, including renal transplants; solid malignancies; rheumatoid conditions; patients with connective tissue disorders; those receiving immunomodulatory therapies; or patients with pre-existing chronic lung conditions or with less typical scenarios, such as Dengue fever [27, 28, 52–58]. In developing countries, where most HIV-infected persons reside, PcP is an emerging disease with high prevalence and is poorly controlled, since the access to cART and PcP chemoprophylaxis is still limited; and possibly also due to lack of PcP diagnostic resources and expertise [59–62]. Recent reports described an increased number of PcP cases in Africa, Asia and South America, with differences in the observed frequencies [61]. These variations are possibly due to differences in study design. Including heterogeneity in the patient population and the diverse laboratory methods used [61, 62].

7.2.2 Circulation and Transmission of P. jirovecii

P. jirovecii is a ubiquitous human-specific pathogen. Initial infection with *P. jirovecii* usually occurs in early childhood; about two-thirds of healthy children have antibodies to *P. jirovecii* by ages 2–4 years [63, 64].

Transmission of *Pneumocystis* can be aerial, and has been suggested based on rodent studies. An article published recently, using the rodent model, shows that the cystic forms (but not the trophic forms) can be transmitted by aerial route from host to host [65]. Also, the detection by molecular methods of *Pneumocystis* DNA in air samples collected in rooms and wards where patients with PcP were hospitalized [66, 67], and case clusters in immunocompromised patients support this transmission route [38, 65, 68, 69]. Vertical transmission of *P. jirovecii* in humans, although controversial, was suggested by some authors [70–73]. *P. jirovecii* is human specific and cannot be acquired from other animal reservoirs [10]; but, since it is accepted to be ubiquitous, the existence of environmental reservoirs of this pathogen cannot be set aside [74].

Pneumocystis pneumonia may probably occur by both de novo acquisition of *P. jirovecii* infection and reactivation of a latent infection [75, 76]. Due to the high seroprevalence of anti-*P. jirovecii* antibodies in children, it was initially assumed that latent organisms were acquired by the host early in childhood [63]. Once acquired, the microorganism remained in the lungs of the immunocompetent host without causing symptoms, since the immune system was able to successfully control the infection, although it was not able to eliminate the microorganism [77, 78]. However, in a situation of immunosuppression, latent infection was reactivated and the disease was manifested. Other studies support the idea that *P. jirovecii* can be acquired de novo, contrary to the theory of latent infection as a primary infection mechanism. Wakefield et al. used sequential sampling and observed that latency is not the rule after an episode of PcP [79]. Other studies showed that genotypes distribution patterns differed at each of the cities where the study was performed and correlated with the patients' place of diagnosis but not by place of birth, suggesting that most cases of PcP do not result from infections acquired early in life [80]. It suggested that at least some infections must be actively acquired from a relatively common source (humans or the environment).

Meanwhile, pulmonary colonization with *P. jirovecii* in patients with diverse levels of immunodeficiency, primary respiratory disorders, or even in the immunocompetent general population, has been reported [18, 79, 81, 82]. *Pneumocystis* colonization occurs in persons without signs and symptoms of acute pneumonia. Because of the low organism burden associated with colonization, polymerase chain reaction (PCR)-based techniques are most often necessary to determine the presence of the organism's DNA [83, 84]. This is an important epidemiological issue, especially in terms of transmission, since colonized populations may be a potential reservoir of the fungus and pass on to others, putting those people at risk especially if they are immunocompromised.

7.2.3 Molecular Epidemiology of P. jirovecii

In the absence of a well-established culture system to isolate and maintain alive *P. jirovecii* organisms, previous efforts to describe differences among isolates have been performed using molecular methodologies. It has been shown that specific single-nucleotide polymorphisms (SNPs) in *Pneumocystis* genes are highly informative markers for comparative analyses between the given genotype and specific clinical or epidemiological data [80, 85–89].

A variety of typing methods have been used for *Pneumocystis* genetic analysis. Single-strand conformation polymorphism [90–92], restriction fragment length polymorphism [93] and type-specific oligonucleotide hybridization [94] were initially used, but have been replaced by DNA sequencing schemes. The multilocus sequence typing approach has become the most frequently used method [95], since it offers many advantages, such as reproducibility, possibility of exchanging data between different laboratories, as well as observation of polymorphisms at many loci at the same time [88]. However, technologies based on Sanger sequencing have difficulty to detect infections with more than one genotype so, amplification of genomic DNA associated with single base extension technology centered on the detection of specific SNPs was developed [96, 97]. This technique detects mixtures of genotypes with a higher sensitivity than technologies based on Sanger sequencing. Recent studies have also proposed the analysis of microsatellite markers, also called short tandem repeat analysis, to genotype *P. jirovecii* [98, 99].

With these techniques, several polymorphic loci have been identified and characterized. Two loci have been most widely studied with the purpose of determining the genetic diversity of this organism: the mitochondrial large subunit ribosomal RNA (*mtLSUrRNA*) and the internal transcribed spacer (*ITS*) regions (ITS1 and ITS2) of the nuclear rRNA operon [88]. The *mtLSUrRNA* gene is also the locus used most widely for PCR-based detection of *P. jirovecii*, as it is present in a large number of copies [100]. *ITS*1 and *ITS*2 show the highest level of sequence variation in *P. jirovecii*, and are considered to be the most discriminatory DNA regions for genotyping studies [101]. Nevertheless, the cytochrome b (*CYB*), the β-tubulin (*β-TUB*), the thymidylate synthase (*TS*), the thioredoxin reductase (*TRR*1), the dihydrofolate reductase (*DHFR*), the dihydropteroate synthase (*DHPS*) and the superoxide dismutase (*SOD*) loci, are also variable regions of the *P. jirovecii* genome suitable to be used in identification and characterization of *P. jirovecii* haplotypes correlated with drug resistance and PcP outcome [86, 88, 89, 96, 97, 102–111]. The major genetic variations in these loci associated with clinical outcome are summarized in Table 7.2.

These specific polymorphic loci have been studied with the purpose of understanding important properties of isolates like biodiversity [80, 85, 91, 92, 115, 116], virulence [87, 89], drug susceptibility [103, 104, 106, 113, 117], geographical

Table 7.2 The major genetic variations of *Pneumocystis jirovecii* associated with clinical outcome to date

Locus	SNP position or Genotype	Clinical outcome	References
DHPS	165, 171	Associated with sulpha drugs prophylaxis failure; Failure of both trimethoprim-sulfamethoxazole treatment and trimethoprim-sulfamethoxazole or dapsone prophylaxis.	[86, 104, 107, 111, 112]
	55, 57	Associated with trimethoprim-sulfamethoxazole resistance.	[105, 106, 109, 112]
DHFR	312, 381	Associated with PcP infection high burden and drug resistance.	[89, 105, 113]
SOD	110, 215	Associated with severity of PcP episodes.	[89, 96, 97, 114]
mtLSUrRNA	85	Associated with high *P. jirovecii* burden levels and unfavourable follow up of infection.	[96, 97]
CYB	SNPs in the Q_o site	Associated with atovaquone prophylaxis failure.	[89, 103, 108]
ITS regions	Genotype Ne	Associated with treatment failure and bad clinical outcome	[110]

distribution of genotypes [80, 114, 117–119], and modes of transmission [67, 68, 71, 73, 75, 80, 82, 120]. Concerning biodiversity, studies demonstrated that patients could be infected with a single genotype, while others could be coinfected by more than one *P. jirovecii* genotype [80, 85, 91, 92], and that new *P. jirovecii* genotypes have been identified in several loci [116]. In terms of virulence and drug resistance, data suggest that these factors may be dependent on multiple *P. jirovecii* genotypes and therefore on the association of multiple polymorphisms that occur in various regions of the genome of the organism. The different associations of polymorphisms, could give the organism distinct characteristics, influencing the presentation and clinical course of PcP [89, 105, 114, 116]. However, studies performed on drug susceptibility strongly suggest that mutations in *P. jirovecii* therapeutic targets are selected by drug pressure [86, 88, 89, 102, 104, 105, 107, 108, 111, 112]. Gene mutations that confer clinical resistance to sulpha/sulphone agents on therapeutic dose are not yet reported [60].

Studies on geographical distribution of genotypes have shown that the isolates could be clustered due to geographic differences but are also dependent on clinical characteristics of the populations studied [88, 118, 119]. On the other hand, studies in transmission of *P. jirovecii* have suggested and supported theories about airborne route [67, 120], the existence of asymptomatic carriers [18, 79, 81, 82], vertical transmission [71, 73], and de novo acquisition of the infection [18, 79, 80, 114]. These advances in the understanding of the genetic diversity of *P. jirovecii* raised important questions, such as whether some organisms are more pathogenic than others and whether resistant or more virulent genotypes spread between hosts. However, despite the wide application of molecular methods to study this organism, inconsistent results in association of specific genotypes to the clinical course or outcome of PcP are common [88]. This is due to the lack of guidance and standardization regarding the appropriateness of different methods and loci to apply in these studies. The standardization of molecular typing guidelines would facilitate comparing results reported by different laboratories and, consequently, the analysis of the correlation between genotypes and a specific clinical picture caused by their invasion may be used to predict the PcP outcome, as well as to facilitate the decision about treatment and prophylaxis of PcP.

7.3 Pathogenesis and Host Immune Response During *P. jirovecii* Infection

7.3.1 Pathogenesis of P. jirovecii *Infection: Lung Injury*

The pathogenesis of *Pneumocystis* pneumonia can be divided into four stages: establishment of the infection, organism proliferation, changes within the alveolar microenvironment and host immune response [121, 122].

P. jirovecii is an atypical fungus exhibiting a highly defined tropism for human lungs, where it exists primarily as an extracellular alveolar pathogen [122, 123]. Thus, the initial event in the establishment of the infection is the preferential attachment of the trophic forms of *Pneumocystis* to type-I epithelial cells and, to a lesser extent, to alveolar type-II epithelial cells, through interdigitation of their cell membranes with those of the host cells [124, 125]. This attachment is facilitated by both fibronectin and vitronectin that are present in the alveolar lung fluid [126–128]. Integrins, which are fibronectin receptors, may also play a role in *Pneumocystis*-epithelial cell interactions [129].

This binding of *Pneumocystis* to cell matrix proteins, triggers altered gene expression that results in increased *Pneumocystis* proliferation and also inhibits the growth of lung epithelial cells [125, 130]. Otherwise, the impaired cellular immunity is generally considered the major predisposing factor for organism proliferation.

Consequently, as infection evolves, the alveolar microenvironment is impaired by direct *Pneumocystis*-mediated lung damage and also by initiation of subsequent host immune and inflammatory response that injures the lung [131]. In the first days of infection, the inflammatory reactions result in cellular alterations with limited lesions. As the infection evolves, characteristic features of the infection begin to appear, such as the presence of an exudate that fills the alveoli uniformly and the observation of focal lesions of necrosis in the alveolar type-I epithelial cells, hypertrophy of the alveolar cells and filling of the alveolar space by injured cells and *Pneumocystis* forms [121]. Although the mechanisms of direct injury are unclear, studies indicate that the release of several *Pneumocystis*-derived proteolytic and glycolytic enzymes may mediate lung collateral damage while performing their intended functions in the life cycle of this pathogen [132–136]. On the other hand, another mechanism by which *Pneumocystis* evades its host is by antigenic variation. This microorganism has a major surface glycoprotein (Msg) represented by a family of related glycoproteins encoded by about 50–80 genes depending on the species [137, 138]. Therefore, *Pneumocystis* is able of expressing one isoform of Msg at a given time of infection and switch to another in the course of infection, which allows it to hide from the host immune system and to proliferate.

Finally, the host immune response elicited by *Pneumocystis* infection is proved to be an important determinant of the ultimate outcome of the infection. Although infection is necessary to cause PcP, certain aspects of the immune response cause or exacerbate disease symptoms. Although all patients that develop clinical PcP are immunocompromised, patients with greater immune competency may better control the number of the organisms, but suffer greater inflammatory injury. Conversely, patients with more severe immunosuppression may suffer fewer inflammatory complications but are more susceptible to direct effects of higher organism burdens. That is why immunocompromised non-HIV-infected patients usually have a more rapid onset of the disease with exacerbated clinical manifestations than those infected with HIV, in whom PcP usually has a subacute presentation with more insidious involvement [123, 139, 140].

Thus, although the specific mechanisms of PcP-related immunopathogenesis are not fully defined, the contribution of inflammation and the immune response to this disease process seems to be clear.

7.3.2 Immunity Against P. jirovecii *Infection*

Because of the inability to reliably culture *Pneumocystis* in vitro, investigations into events associated with immunity against *Pneumocystis* has been relegated to insights gained from animal models. These studies have shown that immunity to *Pneumocystis* is a dynamic interplay between nearly every arm of the immune system. Innate cellular immunity is absolutely critical to providing the initial recognition event that precipitates an immune response and to clearing the pathogenic organism from the host. Adaptive immunity, in turn, is crucial for amplifying the innate cellular response, controlling the invader as well as maintaining long-term memory protection.

In the beginning of the infection, as *Pneumocystis* has tropism for the alveolar spaces of the lung, the alveolar macrophages (AM) are the first line of host defence to control this microorganism. There are many studies showing an inverse correlation between macrophage numbers and severity of PcP [141–143] since AM serve as an important host component to recognize the organism and play a central role in directing the inflammatory and adaptive immune responses to this pathogen. The recognition and uptake of *Pneumocystis* by AM enables them to directly kill both trophic and cystic forms, and occurs through multiple receptor systems, such as the mannose receptor that interacts with gpA/Msg on the surface of *Pneumocystis* and dectin-1 receptors that recognize the β-glucan moieties in the fungal cell wall [144–147]. Once internalized, several studies have shown that *Pneumocystis* is incorporated into phagolysosomes and subsequently degraded by induction of the oxidative burst and hydrogen peroxide generation [143, 145, 147, 148]. In addition to degrading *Pneumocystis*, AM also produce a wide variety of pro-inflammatory mediators that, in addition to assisting in the eradication of this organism by the host and in the activation of the adaptive immune system, also promote inflammatory responses that lead to lung injury [143].

There have been fewer studies on the role of other phagocytes participating in the innate response to *Pneumocystis*; however, dendritic cells have shown ability to protect mice from *Pneumocystis* pneumonia, since these cells can induce anti-*Pneumocystis* IgG antibodies and protect $CD4^+$ T cell-depleted mice from *Pneumocystis* pneumonia [149]. Also, neutrophils are thought to play a potentially negative role in the innate responses against this pathogen, since HIV-infected patients had significantly fewer neutrophils recovered in their bronchoalveolar lavage compared to non-HIV-infected patients, which corroborates with the exacerbated clinical manifestations and inflammatory complications of PcP in non-HIV-infected patients [139, 150]. Even more, episodes of severe inflammation caused by

Pneumocystis infection in the lungs of HIV-infected individuals have been synonymous with the presence of elevated neutrophil counts [151, 152].

When the adaptive immunity is activated, the CD4$^+$ T cells have the central role, having this been well-established with a strong correlation between their diminished numbers in infected individuals and an increased risk of PcP development [22, 32]. While the level of CD4$^+$ T cells clearly correlates with the control of *Pneumocystis* infection, the mechanism by which these cells contribute to clearance of the organism is less clear. Studies with severe combined immunodeficiency (SCID) mice showed that CD4$^+$ T cells play an essential role in the recruitment and activation of effector cells, which are responsible for eliminating the organism. These animals develop severe *Pneumocystis* infection despite the presence of functional neutrophils and macrophages [153, 154]; however, when reconstituted with CD4$^+$ T cells from the spleen, they regain the ability to effectively clear the infection, indicating that CD4$^+$ T cell-mediated activation of macrophages is required for effective host defence against *Pneumocystis* [155, 156]. On the other hand, the specific T-cell subsets required for successful resolution of the infection remain unclear, as authors show the increased recruitment of T helper (Th) lymphocytes, both type 1 (Th1) and type 2 (Th2), to the lungs following experimental *Pneumocystis* infection in mice, besides a predominant frequency of Th2 cells response [157, 158].

Relatively to CD8$^+$ T cells, they work in cohort with CD4$^+$ T cells in the normal, effective immune response to *Pneumocystis* infection. However, there has been much debate on whether their involvement is protective or detrimental, particularly in the setting of CD4$^+$ T-cell deficiency. Authors have shown that in mice selectively depleted of CD4$^+$ T cells, infection with *Pneumocystis* leads to an influx of CD8$^+$ T cells in the lungs, though the mice fail to control the infection [159]. However, mice depleted of both cell types had a higher organism burden than those depleted of CD4$^+$ T cells alone, which suggests that CD8$^+$ T cells have a partial protection role in the context of CD4$^+$ T-cell deficiency or dysfunction [160].

The role of gamma delta (γδ)-T cells in *Pneumocystis* infection has not been studied so extensively. However, recent studies that investigate *Pneumocystis* infection in wild-type mice observed that γδ-T cells were recruited in high numbers to the lungs during *Pneumocystis* infection [161]. They demonstrated that γδ-T cells-deficient mice had augmented resolution of *Pneumocystis* pneumonia correlated with augmented CD8$^+$ T-cell recruitment as well as higher production of interferon gamma (IFN-γ). These results suggest that one role of γδ-T cells may be to influence recruitment of CD8$^+$ T cells to the lungs and possibly modulate their inflammatory effect.

The importance of B cells in protection from *Pneumocystis* infection is evident from experimental animal models of B cell deficiencies [162–164], as well as clinical studies where PcP has been reported following B cell-targeted chemotherapy [165, 166]. In addition to obvious effects on antibody production, the increased susceptibility of these patients may also be due to the fact that B cells can function as critical antigen presenting cells during the infection [164]. Thus, the question arises as to whether B-cell-mediated protection against *Pneumocystis* is primarily antibody mediated, or is produced through B-cell activation of CD4$^+$ T cells. Reports of PcP in patients with genetic mutations affecting immunoglobulin production [167], along with the observations that the majority of healthy adults are seropositive for *P. jirovecii*, rarely presenting symptomatic infection [84], provide strong evidence that anti-*Pneumocystis* humoral immunity is a critical component of protective immunity to this pathogen. On the other hand, studies demonstrated that SCID mice that received $CD4^+$ T cells from normal mice were able to clear *Pneumocystis* infection more efficiently than mice that received $CD4^+$ T cells from B cell-deficient mice [164]. Together, these studies show that antibodies play a crucial role in clearing *Pneumocystis* from the lungs, but also indicate that B cells play a critical role in priming and activating $CD4^+$ T cells, and that this interaction is required for clearance of *Pneumocystis* in the lungs.

In addition to all these cellular immunity participants, various cytokines and chemokines, released by all nucleated cells, participate in the host response and provide a mechanism for intercellular communication that is not dependent on cell contact [122, 147, 168]. They are involved in every aspect of host defence against infection and play a determinant role during resolution of PcP. They also participate in the maintenance of a crucial balance between the clearance of *Pneumocystis* from the lungs and modulate the inflammation response to this pathogen. The main groups of cytokines and chemokines implicated in host defence against *Pneumocystis* are tumour necrosis factor alpha (TNF-α) [169, 170]; interleukin-1 (IL-1), IL-6 and granulocyte macrophage colony-stimulating factor (GM-CSF) [171–173]; IL-10, IL-12 and IL-23-IL-17 axis [174–176]; monocyte chemoattractant protein-1 (MCP-1), lymphotactin and macrophage inflammatory protein 1 and 2 (MIP-1, MIP-2) [177]; and IFN-γ [178, 179].

The interplay of the organism, host inflammatory cells, release of cytokines, generation of toxic metabolites, and involvement of both cellular and humoral immunity is complex. However, the understanding of the implications of the host immune response in the inflammatory process to *Pneumocystis* infection is important because there is strong evidence that pulmonary impairment and mortality are more closely associated with lung inflammation than organism burden [131].

7.4 Clinical Manifestations of PCP

P. jirovecii produces an often-fatal pneumonia in a variety of immunocompromised individuals. It became evident in the HIV-infected population that the infection can be disseminated and, although rare, almost any organ can be involved. The frequency of extrapulmonary *Pneumocystis* in HIV-positive patients seems to be <2.5% [180]. This can occur without concurrent pulmonary involvement and *P. jirovecii* has been found invading lymph nodes, spleen, liver, gastrointestinal tract, pancreas, eye, ear, adrenal and thyroid glands, heart, central nervous system and others [17, 180–182].

The diagnosis of PcP depends on a high index of suspicion. PcP should be suspected in the presence of conditions known to be associated with the disease in HIV-infected and non-HIV-infected patients [183] (Table 7.3).

Table 7.3 Main underlying conditions of *Pneumocystis jirovecii* pneumonia in non-HIV infected patients [183]

Underlying conditions	Percentage of total cases in non-HIV-infected patients
Haematological malignancies	32.5
Solid tumours	18.2
Inflammatory diseases	14.9
Solid organ transplant	12.3
Vasculitis	9.7
Miscellaneous	12.3

Table 7.4 Clinical and laboratory features at presentation of *Pneumocystis jirovecii* pneumonia in HIV-infected and non-HIV-infected patients [185]

	HIV-infected patients	Non-HIV-infected patients
Temperature ≥ 38.0 °C	76%	92%
Dyspnoea	68%	66%
Cough	81%	71%
Sputum	23%	21%
Median duration of symptoms (days)	28	5
Rales	30%	34%
Abnormal chest radiography	96%	92%
Hypoxemia (mmHg)	69	52

Before PcP associated with AIDS, two distinct modes of onset were described. One was the insidious onset of either cough or tachypnea progressing to dyspnea and cyanosis with either no fever or a mildly elevated temperature, always associated with a clinical course of 3 or more weeks, and in some instances lasting over 2 months. This clinical pattern was described in children with congenital immunodeficiency [184]. The other mode of onset was abrupt, with symptoms, usually of fever, often spiking, cough and tachypnea all having their onset within few days of each other. Almost all children and adults on high dose of steroid therapy at the time symptoms began had this clinical picture [184].

In HIV-infected patients, the most common manifestations of PcP are subacute onset of progressive dyspnea, variable temperature elevation, nonproductive cough, and evidence of respiratory distress. In mild cases, the chest is clear, though there may be scattered rales or rhonchi, wheezes, or a change in breath sounds. Acute dyspnea with pleuritic chest pain may indicate the development of a pneumothorax. Oral thrush is a common co-infection. This clinical picture is similar to those described in children with congenital immunodeficiency [184]. Patients who are immunocompromised non-HIV-infected with PcP typically present an abrupt onset of respiratory insufficiency that may correlate with a tapered or increased dosage of immunosuppressant medications [28, 185] (Table 7.4). PcP have a more insidious course in patients with AIDS than in patients with other immunodeficiencies [186].

PcP is often the AIDS-defining illness in patients infected with HIV, occurring most frequently when the $CD4^+$ T cells count <200/mm^3 blood. The mortality rates

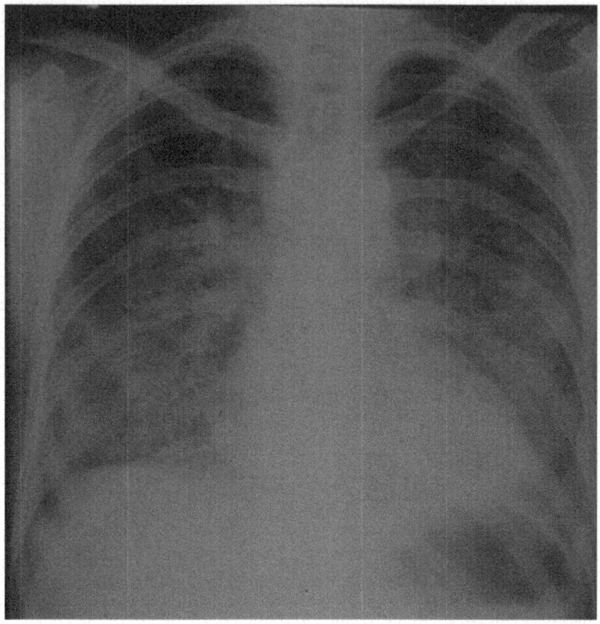

Fig. 7.2 Posteroanterior chest radiograph of a HIV-infected patient with *Pneumocystis jirovecii* pneumonia, demonstrating diffuse bilateral infiltrates

reported in HIV-infected patients for first episodes of PcP is 10–20%, but the rate increases with the need of mechanical ventilation [187]. PcP in non-HIV-infected patients have more neutrophils and fewer *Pneumocystis* organisms in their lungs than patients with PcP associated with AIDS [139]. Despite the availability of effective antimicrobial agents, the mortality rate among patients with PcP in the absence of AIDS is 30–60%, depending on the population risk [28, 188]. Hypoxemia is the most characteristic laboratory abnormality, and elevation of lactate dehydrogenase (LDH) levels to 500 mg/dL is common but nonspecific [189, 190].

Chest radiographies typically demonstrate diffuse, bilateral, symmetrical interstitial infiltrates, spreading from the hilar region to the periphery with time, in a butterfly pattern [191] (Fig. 7.2). A chest radiograph may be normal in patients with early diseases [192]. Atypical radiographic presentation also occurs (nodules, blebs and cysts, asymmetric disease, and upper locations). In the classic diffuse PcP, all lobes of both lungs are generally involved although segmental distribution has been described. Absence of hylar adenopathy and pleural effusion is also characteristic unless PcP is a complication of a basic disease process in which adenopathy and/or pleural effusion are a part. Interstitial and mediastinal emphysema and pneumothorax may occur often with a fatal outcome [193]. In 13–18% of HIV-infected patients with PcP another concurrent cause of pulmonary dysfunction, such as tuberculosis, Kaposi sarcoma, or bacterial pneumonia should be sought [194, 195].

Radiographs of the lungs are often helpful in the diagnosis of a PcP at a time when there are virtually no physical findings and the only complaints are of cough, tachypnea, and fever. Since the haziness is nonspecific, the radiograph is only helpful in the context of clinical findings favouring the development of PcP. Thin-section computed tomography (CT) demonstrating patchy ground-glass attenuation increases the likelihood that a diagnostic study, such as bronchoscopy, will demonstrate *Pneumocystis* in patients with mild-to-moderate symptoms and normal or nonspecific radiograph [196, 197].

In HIV-infected patients, the chest radiographs opacities are discrete and interstitial in distribution; in contrast, in patients with other immunodeficiencies radiographs were alveolar and diffuse [198].

Following therapy, radiographs show gradual clearing after a variable latent period, during which the radiograph may actually appear worse, finally becoming completely normal in 6–8 weeks [184].

7.5 Laboratory Diagnosis of PCP

The presumptive diagnosis of PcP is based on: clinical manifestations, pulmonary function testing, arterial blood gas testing (ABG) at rest and after exercise, and nonspecific radiological and laboratory tests. The history and physical examination are part of the evaluation of the patient with pulmonary symptoms. Over the years it was noticed that the presentation of PcP depends on the underlying disease. Thus, in HIV-infected patients the duration of the clinical manifestations is longer and the diagnosis is more difficult than in patients with other immune deficiencies [186]. Pulmonary manifestations are the most frequent in the natural history of PcP. Taking into account that PcP is an interstitial disease causing serious difficulties in gas exchange occurring in the lungs, a partial pressure of oxygen (PaO2) in peripheral blood $\leq$9.3 kPa (70 mmHg) is indicative of PcP in spite of the fact that 10 to 20% of cases of PcP have normal levels of PaO2 [199]. The measurement of LDH, which increases in the beginning phase of the infection, is a nonspecific laboratory test for PcP diagnosis that can be used as a prognostic tool, and to assess response to PcP therapy [200]. Also, the evaluation of the immune status of the patient is important to determine the risk of developing PcP. This disease occurs most frequently in patients with $CD4^+$ T cells count $\leq$200/mm^3 blood. This information applies to HIV-infected patients and to non-HIV-infected patients with other immunodeficiency, e.g., cancer patients receiving chemotherapy.

All these elements are useful but nonspecific, neither confirming or nor denying the diagnosis of PcP. For many years *P. jirovecii* could not be cultured –until a culture system to propagate *P. jirovecii* in vitro was developed only in 2014 [201]. Since this culture system still needs to be validated, disseminated and shown to be cost-effective for diagnostic purposes, microscopic visualization of cystic or trophic forms in respiratory specimens with cytochemical staining or immunofluorescent staining with monoclonal antibodies (IF-Mab) are the standard procedures to identify this microorganism and diagnose the disease.

7.5.1 *Biological Specimens to Study*

The definitive diagnosis of PcP is determined by cyto-histopathological examination of respiratory specimens. These specimens can be obtained by invasive techniques, such as open-lung biopsy (LB), transbronchial biopsy (TBB), bronchoalveolar lavage fluid (BALF) and bronchial secretions (BS), and by less invasive techniques such as induced sputum (IS), spontaneous sputum (SS), nasopharyngeal aspirate (NA) and oropharyngeal washing (OW). BALF and IS are the most widely used clinical specimens for the diagnosis of PcP.

The diagnosis of lung tissue fragments, allows the observation of the microorganism in more than 95% of cases of infection [202]. The BALF—obtained by fiberoptic bronchoscopy—analysis allows the diagnosis in more than 80% of all patients with PcP, and in more than 95% of patients with HIV co-infection [194, 203]. In turn, induced sputum (IS) that is obtained by inhaling 1.8% saline with the aid of an ultrasonic nebulizer, is widely used for patients with AIDS and PcP, but its utility for patients with other forms of immunodeficiency is less defined. Non-HIV immunocompromised patients with PcP have lower burden of organisms, and sputum induction may consequently have lower diagnostic yield in these patients [204]. This technique allows *Pneumocystis* detection in 30–55% of cases of infection, after nonspecific staining, sometimes difficult to interpret [205]. As an alternative to BALF and IS, upper respiratory samples (NA and OW) have been utilized [83, 206–208] and when studied by PCR technologies, have sensitivity above 75% [206, 209].

The diagnosis of extrapulmonary pneumocystosis is possible by demonstrating *P. jirovecii* cystic or trophic forms in affected tissues. Nonspecific or specific staining method can be utilized, or a PCR method to detect *Pneumocystis* DNA in the affected tissues.

7.5.2 *Staining Methods*

Several staining methods have been described for the diagnosis of PcP. Gomori-methenamine-silver (GMS), Giemsa, or rapid Giemsa-like stains such as Diff-Quik, toluidine blue O (TBO), cresyl echt violet, calcofluor white and IF-MAb. All these methods reveal the characteristic morphology of the cystic and of the trophic forms (Fig. 7.3).

GMS described by Gomori and modified by Grocott and Musto and collaborators [210, 211], has been considered the gold standard technique for the diagnosis of PcP for many years. This reagent selectively stains the *Pneumocystis* cystic form wall that appears dark brown [210] (Fig. 7.3a). TBO stain also has a good affinity for components of the cystic form wall, which stains metachromatically in reddish violet [212] (Fig. 7.3b). GMS, TBO and cresyl echt violet that is similar to TBO, all identify *Pneumocystis* cystic forms but not its other developmental forms. They can

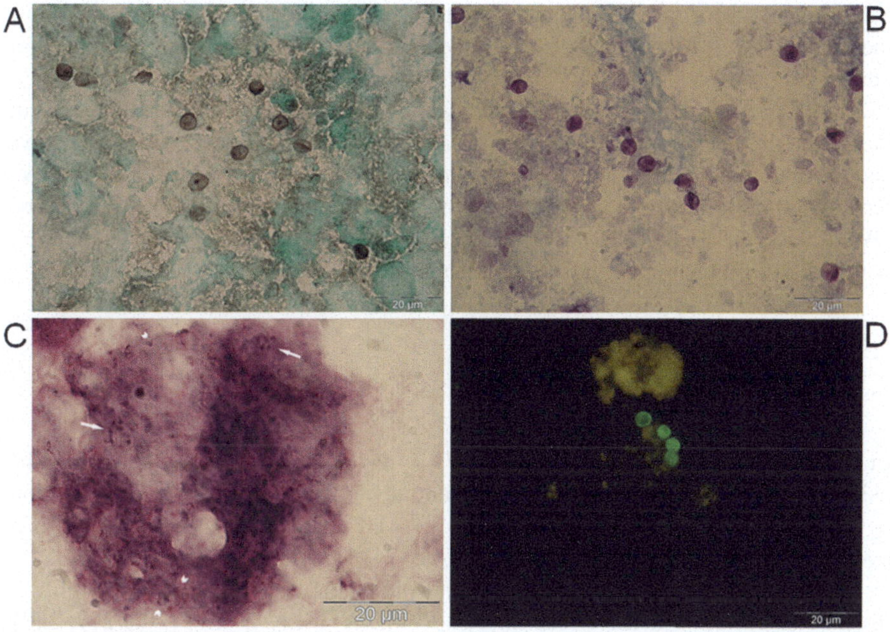

Fig. 7.3 Respiratory specimens with *Pneumocystis* after appropriate staining (magnification ×1000). (**a**) Rat-derived *Pneumocystis* cystic forms stained with Gomori-methenamine-silver (GMS). (**b**) Rat-derived *Pneumocystis* cystic forms stained with toluidine blue O (TBO). (**c**) Clustered trophic and cystic forms of *Pneumocystis jirovecii* in BALF stained with Giemsa. Mature cystic forms (*arrow*), and several trophic forms (*arrowhead*) are quite visible. (**d**) *Pneumocystis jirovecii* cystic forms in BALF stained with immunofluorescent-MAbs anti-*Pneumocystis jirovecii*

be used in any kind of clinical specimens, but they also stain the cell wall of yeasts and other fungi. Therefore, it is required an experienced technician for the reading of the microscopic slides. In turn, calcofluor white is a chemifluorescent agent that nonspecifically binds to β-linked polysaccharide polymers of the cell wall of *Pneumocystis* [213].

Giemsa and Diff-Quik stain the nuclei of all *Pneumocystis* life-cycle stages. They do not stain cystic or sporocytic walls, which appear like a clear peripheral halo around these forms [214] (Fig. 7.3c). These panoptic stains are also used for the diagnosis of other pathogens in the lungs. The reading of the microscopic slides is difficult, especially when dealing with specimens with low fungal burden, requiring the observation by an experienced technician.

The best strategy for a more accurate PcP diagnosis in laboratories that only have these staining methods is the parallel use of a method that stains the cystic form wall and the nuclei of the developing forms, in different smears from the same specimen.

Histologic examination of biopsy specimens for recognition of foamy eosinophilic material in which *P. jirovecii* cystic or trophic forms are embedded can be achieved by performing GMS, Giemsa or Diff-Quik, TBO, or IF-MAb, that show the morphology of the cysts, easier to identify than the trophic forms in paraffin-embedded tissues.

The IF-MAb specific for *P. jirovecii* began to be implemented in the diagnosis of PcP, since their development in 1986 [215, 216]. They are considered more sensitive and specific than other cytochemical staining methods, which makes them the preferred method to apply in specimens with lower diagnostic yield (IS, BS, NA and OW) [205, 216] (Fig. 7.3d). Actually, because of its reliability, IF-MAb is the most commonly used technique in the diagnosis of PcP [217].

7.5.3 Molecular Diagnosis

In the last two decades, the detection of *P. jirovecii* DNA in clinical specimens by using molecular tools has brought important advances in the diagnosis, epidemiology and management of PcP. PCR assays showed high efficiency to amplify *P. jirovecii* DNA in clinical specimens [80, 117, 217–219].

Molecular techniques play an important role when used in association with less invasive or even noninvasive respiratory specimens (IS, NA, OW) [206, 217, 220]. However, the sensitivity of the PCR assays is directly affected by the implicit lower sensitivity of the noninvasive respiratory specimens, due to their lower fungal burden. In BALF, the sensitivity of PCR assays for detection of *P. jirovecii* DNA, such as the *mtLSUrRNA* nPCR, is usually very satisfactory (higher than 95%), with a detection threshold that can reach values of 0.5–1 organism/μL of sample [218]. Yet, in noninvasive specimens, those PCR assays result in lower sensitivity values (less than 80%) [206, 221, 222].

The PCR tools have greatly improved the diagnosis of PcP, allowing the early detection of *P. jirovecii* infection. In fact, PCR assays can detect *P. jirovecii* DNA in respiratory specimens from PcP patients that were negative microscopically [83, 218, 223, 224]. The use of these techniques may lead to a decreased time period from onset of symptoms to treatment, which provides a recognized positive effect on prognosis by avoiding the evolution of early PcP cases to a severe disease with associated respiratory failure and significant mortality [200, 217]. The ambiguity created by the fact that a positive PCR test associated with a negative microscopy may be the result of either PcP or *P. jirovecii* colonization [19, 218]. can be solved by combining the test results with the etiologic features of *P. jirovecii* as opportunistic pathogen and the patient clinical data. Only the careful assessment of the patients' clinical signs and symptoms, radiological and laboratorial complementary tests, in combination with the PCR data analysis, will shed light on the effective state of a specific *P. jirovecii* infection [217].

In clinical practice, colonization or subclinical carriage of *P. jirovecii* is usually identified when the DNA of the pathogen is detected by a PCR assay in respiratory specimens from immunocompromised or immunocompetent persons without clinical manisfestations of PcP [19, 54, 199]. Usually, in these situations, the microscopic examinations are often false negative due to the very low fungal burdens, difficult to detected in those subjects [225].

7.5.3.1 Molecular Techniques for *P. jirovecii* DNA Detection

Several PCR protocols for detection of *P. jirovecii* DNA in clinical specimens have been reported, using different PCR techniques and targeting a variety of *P. jirovecii* genes such as the *mtLSUrRNA*, mitochondrial small subunit rRNA (*mtSSUrRNA*), the *ITS* regions of the rRNA operon, 5S ribosomal RNA (*5S rRNA*), 18S ribosomal RNA (*18S rRNA*), *DHPS*, *DHFR*, *TS*, cell-division-cycle *cdc2* gene, major surface glycoprotein (*MSG*), upstream conserved sequence of the MSG (*UCS*), *KEX*1, *CYB*, *SOD*, α and β tubulin (α- β-*TUB*), *TRR*1 and *arom* [102, 103, 105, 136, 217, 226].

The *mtLSU rRNA* nested-PCR was the most sensitive method among nine PCR assays evaluated for detection of *P. jiroveii* DNA. It produced less false-negative results, and displayed higher degree of concordance with microscopic data than *mtLSU rRNA* single-PCR, *ITS* regions nested-PCR, *DHPS* single- and nested-PCR, *DHFR* nested-PCR, *MSG* heminested-PCR, *18S rRNA* 1-tube nested-PCR and *5S rRNA* real-time quantitative PCR (RT-qPCR) [227]. The fact that several genes used for molecular detection of *P. jirovecii* (e.g. *DHPS*, *DHFR*, *ITS*) are present as single nuclear encoded genes, whereas the mitochondrial *mtLSU rRNA* is a multicopy gene in the *P. jirovecii* genome, contributes largely to the higher successful amplification rates of the *mtLSU rRNA* PCR assays, especially the nested-PCR technique [97, 217, 228]. A bivariate meta-analysis and systematic review assessed the usefulness of PCR-based assays in BALF for diagnosis of PcP. It included 16 studies with a total 1857 BALF samples from 1793 patients published between 1994 and 2012, pointing out for overall pooled sensitivity of 98.3% (95% CI, 91.3%–99.7%) and specificity of 91.0% (95% CI, 82.7%–95.5%), which predicts PCR in BALF as a very accurate method for diagnosis of PcP [229]. Although nested-PCR seems to improve the sensitivity of the detection of *P. jirovecii* in respiratory specimens, the clinical relevance of a positive result remains to be investigated [230]. The lower specificity rates found can be explained by the asymptomatic carriers or colonized patients with positive nested-PCR tests incorrectly diagnosed as having PcP (false positive PcP cases). Again, this leads to the intriguing issue that positive PCR tests associated with negative microscopic examinations may correlate with either PcP or *P. jirovecii* colonization, or even could be an effect of contamination prone process [229]. The use of less invasive samples from the upper respiratory tract in association with highly sensitive and specific molecular tools, such as nested-PCR or RT-qPCR, provide a more cost-effective option than the standard procedures using BALF [223, 231–233].

Molecular strategies for detecting and quantifying *P. jirovecii* burdens were explored by several authors. Several RT-qPCR protocols for detection of *P. jirovecii* DNA in clinical samples have been reported, targeting the MSG multigene family of *P. jirovecii*, *β-TUB*, *KEX1* genes, *mtLSUrRNA* and *cdc2* genes [97, 224, 234–237]. A meta-analysis study assessed the use of RT-qPCR for the diagnosis of PcP in immunocompromised patients, including 10 individual studies from 1990 to 2010 [238]. Overall, the sensitivity of RT-qPCR was 97% (95%CI: 93%–99%) and specificity was 94% (95%CI: 90%–96%). In the subgroup of HIV-infected patients, the sensitivity and specificity were 97% (95%CI: 93%–99%) and 93% (95%CI: 89%–96%), respectively. Regarding *P. jirovecii* DNA detection in BALF, the sensitivity was 98% (95%CI: 94%–99%) and specificity 93% (95%CI: 89%–96%) [238]. Although RT-qPCR demonstrated good diagnostic accuracy, further studies are needed in order to identify any differences in the diagnostic performance of RT-qPCR in HIV-infected and non-HIV-infected immunocompromised patients and in differentiating colonization from disease [217]. In order to achieve more accurate tests, especially in the management of PcP in non-HIV immunocompromised patients, thresholds should be assessed according to underlying diseases and other clinical and radiological parameters [237]. Also, when interpreting the significance of the fungal burden, it is not only important to consider the underlying condition of the patient, but also the quality of the biological specimen under study.

A comparative study of three commercial assays (*Pneumocystis jirovecii* [*carinii*]—FRT PCR Kit [AmpliSens], MycAssay *Pneumocystis* [Myconostica] and real-time PCR *Pneumocystis jirovecii* [Bio-Evolution]) and an in-house PCR was performed [239]. The sensitivity and specificity when testing proven/probable PcP was 100%, 100%, 95% and 83%, 93% and 100%, respectively, and sample concordance between the Amplisens and MycAssay were excellent (Kappa: 0.85) [239]. Also a commercial RT-qPCR for both the detection of *Pneumocystis* and of DHPS point mutations associated with resistance to sulfa-based drugs such as trimethoprim-sulfamethoxazole (TMP-SMX) and dapsone, used for both prophylaxis and treatment of PcP was developed [240].

In conclusion, several studies consistently pointed the nested-PCR and the RT-qPCR assays, especially the ones targeting the *mtLSUrRNA* gene, as the most sensitive and specific molecular tools for detection of *P. jirovecii* [67, 218, 225, 226, 234, 235, 241]. Given the great variety of PcP PCR assays available, it may be wise for diagnostic laboratories to use commercial and standardized tests that have developed an understanding of how to interpret results, in particular low level positives.

PCR-based approaches, namely multilocus sequence typing, can also be used to determine the epidemiology and transmission of infection and to investigate potential outbreaks [80, 121, 242].

7.5.4 New Alternatives for Diagnosis of PcP

7.5.4.1 Usefulness of Blood

The success of microscopy and molecular techniques in the diagnosis of PcP depends on the available resources and technology of the laboratory, the team's experience, as well as the type of biological specimen analyzed. The demand for lesser exclusive technologies and the evidence that invasive techniques to obtain the standard respiratory samples are not always possible to perform and carry associated risks of complications [199], launched the interest in blood and serum specimens. Therefore, these specimens have been tested as an alternative to respiratory specimens since the 1990s. At first, the attention was turned to the detection of *P. jirovecii* in blood samples through molecular methods such as PCR, once a blood-borne phase of the infection was suggested although never demonstrated [243]. However, most studies that experienced this method (5/7) showed low to very low sensitivity (0–30%), depending on the locus in analysis [243–249]. Then, new alternative strategies for the diagnosis of PcP emerged through the measurement of blood biomarkers that reflect the host-pathogen interaction, as noninvasive interventions [250, 251]. Most recently, the interest in the serum antibodies has increased as promising studies using recombinant antigens of *P. jirovecii* and antibody immunodetection techniques, such as immunoenzymatic or immunoblotting assays, have shown potential application in the diagnosis and epidemiological studies of PcP [252–255].

7.5.4.2 Blood Biomarkers

Elevated serum levels of the LDH, beta-d-glucan (BG) and Krebs von den Lungen-6 antigen (KL-6), as well as low serological levels of S-adenosylmethionine (SAM), have been related to PcP and proposed as markers of the disease [223, 250, 251, 256–258]. However, the use of these metabolites in PcP diagnosis is complex because their serum levels are not strictly specific to *P. jirovecii* infection. The measurement of serum BG, the main structural component of the cell wall of *Pneumocystis* cystic forms is the most promising procedure for establishing the serologic diagnosis of PcP [223, 250, 258, 259]. However, BG is also a structural component of fungi and, as such, may be increased in patients with other fungal infections. The combination tests BG/KL-6 was demonstrated to be the most accurate serologic approach for PcP diagnosis, less onerous and minimally invasive, in comparison with the classic diagnostic approach (BALF followed by microscopy or DNA detection) [223, 258].

Two meta-analysis studies performed with a majority of HIV-infected patients estimated that the measurement of BG serological levels for PcP diagnosis presents a high sensitivity (95–96%), a medium specificity (84–86%) and a negative predictive value varying between 98.5% and 98.9% for a PcP prevalence of 20% [259, 260], although recent studies reveal a lower sensitivity for non-HIV-infected patients (85%) [261]. Given the high sensitivity of BG detection in serum, a negative serum BG result can exclude PcP in a patient at risk of the disease, but a positive BG test cannot be used to confirm the diagnosis of PcP. So, a BG positive test alone cannot be considered diagnostic of PcP due to the lack of BG's specificity to *P. jirovecii* infection and also because there is not a consensus threshold for its application in PcP diagnosis [258, 262]. Though, the results of BG tested alone should be interpreted with caution. BG can still contribute to the diagnosis of PcP and to exclude the disease, in combination with PCR of upper respiratory tract specimens—when BALF is not available or difficult to obtain [118], and with other clinical diagnostic criteria indicative of PcP, especially in non-HIV-infected immunocompromised patients.

7.5.4.3 Immunodiagnosis

In the state of the art of immunodiagnosis of PcP, different studies have showed that *P. jirovecii* recombinant antigens are recognized by human serum antibodies, which gives them an enormous potential for application in serological diagnosis of PcP [252–255]. Nowadays, the *Pneumocystis* antigen that has received the most attention is the major surface glycoprotein (Msg), which is highly specific of this pathogen, contains shared and species-specific epitopes, elicits humoral and cellular protective immune responses and plays a central role in the interaction of *Pneumocystis* with its host [263]. Studies that evaluated recombinant fragments covering the entire length of Msg (MsgA, MsgB and MsgC) showed that the carboxyl-terminal domain (MsgC) is the most conserved and reactive region of Msg [252–255]. These data suggest that epitopes that stimulate at least part of the human antibodies against *P. jirovecii* may be located in this region. That was proved recently, in a study where the researchers designed and produced a recombinant synthetic antigen with three antigenic regions of the Msg protein—one from the terminal portion of MsgB and two from MsgC [264]. The authors purified this antigen and used it as an antigenic tool in an ELISA technique, which showed that the IgM anti-*P. jirovecii* levels were statistically increased in patients with PcP ($p = 0.001$), compared with patients without PcP. The ELISA test presented a sensitivity of 100% and a specificity of 80.8%, when associated with the clinical diagnosis of PcP of each patient, showing as an immunodiagnostic tool for PcP diagnosis and a possible improvement in the clinical management of this disease can be near.

7.6 Management of PCP

7.6.1 Treatment

If PcP is left untreated, associated mortality is approximately 100%, especially in non-HIV-infected patients. Mortality from PcP increases with abrupt onset of respiratory failure and delay in PcP diagnosis. Disease can be stratified according to mild, moderate or severe, depending on presenting symptoms, oxygen saturation and chest radiographic changes. Requirement for mechanical ventilation and vasopressors is a poor prognostic feature.

Clinicians should start treatment as soon as possible, even before a definitive diagnosis has been given, because *Pneumocystis* persist in clinical specimens for days or weeks after effective therapy is initiated [265]. Most information on treatment and duration of therapy is extrapolated from HIV guidelines, with limited information in the non-HIV population [266]. TMP-SMX is the preferred agent for treatment because of its low cost, clinical efficacy against PcP, and bioavailability in both intravenous and oral formulations [266]. Oral outpatient therapy with TMP-SMX is highly effective in patients with mild-to-moderate disease. If the patient has severe PcP infection (acute respiratory failure, hemodynamic instability, need of ventilatory support), should be treated with intravenous TMP-SMX, because of the possibility of decreased drug absorption in the critically ill patients [267]. Corticosteroids are used adjunctively with TMP-SMX to reduce the likelihood of respiratory failure and death. Costicosteroids could suppress the acute inflammatory process associated with PcP [266]. If corticosteroids are institute in PcP HIV-infected patients who developed respiratory failure, prednisone 40 mg twice daily for 5 days, 40 mg daily for 5 days, then 20 mg daily for the duration of anti-Pneumocystis therapy is recommended [268]. Among non-HIV immunocompromised patients, corticosteroids use in the treatment of PcP is controversial. However, in severely ill PcP patients corticosteroids could be useful to reduce the duration of mechanical ventilation and ICU admissions [188].

If the patient developed a mild rash while being treated previously with TMP-SMX, desensitization should be first considered [266]. Yet, if the patient develops a severe rash such as Steven–Johnson's syndrome desensitization is not recommended, and TMP-SMX is contraindicated. Alternative therapeutic regimens for mild-to-moderate disease include primaquine plus clindamycin and atovaquone, and alternative therapeutic regimens for patients with moderate-to-severe disease include intravenous pentamidine or intravenous clindamycin plus oral primaquine. Second line PcP therapeutic options have inferior efficacy or similar to that of TMP-SMX but with increased prevalence of adverse effects or increased mortality rates [269–271]. The recommended duration of therapy for PcP is 21 days. Table 7.5 summarizes the various treatment regimens for PcP.

The rate of response to therapy depends on the agent used, number of previous PcP episodes, severity of pulmonary illness, degree of immunodeficiency, timing of initiation of therapy and co-morbidities. Clinical treatment failure is defined as a worsening of or no improvement in lung/respiratory function, evidenced by a decline in oxygen saturation after at least 4–8 days of anti-PcP treatment [23]. Failure attributed to lack of drug efficacy occurs in approximately 10% of those with mild-to-moderate disease. Clinicians should wait at least 4–8 days before switching therapy due to lack of clinical improvement. In the absence of corticosteroid therapy, early and reversible deterioration within the first 3–5 days of therapy is typical. Other concomitant infections must be excluded as the cause of clinical

Table 7.5 Drugs for treatment of *Pneumocystis jirovecii* pneumonia

Drug	Dose	Route	Comments
Trimethoprim–sulfamethoxazole	15–20 mg/kg 75–100 mg/kg daily in divided doses	Oral or intravenous	First choice
Primaquine plus clindamycin	50 mg daily 600 mg three times daily	Oral	Alternate choice
Atovaquone	750 mg two times daily	Oral	Alternate choice
Pentamidine	4 mg/kg daily 600 mg daily	Intravenous Aerosol	Alternate choice

failure [194, 196]. Treatment failure attributed to treatment-limiting toxicities occurs in up to one-third of AIDS patients [272].

7.6.2 Prophylaxis

The most effective way of preventing PcP in HIV-infected people is by immune-reconstitution through the administration of effective antiretroviral therapy. Prophylaxis should be administered until immune reconstitution has been achieved.

HIV-infected adults should receive chemoprophylaxis against PcP if they have $CD4^+$ T cells counts <200/mm^3 or a history of oropharyngeal candidiasis [22–24]. Persons who have a $CD4^+$ T cells percentage of 14% or a history of an AIDS-defining illness, but who do not otherwise qualify, should be considered for prophylaxis [22–24].

For prophylaxis of PcP in non-HIV-infected patients with compromised immune systems there are data related to specific conditions where PcP may have higher prevalence and when prophylaxis could be utilized [58] (Table 7.6). Recommendations for prophylaxis are reviewed in the ECIL guidelines, although especially focused on patients with haematological malignancies and undergoing stem cell transplantation (SCT) [26].

TMP-SMX are the recommended prophylactic agents. This combination therapy at a dose of one double-strength tablet daily confers cross protection against toxoplasmosis and many respiratory bacterial infections [23, 279–282].

TMP-SMX chemoprophylaxis should be continued in patients who have non-life treating adverse reactions. Yet, it should be permanently discontinued in patients with life treating adverse reactions, including possible or definite Stevens–Johnson syndrome or toxic epidermal necrolysis. For patients who cannot tolerate TMP-SMX, alternative prophylactic regimens include dapsone, dapsone plus pyrimethamine plus leucovorin, aerosolized pentamidine, and atovaquone (Table 7.7).

In HIV-infected patients, primary PcP prophylaxis should be discontinued for those patients who have responded to cART with an increase in CD4+ T cells count from <200/mm^3 to ≥200/mm^3 for more than 3 months. Prophylaxis should be reintroduced if CD4+ T cells count decreased to <200/mm^3.

HIV-patients who have a history of PcP should be given chemoprophylaxis for life with TMP-SMX (secondary prophylaxis or maintenance therapy/unless immune reconstitution occurs as a result of cART) [283]. Secondary prophylaxis should be discontinued following recommendations for discontinuation of primary prophylaxis.

Table 7.6 Prophylaxis for *Pneumocystis jirovecii* pneumonia in non-HIV-infected patients

Specific conditions	Comments
Hematopoietic cell transplantation	
Allogeneic stem cell transplant recipients	• PcP prophylaxis until at least 6 months after HCT. For those who continue to receive *immunosuppressive drugs* prophylaxis should be longer than 6 months [273].
Autologous stem cell transplant recipients	• PcP rate is much lower than among allogeneic recipients. PcP prophylaxis should be required in high-risk populations, including underlying lymphoma, leukaemia, or myeloma, especially on intensive treatment with purine analogues or high-dose corticosteroids. Extend PcP prophylaxis should be 3–6 months, or longer for patients with ongoing immunomodulatory therapy post-transplantation [273].
Solid organ transplant recipients	• PcP overall incidence varied in the range of 5–15%, but appeared highest in lung transplant recipients. PcP prophylaxis is recommended for at least 6–12 months' post-transplantation, though longer durations should be considered. Shorter duration of at least 4 months post renal transplant [33, 274].
Acute lymphoblastic leukaemia and prolonged $CD4^+$ counts < 200/μL	• PcP prophylaxis recommended in a broad range of haematological malignancies if the risk is ≥3.5%, including patients on more intensive chemotherapeutic regimens. These patients appear to have a higher risk of PcP associated mortality. Non-neutropenic cancer patients given steroids for prolonged periods of time carry substantial risk of PcP. PcP prophylaxis should continue until $CD4^+$ > 200/μL [275].
Inflammatory bowel disease	• Treatment of IBD with immunomodulators and biologic agents the potential for OI is a key safety concern. PcP prophylaxis is recommended for patients receiving three or more immunomodulatory drugs, including calcineurin inhibitors and/or anti-TNF. The duration of this prophylaxis is currently being reviewed and studied [276].
Primary immunodeficiency diseases, severe protein malnutrition collagen vascular diseases on cytotoxic or immunosuppressive therapy, conditions treated with daily equivalent of 20 mg or more of prednisone for ≥1 month	• PcP prophylaxis in Wegener's granulomatosis when $CD4^+$ < 300/μL. Others AID the risk of infection is related to treatment with systemic steroid, and if the annual risk of PcP is >9% the use of PcP prophylaxis should be recommended if $CD4^+$ < 200/μL [277]. In other immunodepression conditions the risk-benefit ratio of prophylaxis remains unclear. Prophylaxis is warranted when the risk for PcP is ≥3.5%. In children at lower PcP incidence rates [278].

Table 7.7 Drugs for prophylaxis against *Pneumocystis jirovecii* pneumonia

Drug	Dose	Route	Comments
Trimethoprim–sulfamethoxazole	1 double-strength tablet daily or 1 single-strength tablet daily	Oral	First choice
	1 double-strength tablet 3 times per week		Alternate choice
Dapsone	50 mg twice daily or 100 mg daily	Oral	Ensure patient does not have G6PD deficiency
Dapsone plus Pyrimethamine plus Leucovorin	50 mg daily 50 mg weekly 25 mg weekly	Oral	
Dapsone plus Pyrimethamine plus Leucovorin	200 mg weekly 75 mg weekly 25 mg weekly	Oral	
Pentamidine	300 mg monthly	Aerosol	
Atovaquone	1500 mg daily	Oral	Give with high-fat meals, for maximal absorption

G6PD, glucose-6-phosphate dehydrogenase

7.7 Conclusions

Pneumocystis jirovecii is a worldwide pathogen implicated in interstitial pneumonia. Important advances have been made in the past 20 years in understanding this organism, especially with regard to its taxonomy, epidemiology, host specificity, transmission, and pathophysiology. In developed countries, PcP is still a major concern among people infected with HIV, although the incidence of this disease in this population group has dropped dramatically due to successful implementation of anti-PcP prophylaxis and cART. However, we are witnessing an increase and change in the population of patients at risk of PcP, based on immunocompromised non-HIV-infected patients. As the host defence against *Pneumocystis* requires virtually every arm of the immune system to clear the organisms from the lungs, the clinical presentation in HIV-infected patients is different from that in other immunocompromised non-HIV-infected patients. In high-risk patients, clinical presentation and radiology are sufficient for initiating empirical therapy but should not be used as definitive diagnosis that is dependent on specific laboratory diagnostic methodologies, which have improved dramatically. Less invasive samples from the upper respiratory tract in association with highly sensitive and specific molecular methods, such as nested-PCR or RT-qPCR, have been tested intensively in comparison to the standard procedures. Also new diagnostic tools based on the measurement of blood biomarkers of infection and detection of antibodies have been and continue to be developed. Yet, none of the methods developed until now are able to distinguish PcP cases from colonization cases. In addition to diagnosis, molecular based methods can be used for transmission studies, and epidemiological and evolution studies of the organism. The standard first line therapy/prophylaxis for PcP, in HIV- and non-HIV-infected patients, is the combination therapy TMP-SMX. Despite being effective, there are significant prophylactic and treatment failures as well as intolerance or side effects associated with this anti-folate inhibitors combination that may hamper the clinical outcome of the disease. Therefore, optimal strategies for prevention, diagnosis and treament of *P. jirovecii* symptomatic infections are fundamental to providing comprehensive, high-quality care for these patients.

References

1. Chagas C (1909) Nova tripanosomiae humana. Mem Inst Oswaldo Cruz 1:159–218
2. Carini A (1910) Formas de eschizogonia do Trypanosoma lewisi. Soc De Med et Chir de São Paulo. Bull Inst Pasteur IX:973–978
3. Delanoe P, Delanoe M (1912) Sur les rapports des kystes de Carini du poumon des rats avec le Trypanosoma lewisi. C R Hebd Seances Acad Sci 155:658–659
4. Hughes WT (2005) Historical overview. In: Walzer PD, Cushion MT (eds) Pneumocystis pneumonia, 3rd edn. Marcel Dekker, New York, pp 1–37
5. Tang X, Bartlett MS, Smith J et al (1998) Determination of copy number of rRNA genes in Pneumocystis carinii f. sp. hominis. J Clin Microbiol 36:2491–2494
6. Edman JC, Kovacs JA, Masur H et al (1988) Ribosomal RNA sequence shows Pneumocystis carinii to be a member of the fungi. Nature 334:519–522
7. Stringer SL, Hudson K, Blase MA et al (1989) Sequence from ribosomal RNA of Pneumocystis carinii compared to those of four fungi suggests an ascomycetous affinity. J Protozool 36:14S–16S
8. Readhead SA, Cushion MT, Frenkel JK et al (2006) Pneumocystis and Trypanossoma cruzi: nomenclature and typifications. J Eukaryot Microbiol 53:2–11
9. Aliouat-Denis CM, Martinez A, Aliouat EM et al (2009) The pneumocystis life cycle. Mem Inst Oswaldo Cruz 104:419–426
10. Cushion MT (2010) Are members of the fungal genus pneumocystis (a) commensals; (b) opportunists; (c) pathogens; or (d) all of the above? PLoS Pathog 6:e1001009
11. Yoshida Y, Matsumoto Y, Yamada M et al (1984) Pneumocystis carinii: electron microscopic investigation on the interaction of trophozoite and alveolar lining cell. Zentralbl Bakteriol Mikrobiol Hyg A 256:390–399
12. Yoshida Y (1989) Ultrastructural studies of Pneumocystis carinii. J Protozool 36:53–60
13. Vohra PK, Park JG, Sanyal B et al (2004) Expression analysis of PCSTE3, a putative pheromone receptor from the lung pathogenic fungus Pneumocystis carinii. Biochem Biophys Res Commun 319:193–199
14. Matos O (1999) Epidemiologia e diagnóstico da pneumonia por Pneumocystis carinii. Contribuição para a avaliação de técnicas de biologia molecular aplicadas ao diagnóstico da pneumocistose. Dissertation, Universidade Nova de Lisboa, p 246
15. Dei-Cas E, Aliouat EM, Cailliez JC (2005) Pneumocystis cellular structure. In: Walzer PD, Cushion MT (eds) Pneumocystis pneumonia, 3rd edn. Marcel Dekker, New York, pp 61–94
16. Dei-Cas E, Jackson H, Palluault F et al (1991) Ultrastructural observations on the attachment of Pneumocystis carinii in vitro. J Protozool 38:205S–207S
17. Raviglione MC (1990) Extrapulmonary pneumocystosis: the first 50 cases. Rev Infect Dis 12:1127–1138
18. Medrano FJ, Montes-Cano M, Conde M et al (2005) Pneumocystis jirovecii in general population. Emerg Infect Dis 11:245–250
19. Morris A, Lundgren JD, Masur H et al (2004) Current epidemiology of Pneumocystis pneumonia. Emerg Infect Dis 10:1713–1720

20. Huang L, Cattamanchi A, Davis JL et al (2011) International HIV-associated Opportunistic Pneumonias (IHOP) Study; Lung HIV Study. HIV-associated Pneumocystis pneumonia. Proc Am Thorac Soc 8:294–300
21. Miller RF, Huang L, Walzer PD (2013) Pneumocystis pneumonia associated with human immunodeficiency virus. Clin Chest Med 34:229–241
22. Phair J, Muñoz A, Detels R et al (1990) The risk of Pneumocystis carinii pneumonia among men infected with human immunodeficiency virus type 1. Multicenter AIDS Cohort Study Group. N Engl J Med 322:161–165
23. Panel on Opportunistic Infections in HIV-Infected Adults and Adolescents. Guidelines for the prevention and treatment of opportunistic infections in HIV-infected adults and adolescents: recommendations from the Centers for Disease Control and Prevention, the National Institutes of Health, and the HIV Medicine Association of the Infectious Diseases Society of America. http://aidsinfo.nih.gov/contentfiles/lvguidelines/adult_oi.pdf. Accessed 9 Jan 2017
24. Kaplan JE, Hanson DL, Navin TR et al (1998) Risk factors for primary Pneumocystis carinii pneumonia in human immunodeficiency virus-infected adolescents and adults in the United States: reassessment of indications for chemoprophylaxis. J Infect Dis 178:1126–1132
25. Miguez-Burbano MJ, Ashkin D, Rodriguez A et al (2005) Increased risk of Pneumocystis carinii and community-acquired pneumonia with tobacco use in HIV disease. Int J Infect Dis 9:208–217
26. Maertens J, Cesaro S, Maschmeyer G et al (2016) ECIL guidelines for preventing Pneumocystis jirovecii pneumonia in patients with haematological malignancies and stem cell transplant recipientes. J Antimicrob Chemother 71:2397–2404
27. Roblot F, Imbert S, Godet C et al (2004) Risk factors analysis for Pneumocystis jirovecii pneumonia (PCP) in patients with haematological malignancies and pneumonia. Scand J Infect Dis 36:848–854
28. Overgaard UM, Helweg-Larsen J (2007) Pneumocystis jiroveci pneumonia (PCP) in HIV-1-negative patients: a retrospective study 2002-2004. Scand J Infect Dis 39:589–595
29. Cheson BD (1995) Infectious and immunosuppressive complications of purine analog therapy. J Clin Oncol 13:2431–2448
30. Venhuizen AC, Hustinx WN, van Houte AJ et al (2008) Three cases of Pneumocystis jirovecii pneumonia (PCP) during first-line treatment with rituximab in combination with CHOP-14 for aggressive B-cell non-Hodgkin's lymphoma. Eur J Haematol 80:275–276
31. Martin SI, Marty FM, Fiumara K et al (2006) Infectious complications associated with alemtuzumab use for lymphoproliferative disorders. Clin Infect Dis 43:16–24
32. Mansharamani NG, Balachandran D, Vernovsky I et al (2000) Peripheral blood CD4+ T-lymphocyte counts during Pneumocystis carinii pneumonia in immunocompromised patients without HIV infection. Chest 118:712–720
33. Martin SI, Fishman JA, AST Infectious Diseases Community of Practice (2013) Pneumocystis pneumonia in solid organ transplantation. Am J Transplant 13:272–279
34. Hardy AM, Wajszczuk CP, Suffredini AF et al (1984) Pneumocystis carinii pneumonia in renal-transplant recipients treated with cyclosporine and steroids. J Infect Dis 149:143–147
35. Lufft V, Kliem V, Behrend M et al (1996) Incidence of Pneumocystis carinii pneumonia after renal transplantation. Impact of immunosuppression. Transplantation 62:421–423
36. Arend SM, Westendorp RG, Kroon FP et al (1996) Rejection treatment and cytomegalovirus infection as risk factors for Pneumocystis carinii pneumonia in renal transplant recipients. Clin Infect Dis 22:920–925
37. Pliquett RU, Asbe-Vollkopf A, Hauser PM et al (2012) A Pneumocystis jirovecii pneumonia outbreak in a single kidney-transplant center: role of cytomegalovirus co-infection. Eur J Clin Microbiol Infect Dis 31:2429–2437
38. Radisic M, Lattes R, Chapman JF et al (2003) Risk factors for Pneumocystis carinii pneumonia in kidney transplant recipients: a case–control study. Transpl Infect Dis 5:84–93
39. de Boer MG, Bruijnesteijn van Coppenraet LE, Gaasbeek A et al (2007) An outbreak of Pneumocystis jiroveci pneumonia with 1 predominant genotype among renal transplant

recipients: interhuman transmission or a common environmental source? Clin Infect Dis 44:1143–1149
40. Henson JW, Jalaj JK, Walker RW et al (1991) Pneumocystis carinii pneumonia in patients with primary brain tumors. Arch Neurol 48:406–409
41. Varthalitis I, Meunier F (1993) Pneumocystis carinii pneumonia in cancer patients. Cancer Treat Rev 19:387–413
42. Kulke MH, Vance EA (1997) Pneumocystis carinii pneumonia in patients receiving chemotherapy for breast cancer. Clin Infect Dis 25:215–218
43. Barbounis V, Aperis G, Gambletsas E et al (2005) Pneumocystis carinii pneumonia in patients with solid tumors and lymphomas: predisposing factors and outcome. Anticancer Res 25:651–655
44. Godeau B, Coutant-Perronne V, Le Thi HD et al (1994) Pneumocystis carinii pneumonia in the course of connective tissue disease: report of 34 cases. J Rheumatol 21:246–251
45. Kaur N, Mahl TC (2007) Pneumocystis jiroveci (carinii) pneumonia after infliximab therapy: a review of 84 cases. Dig Dis Sci 52:1481–1484
46. Komano Y, Harigai M, Koike R et al (2009) Pneumocystis jiroveci pneumonia in patients with rheumatoid arthritis treated with infliximab: a retrospective review and case-control study of 21 patients. Arthritis Rheum 61:305–312
47. Buchacz K, Baker RK, Palella FJ Jr et al (2010) AIDS-defining opportunistic illnesses in US patients, 1994-2007: a cohort study. AIDS 24:1549–1559
48. Lundberg BE, Davidson AJ, Burman WJ (2000) Epidemiology of Pneumocystis carinii pneumonia in an era of effective prophylaxis: the relative contribution of non-adherence and drug failure. AIDS 14:2559–2566
49. Wolff AJ, O'Donnell AE (2001) Pulmonary manifestations of HIV infection in the era of highly active antiretroviral therapy. Chest 120:1888–1893
50. Kaplan JE, Benson C, Holmes KH et al (2009) Guidelines for prevention and treatment of opportunistic infections in HIV-infected adults and adolescents: recommendations from CDC, the National Institutes of Health, and the HIV Medicine Association of the Infectious Diseases Society of America. MMWR Recomm Rep 58:1–207
51. European Centre for Disease Prevention and Control (ECDC), World Health Organization (WHO) (2015) HIV/AIDS surveillance in Europe 2015. In: Surveillance Report. http://www.euro.who.int/__data/assets/pdf_file/0019/324370/HIV-AIDS-surveillance-Europe-2015.pdf?ua=1. Accessed 15 Feb 2017
52. Nuesch R, Bellini C, Zimmerli W (1999) Pneumocystis carinii pneumonia in human immunodeficiency virus (HIV)-positive and HIV-negative immunocompromised patients. Clin Infect Dis 29:1519–1523
53. Roblot F, Godet C, Le Moal G et al (2002) Analysis of underlying diseases and prognosis factors associated with Pneumocystis carinii pneumonia in immunocompromised HIV-negative patients. Eur J Clin Microbiol Infect Dis 21:523–531
54. Morris A, Sciurba FC, Norris KA (2008) Pneumocystis: a novel pathogen in chronic obstructive pulmonary disease? COPD 5:43–51
55. Mori S, Sugimoto S (2012) Pneumocystis jirovecii infection: an emerging threat to patients with rheumatoid arthritis. Rheumatology 51:2120–2130
56. Maini R, Henderson KL, Sheridan EA et al (2013) Increasing Pneumocystis pneumonia, England, UK, 2000-2010. Emerg Infect Dis 19:386–392
57. Atwal SS, Puranik S, Mehra S et al (2014) Pneumocystis pneumonia – a novel complication in a non-HIV dengue patient. J Clin Diagn Res 8(9):RL01–RL02
58. Avino LJ, Naylor SM, Roecker AM (2016) Pneumocystis jirovecii pneumonia in the non-HIV-infected population. Ann Pharmacother 50:673–679
59. French N, Kaleebu P, Pisani E et al (2006) Human immunodeficiency virus (HIV) in developing countries. Ann Trop Med Parasitol 100:433–454
60. Matos O, Esteves F (2010) Epidemiology and clinical relevance of Pneumocystis jirovecii Frenkel, 1976 dihydropteroate synthase gene mutations. Parasite 17:219–232

61. de Armas RY, Wissmann G, Müller AL et al (2011) Pneumocystis jirovecii pneumonia in developing countries. Parasite 18:219–228
62. Matos O (2012) Pneumocystis jirovecii pneumonia in Africa: impact and implications of highly sensitive diagnostic technologies. N Am J Med Sci 4:486–487
63. Pifer LL, Hughes WT, Stagno S et al (1978) Pneumocystis carinii infection: evidence for high prevalence in normal and immunosuppressed children. Pediatrics 61:35–41
64. Lundgren B, Lebech M, Lind K et al (1993) Antibody response to a major human Pneumocystis carinii surface antigen in patients without evidence of immunosuppression and in patients with suspected atypical pneumonia. Eur J Clin Microbiol Infect Dis 12:105–109
65. Martinez A, Halliez MCM, Moukhtar Aliouat E et al (2013) Growth and airborne transmission of cell-sorted life cycle stages of Pneumocystis carinii. PLoS One 8:e79958
66. Olsson M, Sukura A, Lindberg LA et al (1996) Detection of Pneumocytis carinii DNA by filtration of air. Scand J Infect Dis 28:279–282
67. Wakefield AE (1996) DNA sequences identical to Pneumocystis carinii f. sp. carinii and Pneumocystis carinii f. sp. hominis in samples of air spora. J Clin Microbiol 34:1754–1759
68. Helweg-Larsen J, Tsolaki AG, Miller RF et al (1998) Clusters of Pneumocystis carinii pneumonia: analysis of person-to-person transmission by genotyping. QJM 91:813–820
69. de Boer MG, Kroon FP, le Cessie S et al (2011) Risk factors for Pneumocystis jirovecii pneumonia in kidney transplant recipients and appraisal of strategies for selective use of chemoprophylaxis. Transpl Infect Dis 13:559–569
70. Mortier E, Pouchot J, Bossi P et al (1995) Maternal-fetal transmission of Pneumocystis carinii in human immunodeficiency virus infection. N Engl J Med 332:825
71. Miller RF, Ambrose HE, Novelli V et al (2002) Probable mother-to-infant transmission of Pneumocystis carinii f. sp. hominis Infection. J Clin Microbiol 40:1555–1557
72. Rivero L, de la Horra C, Montes-Cano MA et al (2008) Pneumocystis jirovecii transmission from imunocompetente carriers to infant. Emerg Infect Dis 14:1116–1118
73. Montes-Cano MA, Chabe M, Fontillon-Alberdi M et al (2009) Vertical transmission of Pneumocystis jirovecii in humans. Emerg Infect Dis 15:125–127
74. Dei-Cas E (2000) Pneumocystis infections: the iceberg? Med Mycol 1:23–32
75. Keely SP, Stringer JR, Baughman RP et al (1995) Genetic variation among Pneumocystis carinii hominis isolates in recurrent pneumocystosis. J Infect Dis 172:595–598
76. Stringer JR, Walzer PD (1996) Molecular biology and epidemiology of Pneumocystis carinii infection in AIDS. AIDS 10:561–571
77. Sing A, Roggenkamp A, Autenrieth IB et al (1999) Pneumocystis carinii carriage in immunocompetent patients with primary pulmonary disorders as detected by single or nested PCR. J Clin Microbiol 37:3409–3410
78. Chabé M, Dei-cas E, Creusy C et al (2004) Immunocompetent hosts as a reservoir of Pneumocystis organisms: histological and rt-PCR data demonstrate active replication. Eur J Clin Microbiol Infect Dis 23:89–97
79. Wakefield AE, Lindley AR, Ambrose HE et al (2003) Limited asymptomatic carriage of Pneumocystis jiroveci in human immunodeficiency virus–infected patients. J Infect Dis 187:901–908
80. Beard CB, Carter JL, Keely SP et al (2000) Genetic variation in Pneumocystis carinii isolates from different geographic regions: implications for transmission. Emerg Infect Dis 6:265–272
81. Rabodonirina L, Vanhems P, Couray-Targe S et al (2004) Molecular evidence of interhuman transmission of Pneumocystis pneumonia among renal transplant reciepients hospitalized with HIV-positive patients. Emerg Infect Dis 10:1766–1773
82. Gianella S, Haeberli L, Joos B et al (2010) Molecular evidence of interhuman transmission in an outbreak of Pneumocystis jirovecii pneumonia among renal transplant recipients. Transpl Infect Dis 12:1–10

83. Matos O, Costa MC, Lundgren B et al (2001) Effect of oral washes on the diagnosis of Pneumocystis carinii pneumonia with a low parasite burden and on detection of organisms in subclinical infections. Eur J Clin Microbiol Infect Dis 20:573–575
84. Morris A, Norris KA (2012) Colonization by Pneumocystis jirovecii and its role in disease. Clin Microbiol Rev 25:297–317
85. Latouche S, Ortona E, Mazars E et al (1997) Biodiversity of Pneumocystis carinii hominis: typing with different DNA regions. J Clin Microbiol 35:383–387
86. Kazanjian P, Locke AB, Hossler PA et al (1998) Pneumocystis carinii mutations associated with sulfa and sulfone prophylaxis failures in AIDS patients. AIDS 12:873–878
87. Miller RF, Wakefield AE (1999) Pneumocystis carinii genotypes and severity of pneumonia. Lancet 353:2039–2040
88. Matos O, Esteves F (2010) Pneumocystis jirovecii multilocus gene sequencing: findings and implications. Future Microbiol 5:1257–1267
89. Esteves F, Gaspar J, Marques T et al (2010) Identification of relevant single-nucleotide polymorphisms in Pneumocystis jirovecii: relationship with clinical data. Clin Microbiol Infect 16:878–884
90. Hauser PM, Blanc DS, Bille J et al (1996) Development of a molecular typing method for Pneumocystis carinii sp.f. hominis. J Euk Microbiol 43:34S
91. Hauser PM, Francioli P, Bille J et al (1997) Typing of Pneumocystis carinii f. sp. hominis by single-strand conformation polymorphism of four genomic regions. J Clin Microbiol 35:3086–3091
92. Hauser PM, Blanc DS, Sudre P et al (2001) Genetic diversity of Pneumocystis carinii in HIV-positive and -negative patients as revealed by PCR-SSCP typing. AIDS 15:461–466
93. Ripamonti C, Orenstein A, Kutty G et al (2009) Restriction fragment length polymorphism typing demonstrates substantial diversity among Pneumocystis jirovecii isolates. J Infect Dis 200:1616–1622
94. Lu JJ, Bartlett MS, Smith JW et al (1995) Typing of Pneumocystis carinii f. sp. hominis strains with type-specific oligonucleotide probes derived from nucleotide sequences of internal transcribed spacers of rRNA genes. J Clin Microbial 33:2973–2977
95. Hauser PM (2004) The development of a typing method for an uncultivable microorganism: the example of Pneumocystis jirovecii. Infect Genet Evol 4:199–203
96. Esteves F, Gaspar J, De Sousa B et al (2011) Clinical relevance of multiple single-nucleotide polymorphisms in Pneumocystis jirovecii Pneumonia: development of a multiplex PCR-single-base-extension methodology. J Clin Microbiol 49:1810–1815
97. Esteves F, Gaspar J, de Sousa B et al (2012) Pneumocystis jirovecii multilocus genotyping in pooled DNA samples: a new approach for clinical and epidemiological studies. Clin Microbiol Infect 18:E177–E184
98. Parobek CM, Jiang LY, Patel JC et al (2014) Multilocus microsatellite genotyping array for investigation of genetic epidemiology of Pneumocystis jirovecii. J Clin Microbiol 52:1391–1392
99. Gits-Muselli M, Peraldi MN, de Castro N et al (2015) New short tandem repeat-based molecular typing method for Pneumocystis jirovecii reveals intrahospital transmission between patients from different wards. PLoS One 10:e0125763
100. Wakefield AE, Pixley FJ, Banerji S et al (1990) Amplification of mitochondrial ribosomal RNA sequences from Pneumocystis carinii DNA of rat and human origin. Mol Biochem Parasitol 43:69–76
101. Lee CH, Helweg-Larsen J, Tang X et al (1998) Update on Pneumocystis carinii f. sp. hominis typing based on nucleotide sequence variations in internal transcribed spacer regions of rRNA genes. J Clin Microbiol 36:734–741
102. Mei Q, Gurunathan S, Masur H et al (1998) Failure of co-trimoxazole in Pneumocystis carinii infection and mutations in dihydropteroate synthase gene. Lancet 351:1631–1632

103. Walker DJ, Wakefield AE, Dohn MN et al (1998) Sequence polymorphisms in the Pneumocystis carinii cytochrome b gene and their association with atovaquone prophylaxis failure. J Infect Dis 178:1767–1775
104. Helweg-Larsen J, Benfield TL, Eugen-Olsen J et al (1999) Effects of mutations in Pneumocystis carinii dihydropteroate synthase gene on outcome of AIDS-associated P. carinii pneumonia. Lancet 354:1347–1351
105. Ma L, Borio L, Masur H et al (1999) Pneumocystis carinii dihydropteroate synthase but not dihydrofolate reductase gene mutations correlate with prior trimethoprim-sulfamethoxazole or dapsone use. J Infect Dis 180:1969–1978
106. Armstrong W, Meshnick S, Kanzanjian P (2000) Pneumocystis carinii mutations associated with sulfa and sulfone prophylaxis failures in immunocompromised patients. Microbes Infect 2:61–67
107. Huang L, Beard CB, Creasman J et al (2000) Sulfa or sulfone prophylaxis and geographic region predict mutations in the Pneumocystis carinii dihydropteroate synthase gene. J Infect Dis 182:1192–1198
108. Kazanjian P, Armstrong W, Hossler PA et al (2001) Pneumocystis carinii cytochrome b mutations are associated with atovaquone exposure in patients with AIDS. J Infect Dis 183:819–822
109. Costa MC, Gaspar J, Mansinho K et al (2005) Detection of Pneumocystis jirovecii dihydropteroate synthase polymorphisms in patients with Pneumocystis pneumonia. Scand J Infect Dis 37:766–771
110. Matos O, Lee CH, Jin S et al (2003) Pneumocystis jiroveci in Portuguese immunocompromised patients: association of specific ITS genotypes with treatment failure, bad clinical outcome and childhood. Infect Genet Evol 3:281–285
111. Crothers K, Beard CB, Turner J et al (2005) Severity and outcome of HIV-associated Pneumocystis pneumonia containing Pneumocystis jirovecii dihydropteroate synthase gene mutations. AIDS 19:801–805
112. Kazanjian P, Armstrong W, Hossler PA et al (2000) Pneumocystis carinii mutations are associated with duration of sulfa or sulfone prophylaxis exposure in AIDS patients. J Infect Dis 182:551–557
113. Costa MC, Esteves F, Antunes F et al (2006) Genetic characterization of the dihydrofolate reductase gene of Pneumocystis jirovecii isolates from Portugal. J Antimicrob Chemother 58:1246–1249
114. Miller RF, Lindley AR, Ambrose HE et al (2003) Genotypes of Pneumocystis jiroveci isolates obtained in Harare, Zimbabwe, and London, United Kingdom. Antimicrob Agents Chemother 47:3979–3981
115. Costa MC, Esteves F, Antunes F et al (2006) Multilocus genotyping of Pneumocystis jirovecii in immunocompromised patients: preliminary results. J Eukaryot Microbiol 53:S104–S105
116. Esteves F, Gaspar J, Tavares A et al (2010) Population structure of Pneumocystis jirovecii isolated from immunodeficiency virus-positive patients. Infect Genet Evol 10:192–199
117. Esteves F, Montes-Cano MA, de la Horra C et al (2008) Pneumocystis jirovecii multilocus genotyping profiles in patients from Portugal and Spain. Clin Microbiol Infect 14:356–362
118. Alanio A, Gits-Muselli M, Calderon E et al (2016) European study on Pneumocystis jirovecii short tandem repeats genotyping reveals wide population diversity with geographic specificities. J Med Mycol 26:e5–e6
119. Esteves F, de Sousa B, Calderon EJ et al (2016) Multicentre study highlighting clinical relevance of new high-throughput methodologies in molecular epidemiology of Pneumocystis jirovecii pneumonia. Clin Microbiol Infect 22:566.e9–566.e19
120. Olsson M, Lidman C, Latouche S et al (1998) Identification of Pneumocystis carinii f. sp. hominis gene sequences in filtered air in hospital environments. J Clin Microbiol 36:1737–1740
121. Walzer PD (1994) Pathogenic mechanisms. In: Walzer PD (ed) Pneumocystis carinii Pneumonia, 2nd edn. Marcel Dekker, New York, pp 251–265

122. Thomas CF Jr, Limper AH (2007) Current insights into the biology and pathogenesis of Pneumocystis pneumonia. Nat Rev Microbiol 5:298–308
123. Gigliotti F, Wright TW (2005) Immunopathogenesis of Pneumocystis carinii pneumonia. Expert Rev Mol Med 7:1–16
124. Walzer PD (1986) Attachment of microbes to host cells: relevance of Pneumocystis carinii. Lab Investig 54:589–592
125. Skalski JH, Kottom TJ, Limper AH (2015) Pathology of Pneumocystis pneumonia: life cycle, cell wall and cell signal transduction. FEMS Yeast Res 15:fov046
126. Pottratz ST, Martin WJ (1990) Role of fibronectin in Pneumocystis carinii attachment to cultured lung cells. J Clin Invest 85:351–356
127. Pottratz ST, Paulsrud J, Smith JS et al (1991) Pneumocystis carinii attachment to cultured lung cells by pneumocystis gp 120, a fibronectin binding protein. J Clin Invest 88:403–407
128. Limper AH, Standing JE, Hoffman OA et al (1993) Vitronectin binds to Pneumocystis carinii and mediates organism attachment to cultured lung epithelial cells. Infect Immun 61:4302–4309
129. Pottratz ST, Weir AL, Wisniowski PE (1994) Pneumocystis carinii attachment increases expression of fibronectin-binding integrins on cultured lung cells. Infect Immun 62:5464–5469
130. Limper AH, Martin WJ 2nd (1990) Pneumocystis carinii: inhibition of lung cell growth mediated by parasite attachment. J Clin Invest 85:391–396
131. Wright TW, Gigliotti F (2005) Mechanisms of lung injury during Pneumocystis pneumonia. In: Walzer PD, Cushion MT (eds) Pneumocystis pneumonia, 3rd edn. Marcel Dekker, New York, pp 327–348
132. Hayes DJ, Stubberfield CR, McBride JD et al (1991) Alterations in cysteine proteinase content of rat lung associated with development of Pneumocystis carinii infection. Infect Immun 59:3581–3588
133. Breite M, Bailey AM, Martin WJ (1993) Pneumocystis carinii chymase is capable of altering epithelial cell permeability. Am Rev Respir Dis 147:A33
134. Sukura A, Konttinen YT, Sepper R et al (1995) Collagenases and the serine proteinases elastase and cathepsin G in steroid-induced Pneumocystis carinii pneumonia. J Clin Microbiol 33:829–834
135. Fox D, Smulian AG (2001) Plasminogen-binding activity of enolase in the opportunistic pathogen Pneumocystis carinii. Med Mycol 39:495–507
136. Kutty G, Kovacs JA (2003) A single-copy gene encodes Kex1, a serine endoprotease of Pneumocystis jiroveci. Infect Immun 71:571–574
137. Stringer JR (2007) Antigenic variation in Pneumocystis. J Eukaryot Microbiol 54:8–13
138. Keely SP, Stringer JR (2009) Complexity of the MSG gene family of Pneumocystis carinii. BMC Genomics 10:367
139. Limper AH, Offord KP, Smith TF et al (1989) Pneumocystis carinii pneumonia. Differences in lung parasite number and inflammation in patients with and without AIDS. Am Rev Respir Dis 140:1204–1209
140. Gigliotti F, Limper AH, Wright T (2014) Pneumocystis. Cold Spring Hard Perspect Med 4:a019828
141. Fleury J, Escudier E, Pocholle MJ et al (1985) Cell population obtained by bronchoalveolar lavage in Pneumocystis carinii pneumonitis. Acta Cytol 29:721–726
142. Young JA, Stone JW, Mcgonigle RJ et al (1986) Diagnosing Pneumocystis carinii pneumonia by cytological examination of bronchoalveolar lavage fluid: report of 15 cases. J Clin Pathol 39:945–949
143. Limper AH, Hoyte JS, Standing JE (1997) The role of alveolar macrophages in Pneumocystis carinii degradation and clearance from the lung. J Clin Invest 99:2110–2117
144. Ezekowitz RA, Williams DJ, Koziel H et al (1991) Uptake of Pneumocystis carinii mediated by the macrophage mannose receptor. Nature 351:155–158

145. Steele C, Marrero L, Swain S et al (2003) Alveolar macrophage-mediated killing of Pneumocystis carinii f. sp. muris involves molecular recognition by the Dectin-1 β-glucan receptor. J Exp Med 198:1677–1688
146. Brown GD, Gordon S (2005) Immune recognition of fungal beta-glucans. Cell Microbiol 7:471–479
147. Kelly MN, Shellito JE (2010) Current understanding of Pneumocystis immunology. Future Microbiol 5:43–65
148. Laursen AL, Moller B, Rungby J et al (1994) Pneumocystis carinii – induced activation of the respiratory burst in human monocytes and macrophages. Clin Exp Immunol 98:196–202
149. Zheng M, Shellito JE, Marrero L et al (2001) CD4+ T-cell independent vaccination against Pneumocystis carinii in mice. J Clin Invest 108:1469–1474
150. Monnet X, Vidal-Petiot E, Osman D et al (2008) Critical care management and outcome of severe Pneumocystis pneumonia in patients with and without HIV infection. Crit Care 12:R28
151. Jensen BN, Lisse IM, Gerstoft J et al (1991) Cellular profiles in bronchoalveolar lavage fluid of HIV-infected patients with pulmonary symptoms: relation to diagnosis and prognosis. AIDS 5:527–533
152. Sadaghdar H, Huang ZB, Eden E (1992) Correlation of bronchoalveolar lavage findings to severity of Pneumocystis carinii pneumonia in AIDS. Evidence for the development of high-permeability pulmonary edema. Chest 102:63–69
153. Roths JB, Marshall JD, Allen RD et al (1990) Spontaneous Pneumocystis carinii pneumonia in immunodeficient mutant SCID mice. Natural history and pathobiology. Am J Pathol 136:1173–1186
154. Wright TW, Gigliotti F, Finkelstein JN et al (1999) Immune-mediated inflammation directly impairs pulmonary function, contributing to the pathogenesis of Pneumocystis carinii pneumonia. J Clin Invest 104:1307–1317
155. Harmsen AG, Stankiewicz M (1990) Requirement for CD4+ cells in resistance to Pneumocystis carinii pneumonia in mice. J Exp Med 172:937–945
156. Beck JM, Harmsen AG (1998) Lymphocytes in host defense against Pneumocystis carinii. Semin Respir Infect 13:330–338
157. Shellito JE, Tate C, Ruan S et al (2000) Murine CD4 T lymphocyte subsets and host defense against Pneumocystis carinii. J Infect Dis 181:2011–2017
158. Shipley TW, Kling HM, Morris A et al (2010) Persistent Pneumocystis colonization leads to the development of chronic obstructive pulmonary disease in a nonhuman primate model of AIDS. J Infect Dis 202:302–312
159. Beck JM, Warnock ML, Curtis JL et al (1991) Inflammatory responses to Pneumocystis carinii in mice selectively depleted of helper T lymphocytes. Am J Respir Cell Mol Biol 5:186–197
160. Beck JM, Newbury RL, Palmer BE et al (1996) Role of CD8 lymphocytes in host defense against Pneumocystis carinii in mice. J Lab Clin Med 128:477–487
161. Steele C, Zheng M, Young E et al (2002) Increased host resistance against Pneumocystis carinii pneumonia in gamma-delta T-cell-deficient mice: protective role of gamma interferon and CD8 T cells. Infect Immun 70:5208–5215
162. Marcotte H, Levesque D, Delanay K et al (1996) Pneumocystis carinii infection in transgenic B cell-deficient mice. J Infect Dis 173:1034–1037
163. Lund FE, Schuer K, Hollifield M et al (2003) Clearance of Pneumocystis carinii in mice is dependent on B cells but not on P. carinii-specific antibody. J Immunol 171:1423–1430
164. Lund FE, Hollifield M, Schuer K et al (2006) B cells are required for generation of protective effector and memory CD4 cells in response to Pneumocystis lung infection. J Immunol 176:6147–6154
165. Hugle B, Solomon M, Harvey E et al (2010) Pneumocystis jiroveci pneumonia following rituximab treatment in Wegener's granulomatosis. Arthritis Care Res 62:1661–1664

166. Kurokawa T, Kaya H, Yoshida T (2010) Two cases of Pneumocystis jiroveci pneumonia with non-Hodgkin's lymphoma after CHOP-based chemotherapy containing rituximab. J Clin Exp Hematop 50:159–162
167. Milledge J, Kakakios A, Gillis J et al (2003) Pneumocystis carinii pneumonia as a presenting feature of X-linked hyper-IgM syndrome. J Paediatr Child Health 39:704–706
168. Steele C, Shellito JE, Kolls JK (2005) Immunity against the opportunistic fungal pathogen Pneumocystis. Med Mycol 43:1–19
169. Hoffman OA, Standing JE, Limper AH (1993) Pneumocystis carinii stimulates tumor necrosis factor-alpha release from alveolar macrophages through a beta-glucan-mediated mechanism. J Immunol 150:3932–3940
170. Wright TW, Pryhuber GS, Chess PR et al (2004) TNF receptor signalling contributes to chemokine secretion, inflammation, and respiratory deficits during Pneumocystis pneumonia. J Immunol 172:2511–2521
171. Chen W, Havell EA, Moldawer LL et al (1992) Interleukin 1: an important mediator of host resistance against Pneumocystis carinii. J Exp Med 176:713–718
172. Chen W, Havell EA, Gigliotti F et al (1993) Interleukin-6 production in a murine model of Pneumocystis carinii pneumonia: relation to resistance and inflammatory response. Infect Immun 61:97–102
173. Paine R 3rd, Preston AM, Wilcoxen S et al (2000) Granulocyte-macrophage colony-stimulating factor in the innate immune response to Pneumocystis carinii pneumonia in mice. J Immunol 164:2602–2609
174. Qureshi MH, Harmsen AG, Garvy BA (2003) IL-10 modulates host responses and lung damage induced by Pneumocystis carinii infection. J Immunol 170:1002–1009
175. Ruan S, Mckinley L, Zheng M et al (2008) Interleukin-12 and host defense against murine Pneumocystis pneumonia. Infect Immun 76:2130–2137
176. Rudner XL, Happel KI, Young EA et al (2007) Interleukin-23 (IL-23)–IL-17 cytokine axis in murine Pneumocystis carinii infection. Infect Immun 75:3055–3061
177. Wright TW, Johnston CJ, Harmsen AG et al (1999) Chemokine gene expression during Pneumocystis carinii-driven pulmonary inflammation. Infect Immun 67:3452–3460
178. Ishimine T, Kawakami K, Nakamoto A et al (1995) Analysis of cellular response and gamma interferon synthesis in bronchoalveolar lavage fluid and lung homogenate of mice infected with Pneumocystis carinii. Microbiol Immunol 39:49–58
179. Levy DE, Marie I, Prakash A (2003) Ringing the interferon alarm: differential regulation of gene expression at the interface between innate and adaptive immunity. Curr Opin Immunol 15:52–58
180. Hof H (2012) Pneumocystis jirovecii: a peculiar fungus posing particular problems for therapy and prophylaxis. Mycosis 55:1–7
181. Mofenson LM, Brady MT, Danner SP et al (2009) Guidelines for the Prevention and Treatment of Opportunistic Infections among HIV-exposed and HIV-infected children: recommendations from CDC, the National Institutes of Health, the HIV Medicine Association of the Infectious Diseases Society of America, the Pediatric Infectious Diseases Society, and the American Academy of Pediatrics. MMWR Recomm Rep 58(RR-11):1–166
182. Ng VL, Yajko DM, Hadley WK (1997) Extrapulmonary pneumocystosis. Clin Microbiol Rev 10:401–418
183. Fillatre P, Decaux O, Jouneau S et al (2014) Incidence of Pneumocystis jiroveci pneumonia among groups at risk in HIV-negative patients. Am J Med 127:1242.e11–1242.e17
184. Burke BA, Good RA (1973) Pneumocystis carinii infection. Medicine 52:23–51
185. Kovacs JA, Hiemenz JW, Macher AM et al (1984) Pneumocystis carinii pneumonia: a comparison between patients with the acquired immunodeficiency syndrome and patients with other immunodeficiencies. Ann Intern Med 100:663–671
186. Baughman RP (1994) Current methods of diagnosis. In: Walzer PD (ed) Pneumocystis carinii Pneumonia, 2nd edn. Marcel Dekker, New York, pp 381–401

187. Curtis JR, Yarnold PR, Schwartz DN et al (2000) Improvements in outcomes of acute respiratory failure for patients with human immunodeficiency virus-related Pneumocystis carinii pneumonia. Am J Respir Crit Care Med 162:393–398
188. Pareja JG, Garland R, Koziel H (1998) Use of adjunctive corticosteroids in severe adult non-HIV Pneumocystis carinii pneumonia. Chest 113:1215–1224
189. Smith DE, McLuckie A, Wyatt J et al (1988) Severe exercise hypoxaemia with normal or near normal X-rays: a feature of Pneumocystis carinii infection. Lancet 2:1049–1051
190. Zaman MK, Wooten OJ, Suprahmanya B et al (1988) Rapid noninvasive diagnosis of Pneumocystis carinii from induced liquefied sputum. Ann Intern Med 109:7–10
191. Selwyn PA, Pumerantz AS, Durante A et al (1998) Clinical predictors of Pneumocystis carinii pneumonia, bacterial pneumonia and tuberculosis in HIV-infected patients. AIDS 12:885–893
192. Opravil M, Marincek B, Fuchs WA et al (1994) Shortcomings of chest radiography in detecting Pneumocystis carinii pneumonia. J Acquir Immune Defic Syndr 7:39–45
193. Metersky ML, Colt HG, Olson LK et al (1995) Aids-related spontaneous pneumothorax: risk factors and treatment. Chest 108:946–951
194. Baughman RP, Dohn MN, Frame PT (1994) The continuing utility of bronchoalveolar lavage to diagnose opportunistic infection in AIDS patients. Am J Med 97:515–522
195. Stover DE, Zaman MB, Hajdu SI et al (1984) Bronchoalveolar lavage in the diagnosis of diffuse pulmonary infiltrates in the immunosuppressed host. Ann Intern Med 101:1–7
196. Gruden JF, Huang L, Turner J et al (1997) High-resolution CT in the evaluation of clinically suspected Pneumocystis carinii pneumonia in AIDS patients with normal, equivocal, or nonspecific radiographic findings. AJR Am J Roentgenol 169:967–975
197. Hidalgo A, Falcó V, Mauleón S et al (2003) Accuracy of high-resolution CT in distinguishing between Pneumocystis carinii pneumonia and non- Pneumocystis carinii pneumonia in AIDS patients. Eur Radiol 13:1179–1184
198. Collazos J, Beneitez C, Perez-Barba C et al (1985) Pneumocystis carinii Pneumonia. Ann Intern Med 102:137–138
199. Barry SM, Johnson MA (2001) Pneumocystis carinii pneumonia: a review of current issues in diagnosis and management. HIV Med 2:123–132
200. Huang L (2005) Clinical presentation and diagnosis of pneumocystis pneumonia in HIV-infected patients. In: Walzer PD, Cushion MT (eds) Pneumocystis pneumonia, 3rd edn. Marcel Dekker, New York, pp 349–406
201. Schildgen V, Mai S, Khalfaoui S et al (2014) Pneumocystis jirovecii can be productively cultured in differentiated CuFi-8 airway cells. MBio 5:e01186–e01114
202. Rossiter SJ, Miller DC, Churg AM et al (1979) Open lung biopsy in the immunosuppressed patient. Is it really beneficial? J Thorac Cardiovasc Surg 77:338–345
203. Fishman JA (2002) Pneumocystis carinii and parasitic infection in the immunocompromised host. In: Rubin RH, Lowell SY (eds) Clinical approach to infection in the compromised host. Kluwer Academic/Plenum, New York, pp 265–334
204. LaRocque RC, Katz JT, Perruzzi P et al (2003) The utility of sputum induction for diagnosis of Pneumocystis pneumonia in immunocompromised patients without human immunodeficiency virus. Clin Infect Dis 37:1380–1383
205. Kovacs JA, Ng VL, Masur H et al (1988) Diagnosis of Pneumocystis carinii pneumonia: improved detection in sputum with use of monoclonal antibodies. New Engl J Med 318:589–593
206. Wakefield AE, Miller RF, Guiver LA et al (1993) Oropharyngeal samples for detection of Pneumocystis carinii by DNA amplification. Int J Med 86:401–406
207. Helweg-Larsen J, Jensen JS, Lundgren B (1997) Non-invasive diagnosis of Pneumocystis carinii pneumonia by PCR on oral washes. Lancet 350:1363
208. Samuel CM, Whitelaw A, Corcoran C et al (2011) Improved detection of Pneumocystis jirovecii in upper and lower respiratory tract specimens from children with suspected pneumocystis pneumonia using real-time PCR: a prospective study. BMC Infect Dis 11:329

209. To KK, Wong SC, Xu T et al (2013) Use of nasopharyngeal aspirate for diagnosis of Pneumocystis pneumonia. J Clin Microbiol 51:1570–1574
210. Grocott RG (1955) A stain for fungi in tissue sections and smears using Gomori's methenamine-silver nitrate technic. Am J Clin Pathol 25:975–979
211. Musto L, Flanigan M, Elbadawi A (1982) Ten-minute silver stain for Pneumocystis carinii and fungi in tissue sections. Arch Pathol Lab Med 106:292–294
212. Gosey LL, Howard RM, Witebsky FG et al (1985) Advantages of a modified toluidine blue O stain and bronchoalveolar lavage for the diagnosis of Pneumocystis carinii pneumonia. J Clin Microbiol 22:803–807
213. Baselski VS, Robison MK, Pifer LW et al (1990) Rapid detection of Pneumocystis carinii in bronchoalveolar lavage samples by using Cellufluor staining. J Clin Microbiol 28:393–394
214. Holten-Andersen W, Kolmos HJ (1989) Comparison of methenamine silver nitrate and Giemsa stain for detection of Pneumocystis carinii in bronchoalveolar lavage specimens from HIV infected patients. APMIS 97:745–747
215. Kovacs JA, Gill V, Swan JC et al (1986) Prospective evaluation of a monoclonal antibody in diagnosis of Pneumocystis carinii pneumonia. Lancet 2:1–3
216. Kovacs JA, Gill VJ, Meshnick S et al (2001) New insights into transmission, diagnosis, and drug treatment of Pneumocystis carinii pneumonia. J Am Med Assoc 286:2450–2460
217. Matos O, Esteves F (2016) Laboratory diagnosis of Pneumocystis jirovecii pneumonia. In: Kon K, Rai M (eds) Microbiology of respiratory system infection. Elsevier, Amsterdam, pp 185–210
218. Durand-Joly I, Chabé M, Soula F et al (2005) Molecular diagnosis of Pneumocystis pneumonia (PcP). FEMS Immunol Med Microbiol 45:405–410
219. Azoulay E, Bergeron A, Chevret S et al (2009) Polymerase chain reaction for diagnosing Pneumocystis pneumonia in non-HIV immunocompromised patients with pulmonary infiltrates. Chest 135:655–661
220. Respaldiza N, Montes-Cano MA, Friaza V et al (2006) Usefulness of oropharyngeal washings for identifying Pneumocystis jirovecii carriers. J Eukaryot Microbiol 53:100–101
221. Tsolaki AG, Miller RF, Wakefield AE (1999) Oropharyngeal samples for genotyping and monitoring response to treatment in AIDS patients with Pneumocystis carinii pneumonia. J Med Microbiol 48:897–905
222. Larsen HH, Huang L, Kovacs JA et al (2004) A prospective, blinded study of quantitative touch-down polymerase chain reaction using oral-wash samples for diagnosis of Pneumocystis pneumonia in HIV-infected patients. J Infect Dis 189:1679–1683
223. Esteves F, Calé SS, Badura R et al (2015) Diagnosis of Pneumocystis pneumonia: evaluation of four serologic biomarkers. Clin Microbiol Infect 21:379.e1–379.e10
224. Rohner P, Jacomo V, Studer R et al (2009) Detection of Pneumocystis jirovecii by two staining methods and two quantitative PCR assays. Infection 37:261–265
225. Vidal S, de la Horra C, Martín J et al (2006) Pneumocystis jirovecii colonisation in patients with interstitial lung disease. Clin Microbiol Infect 12:231–235
226. Wakefield AE, Pixley FJ, Banerji S (1990) Detection of Pneumocystis carinii with DNA amplification. Lancet 336:451–453
227. Totet A, Pautard JC, Raccurt C et al (2003) Genotypes at the internal transcribed spacers of the nuclear rRNA operon of Pneumocystis jiroveci in nonimmunosuppressed infants without severe pneumonia. J Clin Microbiol 41:1173–1180
228. Olsson M, Strålin K, Holmberg H (2001) Clinical significance of nested polymerase chain reaction and immunofluorescence for detection of Pneumocystis carinii pneumonia. Clin Microbiol Infect 7:492–497
229. Fan LC, Lu HW, Cheng KB et al (2013) Evaluation of PCR in bronchoalveolar lavage fluid for diagnosis of Pneumocystis jirovecii pneumonia: a bivariate meta-analysis and systematic review. PLoS One 8:e73099
230. Moonens F, Liesnard C, Brancart F et al (1995) Rapid simple and nested polymerase chain reaction for the diagnosis of Pneumocystis carinii pneumonia. Scand J Infect Dis 27:358–362

231. Thomas CF, Limper AH (2004) Pneumocystis pneumonia. N Engl J Med 350:2487–2498
232. Fillaux J, Malvy S, Alvarez M et al (2008) Accuracy of a routine real-time PCR assay for the diagnosis of Pneumocystis jirovecii pneumonia. J Microbiol Methods 75:258–261
233. Harris JR, Marston BJ, Sangrujee N et al (2011) Cost-effectiveness analysis of diagnostic options for Pneumocystis pneumonia (PCP). PLoS One 6:e23158
234. Botterel F, Cabaret O, Foulet F et al (2012) Clinical significance of quantifying Pneumocystis jirovecii DNA by using real-time PCR in bronchoalveolar lavage fluid from immunocompromised patients. J Clin Microbiol 50:227–231
235. McTaggart LR, Wengenack NL, Richardson SE (2012) Validation of the mycassay Pneumocystis kit for detection of Pneumocystis jirovecii in bronchoalveolar lavage specimens by comparison to a laboratory standard of direct immunofluorescence microscopy, real-time PCR, or conventional PCR. J Clin Microbiol 50:1856–1859
236. Larsen HH, Masur H, Kovacs JA et al (2002) Development and evaluation of a quantitative, touchdown, real-time PCR assay for diagnosing Pneumocystis carinii pneumonia. J Clin Microbiol 40:490–494
237. Montesinos I, Brancart F, Schepers K et al (2015) Comparison of 2 real-time PCR assays for diagnosis of Pneumocystis jirovecii pneumonia in human immunodeficiency virus (HIV) and non-HIV immunocompromised patients. Diagn Microbiol Infect Dis 82:143–147
238. Summah H, Zhu YG, Falagas ME et al (2013) Use of real-time polymerase chain reaction for the diagnosis of Pneumocystis pneumonia in immunocompromised patients: a meta-analysis. Chin Med J 126:1965–1973
239. Sasso M, Chastang-Dumas E, Bastide S et al (2016) Performances of four real-time PCR assays for the diagnosis of Pneumocystis jirovecii Pneumonia. J Clin Microbiol 54:625–630
240. Montesinos I, Delforge M-L, Ajjaham F et al (2017) Evaluation of a new commercial real-time PCR assay for diagnosis of Pneumocystis jirovecii pneumonia and identification of dihydropteroate synthase (DHPS) mutations. Diagn Microbiol Infect Dis 87(1):32–36
241. Lu JJ, Lee CH (2008) Pneumocystis pneumonia. J Formos Med Assoc 107:830–842
242. Depypere M, Saegeman V, Lagrou K (2016) Typing of Pneumocystis jirovecii by multiclocus sequencing: evidence of outbreak? Eur J Clin Microbiol Infect Dis 35:911–916
243. Schluger N, Sepkowitz K, Armstrong D et al (1991) Detection of Pneumocystis carinii in serum of AIDS patients with Pneumocystis pneumonia by the polymerase chain reaction. J Protozool 38:240S–242S
244. Lipschik GY, Gill VJ, Lundgren JD et al (1992) Improved diagnosis of Pneumocystis carinii infection by polymerase chain reaction on induced sputum and blood. Lancet 340:203–206
245. Roux P, Lavrard I, Poirot JL et al (1994) Usefulness of PCR for detection of Pneumocystis carinii DNA. J Clin Microbiol 32:2324–2326
246. Atzori C, Lu JJ, Jiang B et al (1995) Diagnosis of Pneumocystis carinii pneumonia in AIDS patients by using polymerase chain reactions on serum specimens. J Infect Dis 172:1623–1626
247. Tamburrini E, Mencarini P, Visconti E et al (1996) Detection of Pneumocystis carinii DNA in blood by PCR is not of value for diagnosis of P. carinii pneumonia. J Clin Microbiol 34:1586–1588
248. Matos O, Lundgren B, Caldeira L et al (1999) Evaluation of a nested PCR for detection of Pneumocystis carinii in serum from imunocompromised patients. J Eukaryot Microbiol 46:104S–105S
249. Rabodonirina M, Cotte L, Boibieux A et al (1999) Detection of Pneumocystis carinii DNA in blood specimens from human immunodeficiency virus-infected patients by nested PCR. J Clin Microbiol 37:127–131
250. Tasaka S, Hasegawa N, Kobayashi S et al (2007) Serum indicators for the diagnosis of Pneumocystis pneumonia. Chest 131:1173–1180
251. Finkelman MA (2010) Pneumocystis jirovecii infection: Cell wall (1-3)-D-glucan biology and diagnostic utility. Crit Rev Microbiol 36:271–281

252. Daly KR, Fichtenbaum CJ, Tanaka R et al (2002) Serologic responses to epitopes of the major surface glycoprotein of Pneumocystis jirovecii differ in human immunodeficiency virus-infected and uninfected persons. J Infect Dis 186:644–651
253. Daly KR, Koch J, Levin L et al (2004) Enzyme-linked immunosorbent assay and serologic responses to Pneumocystis jirovecii. Emerg Infect Dis 10:848–854
254. Djawe K, Huang L, Daly KR et al (2010) Serum antibody levels to the Pneumocystis jirovecii major surface glycoprotein in the diagnosis of P. jirovecii pneumonia in HIV+ patients. PLoS One 5:e14259
255. Gingo MR, Lucht L, Daly KR et al (2011) Serologic responses to Pneumocystis proteins in human immunodeficiency virus patients with and without Pneumocystis jirovecii pneumonia. J Acquir Immune Defic Syndr 57:190–196
256. Skelly MJ, Holzman RS, Merali S (2008) S-adenosylmethionine levels in the diagnosis of Pneumocystis carinii pneumonia in patients with HIV infection. Clin Infect Dis 46:467–471
257. Nakamura H, Tateyama M, Tasato D et al (2009) Clinical utility of serum β-d-glucan and KL-6 levels in Pneumocystis jirovecii pneumonia. Intern Med 48:195–202
258. Esteves F, Lee CH, de Sousa B et al (2014) (1-3)-beta-D-glucan in association with lactate dehydrogenase as biomarkers of Pneumocystis pneumonia (PcP) in HIV-infected patients. Eur J Clin Microbiol Infect Dis 33:1173–1180
259. Onishi A, Sugiyama D, Kogata Y et al (2012) Diagnostic accuracy of serum 1,3-beta-d-glucan for Pneumocystis jiroveci pneumonia, invasive candidiasis, and invasive aspergillosis: systematic review and meta-analysis. J Clin Microbiol 50:7–15
260. Karageorgopoulos DE, Qu JM, Korbila IP et al (2013) Accuracy of beta-d-glucan for the diagnosis of Pneumocystis jirovecii pneumonia: a meta-analysis. Clin Microbiol Infect 19:39–49
261. Li WJ, Guo YL, Liu TJ et al (2015) Diagnosis of pneumocystis pneumonia using serum (1-3)-β-Glucan: a bivariate meta-analysis and systematis review. Thorac Dis 7:2214–2225
262. Salerno D, Mushatt D, Myers L et al (2014) Serum and BAL beta-D-glucan for the diagnosis of Pneumocystis pneumonia in HIV positive patients. Respir Med 108:1688–1695
263. Stringer JR (2005) Surface antigens. In: Walzer PD, Cushion MT (eds) Pneumocystis pneumonia, 3rd edn. Marcel Dekker, New York, pp 95–126
264. Tomás AL, Cardoso F, Esteves F et al (2016) Serological diagnosis of pneumocystosis: production of a synthetic recombinant antigen for immunodetection of Pneumocystis jirovecii. Sci Rep. doi:10.1038/srep36287
265. Roger PM, Vandenbos F, Pugliese P et al (1998) Persistence of Pneumocystis carinii after effective treatment of P. carinii pneumonia is not related to relapse or survival among patients infected with human immunodeficiency virus. Clin Infect Dis 26:509–510
266. Cooley L, Dendle C, Wolf J et al (2014) Consensus guidelines for diagnosis, prophylaxis and management of Pneumocystis jirovecii pneumonia in patients with haematological and solid malignancies. Intern Med J 44:1350–1363
267. Masur H (1992) Prevention and treatment of Pneumocystis carinii pneumonia. N Engl J Med 327:1853–1860
268. Bozzette SA, Sattler FR, Chiu J et al (1990) California collaborative treatment group: a controlled trial of early adjunctive treatment with corticosteroids for Pneumocystis carinii pneumonia in the acquired immunodeficiency syndrome. N Engl J Med 323:1451–1457
269. Klein NC, Duncanson FP, Lenox TH et al (1992) Trimethoprim–sulfamethoxazole versus pentamidine for Pneumocystis carinii pneumonia in AIDS patients: results of a large prospective randomized treatment trial. AIDS 6:301–305
270. Toma E, Thorne A, Singer J et al (1998) The CTN-PCP Study Group. Clindamycin with primaquine vs trimethoprim-sulfamethoxazole therapy for mild and moderately severe Pneumocystis carinii pneumonia in patients with AIDS: a multicenter, double-blind, randomized trial (CTN 004). Clin Infect Dis 27:524–530

271. Hughes W, Leoung G, Kramer F et al (1993) Comparison of atovaquone (566C80) with trimethoprim-sulfamethoxazole to treat Pneumocystis carinii pneumonia in patients with AIDS. N Engl J Med 328:1521–1527
272. Safrin S, Finkelstein DM, Feinberg J et al (1996) Comparison of three regimens for treatment of mild to moderate Pneumocystis carinii pneumonia in patients with AIDS. A double-blind, randomized, trial of oral trimethoprim-sulfamethoxazole, dapsone-trimethoprim, and clindamycin-primaquine. ACTG 108 Study Group. Ann Intern Med 124:792–802
273. Tomblyn M, Chiller T, Einsele H et al (2009) Guidelines for preventing infectious complications among hematopoietic cell transplantation recipients: a global perspective. Biol Blood Marrow Transplant 15:1143–1238
274. Wang EH, Partovi N, Levy RD et al (2012) Pneumocystis pneumonia in solid organ transplant recipients: not yet an infection of the past. Transpl Infect Dis 14:519–525
275. Neumann S, Krause SW, Maschmeyer G et al (2013) Primary prophylaxis of bacterial infections and Pneumocystis jirovecii pneumonia in patients with hematological malignancies and solid tumors: guidelines of the Infectious Diseases Working Party (AGIHO) of the German Society of Hematology and Oncology (DGHO). Ann Hematol 92:433–442
276. Rahier JF, Ben-Horin S, Chowers Y et al (2009) European evidence-based consensus on the prevention, diagnosis and management of opportunistic infections in inflammatory bowel disease. J Crohns Colitis 3:47–91
277. Sowden E, Carmichael AJ (2004) Autoimmune inflammatory disorders, systemic corticosteroids and pneumocystis pneumonia: a strategy for prevention. BMC Infect Dis 4:42
278. Green H, Paul M, Vidal L et al (2007) Prophylaxis of Pneumocystis pneumonia in immunocompromised non-HIV-infected patients: systematic review and meta-analysis of randomized controlled trials. Mayo Clin Proc 82:1052–1059
279. Centers for Disease Control and Prevention (1989) Guidelines for prophylaxis against Pneumocystis carinii pneumonia for persons infected with human immunodeficiency virus. MMWR Morb Mortal Wkly Rep 38:1–9
280. Bozzette SA, Finkelstein DM, Spector SA et al (1995) A randomized trial of three antipneumocystis agents in patients with advanced human immunodeficiency virus infection. NIAID AIDS Clinical Trials Group. N Engl J Med 332:693–699
281. Schneider MME, Hoepelman AIM, Eeftinck Schattenkerk JKM et al (1992) A controlled trial of aerosolized pentamidine or trimethoprim–sulfamethoxazole as primary prophylaxis against Pneumocystis carinii pneumonia in patients with human immunodeficiency virus infection. N Engl J Med 327:1836–1841
282. Schneider MME, Nielsen TL, Nelsing S et al (1995) Efficacy and toxicity of 2 doses of trimethoprim-sulfamethoxazole as primary prophylaxis against Pneumocystis carinii pneumonia in patients with human immunodeficiency virus. J Infect Dis 171:1632–1636
283. Masur H, Kaplan JE, Holmes KK (2002) Guidelines for preventing opportunistic infections among HIV-infected persons – 2002. Recommendations of the U.S. Public Health Service and the Infectious Diseases Society of America. Ann Intern Med 137:435–478

Chapter 8
Coccidioides and Coccidioidomycosis

Marcus M. Teixeira and Bridget M. Barker

Abstract *Coccidioides immitis* and *C. posadasii* are the causative agents of coccidioidomycosis, or San Joaquin Valley Fever. The disease was first described in 1892 in Argentina, and its distribution and frequency investigated into the mid-1900s. National reporting has measured a recent increase in disease burden in the United States. All mammals appear to be susceptible to infection; however disease manifestation varies from asymptomatic to lethal. Several thousand cases of the disease are reported annually in the United States. We discuss the epidemiology, diagnostics, virulence factors, and treatment options for coccidioidomycosis.

8.1 History and Taxonomy

Coccidioidomycosis or Valley fever is a deep systemic mycosis reported in arid and semi-arid areas of the Americas. The disease is caused by the geo-zoo-anthropophylic fungal species nested in the *Coccidioides* genus [1–3]. The history of coccidioidomycosis began in 1892 when Alejandro Posadas evaluated the cavalry soldier Domingos Ezcurra [4]. He had disseminated verrucous lesions after suffering from an unknown disease for a year while battling on the border of the Chaco region of Argentina. He also presented lymphadenopathy, and several attempts to use treatments and surgical interventions were performed, but did not halt the progression of the disease. Domingos Ezcurra's death in 1889 and a complete autopsy demonstrated lesions in the skin, lymph nodes, lungs, liver, and spleen. Histopathological investigation of

M.M. Teixeira, Ph.D.
Division of Pathogen Genomics, Translational Genomics Research Institute (TGen-North), Flagstaff, AZ, USA

Pathogen and Microbiome Institute, Northern Arizona University, Flagstaff, AZ, USA

B.M. Barker, Ph.D. (✉)
Division of Pathogen Genomics, Translational Genomics Research Institute (TGen-North), Flagstaff, AZ, USA

Pathogen and Microbiome Institute, Northern Arizona University, Flagstaff, AZ, USA

Department of Medicine, Valley Fever Center for Excellence, University of Arizona, Tucson, AZ, USA
e-mail: bridget.barker@nau.edu

H.M. Mora-Montes, L.M. Lopes-Bezerra (eds.), *Current Progress in Medical Mycology*, DOI 10.1007/978-3-319-64113-3_8

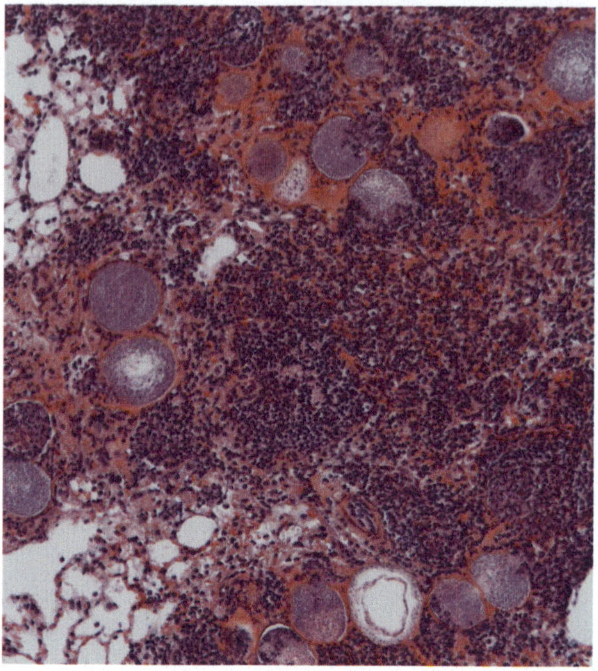

Fig. 8.1. CD-1 mice infected with 10^6 arthroconidia instilled intranasally show significant infection and lung damage by day 5. H&E staining reveals large spherules and massive influx of host immune cells

involved organs revealed the presence of epitheliod granulomas, infiltrates of mononuclear cells as well the presence of spherules as shown in Fig. 8.1 [5, 6].

For more than a century, *Coccidioides immitis* was considered a singles species; however, molecular systematics revealed that at least two species complexes, containing distinct populations with limited gene flow [7–10]. *Coccidioides posadasii* was proposed by [11] in a tribute to Alejandro Posadas, the discoverer of coccidioidomycosis [11]. *C. immitis* is prevalent in central/southern California, Eastern Washington and north Mexico while *C. posadasii* is endemic in Arizona with a discontinuous distribution extending into Utah, Texas, Mexico, Central and South America [12, 13]. Recent comparative genomic and population genetic studies suggested that *C. immitis* is composed of three subpopulations, divided into San Diego/Mexico, San Joaquin Valley and Washington. These studies also proposed that *C. posadasii* is divided into Arizona, Mexico, Guatemala, and Texas/South America populations [14, 15]. It is likely that *C. immitis* and *C. posadasii* co-occur in nature, as both species have been found in patients from both San Diego/USA and Mexico, and hybrid strains have been discovered [15, 16].

8.2 Clinical Aspects of Disease

The disease affects both healthy and immune compromised human populations, as well as many other mammals [1, 17, 18]. Populations at greatest risk for severe disease are persons with underlying immunocompromising conditions. These include HIV/AIDS, immunomodulatory therapies for cancer, organ transplant, and autoimmune diseases [19–25]. Males more than females, African American and Filipino more than European ancestry, and pregnancy also have greater risk of dissemination [26–30]. Mutations of IL-12rβ1, IFN-γr1, STAT1, and STAT3 have also been associated with disseminated disease [31].

Coccidioides sp. infections primarily affect the lungs. Most commonly, infection is initiated by the inhalation of infectious propagules of the fungus (arthroconidia), but can be a result of dermal invasion [32]. The disease is asymptomatic in approximately 60% of infections. The remaining cases are usually characterized by a self-limited respiratory illness [33–37]. The disease may be confounded with several other upper respiratory infections and requires specific laboratory tests including culture and serology, along with associated travel history or inhabitation of endemic areas of the disease [38–40].

For those with symptoms, the most common forms are acute and sub-acute disease, which manifests as mild respiratory illness (often flu-like) with fever, cough, and is frequently associated with rib cage pain and/or fatigue [39, 41]. In the majority of acute pulmonary cases the use of antifungal drugs is not required [38, 39]. However, pneumonia may be observed in those who develop a severe acute respiratory disease and chemotherapy may be necessary for disease management. Clinical conditions in those patients include intense and elevated fever, dyspnea, hypoxemia, and are frequently observed in immunocompromised individuals [2, 3, 38, 39].

Several factors may contribute to those cases, including the inoculum, pathogen variation, and host genetic background [42–46]. Skin lesions are the most frequent extra-pulmonary manifestation and are present in 10–50% of cases [47, 48]. Lesions frequently observed include papules, nodules, pustules, abscesses, warts or ulcerations. Lymphohematogenous disease is associated with high morbidity and mortality rates, and may require life-long antifungal therapy in most cases. One percent of cases are diagnosed as disseminated coccidioidomycosis; however, in susceptible populations the risk of dissemination can reach 75% [31]. Dissemination usually onsets within a few weeks to months following the primary infection [49]. A widespread septic fungemia may occur in some cases and is often fatal without treatment [2, 33, 50, 51].

8.3 Epidemiology

Coccidioidomycosis disease range exists across arid to semi-arid regions of the Americas. The disease is endemic to many desert biomes across the western United States, Mexico, Guatemala, Venezuela, Colombia, Paraguay, Argentina, and Brazil [1, 3, 52–55]. Those areas are generally characterized by having elevated temperatures and low annual rainfall with sporadic seasonal rainy periods. During seasonal rains, is believed that the fungus grows in the filamentous form. Upon maturation, arthroconidia are produced asexually and become airborne during drought, wind storms, and haboobs [32, 56–59].

This theory of "grow and blow" may be responsible for the annual peaks of infection in the western USA during summer months [60]. Soil perturbation by natural phenomenon such as windstorms and earthquakes; or by anthropogenic actions such as agriculture, excavation of archeological sites, deforestation, or hunting armadillos, which are known risk factors for coccidioidomycosis [2, 61–67]. In addition, microenvironmental conditions such as soil type and available nutrients may be key factors for fungal development and may explain the discontinuous distribution of the fungus in specific endemic areas [59, 68, 69]. Upon inhalation of an arthroconidium into a mammal lung, both *Coccidioides immitis* and *C. posadasii* will shift into a parasitic spherule cycle. Spherules became mature in tissue initiating an encasement for multiple uninucleate endospores that after rupture can disperse throughout the body (Fig. 8.2; [32]).

The genus *Coccidioides* is nested within the Onygenales order (Ascomycote) and shares with *Blastomyces*, *Paracoccidioides*, *Histoplasma,* and Dermatophytes the ability to degrade animal derived material such as keratin [70]. Genomic analyses show that these human fungal pathogens co-inherited genes that allow the degradation of proteins present skin, carcasses, bones, and other animal tissues [71–73]. *Coccidioides* was found associated with different wild rodents in Arizona and California, and was isolated from armadillos in Brazil, which suggests that burrow-

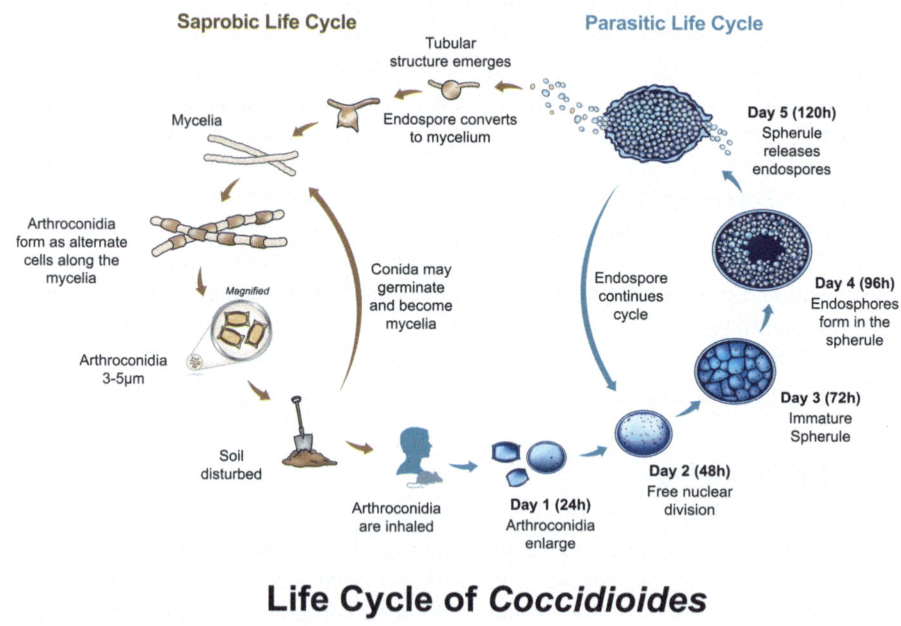

Fig. 8.2. Upon inhalation of an arthroconidium into a mammalian lung, both *Coccidioides immitis* and *C. posadasii* will shift into a parasitic spherule cycle. Depending on host and fungal factors, the spherules became mature within 5–10 days in tissue initiating an encasement for multiple uninucleate endospores, which are released after rupture and can disseminate throughout the body [32]

forming mammals may be direct (i.e. infection) or indirect (i.e. thriving in the burrow environment) hosts and responsible for the maintenance and dispersion of the fungus [62, 74–77].

All mammals are susceptible to infection. Primates (humans and other primates) and canids (wolves and domestic dogs) appear to have more severe disease on average than other mammals [18, 78–85]. Agricultural animals in the endemic region; including cattle, bison, pigs, sheep, and goats, have high rates of infection/exposure, but do not appear to have frequent pulmonary symptoms or severe disease [86–88]. Equine disease is rare, but severe and complicated to treat [89–91]. Infection of lizards and snakes is thought to be rare, but is reported [92, 93]. Bats, sea mammals, cats, zoo animals, and many other animals, all seem to have some level of infectivity, but disease rarely manifests [92, 94–99]. Birds do not appear to be infected, as we can find no reported cases.

In the 1990s in the United States, cases became reportable to the CDC and at the state level in California, Arizona, Nevada, New Mexico, and Utah [100]. The reported annual incidence of coccidioidomycosis in USA is estimated to be 150,000 cases annually. Since being tracked, the reported rates have increased to a high of 25,000 cases in 2011 [2, 101]. There were changes in reporting at this time in Arizona and the cause of the rise that year is still debated. From 2012 to 2014 the reports declined; however, a recent increase was reported in 2015 [102, 103]. Therefore, residents of and recent travelers to regions where *Coccidioides* is endemic that are diagnosed with community acquired pneumonia should be evaluated routinely for coccidioidomycosis as a possible etiologic agent [40, 104]. It was thought that dust storms and large-scale soil disturbance led to increased disease, but the results are mixed. The Northridge earthquake resulted in spike in the region near the earthquake origin; however, no increase in disease was observed after haboobs in Arizona [56, 59, 60, 65, 67, 105].

8.4 Virulence Factors

The species from the genus *Coccidioides* are dimorphic fungi and the shift to spherule cell types promotes a gene reprograming, which allows the fungi to proliferate or stay quiescent inside mammalian tissues [106–108]. Auxotrophic mutants of *Coccidioides* have been reported to decrease virulence or are avirulent in a murine model, but few virulence factors have been investigated for *Coccidioides* to date [109, 110].

Upon infection, the arthroconidia differentiate, resulting in the formation of specialized endosporulating spherules (Figs. 8.1 and 8.3). This morphological shift is one mechanism that allows *Coccidioides* to survive in harsh conditions imposed by the immune system even in immunocompetent hosts [111–113].

Three functional genomic studies based on transcriptomic analysis have identified genes and pathways expressed differentially in the spherule phase compared to the saprophytic stage of both *C. immitis* and *C. posadasii*. These studies revealed an

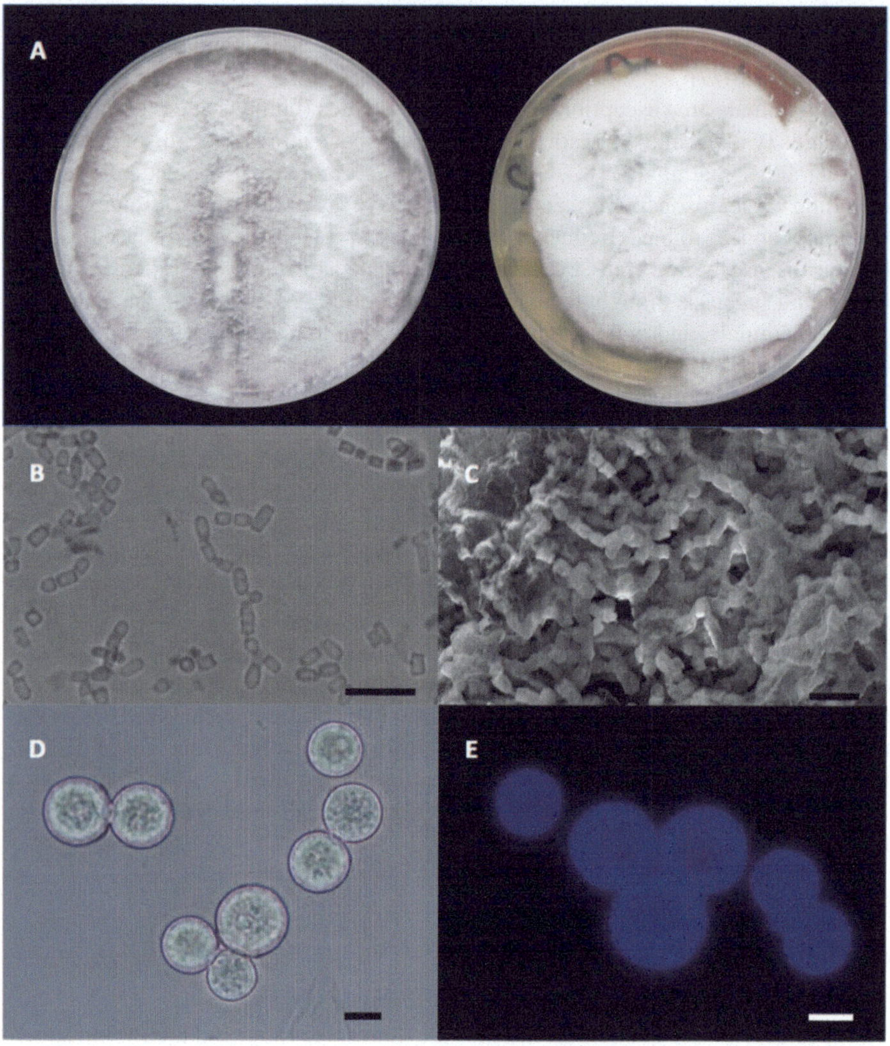

Fig. 8.3 (**a**) Optimal fungal growth is between 28 and 37 °C and white/beige cotonous mycelial colonies are expected to appear between days 2 and 7 (or potentially longer) surrounding the cultured biomaterial. Older *Coccidioides* colonies can also present dark pigmentation and a wrinkled surface. (**b**) Microscopically, the filamentous form of *Coccidioides* sp. is characterized by hyaline and septate hyphae harboring barrel-shaped arthroconidia ranging from 2 to 5 um. (**c**) Using Scanning Electron Microscopy (SEM) the barrel shape of the conidia can be easily observed. (**d–e**) Spherules and endospores cells can be visualized from BALF's and other biofluids. Direct mycological examinations are usually prepared using potassium hydroxide (**d**), and/or calcofluor white labeling (**e**) *Black bars* represent 5 μM

upregulation of spherule-specific genes controlling the dynamic and complex morphological switch that contributes to the development of infection, which codify for cell wall components (chitin and glucan modeling), oxidative stress, pH regulation, lipid metabolism, sugar transporters, and proteases [106, 108, 114]. Recently, the proteomic composition of spherulin (lysate of the spherule phase) was characterized and revealed the presence of important metabolic enzymes and lectin-binding glycoproteins of the infective phase of the fungus [115]. Structural and functional genomic studies of *C. immitis* and *C. posadasii* coupled with efficient gene disruption and insertional mutagenesis methods allowed the deletion and validation of potential virulence factors, listed below.

SOWgp. The outer cell wall portion of spherules is coated with several immunomodulatory proteins [116–119]. A specific glycoprotein (SOWgp) was shown to be dominant in the endospore surface (early spherule phase) and plays an important role in the cellular and humoral immunity in patients infected with *Coccidioides* [120–122]. The function of this gene was determined by targeted gene replacement and bioinformatics methods. The SOWgp gene encodes for a 422-amino acid protein harboring six blocks of proline-rich sites [123, 124]. The specific expression of this gene in the beginning of the parasitic phase of *Coccidioides* was first demonstrated by specific RNA hybridization and later confirmed by RNA-seq experiments [108]. The adhesin function of this antigen was determined by the ability of the recombinant SOWgp to bind to extracellular matrices (laminin, fibronectin, and collagen type IV). The disruption of SOWgp gene in the wild type *C. posadasii* C735 strain resulted in a lack of adhesion and a significant reduction in the virulence in a murine model of infection [123].

Mep1. Another important protein identified in the lysates of the parasitic phase of *Coccidioides* is the Mep1 protease. Early proteomic analysis revealed that spherule glycoproteins were naturally cleaved [125, 126]. It was shown that Mep1 cleaves the major *Coccidioides* spherule antigen SOWgp, thus protecting the pathogen from phagocytosis by innate immune cells [125]. This study revealed that SOWgp may play a role as a pathogen-associated molecular pattern (PAMP) molecule. Mep1 belongs to the metalloprotease subfamily M43B, containing 276 amino acids and is potentially secreted to outer cell space. Two conserved zinc-biding sites were identified being important to the structural stability of Mep1 [127]. The Mep1 gene was found to be highly expressed in the parasitic phase of *Coccidioides* and this was later confirmed by RNA-seq experiments. In vivo studies showed that the gene disruption of Mep1 in the C735 strain increased C57BL/6 mice survival, compared to infection with the native strain. This effect was more evident when mice were previously challenged with SOWgp antibodies. The experiments proved that the exposure of SOWgp could increase the phagocytic activity of alveolar macrophages allowing the fungal proliferation in the host [122, 123, 127].

Cps1. The Cps1 gene encodes for an AMP-binding protein, and may be part of non-ribosomal peptide synthase (NRPS) complex conserved among Ascomycetous fungi [128]. The novel peptide synthetase was initially discovered in the plant pathogen *Cochliobolus heterostrophus*, shown to be necessary for virulence, and the depleted Cps1 strain was protective in plant infection models [128]. The gene knockout of the homologous Cps1 copy in the *C. posadasii* genome produced a reduction in the virulence effect as compared to the wild type strain. The Cps1 mutant was not able to produce disease, or persist in tissue in mice models. The role of Cps1 was not identified; however, RNA-seq experiments demonstrated that this transcript regulates at least 33 other *Coccidioides posadasii* genes [129].

Urease/Ugh. Urease acts in the conversion of urea into ammonia in almost all fungal species and is responsible for pH regulation during infection [130]. Urease was previously shown to be released from spherules during endosporulation. The urease gene of *Coccidioides* was initially characterized in 1997 and encodes for a 2517-bp ORF that translates to protein of 91.5 kDa containing 839 amino acid residues [131]. Despite the lack of signal peptide the protein is carried from the cytosol to central vacuole via vesicles of spherules and promotes the alkalization of the infection site [132]. The loss of function of urease gene resulted in a low virulent strain, allowing the clearance of *Coccidioides* infection more efficiently than native or reconstituted strains in murine infection experiments. Finally, immunization of mice with recombinant *Coccidioides* urease was protective against arthroconidia challenge by stimulating the T helper type 1 cells (Th1) response [133]. Fungi also metabolize ureidoglycolate into glyoxylate, generating CO_2 and ammonia. This process is mediated by the enzyme ureidoglycolate hydrolase and the lack of the Ugh gene in *Coccidioides* significantly decreases the virulence of this pathogen [134].

Chitin synthase and Chitinases. Chitin is an indispensable molecule of the cell wall of fungal pathogens, and is metabolized in *Coccidioides* by a subset of seven different classes of enzymes (CpCHS1-7), which may play an important role in fungal morphogenesis [116, 135]. Gene expression comparisons between spherule and mycelium revealed that transcripts of CpCHS1 and CpCHS4 are more abundant in the pathogenic phase [135]. Nikkomycin Z seems to be effective against *Coccidioides* by inactivating chitin synthases of the pathogenic phase of the fungus [136–138].

Chitinases also play an important role in spherule formation by cleaving the septal wall complex during endosporulation [116]. Two chitinases (CTS2 and CTS3) out of eight tested were found to be highly expressed during spherule development [139, 140]. These two genes are located on different chromosomes of *C. posadasii* and were disrupted to observe effects on virulence in a murine model of coccidioidomycosis. A third gene (ARD1) located next to the CTS3 was also disrupted during the homologous recombination of the insert creating a triple deletion strain [140]. This *C. posadasii* mutant strain has the ability to grow in the filamentous phase, producing hyphae and arthroconidia, but lacks the ability to endosporulate in vitro or within the host, and does not induce disease. This strain was later used in immunization trials in susceptible BALB/c mice. Mice immunized with the attenuated strain were protected against a secondary infection with wild type *C. posadasii* [140].

8.5 Immunopathology

Whereas adaptive immunity to infection has been studied in the development of vaccines, the humoral response is less well characterized. For most fungal infections, phagocytes such as neutrophils, macrophages and dendritic cells are involved in the initial control of infection [141]. For *Coccidioides*, these cell types have been implicated in the control of disease.

Human polymorphonuclear neutrophils (PMNs) have phagocytic activity against *C. immitis* in in vitro infection models when exposed to *Coccidioides* antigen, such as spherulin or coccidioidin [142, 143]. This process is also efficient for early released endospores, but not in mature spherules, primarily due to size [144]. Recent work showing frustrated phagocytosis reveals strong interactions and adherence, but lack of phagocytosis for large spherules [145]. Although PMN readily engulf arthroconidia and small endospores, they do not appear to be efficient in killing cells, as less than 20% mortality was observed in studies [146–148]. PMN infiltrates are usually observed in histopathological findings and these cell types appear to be rapidly recruited in the lungs in experimental pulmonary coccidioidomycosis [140, 149]. This is characterized by an intense inflammatory response in lungs that may contribute to tissue damage and granuloma formation that occurs during fungal infection (Fig. 8.1; [149, 150]).

Coccidioides endospores and arthroconidia are also phagocytized by alveolar and peritoneal macrophages from naive macaques, humans, and mice [146, 147, 151, 152]. In these studies, both cell types were phagocytosed, however, killing was less than 1% for unstimulated cells. For alveolar and peritoneal macrophages from BALB/c mice incubated with gamma-interferon (IFN-γ), a reduction of 50% of recovered CFU was observed [146]. In addition, a few studies suggest that lysosome fusion is inhibited by the fungus in naive lymphocytes; however, lysosomal fusion was observed in lymphocytes that were harvested from mice immunized with formalin-killed spherules [147].

Comparisons between more (DBA/2) and less (BALB/c) resistant mouse strains have revealed the resistance to coccidioidomycosis may be associated with higher pro-inflammatory cytokines tumor necrosis factor alpha (TNF-α), interleukin-1α (IL-1α), and interleukin -6 (IL-6) [153]. TNF-α is also induced in macrophages in response to exposure to formalin-killed spherules [154]. Additionally, IFN-γ is important for resistance to *Coccidioides* infection in DBA/5 mice, whereas higher interleukin-4 (IL-4) appears to increase fungal burden in BALB/c mouse lungs [155]. Thus, while the predominant Th1 protective response is initiated in DBA/5 mice via IFN-γ, BALB/c mice generate a T helper type 2 (Th2) response, which can be ameliorated by recombinant interleukin-12 (IL-12) [156].

Interestingly, the presence of SOWgp on the surface of spherules triggers the Th2 response in naive host cells, which produce IL-6 and IL-10, but SOWgp antigen induces a Th1 response in activated cells [122]. SOWgp is an immunomodulatory molecule that is expressed in endospores. In order to avoid the recognition by macrophages, the fungus secretes the metalloproteinase, Mep1, that hydrolyses surface-bound SOWgp and consequently reduces phagocytosis [127].

Coccidioides antigens also induce antibody responses during infection. However, high antibody titers typically correlate with a poor clinical outcome [157]. Certain antibody subsets can act as opsonins against both arthroconidia and endospores [158]. Both T27K *Coccidioides* antigen derived from spherules and heat-killed spherules interact with immature dendritic cells (DCs) that function to capture antigen, activating lymphocytes [159]. This suggests that DCs may play a critical early role in the development of adaptive immunity and T cell activation [160].

Pattern recognition receptors (PRRs) recognize *Coccidioides* specific proteins, polysaccharides, or nucleic acid motifs, and activate various inflammatory and antimicrobial responses [161, 162]. Dectin-1, Toll-like receptor 2 (TLR-2) and IL-1βR are all PRRs involved in *Coccidioides* infection [161, 163, 164]. In the absence of Dectin-1, several cytokines are differentially expressed 14 days P.I., which are required for the protective Th1 response [162]. However, other common antifungal defenses shown to be important for *Candida*, *Aspergillus*, and *Cryptococcus*, such as Dectin-2, mannose receptor (MR), and reactive oxygen species, do not appear to play a role for coccidioidomycosis, so it is possible that *Coccidioides* may have alternative strategies to evade the host immune response [165–171].

Early studies showed that T-cells are crucial in conferring protection against early infection with *Coccidioides* in mice models [172]. Th1 responses have generally been correlated with protection in murine models of coccidioidomycosis [133], whereas a predominant Th2 response is thought to lead to more severe disease [156, 173]. This is a general trend that is observed for many fungi [174–180]. However, what has become apparent is the fact that ~~is it~~ it is the *balance* among the Th1/Th2/Th17 response that mitigates the protective response to infection [163]. Loss of either the Th1 or Th2 response did not affect vaccine-mediated protection; however, the loss of a Th17 response resulted in more vaccinated mice succumbing to infection. T helper type 17 cells (Th17) are activated in response to early stages of *Coccidioides* infection and further contribute to the recruitment of innate cells to the lungs. IL-17 is produced by Th17 cells, and induces neutrophil recruitment and is Dectin-1 dependent for both *Coccidioides* and *Blastomyces* [181]. These cytokines also stimulate innate immune cells to increase the release of IL-1β, TNF-α, and IL-6 [163]. Additionally, the roles of Myd88 and CARD9 are being investigated, and appear to be essential for the development of the protective Th17 response [182, 183]. Taken in total, it is clear that the Th1/Th17 response together provide protection against the development of disease, but many questions remain. Recent work shows that the strain type that is used for infection can result in variable immune response [184]. Thus, disparate results between research groups could be due in part to use of different wild type strains.

8.6 Diagnostics

Differential diagnostics for coccidioidomycosis are extremely important because acute disease can manifest as a community acquired pneumonia [40]. Moreover, delays in the diagnosis of coccidioidomycosis or inadequate or unnecessary therapy may allow the extra-pulmonary proliferation of the fungus, and may require longer antifungal treatments for infected patients [38, 185]. Although coccidioidomycosis can be caused by two different species, there are no known differences in the clinical diagnostics between patients infected with *C. immitis* and *C. posadasii.* We will discuss the most common approaches to detect and identify the causative agents of disease.

Microscopy and Culture. *Coccidioides* is a Risk Group 3 microorganism and any culture of this fungus should be manipulated in a BSL3 environment as laboratory acquired infections have been reported [186]. Cultures of *Coccidioides* spp. can be obtained by direct plating tissue biopsies, BALF or biofluids on Mycosel, Sabouraud dextrose agar, blood agar, or any common agar used in a medical mycology laboratory [2, 185, 187]. Optimal fungal growth is between 28 and 37 °C and white/beige cotonous mycelial colonies are expected to appear between days 2–7 (or potentially longer) surrounding the cultured biomaterial. Older *Coccidioides* colonies can also present dark pigmentation and a wrinkled surface (Figure 8.3a). Microscopically, the filamentous form of *Coccidioides* sp. is characterized by hyaline and septate hyphae harboring barrel-shaped arthroconidia ranging from 2 to 5 um (Fig. 8.3b,c). The use of a *Coccidioides*-specific fluorescent DNA probe (AccuProbe) can be used to rapidly identify suspected colonies; however, false-negative results in samples fixed with formaldehyde have been seen [187–189]. Coccidioidal spherules and endospores cells can be visualized from BALF's and other biofluids. Direct mycological examinations are usually prepared using potassium hydroxide, calcofluor, and Papanicolaou (Figure 8.3d, e). Fixed biopsies can also be used for histopathological diagnostics and the observation of spherules is commonly achieved using routinely stains such as Hematoxylin-Eosin or Gomori Methenamine-Silver [185, 190].

Serological testing. Immunodiagnostics of coccidioidomycosis are widely used in clinical practice, especially in endemic areas of the disease. These methods include tube precipitin, complement fixation, immunodiffusion, agar gel precipitin-inhibition test, counterimmunoelectrophoresis, latex particle agglutination, fluorescent antibody, radioimmunoassay and ELISA [2, 185, 190–192]. Immunoglobulin (Ig)M positive reactions are reported to be positive during early and acute infection while after 2–3 weeks of illness the IgG reaction becomes positive and values are usually followed in a patient as a disease prognostic. In most asymptomatic cases both IgM and IgG are initially detected but titers may drop to negative after infection resolution [33, 191]. Tube precipitin and latex particle agglutination methods were largely used to detect IgM, whereas complement fixation was employed to detect late and chronic/disseminated cases of the disease [193]. A hallmark serological inquiry utilizing the enzyme immunoassay method based on the combination of

IgM and IgG revealed specificity and sensitivity 98.5% and 95.5% respectively and so far, is the most common method employed for the diagnostics and prognostics of coccidioidomycosis [194].

Some studies have discussed difficulties in the antibody detection at early timepoints of infection, as well as in immunosuppressed patients [195]. An alternative approach is the detection of fungal antigens in biofluids via immunoassay. Recently, studies using antibodies against coccidioidal galactomannan were revealed to be efficient in the rapid diagnosis of severe coccidioidomycosis [196]. A commercial version, MiraVista® *Coccidioides* quantitative antigen test, is available. Cross-reactivity with other endemic mycosis such as blastomycosis, histoplasmosis, and paracoccidioidomycosis was already reported, so further diagnostics may be required and interpretation of results should consider the possibility of infection with other etiologic agents [197].

Nucleic acid detection. About 90 *Coccidioides* sp. genomes have been sequenced to date, and are available in public databases [14, 72]. This allows the scientific community to develop sophisticated molecular tools to detect the fungal DNA during infection. rDNA (ITS1-5.8S-ITS2) sequencing is the most common fungal barcode and still used for fungal detection and genotyping. Specific *Coccidioides* rDNA primers were developed and tested by PCR in serum samples of 94 patients, as well as from paraffin embedded tissues revealing a low positivity [198, 199]. The ITS2 region was therefore targeted for the *Coccidioides* detection via real-time PCR. Sixteen samples out of 20 positive cultures were positive for *Coccidioides* and few samples with negative histopathology were positive suggesting being a potential useful diagnostic assay for the disease [200]. Other authors revealed the conventional and nested PCR methods targeting the antigen-2/proline rich antigen gene reveled to be promising in clinical practice [201, 202]. Recently a molecular assay based on real time qPCR was tested in clinical and environmental samples, which seems to accurately and sensitively detect both *C. posadasii* and *C. immitis* from a variety of sample types [13]. This assay uses a set of variable flanking primers and an internal probe targeting a specific *Coccidioides* LTR transposon and will be commercially available in the near future.

8.7 Therapeutics

Most coccidioidomycosis infections are self-limited without antifungal treatment, although this process may take many weeks or many months to completely resolve [39]. For those patients, periodic reassessment of symptoms and radiographic findings are important to follow-up the disease resolution without chemotherapy [2, 38, 39]. The genetic background of the patient (e.g. Filipino or African descent), immunosuppression status (e.g. HIV-infected, transplanted or under corticosteroid therapy), pregnancy or those with preexisting disease (e.g. diabetes mellitus or cardiopulmonary disease) may need early antifungal treatment upon diagnosis of coccidioidomycosis [19, 38, 39]. A recent finding regarding *Coccidioides* ssp.

antifungal susceptibility shows variability among isolates, and azole resistant strains have been detected [203].

A small percentage of patients progress to a chronic pneumonia or develop extra-pulmonary dissemination involving the skin, bones and the meninges. In those cases, antifungal therapy is used to control the *Coccidioides* infection [38, 39]. Coccidioidomycosis clinical presentations vary widely and the proper management strategies must be specific to a patient. Commonly used indicators of severe disease that requires antifungal treatment include: weight loss of >10%, periodic night sweats (longer than 3 weeks), pulmonary infiltrates compromising more than one-half of one lung or portions of both lungs, prominent or persistent hilar adenopathy, high anticoccidiodial antibody concentrations in excess of 1:16, inability to work, symptoms that persist for >2 months, or age > 55 years [38, 39].

Azoles. The most common class of drugs used in the treatment of coccidioidomycosis include oral azoles: fluconazole, itraconazole, and ketoconazole. Fluconazole is preferred due to high bioavailability and low adverse drug reaction [38, 39]. According to the 2016 Infectious Diseases of America treatment guidelines for coccidioidomycosis, the advised doses and administration routes for azole treatment are: ketoconazole (400 mg every day administered orally), fluconazole (400–800 mg/day administered orally or intravenously), and itraconazole (200 mg twice per day or 3 times per day administered orally). The use of azole treatment is primarily recommended in unresolved acute infections that evolve to chronic pneumonia or disseminated infections including meningitis. Risk groups listed above that have a current acute infection identified by recommended diagnostic methods are advised to start azole antifungal therapy. The duration of azole therapy is unclear but is commonly recommended up to 6 months, with periodic follow-up visits every trimester for up to 2 years after clinical cure [38, 39]. More recently, the newer class of azoles, voriconazole and posaconazole were tested in small clinical trials. Posaconazole has been used in disseminated nonmeningeal coccidioidomycosis that previously failed to respond to azole and/or AmB therapy as well as in patients with short-term previous treatment. Voriconazole was shown to be effective in cases of azole resistance. At least 6–12 months of therapy is recommended for both Voriconazole and Posaconazole [209–213].

Amphotericin B. Deoxycholate and liposomal formulations of amphotericin B are effective, and an alternative for coccidioidomycosis treatment in the case of azole failure [209]. It is recommended in patients with rapidly progressing disease characterized by respiratory failure or rapid dissemination, and also co-diagnosis with HIV [19, 38, 39]. Again, the 2016 Infectious Diseases of America treatment guidelines advise doses and administration routes for amphotericin B. Amphotericin B deoxycholate (0.5–1.5 mg/kg per day or alternate day administered intravenously), and lipid formulations of amphotericin B (2.0–5.0 mg/kg or greater per day administered intravenously) are both effective. The combination of amphotericin B and azole therapy in severe cases was shown to improve the outcome of patients. Pregnant women diagnosed with valley fever should undergo amphotericin B treatment due teratogenicity of azoles. Administration of amphotericin B may be reduced over time as patient outcome improves [38, 39]. The deoxycholate formulation of amphotericin B was show to be ineffective against CNS infections and this formulation seems to be more effective in treating disease [38, 210–213].

Novel agents. Several other molecules have been tested for in vitro and in vivo effects against the ethological agents of coccidioidomycosis. Nikkomycin Z (NikZ) is a natural inhibitor of chitin synthase, an essential structural component for fungal cell wall integrity [136, 214, 215]. This molecule has a strong in vitro activity against *C. immitis* and *C. posadasii* [136]. In vivo studies showed that NikZ is efficient in treating canine coccidioidomycosis, treating induced disease in murine models, and has undergone phase 1 studies in humans [137, 138]. NikZ has demonstrated sterilization of the lungs of experimentally infected mice. If confirmed in clinical trials, this would be a major advance for treatment of this disease. Recently, the effective dosage of this drug was determined in murine models of coccidioidomycosis infection for phase 2 and 3 clinical trials, and shows promise in treating this important fungal infection [138]. Recently, a new compound that inhibits CYP51 named VT-1161 showed antifungal activity in vitro against *Coccidioides* ssp., and was demonstrated to be effective in murine central nervous system models of the disease [216].

8.8 Vaccination Strategies

The population living or visiting endemic areas in the western USA and other desert areas of Latin America are at high risk, and are potential candidates for vaccination against coccidioidomycosis. At least 20 million people reside in or visit the endemic areas of the southwestern USA [158, 217, 218]. A majority of people resolve coccidioidal infections without complications, and acquire life-long resistance to a second infection [219, 220]. Thus, there is a strong public health argument for the value of the development of a protective vaccine.

The initial attempt for immunization against coccidioidomycosis was done in a five-year clinical trial using formalin-inactivated spherules. However, this study failed to produce a significant reduction in incidence or severity of disease in phase 3 trials and high levels of skin reaction among participants were observed when protective concentrations of the vaccine were administered [221]. A second study used auxotrophic *Coccidioides* mutants with attenuated virulence, and mice immunized with those strains showed protection against coccidioidomycosis [110].

One approach to developing vaccine strains is to use targeted gene deletion on putative virulence factors. One study generated an avirulent mutant deficient for two chitin synthase genes (CTS2 and CTS3), which are important for cell wall integrity of *Coccidioides posadasii* [140]. The mutant strain was avirulent and did not induce disease in a murine model due the inability of spherules to mature and produce endospores. The protective nature of this knockout strain was tested, and results indicated a protective response against the disease; however, fungal clearance was not achieved in recovered tissues after infection with wild type strains and tested animals presented with an exacerbated inflammatory response [140]. Similar results were obtained using another avirulent strain of *C. posadasii*. This mutant lacked a single chitin synthase gene CHS5, grew atypically, had a slow development of spherules in vitro, and did not convert into the parasitic phase in vivo [222, 223]. Immunized mice survived beyond 45 days post-infection, compared to control mice; however, viable cells of *Coccidioides* were still recovered from the organs. However, one recent vaccine candidate has shown clearance and protection in a murine model of infection. Immunizations in both C57BL/6 or BALB/c mice with a live and attenuated *C. posadasii* strain lacking the CPS1 gene resulted in a survival of over 95% of mice infected with lethal doses. Few residual fungal colonies were recovered from the lungs, and this live attenuated strain is a strong candidate for coccidioidomycosis vaccination [129].

Purified antigens have also been tested as vaccine candidates and provided important advances in the knowledge of the immunopathology of these fungal pathogens. *C. immitis* purified urease and the urease-DNA ectopically expressed by a mammalian plasmid promotes an immunoprotective Th1 response against coccidioidomycosis in a murine model [133]. The purified recombinant *C. posadasii* Gel1 protein also conferred protection to C57BL/6 mice against coccidioidomycosis. However, a significant decrease in the level of Th1 that is fundamental to immune response to coccidioidomycosis was observed [224]. The proline-rich antigen combined with a coccidioidal antigen 2 (Ag2/PRA), which codify for a recombinant 194 amino acid protein, has been shown to be protective against *Coccidioides* infection. Recombinant Ag2/PRA confers protection from *Coccidioides* infection by inducing a T-cell $CD4^{+}$-type response and is a potent vaccine candidate for this important disease [33, 225]. Other antigens such as SOW, ELI1, HSP60, aspartyl protease, and *Coccidioides*-specific antigen (CSA) have been used to immunize mice and showed significant reduction of fungal burden, but none promote complete clearance of the pathogen [33, 158, 217]. Finally, fungal glucans purified from *S. cerevisiae,* a potent immunomodulator that binds to Dectin-1, has been shown to stimulate immunity in CD1 mice infected with *C. posadasii*. This same vaccine was effective against Aspergillosis, which leaves open the possibility of a pan-fungal vaccine [226].

There are several candidates for the vaccine of coccidioidomycosis; however, none of those compounds or attenuated strains is commercially available. Cell-mediated immunity is critical for defense; however, the role of the humoral response in coccidioidomycosis is still a gap that impairs vaccine development. Moreover, considering that disease may disproportionately affect a naive elderly population, new adjuvants that increase the priming of the immune system and improve the performance of vaccines are needed [227].

Acknowledgements We wish to thank Dr. John N Galgiani and Dr. Neil Ampel, University of Arizona Valley Fever Center for Excellence, University of Arizona College of Medicine for critical comments in the preparation of this manuscript.

References

1. Brown J, Benedict K, Park BJ, Thompson GR 3rd (2013) Coccidioidomycosis: epidemiology. Clin Epidemiol 5:185–197
2. Nguyen C, Barker BM, Hoover S, Nix DE, Ampel NM, Frelinger JA, Orbach MJ, Galgiani JN (2013) Recent advances in our understanding of the environmental, epidemiological, immunological, and clinical dimensions of coccidioidomycosis. Clin Microbiol Rev 26:505–525
3. Stockamp NW, Thompson GR 3rd (2016) Coccidioidomycosis. Infect Dis Clin N Am 30:229–246
4. Posadas A (1892) Un Nuevo Caso De Micosis Fungoide Con Psorospermias. Círculo Médico Argentino 52:11
5. Hirschmann JV (2007) The early history of coccidioidomycosis: 1892-1945. Clin Infect Dis 44:1202–1207
6. Negroni R (2011) Historia Del Descubrimiento De La Coccidioidomicosis. Rev Argent Dermatol 92(3):1118
7. Burt A, Carter DA, Koenig GL, White TJ, Taylor JW (1996) Molecular markers reveal cryptic sex in the human pathogen Coccidioides Immitis. Proc Natl Acad Sci U S A 93:770–773
8. Burt A, Dechairo BM, Koenig GL, Carter DA, White TJ, Taylor JW (1997) Molecular markers reveal differentiation among isolates of Coccidioides Immitis from California, Arizona and Texas. Mol Ecol 6:781–786
9. Koufopanou V, Burt A, Szaro T, Taylor JW (2001) Gene genealogies, cryptic species, and molecular evolution in the human pathogen Coccidioides Immitis and relatives (Ascomycota, Onygenales). Mol Biol Evol 18:1246–1258
10. Koufopanou V, Burt A, Taylor JW (1997) Concordance of gene genealogies reveals reproductive isolation in the pathogenic fungus Coccidioides Immitis. Proc Natl Acad Sci U S A 94:5478–5482
11. Fisher MC, Koenig GL, White TJ, Taylor JW (2002) Molecular and phenotypic description of Coccidioides Posadasii Sp. Nov., previously recognized as the non-California population of Coccidioides Immitis. Mycologia 94:73–84
12. Fisher MC, Koenig GL, White TJ, San-Blas G, Negroni R, Alvarez IG, Wanke B, Taylor JW (2001) Biogeographic range expansion into South America by Coccidioides Immitis mirrors new world patterns of human migration. Proc Natl Acad Sci U S A 98:4558–4562
13. Litvintseva AP, Marsden-Haug N, Hurst S, Hill H, Gade L, Driebe EM, Ralston C, Roe C, Barker BM, Goldoft M et al (2015) Valley fever: finding new places for an old disease: Coccidioides Immitis found in Washington state soil associated with recent human infection. Clin Infect Dis 60:E1–E3
14. Engelthaler DM, Roe CC, Hepp CM, Teixeira M, Driebe EM, Schupp JM, Gade L, Waddell V, Komatsu K, Arathoon E et al (2016) Local population structure and patterns of western hemisphere dispersal for Coccidioides Spp., the fungal cause of valley fever. MBio 7(2):e00550
15. Teixeira MM, Barker BM (2016) Use of population genetics to assess the ecology, evolution, and population structure of Coccidioides. Emerg Infect Dis 22:1022–1030
16. Neafsey DE, Barker BM, Sharpton TJ, Stajich JE, Park DJ, Whiston E, Hung CY, Mcmahan C, White J, Sykes S et al (2010) Population genomic sequencing of Coccidioides fungi reveals recent hybridization and transposon control. Genome Res 20:938–946
17. Del Rocio Reyes-Montes M, Perez-Huitron MA, Ocana-Monroy JL, Frias-De-Leon MG, Martinez-Herrera E, Arenas R, Duarte-Escalante E (2016) The habitat of Coccidioides Spp. and the role of animals as reservoirs and disseminators in nature. BMC Infect Dis 16:550
18. Shubitz LF (2007) Comparative aspects of coccidioidomycosis in animals and humans. Ann N Y Acad Sci 1111:395–403
19. Ampel NM (2005) Coccidioidomycosis in persons infected with Hiv type 1. Clin Infect Dis 41:1174–1178

20. Bergstrom L, Yocum DE, Ampel NM, Villanueva I, Lisse J, Gluck O, Tesser J, Posever J, Miller M, Araujo J et al (2004) Increased risk of coccidioidomycosis in patients treated with tumor necrosis factor alpha antagonists. Arthritis Rheum 50:1959–1966
21. Blair JE, Logan JL (2001) Coccidioidomycosis in solid organ transplantation. Clin Infect Dis 33:1536–1544
22. Borchers AT, Gershwin ME (2010) The immune response in coccidioidomycosis. Autoimmun Rev 10:94–102
23. Dierberg KL, Marr KA, Subramanian A, Nace H, Desai N, Locke JE, Zhang S, Diaz J, Chamberlain C, Neofytos D (2012) Donor-derived organ transplant transmission of coccidioidomycosis. Transpl Infect Dis 14:300–304
24. Mertz LE, Blair JE (2007) Coccidioidomycosis in rheumatology patients: incidence and potential risk factors. Ann N Y Acad Sci 1111:343–357
25. Santelli AC, Blair JE, Roust LR (2006) Coccidioidomycosis in patients with diabetes mellitus. Am J Med 119:964–969
26. Durry E, Pappagianis D, Werner SB, Hutwagner L, Sun RK, Maurer M, Mcneil MM, Pinner RW (1997) Coccidioidomycosis in Tulare County, California, 1991: Reemergence of an endemic disease. J Med Vet Mycol 35:321–326
27. Leake JA, Mosley DG, England B, Graham JV, Plikaytis BD, Ampel NM, Perkins BA, Hajjeh RA (2000) Risk factors for acute symptomatic Coccidioidomycosis among elderly persons in Arizona, 1996-1997. J Infect Dis 181:1435–1440
28. Pappagianis D, Lindsay S, Beall S, Williams P (1979) Ethnic background and the clinical course of coccidioidomycosis. Am Rev Respir Dis 120:959–961
29. Rosenstein NE, Emery KW, Werner SB, Kao A, Johnson R, Rogers D, Vugia D, Reingold A, Talbot R, Plikaytis BD et al (2001) Risk factors for severe pulmonary and disseminated coccidioidomycosis: Kern County, California, 1995-1996. Clin Infect Dis 32:708–715
30. Ruddy BE, Mayer AP, Ko MG, Labonte HR, Borovansky JA, Boroff ES, Blair JE (2011) Coccidioidomycosis in African Americans. Mayo Clin Proc 86:63–69
31. Odio CD, Marciano BE, Galgiani JN, Holland SM (2017) Risk factors for disseminated coccidioidomycosis, United States. Emerg Infect Dis 23(2):308
32. Lewis ER, Bowers JR, Barker BM (2015a) Dust devil: the life and times of the fungus that causes valley fever. PLoS Pathog 11:E1004762
33. Cox RA, Magee DM (2004) Coccidioidomycosis: host response and vaccine development. Clin Microbiol Rev 17:804–839
34. De Perio MA, Niemeier RT, Burr GA (2015) Coccidioides exposure and coccidioidomycosis among prison employees, California, United States. Emerg Infect Dis 21:1031–1033
35. Drips W Jr, Smith CE (1964) Epidemiology of coccidioidomycosis. A contemporary military experience. JAMA 190:1010–1012
36. Smith CE, Saito MT, Simons SA (1956) Pattern of 39,500 serologic tests in coccidioidomycosis. J Am Med Assoc 160:546–552
37. Wack EE, Ampel NM, Sunenshine RH, Galgiani JN (2015) The return of delayed-type hypersensitivity skin testing for coccidioidomycosis. Clin Infect Dis 61:787–791
38. Ampel NM (2015) The treatment of coccidioidomycosis. Rev Inst Med Trop Sao Paulo 57(Suppl 19):51–56
39. Galgiani JN, Ampel NM, Blair JE, Catanzaro A, Geertsma F, Hoover SE, Johnson RH, Kusne S, Lisse J, Macdonald JD et al (2016) 2016 infectious diseases society of America (Idsa) clinical practice guideline for the treatment of coccidioidomycosis. Clin Infect Dis 63:E112–E146
40. Valdivia L, Nix D, Wright M, Lindberg E, Fagan T, Lieberman D, Stoffer T, Ampel NM, Galgiani JN (2006) Coccidioidomycosis as a common cause of community-acquired pneumonia. Emerg Infect Dis 12:958–962
41. Blair JE, Chang YH, Cheng MR, Vaszar LT, Vikram HR, Orenstein R, Kusne S, Ho S, Seville MT, Parish JM (2014) Characteristics of patients with mild to moderate primary pulmonary coccidioidomycosis. Emerg Infect Dis 20:983–990

42. Berman RJ, Friedman L, Roessler WG, Smith CE (1956) The virulence and infectivity of twenty-seven strains of Coccidioides Immitis. Am J Hyg 64:198–210
43. Friedman L, Smith CE (1957) The comparison of four strains of Coccidioides Immitis with diverse histories. Mycopathol Mycol Appl 8:47–53
44. Friedman L, Smith CE, Gordon LE (1955) The assay of virulence of Coccidioides in white mice. J Infect Dis 97:311–316
45. Hugenholtz PG, Reed RE, Maddy KT, Trautman RJ, Barger JD (1958) Experimental coccidioidomycosis in dogs. Am J Vet Res 19:433–437
46. Kirkland TN, Fierer J (1983) Inbred mouse strains differ in resistance to lethal Coccidioides Immitis infection. Infect Immun 40:912–916
47. Blair JE (2007) State-of-the-art treatment of coccidioidomycosis: skin and soft-tissue infections. Ann N Y Acad Sci 1111:411–421
48. Garcia Garcia SC, Salas Alanis JC, Flores MG, Gonzalez Gonzalez SE, Vera Cabrera L, Ocampo Candiani J (2015) Coccidioidomycosis and the skin: a comprehensive review. An Bras Dermatol 90:610–619
49. Mischel PS, Vinters HV (1995) Coccidioidomycosis of the central nervous system: Neuropathological and Vasculopathic manifestations and clinical correlates. Clin Infect Dis 20:400–405
50. Keckich DW, Blair JE, Vikram HR (2010) Coccidioides Fungemia in six patients, with a review of the literature. Mycopathologia 170:107–115
51. Wang CY, Jerng JS, Ko JC, Lin MF, Hsiao CH, Lee LN, Hsueh PR, Kuo SH (2005) Disseminated coccidioidomycosis. Emerg Infect Dis 11:177–179
52. Bonifaz A, Vazquez-Gonzalez D, Perusquia-Ortiz AM (2011) Endemic systemic mycoses: coccidioidomycosis, histoplasmosis, paracoccidioidomycosis and blastomycosis. J Dtsch Dermatol Ges 9:705–714. Quiz 715
53. Colombo AL, Tobon A, Restrepo A, Queiroz-Telles F, Nucci M (2011) Epidemiology of endemic systemic fungal infections in Latin America. Med Mycol 49:785–798
54. Schwartz IS, Kenyon C, Thompson GR (2016) Endemic mycoses: What's new about old diseases? Curr Clin Microbiol Rep 3:71–80
55. Sifuentes-Osornio J, Corzo-Leon DE, Ponce-De-Leon LA (2012) Epidemiology of invasive fungal infections in Latin America. Curr Fungal Infect Rep 6:23–34
56. Comrie AC (2005) Climate factors influencing coccidioidomycosis seasonality and outbreaks. Environ Health Perspect 113:688–692
57. Pappagianis D (1988) Epidemiology of coccidioidomycosis. Curr Top Med Mycol 2:199–238
58. Smith CE, Beard RR et al (1946) Effect of season and dust control on coccidioidomycosis. J Am Med Assoc 132:833–838
59. Sorooshian A, Wonaschutz A, Jarjour EG, Hashimoto BI, Schichtel BA, Betterton EA (2011) An aerosol climatology for a rapidly growing arid region (southern Arizona): major aerosol species and remotely sensed aerosol properties. J Geophys Res Atmos 116:16
60. Comrie AC, Glueck MF (2007) Assessment of climate-coccidioidomycosis model: model sensitivity for assessing climatologic effects on the risk of acquiring coccidioidomycosis. Ann N Y Acad Sci 1111:83–95
61. Centers for Disease Control and Prevention (2001) Coccidioidomycosis in workers at an Archeologic site—dinosaur National Monument, Utah, June-July 2001. MMWR Morb Mortal Wkly Rep 50:1005–1008
62. Eulalio KD, De Macedo RL, Cavalcanti MA, Martins LM, Lazera MS, Wanke B (2001) Coccidioides Immitis isolated from armadillos (Dasypus Novemcinctus) in the state of Piaui, Northeast Brazil. Mycopathologia 149:57–61
63. Flynn NM, Hoeprich PD, Kawachi MM, Lee KK, Lawrence RM, Goldstein E, Jordan GW, Kundargi RS, Wong GA (1979) An unusual outbreak of windborne coccidioidomycosis. N Engl J Med 301:358–361
64. Johnson L, Gaab EM, Sanchez J, Bui PQ, Nobile CJ, Hoyer KK, Peterson MW, Ojcius DM (2014) Valley fever: danger lurking in a dust cloud. Microbes Infect 16:591–600

65. Pappagianis D, Einstein H (1978) Tempest from Tehachapi takes toll or Coccidioides conveyed aloft and afar. West J Med 129:527–530
66. Schneider E, Hajjeh RA, Spiegel RA, Jibson RW, Harp EL, Marshall GA, Gunn RA, Mcneil MM, Pinner RW, Baron RC et al (1997) A coccidioidomycosis outbreak following the Northridge, Calif, earthquake. JAMA 277:904–908
67. Williams PL, Sable DL, Mendez P, Smyth LT (1979) Symptomatic coccidioidomycosis following a severe natural dust storm. An outbreak at the naval Air Station, Lemoore, Calif. Chest 76:566–570
68. Barker B, Tabor JA, Shubitz LF, Perrill R, Orbach MJ (2012) Detection and phylogenetic analysis of Coccidioides Posadasii in Arizona soil samples. Fungal Ecol 5:13
69. Barker BM, Jewell KA, Kroken S, Orbach MJ (2007) The population biology of Coccidioides: epidemiologic implications for disease outbreaks. Ann N Y Acad Sci 1111:147–163
70. Untereiner WA, Scott JA, Naveau FA, Sigler L, Bachewich J, Angus A (2004) The Ajellomycetaceae, a new family of vertebrate-associated Onygenales. Mycologia 96:812–821
71. Burmester A, Shelest E, Glockner G, Heddergott C, Schindler S, Staib P, Heidel A, Felder M, Petzold A, Szafranski K et al (2011) Comparative and functional genomics provide insights into the pathogenicity of Dermatophytic fungi. Genome Biol 12:R7
72. Sharpton TJ, Stajich JE, Rounsley SD, Gardner MJ, Wortman JR, Jordar VS, Maiti R, Kodira CD, Neafsey DE, Zeng Q et al (2009) Comparative genomic analyses of the human fungal pathogens Coccidioides and their relatives. Genome Res 19:1722–1731
73. Whiston E, Taylor JW (2015) Comparative Phylogenomics of pathogenic and nonpathogenic species. G3 (Bethesda) 6:235–244
74. Brillhante RS, Moreira Filho RE, Rocha MF, Castelo-Branco Dde S, Fechine MA, Lima RA, Picanco YV, Cordeiro Rde A, Camargo ZP, Queiroz JA et al (2012) Coccidioidomycosis in armadillo hunters from the state of Ceara, Brazil. Mem Inst Oswaldo Cruz 107:813–815
75. Catalan-Dibene J, Johnson SM, Eaton R, Romero-Olivares AL, Baptista-Rosas RC, Pappagianis D, Riquelme M (2014) Detection of Coccidioidal antibodies in serum of a small rodent community in Baja California, Mexico. Fungal Biol 118:330–339
76. Cordeiro RA, Brilhante RS, Rocha MF, Fechine MA, Camara LM, Camargo ZP, Sidrim JJ (2006) Phenotypic characterization and ecological features of Coccidioides Spp. from Northeast Brazil. Med Mycol 44:631–639
77. Sharapov VM (1971) Mycotic elements in the lungs of wild mammals resembling spherules of Coccidiodes Immitis. Vestn Dermatol Venerol 45:50–52
78. Ajello L, Budurin AA, Maddy KT, Moore JC, Reed RE (1956) Ecological and Epizootiological studies on canine coccidioidomycosis. J Am Vet Med Assoc 129:485–490
79. Beaman L, Holmberg C, Henrickson R, Osburn B (1980) The incidence of coccidioidomycosis among nonhuman primates housed outdoors at the California primate research Center. J Med Primatol 9:254–261
80. Blundell GP, Castleberry MW, Lowe EP, Converse JL (1961) The pathology of Coccidioides Immitis in the Macaca Mulatta. Am J Pathol 39:613–630
81. Grayzel SE, Martinez-Lopez B, Sykes JE (2016) Risk factors and spatial distribution of canine coccidioidomycosis in California, 2005-2013. Transbound Emerg Dis 64(4):1110
82. Hoffman K, Videan EN, Fritz J, Murphy J (2007) Diagnosis and treatment of ocular coccidioidomycosis in a female captive chimpanzee (Pan Troglodytes): a case study. Ann N Y Acad Sci 1111:404–410
83. Reed RE, Converse JL (1966) The seasonal incidence of canine coccidioidomycosis. Am J Vet Res 27:1027–1030
84. Rosenberg DP, Gleiser CA, Carey KD (1984) Spinal coccidioidomycosis in a baboon. J Am Vet Med Assoc 185:1379–1381
85. Straub M, Trautman RJ, Greene JW (1961) Coccidioidomycosis In 3 Coyotes. Am J Vet Res 22:811–813
86. Jessup D, Kock N, Berbach M (1989) Coccidioidomycosis in a desert bighorn sheep (Ovis Canadensis Nelsoni) from California. J Zoo Wildl Med 20:3

87. Morrow W (2006) Holocene coccidioidomycosis: valley fever in early Holocene bison (bison Antiquus). Mycologia 98:669–677
88. Prchal CJ (1948) Coccidioidomycosis Of Cattle In Arizona. J Am Vet Med Assoc 112:461–465
89. Higgins JC, Leith GS, Pappagianis D, Pusterla N (2006) Treatment of Coccidioides Immitis pneumonia in two horses with fluconazole. Vet Rec 159:349–351
90. Maleski K, Magdesian KG, Lafranco-Scheuch L, Pappagianis D, Carlson GP (2002) Pulmonary coccidioidomycosis in a neonatal foal. Vet Rec 151:505–508
91. Ziemer EL, Pappagianis D, Madigan JE, Mansmann RA, Hoffman KD (1992) Coccidioidomycosis in horses: 15 cases (1975-1984). J Am Vet Med Assoc 201:910–916
92. Reed, R.E., Ingram, K.A., Reggiardo, C., Shupe, M.R. (1996). Coccidioidomycosis in domestic and wild animals. Paper Presented at: 5th International conference on coccidioidomycosis (National Foundation For Infectious Disease)
93. Timm KI, Sonn RJ, Hultgren BD (1988) Coccidioidomycosis in a Sonoran gopher snake, Pituophis Melanoleucus Affinis. J Med Vet Mycol 26:101–104
94. Cordeiro Rde A, e Silva KR, Brilhante RS, Moura FB, Duarte NF, Marques FJ, Filho RE, De Araujo RW, Bandeira Tde J, Rocha MF et al (2012) Coccidioides Posadasii infection in bats, Brazil. Emerg Infect Dis 18:668–670
95. Greene RT, Troy GC (1995) Coccidioidomycosis in 48 cats: a retrospective study (1984-1993). J Veterin Intern Med 9:86–91
96. Huckabone SE, Gulland FM, Johnson SM, Colegrove KM, Dodd EM, Pappagianis D, Dunkin RC, Casper D, Carlson EL, Sykes JE et al (2015) Coccidioidomycosis and other systemic mycoses of marine mammals stranding along the Central California, Usa coast: 1998-2012. J Wildl Dis 51:295–308
97. Krutzsch PH, Watson RH (1978) Isolation of Coccidioides Immitis from bat guano and preliminary findings on laboratory infectivity of bats with Coccidioides Immitis. Life Sci 22:679–684
98. Reed RE, Migaki G, Cummings JA (1976) Coccidioidomycosis in a California Sea lion (Zalophus Californianus). J Wildl Dis 12:372–375
99. Reidarson TH, Griner LA, Pappagianis D, Mcbain J (1998) Coccidioidomycosis in a bottlenose dolphin. J Wildl Dis 34:629–631
100. Centers for Disease Control and Prevention (2013) Increase in reported coccidioidomycosis—United States, 1998-2011. MMWR Morb Mortal Wkly Rep 62:217–221
101. Galgiani JN (2007) Coccidioidomycosis: changing perceptions and creating opportunities for its control. Ann N Y Acad Sci 1111:1–18
102. Thompson GR 3rd, Stevens DA, Clemons KV, Fierer J, Johnson RH, Sykes J, Rutherford G, Peterson M, Taylor JW, Chaturvedi V (2015) Call for a California coccidioidomycosis consortium to face the top ten challenges posed by a recalcitrant regional disease. Mycopathologia 179:1–9
103. Twarog M, Thompson GR 3rd (2015) Coccidioidomycosis: recent updates. Semin Respir Crit Care Med 36:746–755
104. Panackal AA, Hajjeh RA, Cetron MS, Warnock DW (2002) Fungal infections among returning Travelers. Clin Infect Dis 35:1088–1095
105. Sunenshine RH, Anderson S, Erhart L, Vossbrink A, Kelly PC, Engelthaler D, Komatsu K (2007) Public health surveillance for coccidioidomycosis in Arizona. Ann N Y Acad Sci 1111:96–102
106. Johannesson H, Kasuga T, Schaller RA, Good B, Gardner MJ, Townsend JP, Cole GT, Taylor JW (2006) Phase-specific gene expression underlying morphological adaptations of the dimorphic human pathogenic fungus, Coccidioides Posadasii. Fungal Genet Biol 43:545–559
107. Pappagianis D, Smith CE, Kobayashi GS (1956) Relationship of the in vivo form of Coccidioides Immitis to virulence. J Infect Dis 98:312–319
108. Whiston E, Zhang Wise H, Sharpton TJ, Jui G, Cole GT, Taylor JW (2012) Comparative Transcriptomics of the Saprobic and parasitic growth phases in Coccidioides Spp. PLoS One 7:E41034

109. Hung CY, Xue J, Cole GT (2007) Virulence mechanisms of Coccidioides. Ann N Y Acad Sci 1111:225–235
110. Walch HA, Kalvoda A (1971) Immunization of mice with induced mutants of Coccidioides Immitis. I. Characterization of mutants and preliminary studies of their use as viable vaccines. Sabouraudia 9:173–184
111. Boyce KJ, Andrianopoulos A (2015) Fungal dimorphism: the switch from hyphae to yeast is a specialized morphogenetic adaptation allowing colonization of a host. FEMS Microbiol Rev 39:15
112. Gauthier GM (2015) Dimorphism in fungal pathogens of mammals, plants, and insects. PLoS Pathog 11:E1004608
113. Rappleye CA, Goldman WE (2006) Defining virulence genes in the dimorphic fungi. Annu Rev Microbiol 60:281–303
114. Viriyakosol S, Singhania A, Fierer J, Goldberg J, Kirkland TN, Woelk CH (2013b) Gene expression in human fungal pathogen Coccidioides Immitis changes as Arthroconidia differentiate into spherules and mature. BMC Microbiol 13:121
115. Grys TE, Kaushal S, Chowdhury Y, Dasari S, Mitchell NM, Magee DM, Blair JE, Colby TV, Lake DF (2016) Total and Lectin-binding proteome of Spherulin from Coccidioides Posadasii. J Proteome Res 15:3463–3472
116. Cole GT, Hung CY (2001) The parasitic Cell Wall of Coccidioides Immitis. Med Mycol 39(Suppl 1):31–40
117. Cole GT, Kirkland TN, Franco M, Zhu S, Yuan L, Sun SH, Hearn VM (1988a) Immunoreactivity of a Surface Wall fraction produced by spherules of Coccidioides Immitis. Infect Immun 56:2695–2701
118. Cole GT, Seshan KR, Franco M, Bukownik E, Sun SH, Hearn VM (1988b) Isolation and morphology of an Immunoreactive Outer Wall fraction produced by spherules of Coccidioides Immitis. Infect Immun 56:2686–2694
119. Levine HB, Cobb JM, Smith CE (1961) Immunogenicity of spherule-endospore vaccines of Coccidioides Immitis for mice. J Immunol 87:218–227
120. Cole GT, Kirkland TN (1993) Identification of antigens of Coccidioides Immitis which stimulated immune T lymphocytes. Arch Med Res 24:281–291
121. Cole GT, Kruse D, Seshan KR (1991) Antigen complex of Coccidioides Immitis which elicits a precipitin antibody response in patients. Infect Immun 59:2434–2446
122. Hung CY, Ampel NM, Christian L, Seshan KR, Cole GT (2000b) A major cell surface antigen of Coccidioides Immitis which elicits both Humoral and cellular immune responses. Infect Immun 68:584–593
123. Hung CY, Yu JJ, Seshan KR, Reichard U, Cole GT (2002) A parasitic phase-specific Adhesin of Coccidioides Immitis contributes to the virulence of this respiratory fungal pathogen. Infect Immun 70:3443–3456
124. Pan S, Cole GT (1995) Molecular and biochemical characterization of a Coccidioides Immitis-specific antigen. Infect Immun 63:3994–4002
125. Cole GT, Zhu SW, Hsu LL, Kruse D, Seshan KR, Wang F (1992) Isolation and expression of a gene which encodes a wall-associated proteinase of Coccidioides Immitis. Infect Immun 60:416–427
126. Yuan L, Cole GT, Sun SH (1988) Possible role of a proteinase in Endosporulation of Coccidioides Immitis. Infect Immun 56:1551–1559
127. Hung CY, Seshan KR, Yu JJ, Schaller R, Xue J, Basrur V, Gardner MJ, Cole GT (2005b) A metalloproteinase of Coccidioides Posadasii contributes to evasion of host detection. Infect Immun 73:6689–6703
128. Lu SW, Kroken S, Lee BN, Robbertse B, Churchill AC, Yoder OC, Turgeon BG (2003) A novel class of gene controlling virulence in plant pathogenic Ascomycete fungi. Proc Natl Acad Sci U S A 100:5980–5985

129. Narra HP, Shubitz LF, Mandel MA, Trinh HT, Griffin K, Buntzman AS, Frelinger JA, Galgiani JN, Orbach MJ (2016) A Coccidioides Posadasii Cps1 deletion mutant is Avirulent and protects mice from lethal infection. Infect Immun 84:3007–3016
130. Rutherford JC (2014) The emerging role of urease as a general microbial virulence factor. PLoS Pathog 10:E1004062
131. Yu JJ, Smithson SL, Thomas PW, Kirkland TN, Cole GT (1997) Isolation and characterization of the urease gene (Ure) from the pathogenic fungus Coccidioides Immitis. Gene 198:387–391
132. Mirbod F, Schaller RA, Cole GT (2002) Purification and characterization of urease isolated from the pathogenic fungus Coccidioides Immitis. Med Mycol 40:35–44
133. Li K, Yu JJ, Hung CY, Lehmann PF, Cole GT (2001a) Recombinant urease and urease DNA of Coccidioides Immitis elicit an Immunoprotective response against coccidioidomycosis in mice. Infect Immun 69:2878–2887
134. Wise HZ, Hung CY, Whiston E, Taylor JW, Cole GT (2013) Extracellular ammonia at sites of pulmonary infection with Coccidioides Posadasii contributes to severity of the respiratory disease. Microb Pathog 59-60:19–28
135. Mandel MA, Galgiani JN, Kroken S, Orbach MJ (2006) Coccidioides Posadasii contains single chitin synthase genes corresponding to classes I to vii. Fungal Genet Biol 43:775–788
136. Hector RF, Zimmer BL, Pappagianis D (1990b) Evaluation of Nikkomycins X and Z in murine models of coccidioidomycosis, Histoplasmosis, and Blastomycosis. Antimicrob Agents Chemother 34:587–593
137. Shubitz LF, Roy ME, Nix DE, Galgiani JN (2013a) Efficacy of Nikkomycin Z for respiratory coccidioidomycosis in naturally infected dogs. Med Mycol 51:747–754
138. Shubitz LF, Trinh HT, Perrill RH, Thompson CM, Hanan NJ, Galgiani JN, Nix DE (2014) Modeling Nikkomycin Z dosing and pharmacology in murine pulmonary coccidioidomycosis preparatory to phase 2 clinical trials. J Infect Dis 209:1949–1954
139. Pishko EJ, Kirkland TN, Cole GT (1995) Isolation and characterization of two Chitinase-encoding genes (Cts1, Cts2) from the fungus Coccidioides Immitis. Gene 167:173–177
140. Xue J, Chen X, Selby D, Hung CY, Yu JJ, Cole GT (2009a) A genetically engineered live attenuated vaccine of coccidioides Posadasii protects Balb/C mice against coccidioidomycosis. Infect Immun 77:3196–3208
141. Skeldon A, Saleh M (2011) The Inflammasomes: molecular effectors of host resistance against bacterial, viral, parasitic, and fungal infections. Front Microbiol 2:15
142. Galgiani JN (1985) Studies of the effects of Spherulin from Coccidioides Immitis on human Polymorphonuclear leukocytes. Mycopathologia 90:107–111
143. Galgiani JN, Hayden R, Payne CM (1982) Leukocyte effects on the dimorphism of Coccidioides Immitis. J Infect Dis 146:56–63
144. Galgiani JN (1986) Inhibition of different phases of Coccidioides Immitis by human neutrophils or hydrogen peroxide. J Infect Dis 153:217–222
145. Lee CY, Thompson GR 3rd, Hastey CJ, Hodge GC, Lunetta JM, Pappagianis D, Heinrich V (2015) Coccidioides endospores and spherules draw strong chemotactic, adhesive, and phagocytic responses by individual human neutrophils. PLoS One 10:E0129522
146. Beaman L (1987) Fungicidal activation of murine macrophages by recombinant gamma interferon. Infect Immun 55:2951–2955
147. Beaman L, Benjamini E, Pappagianis D (1981) Role of lymphocytes in macrophage-induced killing of Coccidioides Immitis in vitro. Infect Immun 34:347–353
148. Galgiani JN (1995) Differences in oxidant release by human Polymorphonuclear leukocytes produced by stimulation with different phases of Coccidioides Immitis. J Infect Dis 172:199–203
149. Shubitz LF, Dial SM, Perrill R, Casement R, Galgiani JN (2008) Vaccine-induced cellular immune responses differ from innate responses in susceptible and resistant strains of mice infected with Coccidioides Posadasii. Infect Immun 76:5553–5564

150. Capone D, Marchiori E, Wanke B, Dantas KE, Cavalcanti MA, Deus Filho A, Escuissato DL, Warszawiak D (2008) Acute pulmonary coccidioidomycosis: Ct findings from 15 patients. Br J Radiol 81:721–724
151. Beaman L, Benjamini E, Pappagianis D (1983) Activation of macrophages by Lymphokines: enhancement of Phagosome-lysosome fusion and killing of Coccidioides Immitis. Infect Immun 39:1201–1207
152. Beaman L, Holmberg CA (1980) In vitro response of alveolar macrophages to infection with Coccidioides Immitis. Infect Immun 28:594–600
153. Cox RA, Magee DM (1995) Production of tumor necrosis factor alpha, interleukin-1 alpha, and interleukin-6 during murine coccidioidomycosis. Infect Immun 63:4178–4180
154. Slagle DC, Cox RA, Kuruganti U (1989) Induction of tumor necrosis factor alpha by spherules of Coccidioides Immitis. Infect Immun 57:1916–1921
155. Magee DM, Cox RA (1995) Roles of gamma interferon and interleukin-4 in genetically determined resistance to Coccidioides Immitis. Infect Immun 63:3514–3519
156. Magee DM, Cox RA (1996) Interleukin-12 regulation of host Defenses against Coccidioides Immitis. Infect Immun 64:3609–3613
157. Oldfield EC 3rd, Bone WD, Martin CR, Gray GC, Olson P, Schillaci RF (1997) Prediction of relapse after treatment of coccidioidomycosis. Clin Infect Dis 25:1205–1210
158. Yoon HJ, Clemons KV (2013) Vaccines Against Coccidioides. Korean J Intern Med 28:403–407
159. Dionne SO, Podany AB, Ruiz YW, Ampel NM, Galgiani JN, Lake DF (2006) Spherules derived from Coccidioides Posadasii promote human dendritic cell maturation and activation. Infect Immun 74:2415–2422
160. Richards JO, Ampel NM, Galgiani JN, Lake DF (2001) Dendritic cells pulsed with Coccidioides Immitis lysate induce antigen-specific naive T cell activation. J Infect Dis 184:1220–1224
161. Viriyakosol S, Fierer J, Brown GD, Kirkland TN (2005b) Innate immunity to the pathogenic fungus Coccidioides Posadasii is dependent on toll-like receptor 2 and Dectin-1. Infect Immun 73:1553–1560
162. Viriyakosol S, Jimenez Mdel P, Gurney MA, Ashbaugh ME, Fierer J (2013a) Dectin-1 is required for resistance to coccidioidomycosis in mice. MBio 4:E00597–E00512
163. Hung CY, Gonzalez A, Wuthrich M, Klein BS, Cole GT (2011a) Vaccine immunity to coccidioidomycosis occurs by early activation of three signal pathways of T helper cell response (Th1, Th2, and Th17). Infect Immun 79:4511–4522
164. Hung CY, Jimenez-Alzate Mdel P, Gonzalez A, Wuthrich M, Klein BS, Cole GT (2014) Interleukin-1 receptor but not toll-like receptor 2 is essential for Myd88-dependent Th17 immunity to Coccidioides infection. Infect Immun 82:2106–2114
165. Gazi U, Rosas M, Singh S, Heinsbroek S, Haq I, Johnson S, Brown GD, Williams DL, Taylor PR, Martinez-Pomares L (2011) Fungal recognition enhances mannose receptor shedding through Dectin-1 engagement. J Biol Chem 286:7822–7829
166. Gonzalez A, Hung CY, Cole GT (2011) Absence of phagocyte Nadph oxidase 2 leads to severe inflammatory response in lungs of mice infected with Coccidioides. Microb Pathog 51:432–441
167. Lam JS, Mansour MK, Specht CA, Levitz SM (2005) A model vaccine exploiting fungal Mannosylation to increase antigen immunogenicity. J Immunol 175:7496–7503
168. Li H, Barker BM, Grahl N, Puttikamonkul S, Bell JD, Craven KD, Cramer RA Jr (2011) The small Gtpase Raca mediates intracellular reactive oxygen species production, polarized growth, and virulence in the human fungal pathogen Aspergillus Fumigatus. Eukaryot Cell 10:174–186
169. Robinson MJ, Osorio F, Rosas M, Freitas RP, Schweighoffer E, Gross O, Verbeek JS, Ruland J, Tybulewicz V, Brown GD et al (2009) Dectin-2 is a Syk-coupled pattern recognition receptor crucial for Th17 responses to fungal infection. J Exp Med 206:2037–2051

170. Viriyakosol S, Jimenez Mdel P, Saijo S, Fierer J (2014) Neither Dectin-2 nor the mannose receptor is required for resistance to Coccidioides Immitis in mice. Infect Immun 82:1147–1156
171. Wang H, Lebert V, Hung CY, Galles K, Saijo S, Lin X, Cole GT, Klein BS, Wuthrich M (2014) C-type Lectin receptors differentially induce Th17 cells and vaccine immunity to the endemic mycosis of North America. J Immunol 192:1107–1119
172. Beaman LV, Pappagianis D, Benjamini E (1979) Mechanisms of resistance to infection with Coccidioides Immitis in mice. Infect Immun 23:681–685
173. Abuodeh RO, Shubitz LF, Siegel E, Snyder S, Peng T, Orsborn KI, Brummer E, Stevens DA, Galgiani JN (1999) Resistance to Coccidioides Immitis in mice after immunization with recombinant protein or a DNA vaccine of a Proline-rich antigen. Infect Immun 67:2935–2940
174. Brown GD, Denning DW, Gow NA, Levitz SM, Netea MG, White TC (2012a) Hidden killers: human fungal infections. Sci Transl Med 4:165rv113
175. Brown GD, Denning DW, Levitz SM (2012b) Tackling human fungal infections. Science 336:647
176. Brown GD, Netea MG (2012) Exciting developments in the immunology of fungal infections. Cell Host Microbe 11:422–424
177. Brown JC, Nelson J, Vandersluis B, Deshpande R, Butts A, Kagan S, Polacheck I, Krysan DJ, Myers CL, Madhani HD (2014) Unraveling the biology of a fungal meningitis pathogen using chemical genetics. Cell 159:1168–1187
178. Drummond RA, Gaffen SL, Hise AG, Brown GD (2015) Innate defense against fungal pathogens. Cold Spring Harb Perspect Med 5
179. Netea MG, Brown GD (2012) Fungal infections: the next challenge. Curr Opin Microbiol 15:403–405
180. Vautier S, Maccallum DM, Brown GD (2012) C-type Lectin receptors and cytokines in fungal immunity. Cytokine 58:89–99
181. Wuthrich M, Gern B, Hung CY, Ersland K, Rocco N, Pick-Jacobs J, Galles K, Filutowicz H, Warner T, Evans M et al (2011) Vaccine-induced protection against 3 systemic mycoses endemic to North America requires Th17 cells in mice. J Clin Invest 121:554–568
182. Hung CY, Castro-Lopez N, Cole GT (2016) Card9- and Myd88-mediated gamma interferon and nitric oxide production is essential for resistance to subcutaneous Coccidioides Posadasii infection. Infect Immun 84:1166–1175
183. Wang H, Li M, Hung CY, Sinha M, Lee LM, Wiesner DL, Lebert V, Lerksuthirat T, Galles K, Suresh M et al (2016) Myd88 shapes vaccine immunity by extrinsically regulating survival of Cd4+ T cells during the contraction phase. PLoS Pathog 12:E1005787
184. Lewis ER, David VR, Doyle AL, Rajabi K, Kiefer JA, Pirrotte P, Barker BM (2015b) Differences in host innate responses among Coccidioides isolates in a murine model of pulmonary coccidioidomycosis. Eukaryot Cell 14:1043–1053
185. Ampel NM (2010) The diagnosis of coccidioidomycosis. F1000 med rep 2
186. Stevens DA, Clemons KV, Levine HB, Pappagianis D, Baron EJ, Hamilton JR, Deresinski SC, Johnson N (2009) Expert opinion: what to do when there is Coccidioides exposure in a laboratory. Clin Infect Dis 49:919–923
187. Sutton DA (2007) Diagnosis of coccidioidomycosis by culture: safety considerations, traditional methods, and susceptibility testing. Ann N Y Acad Sci 1111:315–325
188. Gromadzki SG, Chaturvedi V (2000) Limitation of the Accuprobe Coccidioides Immitis culture identification test: false-negative results with formaldehyde-killed cultures. J Clin Microbiol 38:2427–2428
189. Padhye AA, Smith G, Standard PG, Mclaughlin D, Kaufman L (1994) Comparative evaluation of Chemiluminescent DNA probe assays and exoantigen tests for rapid identification of Blastomyces Dermatitidis and Coccidioides Immitis. J Clin Microbiol 32:867–870
190. Saubolle MA, Mckellar PP, Sussland D (2007) Epidemiologic, clinical, and diagnostic aspects of coccidioidomycosis. J Clin Microbiol 45:26–30

191. Pappagianis D (2001) Serologic studies in coccidioidomycosis. Semin Respir Infect 16:242–250
192. Wieden MA, Galgiani JN, Pappagianis D (1983) Comparison of Immunodiffusion techniques with standard complement fixation assay for quantitation of Coccidioidal antibodies. J Clin Microbiol 18:529–534
193. Pappagianis D, Zimmer BL (1990) Serology of coccidioidomycosis. Clin Microbiol Rev 3:247–268
194. Kaufman L, Sekhon AS, Moledina N, Jalbert M, Pappagianis D (1995) Comparative evaluation of commercial premier Eia and Microimmunodiffusion and complement fixation tests for Coccidioides Immitis antibodies. J Clin Microbiol 33:618–619
195. Blair JE, Coakley B, Santelli AC, Hentz JG, Wengenack NL (2006) Serologic testing for symptomatic coccidioidomycosis in immunocompetent and immunosuppressed hosts. Mycopathologia 162:317–324
196. Durkin M, Connolly P, Kuberski T, Myers R, Kubak BM, Bruckner D, Pegues D, Wheat LJ (2008) Diagnosis of coccidioidomycosis with use of the coccidioides antigen enzyme immunoassay. Clin Infect Dis 47:E69–E73
197. Malo J, Holbrook E, Zangeneh T, Strawter C, Oren E, Robey I, Erickson H, Chahal R, Durkin M, Thompson C et al (2017) Enhanced antibody detection and diagnosis of coccidioidomycosis with the Miravista Igg and Igm detection enzyme immunoassay. J Clin Microbiol 55(3):893–901
198. Johnson SM, Simmons KA, Pappagianis D (2004) Amplification of Coccidioidal DNA in clinical specimens by Pcr. J Clin Microbiol 42:1982–1985
199. Kishi K, Fujii T, Takaya H, Miyamoto A, Kurosaki A, Kohno T, Yoshimura K (2008) Pulmonary coccidioidomycosis found in healthy Japanese individuals. Respirology 13:252–256
200. Binnicker MJ, Buckwalter SP, Eisberner JJ, Stewart RA, Mccullough AE, Wohlfiel SL, Wengenack NL (2007) Detection of Coccidioides species in clinical specimens by real-time Pcr. J Clin Microbiol 45:173–178
201. Bialek R, Kern J, Herrmann T, Tijerina R, Cecenas L, Reischl U, Gonzalez GM (2004) Pcr assays for identification of Coccidioides Posadasii based on the nucleotide sequence of the antigen 2/Proline-rich antigen. J Clin Microbiol 42:778–783
202. De Aguiar Cordeiro R, Nogueira Brilhante RS, Gadelha Rocha MF, Araujo Moura FE, Pires De Camargo Z, Costa Sidrim JJ (2007) Rapid diagnosis of coccidioidomycosis by nested Pcr assay of sputum. Clin Microbiol Infect 13:449–451
203. Thompson GR 3rd, Barker BM, Wiederhold NP (2017) Large scale evaluation of in vitro amphotericin B, Triazole, and Echinocandin activity against Coccidioides species from United States institutions. Antimicrob Agents Chemother 61:e02634
204. Kim MM, Vikram HR, Kusne S, Seville MT, Blair JE (2011) Treatment of refractory coccidioidomycosis with voriconazole or posaconazole. Clin Infect Dis 53:1060–1066
205. Levy ER, Mccarty JM, Shane AL, Weintrub PS (2013) Treatment of Pediatric refractory coccidioidomycosis with combination voriconazole and caspofungin: a retrospective case series. Clin Infect Dis 56:1573–1578
206. Prabhu RM, Bonnell M, Currier BL, Orenstein R (2004) Successful treatment of disseminated nonmeningeal coccidioidomycosis with Voriconazole. Clin Infect Dis 39:E74–E77
207. Stevens DA, Rendon A, Gaona-Flores V, Catanzaro A, Anstead GM, Pedicone L, Graybill JR (2007) Posaconazole therapy for chronic refractory coccidioidomycosis. Chest 132:952–958
208. Wollina U, Hansel G, Vennewald I, Schonlebe J, Tintelnot K, Seibold M, Kittner T (2009) Successful treatment of relapsing disseminated coccidioidomycosis with cutaneous involvement with Posaconazole. J Dtsch Dermatol Ges 7:46–49
209. Johnson RH, Einstein HE (2007) Amphotericin B and coccidioidomycosis. Ann N Y Acad Sci 1111:434–441
210. Antony SJ, Jurczyk P, Brumble L (2006) Successful use of combination antifungal therapy in the treatment of Coccidioides meningitis. J Natl Med Assoc 98:940–942

211. Bajema KL, Dalesandro MF, Fredricks DN, Ramchandani M (2017) Disseminated coccidioidomycosis presenting with intramedullary spinal cord abscesses: management challenges. Med Mycol Case Rep 15:1–4
212. Clemons KV, Capilla J, Sobel RA, Martinez M, Tong AJ, Stevens DA (2009) Comparative efficacies of lipid-Complexed amphotericin B and liposomal amphotericin B against Coccidioidal meningitis in rabbits. Antimicrob Agents Chemother 53:1858–1862
213. Mathisen G, Shelub A, Truong J, Wigen C (2010) Coccidioidal meningitis: clinical presentation and management in the fluconazole era. Medicine (Baltimore) 89:251–284
214. Tariq VN, Devlin PL (1996) Sensitivity of fungi to Nikkomycin Z. Fungal Genet Biol 20:4–11
215. Yadan JC, Gonneau M, Sarthou P, Le Goffic F (1984) Sensitivity to Nikkomycin Z in Candida Albicans: role of peptide permeases. J Bacteriol 160:884–888
216. Shubitz LF, Trinh HT, Galgiani JN, Lewis ML, Fothergill AW, Wiederhold NP, Barker BM, Lewis ER, Doyle AL, Hoekstra WJ et al (2015) Evaluation of Vt-1161 for treatment of coccidioidomycosis in murine infection models. Antimicrob Agents Chemother 59:7249–7254
217. Clemons KV, Stevens DA (2011) Vaccines against coccidioidomycosis: past and current developments. J Inv Fungal Infect 5:11
218. Cole GT, Hurtgen BJ, Hung CY (2012) Progress toward a human vaccine against coccidioidomycosis. Curr Fungal Infect Rep 6:235–244
219. Cox RA, Vivas JR, Gross A, Lecara G, Miller E, Brummer E (1976) In vivo and in vitro cell-mediated responses in coccidioidomycosis. I. Immumologic responses of persons with primary, asymptomatic infections. Am Rev Respir Dis 114:937–943
220. Galgiani JN, Yam P, Petz LD, Williams PL, Stevens DA (1980) Complement activation by Coccidioides Immitis: in vitro and clinical studies. Infect Immun 28:944–949
221. Pappagianis D (1993) Evaluation of the protective efficacy of the killed Coccidioides Immitis spherule vaccine in humans. The valley fever vaccine study group. Am Rev Respir Dis 148:656–660
222. Cole GT, Xue JM, Okeke CN, Tarcha EJ, Basrur V, Schaller RA, Herr RA, Yu JJ, Hung CY (2004) A vaccine against coccidioidomycosis is justified and attainable. Med Mycol 42:189–216
223. Mcginnis MR, Smith MB, Hinson E (2006) Use of the Coccidioides Posadasii Deltachs5 strain for quality control in the Accuprobe culture identification test for Coccidioides Immitis. J Clin Microbiol 44:4250–4251
224. Xue J, Hung CY, Yu JJ, Cole GT (2005) Immune response of vaccinated and non-vaccinated mice to Coccidioides Posadasii infection. Vaccine 23:3535–3544
225. Herr RA, Hung CY, Cole GT (2007) Evaluation of two homologous Proline-rich proteins of Coccidioides Posadasii as candidate vaccines against coccidioidomycosis. Infect Immun 75:5777–5787
226. Clemons KV, Antonysamy MA, Danielson ME, Michel KS, Martinez M, Chen V, Stevens DA (2015) Whole Glucan particles as a vaccine against systemic coccidioidomycosis. J Med Microbiol 64:1237–1243
227. Cole GT, Hung CY, Sanderson SD, Hurtgen BJ, Wuthrich M, Klein BS, Deepe GS, Ostroff GR, Levitz SM (2013) Novel strategies to enhance vaccine immunity against coccidioidomycosis. PLoS Pathog 9:E1003768

Chapter 9
Paracoccidioides spp. and Paracoccidioidomycosis

Gustavo A. Niño-Vega, Emma Camacho, Ángela Restrepo Moreno, Angela María Tobón, Beatriz L. Gómez, Marcus M. Teixeira, and Bridget M. Barker

Abstract Paracoccidioidomycosis, the most prevalent deep mycosis of Latin America, is caused by *Paracoccidioides* spp., pleomorphic fungi which are present as multibudding yeast cells in infected tissues. Due to their importance to the region, these fungi and their related disease have been the subjects of intense research for

G.A. Niño-Vega, Ph.D. (✉)
Departamento de Biología, División de Ciencias Naturales y Exactas, Universidad de Guanajuato, Guanajuato, Guanajuato, Mexico

Instituto Venezolano de Investigaciones Científicas, Centro de Microbiología y Biología Celular, Caracas, Venezuela
e-mail: gustavo.nino@ugto.mx

E. Camacho, Ph.D.
Department of Molecular Microbiology and Immunobiology, Johns Hopkins Bloomberg School of Public Health, Johns Hopkins University,
615 N Wolfe St W5704, Baltimore, MD 21210, USA
e-mail: ecamach2@jhu.edu

Á.R. Moreno, Ph.D.
Corporación para Investigaciones Biológicas (CIB), Medellín, Colombia
e-mail: angelares@une.net.co

A.M. Tobón
Facultad de Medicina, Centro de Estudios en Salud (CES), Medellín, Colombia

Corporación para Investigaciones Biológicas (CIB), Medellín, Colombia
e-mail: angelamtobon@hotmail.com

B.L. Gómez, B.Sc., Ph.D.
School of Medicine and Health Sciences, University of El Rosario, Bogotá, Colombia

Corporación para Investigaciones Biológicas (CIB), Medellín, Colombia
e-mail: beatrizlgomez@hotmail.com

M.M. Teixeira, Ph.D. • B.M. Barker, Ph.D.
Translational Genomics Research Institute, Northern Arizona Center For Valley Fever Research, Flagstaff, AZ, USA
e-mail: marcus.teixeira@nau.edu; bridget.barker@nau.edu

H.M. Mora-Montes, L.M. Lopes-Bezerra (eds.), *Current Progress in Medical Mycology*, DOI 10.1007/978-3-319-64113-3_9

over 50 years. The present chapter is an attempt to summarize the findings of those decades of research and clinical findings, with an update on the latest advances on taxonomy studies, species distribution and reproductive modes, diagnosis trends, virulence factors, as well as clinical manifestations, and updated treatment of the disease.

9.1 Etiologic Agents

For almost 100 years, since its first description in 1908 by Adolpho Lutz in Brazil [1], *Paracoccidioides brasiliensis* was considered the only causative agent of paracoccidioidomycosis (PCM), a human systemic mycosis geographically confined to Latin America, where it constitutes one of the most prevalent deep mycoses. However, in 2006, the first report on molecular phylogenetic data revealed the presence of cryptic species, namely *P. brasiliensis* S1, PS2, and PS3 [2]. To date, a fourth species within the *P. brasiliensis* species complex, PS4, and a newly identified species within the *Paracoccidioides* genus, *Paracoccidioides lutzii*, have been reported [3–6] (Fig. 9.1).

Paracoccidioides spp. are pleomorphic fungi that depend on the temperature of incubation to establish a specific cell morphology. In infected tissues, and in vitro cultures at 37 °C, these fungi grow in the form of spherical-to-oval cells of different sizes, which can go from a few nanometers in diameter for young, recently separated buds, to up to 30 μm in mature yeast (Y) cells [7]. Yeast cells multiply by multipolar budding and are multinucleate. Buds connect to the mother cell by narrow necks, giving the whole structure an appearance of a ship's wheel (Fig. 9.2), a shape considered

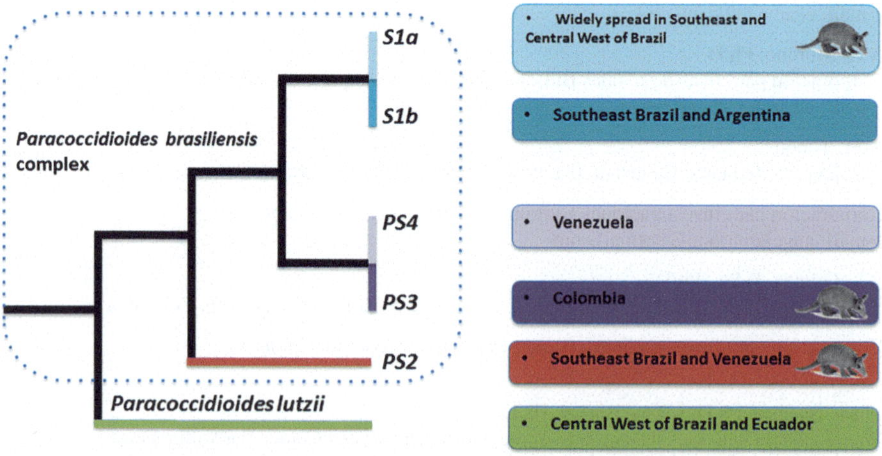

Fig. 9.1 Phylogenetic relationships of *Paracoccidioides* spp. Colored boxes indicate species geographic distribution

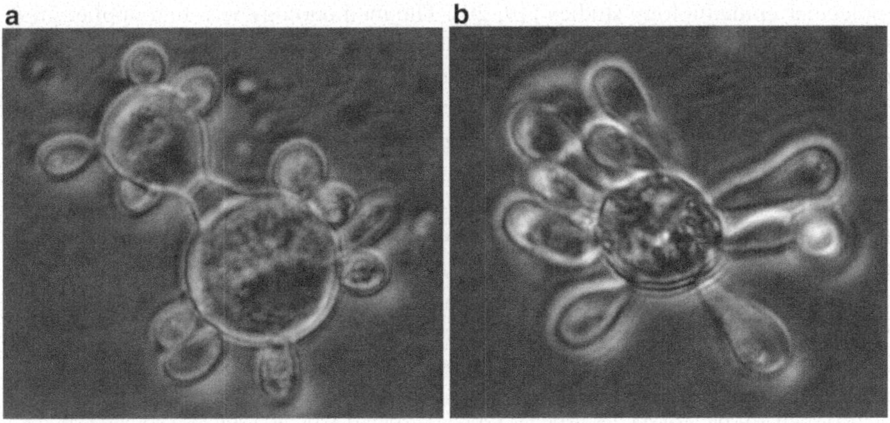

Fig. 9.2 Multibudding yeast phase of (**a**) *P. brasiliensis* and (**b**) *P. lutzii*. Cells were grown at 37 °C on YPD supplemented with 5% horse serum. Magnification ×400. Photos courtesy of Sabrina Rodríguez and Laura Barreto, IVIC, Caracas, Venezuela

as a diagnostic characteristic of *Paracoccidioides* spp. The mycelial (M) form grows slowly at room temperature. Microscopic observations show septate and freely branching hyphae, 1–3 µm in width [8], with the appearance of interwoven threads. Several types of conidia [intercalary, septate, pedunculate] are formed after 2 months of growth under conditions of nutritional deprivation [9, 10]. Conidia can germinate and produce Y cells at 37 °C, with the capacity to infect mice, or may produce germ tubes and branching mycelia if kept at 20–23 °C.

9.2 Taxonomy, Species Distribution, and Reproductive Modes of *Paracoccidioides* spp.

P. brasiliensis [1] was considered a single species for almost 100 years. With the advances in the molecular biology field and fungal genotyping techniques during the 90s, many authors demonstrated that the *Paracoccidioides* genus has an extensive genetic diversity deciphered by karyotyping profiles, RAPD, RFLP, and microsatellites typing methods [11–17]. Those molecular methods were very important for an initial assessment of the population diversity of this pathogen, and provided evidence for geographical and genetic isolation [14], and varied ploidy [11, 17]. The Sanger sequencing technology in the late 90s brought to the mycologist an opportunity to analyze clinical isolates at a nucleotide base level across multiple genomic loci. This allowed the emergence of an operational method proposed by [18] in which fungal species could be delineated by phylogenetic concordance and nondiscordance of multiple genes (GCPSR – Genealogical Concordance for Phylogenetic Species Recognition). This new concept is extensively applied in the mycology field as a tool for species discrimination as well as giving support for molecular epidemiology studies [19, 20]. The methodology was first applied in *P. brasiliensis* by [2] in a set of 65 isolates through eight gene loci, and at least three distinct phylogenetic species were proposed: ***P. brasiliensis* S1** – widely distributed species in Brazil, Uruguay, Peru, and Argentina; ***P. brasiliensis* PS2** - limited to southeast Brazil and Venezuela and ***P. brasiliensis* PS3** - geographically restricted to Colombia. This study was extremely important for species delineation and eco-epidemiology in *Paracoccidioides*; however, no differentiation between soil, veterinarian or human isolates was found. A fourth phylogenetic species was later proposed to be restricted to Venezuela, being named as ***P. brasiliensis* PS4** (Fig. 9.1) [3, 4]. One of the most studied isolates in the paracoccidioidomycosis field, Pb01, was not included in the dataset of Matute et al. [2], which later was found to present a unique multilocus phylogenetic composition, when compared with the three reported *P. brasiliensis* species S1, PS2, and PS3 [5]. More isolates sharing a genetic profile with Pb01 (*Pb01-like*) were later identified based on the same techniques, revealing a great genetic isolation from the previous identified *P. brasiliensis* complex [6]. In addition, *Pb01-like* strains displayed exclusive elongate rod-shaped conidia beyond the typical bicorn cocked hat- and barrel-shaped conidia produced by *P. brasiliensis* [21]. The strains that composed the *Pb01-like* clade were mostly recovered from Central-North areas of Brazil and from Ecuador. Recently the phylogenetic species *Pb01-like* was ranked as a new species apart from the *P. brasiliensis* species complex and named ***Paracoccidioides lutzii,*** as a tribute to the discoverer of PCM, Adolfo Lutz (Fig. 9.1) [3, 22].

All *Paracoccidioides* species described so far cause both, acute and chronic disease; however, some relevant clinical features and phenotypes were observed in those patients infected with *P. lutzii* and *P. brasiliensis* species complex [23]. (1) Serology-based diagnostics of PCM must be performed with species-specific antigens, since most sera recovered from patients infected with *P. lutzii* are not recognized by *P. brasiliensis* antigens and vice-versa [3, 24–26]. (2) Most of PCM infections reported in endemic areas of *P. lutzii* show frequent lymphatic-abdominal clinic manifestation [27, 28]. (3) A preliminary study reported that patients infected with *P. lutzii* isolates have a better response to trimethoprim-sulfamethoxazole therapy than those patients infected with *P. brasiliensis* [17]. (4) *P. brasiliensis* has lower interaction capabilities with hemocytes, when compared to the *P. lutzii* infection model in *Galleria mellonella*, which would indicate different phagocytosis mechanisms [29]. (5) *P. brasiliensis* displays higher adhesion and adhesin expression levels, and greater virulence than *P. lutzii,* in the infection model *G. mellonella* [30]. With the exception of serological achievements, those studies were based on few isolates, and studies on virulence models require more isolates from each group in order to improve knowledge of infection dynamics of both pathogens.

In recent years, next-generation sequencing (NGS) has decreased the cost of sequencing whole genomes and increased the amount of generated data per sequencing run. This allows an increase in the power of genetic discrimination between microorganisms by analysis of thousands of loci instead of just a few. The genomes of *P. brasiliensis* S1 (Pb18), PS2 (Pb3) and *P. lutzii* (Pb01) were fully sequenced, revealing unique adaptations of the *Paracoccidioides* lineage as well as reinforcing the genetic divergences between those two species [31]. Recently, the genomes were carefully reassembled and re-annotated generating high confidence reference genomes for both, the *P. brasiliensis* complex and *P. lutzii*, providing support for genealogical concordance based on whole genome sequence typing [32]. Furthermore, a whole-genome sequence study including 36 *Paracoccidioides* isolates covering the five phylogenetic species of this genus was performed. Gene order was highly conserved across the major lineages of *Paracoccidioides* and few chromosomal rearrangements were found. The phylogenetic species *P. brasiliensis* S1 was split into two clades: S1a and S1b, which were deciphered by increasing the number of genomic loci [33].

With the increase of genomic information, new molecular markers specific for the *Paracoccidioides* genus were developed, enabling the identification of the fungus presence in the air and soil within the endemic regions [34, 35]. ITS sequence analysis of some *P. brasiliensis* S1a and PS2 samples recovered from soil and armadillos, showed that they cluster with clinical samples [21, 32]. However, ITS amplicon Sanger sequencing revealed additional genotypes of *Paracoccidioides* found in soil and aerosol that had not been previously reported in clinical samples [35]. These genotypes may represent noninfectious or low virulence strains that may be present in the environment. Once exposed to these low virulence strains, a mammalian host could be naturally vaccinated and acquire humoral response against other *Paracoccidioides* genotypes. Isolation of *Paracoccidioides* from the environment is laborious and has been mainly unsuccessful in many published attempts. Thus, molecular markers might be a better choice for environmental studies in the PCM field [35]. Moreover, the recently published molecular probe for *P. brasiliensis* and *P. lutzii* may be useful for diagnostics in clinical practices, since it presents a great specificity in in vitro assays [35].

Population genetics of the genus *Paracoccidioides* revealed different breed strategies [3]. For example, the species *P. brasiliensis* PS3 has little genetic variation and a genetic constitution compatible with a clonal microorganism [2]. Phylogeographic analysis suggests that PS3 species emerged from a migration event from *P. brasiliensis* S1b, which kept stable after the Andean uplift [21]. The remaining species (S1a, PS2 and *P. lutzii*) show much more variation and wider distribution. Moreover, Most Recent Common Ancestor (MRCA) estimations supported S1b as the early-diverging *P. brasiliensis* population, with the later emergence of PS2, S1a, and PS3 [33]. Population genetic studies reveals that those species are recombining and sexual reproduction could be the source of genetic and phenotypic variation of those populations [2, 21, 22]. The genus *Paracoccidioides* was considered for many years asexual, since no sexual structures had been reported. However, comparative genomic analysis revealed that the species from the genus *Paracoccidioides* harbor a genomic architecture shared with the close Ajellomycetacean species *H. capsulatum* and *B. dermatitidis* [36, 37], both with a defined sexual stage [38]. The ***MA***ting-***T***ype (*MAT*) locus that regulates the sexual cycle in many Ascomycetes is present in *Paracoccidioides* species in chromosomal synteny with the sexual counterparts of *H. capsulatum* and *B. dermatitidis* [38]. A ratio of 1:1 between both *MAT1-1* and *MAT1-2* idiomorphs was identified, which is compatible with sexual recombining species [29, 39]. Moreover, a series of meiotic regulators were found in all *Paracoccidioides* genomes, which are conserved with the well characterized *Aspergillus* and *Saccharomyces* species [38, 40]. Finally, sexual structures were reported for both *P. brasiliensis* and *P. lutzii* by co-incubating different mating-type strains under soil extract agar. Those authors revealed the presence of young ascocarps (immature gymnothecia) by configuring arrays of intercoiling hyphae [38]. None mature gymnothecia or ascospores have been observed so far. Also, nuclear migration in intercoiling hyphae sections and parasexual cycle should not be discarded, since mature sexual structures were not observed. Aneuploidy was already reported for some *Paracoccidioides* isolates which also indicate mitotic recombination followed by haploidization [41].

9.3 Epidemiology

PCM is a systemic disease affecting Latin America, where it is diagnosed from Mexico (23°N) to Argentina (34° S), with Brazil accounting for more than 80% of all reported cases, followed by Venezuela, Colombia, Ecuador, Bolivia and Argentina. No cases have been reported in Chile, Surinam, Nicaragua or Belize, and with the exception of few isolated reports, the mycosis has neither been observed in Trinidad, Granada or Guadeloupe. Additionally, in endemic countries, the mycosis is distributed irregularly throughout the territory, preferring sites where humidity is high and forests abound [42–44]. Mention should be made of the non-autochthonous patients - less than 100 - reported in Europe, Japan or the United States. All those patients had previously resided or visited the Latin America endemic regions for a mean period of 14 years before the mycosis became manifested. These observations revealed the ability of the etiologic agents to become latent once in the host with no loss of viability such as exemplified by the patient diagnosed in Japan by Kurai et al. [45].

The mycosis is caused by two species of the genus *Paracoccidioides,* namely, *P. brasiliensis* and *P. lutzii*, with the former predominating within the endemic region (Fig. 9.1) [21, 46, 47].

Based on Brazilian statistics, the incidence of PCM was estimated in 10–30 cases per 100,000 inhabitants per year [48]. Nonetheless, as an example, variations were recorded in the state of Paraná where 102 cases were diagnosed in a period of 18 months [49]. Additionally, another place with high incidence of PCM appears to be the Brazilian state of Rondônia where 39.1 cases/100,000 inhabitants were informed from 1997 to 2012 [50].

As for mortality, a study in the Brazilian state of Paraná estimated the annual death toll to be 3.48 per million inhabitants, thus making PCM the fifth cause of death among the chronic infectious diseases of the low respiratory tract [51]. During the period 1998–2006, a Brazilian study was performed covering 35% of the country, which represented 27% of the 5560 Brazilian municipalities - including the larger part of the endemic PCM area. The study constituted a population estimated at

4.3 per one million inhabitants, and analyzed 6732 hospitalization events attributable to PCM (82% male). They reported an estimated global mortality of 1.4 per one million inhabitants with 60% of deaths being attributed to the mycosis itself and the remaining to the fibrotic sequelae [52]. Another retrospective Brazilian study analyzed 3,583 death certificates attributed to various systemic mycoses. It found that PCM was the most frequent cause of death with annual tolls of 53.9% and 44.6% at the beginning and end of the study, respectively. Additionally, the association with AIDS was found to be detrimental [53].

Studies in groups of patients have shown that work with soil and certain crops, common in rural Latin American fields (coffee, cotton, tobacco, sugar cane), predominated as the most common occupation (<60%). Other jobs mentioned were masonry and lumberjacking [44, 54–56].

Even though the precise micro-niche of the etiologic agent has not been fully recognized, it is accepted that PCM is acquired by inhalation of the fungus' conidia [57]. Most individuals exposed to the fungal propagules develop an asymptomatic pulmonary infection that disseminates to regional lymph nodes reaching, at times, deep organs [44, 55, 58].

In infected individuals *P. brasiliensis* may remain latent but viable for long periods of time, as demonstrated by patients diagnosed in non-endemic countries [45, 55]. Once revitalized, the fungus may spread to other organs and tissues through blood and lymphatic systems [55, 59]. If this occurs, reactivation of the disease process may take place with dissemination to lymph nodes, skin, mucosae, larynx, spleen, liver, bones, adrenals, central nervous system and other organs [58, 60].

A series of cases have revealed that, at time of diagnosis, 70% of the patients were older than 30 years of age, 20% were adolescents with the remaining 10% corresponding to children with most cases (58.2%) being observed in patients aged between 40 and 59 years [42, 43, 56, 58, 61]. There is predominance of the mycosis in males over females (13 to 1), although in certain countries (Argentina, Colombia) this relation can be as high as 70 males per 1 female [42, 55, 62] Among children and youngsters aged less than 14 years, no gender differences are noticed, probably because 17 beta-estradiol, the hormone known to hinder the transformation of the infectious air-borne, mycelial-derived conidia into the yeast tissue forms, has not yet been expressed as it occurs in adult patients [63].

No race predilection has been noticed, except that the European and Asiatic emigrants established in the PCM endemic regions, tend to develop a more severe disease [46, 56, 58].

Among the risk factors to develop PCM, agriculture, as mentioned before, has been most frequently mentioned with over 70% of the patients being farmers [56, 58, 60]. Construction work and masonry have also been mentioned [42, 58, 61]. As noted above, male gender is also a risk factor as men develop PCM much more frequently than females [55, 58, 62] even if infection - as shown by delayed type hypersensitivity to the paracoccidioidin test - is equal in both genders [64]. Young age is related to the severity of the mycosis but not with an increased number of cases [65].

9.4 Clinical Manifestations.

PCM has been divided into four categories—two devoid of clinical manifestations and another two directly related to disease processes and largely dependent on age and host immune status:

Subclinical infection: Considered to be a latent, clinically asymptomatic process lasting for 1 month to years after the initial fungal contact, and which might evolve towards a progressive, clinically manifested disease [66].

Residual form: Characterized by the presence of sequelae originating in the previously infected fibrous tissues, mainly in the lungs, but also reported in adrenal glands and mucosae.

Acute/subacute disease: An overt and severe process evidenced by involvement of multiple organs, with lymph node, liver, and spleen hypertrophy and manifestation also of skin lesions, the latter often multiple and widely distributed throughout the body. The lungs do not often reveal radiographic abnormalities, nor are the clinical manifestations obvious, even though *P. brasiliensis* is frequently detected in respiratory specimens. This form is usually diagnosed in children, young adults, and AIDS patients, as well as in those with other immune alterations.

Chronic progressive disease: The most common (90%) of the clinical presentations, diagnosed in older patients and characterized by important lung involvement and frequent lesions in the mucosae, skin, adrenal glands, and other sites.

In any of the clinically manifested forms, constitutional symptoms (fever, asthenia, general malaise, weight loss) are regularly observed [67].

9.4.1 Physiopathology

Lungs are the primary site of infection, whose clinical manifestations are cough, expectoration, hemoptoic sputum, chest pain, and some degree of dyspnea are regularly noticed. Auscultation reveals minimal abnormalities compared to radiographic findings, with a clear dissociation between symptoms and imaging studies [57, 68].

Chest radiograph images reveal mostly interstitial infiltrates (64%), followed by mixed lesions consisting of nodular and alveolar infiltrates that are occasionally confluent, frequently bilateral, symmetrical, and located preferentially in the central and lower fields [57, 68, 69]. High-resolution computed tomography demonstrates abnormal findings in more than 90% of the patients with chronic pulmonary disease, with the most common abnormalities being interlobular septal thickening (88%), peribronchovascular interstitial thickening (78%), centrilobular opacities (63%), intralobular lines (59%), ground-glass opacities (34%), cavities (17%), and airspace consolidation (12%) [70]. Gallium-67 scans reveal pulmonary lesions in practically all patients [71]. At diagnosis, one third of patients with long-lasting disease have serious pulmonary sequelae fibrosis (32%), bullae (27%), as well as indirect signs of pulmonary hypertension with right ventricle enlargement and even cor pulmonale [57, 68, 69].

The oral mucosa is involved in 50% of the cases, with lesions regularly localized in the gingiva and the palate and, with extension to the neighboring skin on the lips and nose. The larynx, pharynx, and gastrointestinal mucosa are also affected. Lesions are infiltrative, edematous, ulcerated, and have a granulomatous base, mulberry-like in appearance. They are extremely painful and hinder the patient's food intake; upon healing, they produce fibrous scarring [72, 73]. This compromise is the principal difference with tuberculosis, although nowadays the oral presentation of TB is rare, with a prevalence ranging from 0.1 to 0.4% [74, 75].

Cutaneous lesions are observed in patients with either the adult form (25%) or the acute/subacute presentation (50%). Skin abnormalities may represent hematogenous dissemination from the primary pulmonary focus or secondary extension from contiguous mucocutaneous lesions. They tend to ulcerate and show a thick well-delimited external border with a granulomatous base, accompanied by infiltration of the neighboring tissues. In AIDS patients, lesions are multiple and rather small, flat, covered by crusts, and with no inflammation in the surrounding tissues [73, 75].

All **lymph node chains** may be involved, with predilection for cervical, axillary, mediastinal, and mesenteric nodes. If mucosal or skin lesions exist, enlargement of the corresponding draining lymph nodes is noted, and they may form fistulas. Lymphatic hypertrophy can cause complications by compression of contiguous structures [67, 68].

In endemic countries, this mycosis is one of the main causes of adrenal insufficiency. Involvement is evidenced by gland hypertrophy with damage to both the cortex and the medulla; in addition, 10% of the patients may present Addison's syndrome. When cortisol is measured at basal and post–adrenocorticotropic hormone stimulation times, diminished adrenal reserve is detected in 33–40% of the patients. Autopsy studies, however, have shown that adrenal involvement may occur in up to 85% of cases. Prompt initiation of antifungal therapy may restore adrenal function [76].

Other lesions: PCM also involves the central nervous system in 9.65–25.45%, with formation of cerebral and meningeal lesions and spine compromise with spondylodiscitis. Lesions are more frequent in the brain hemispheres (69%), in 65% there are multiple granuloma characterized by hypodense images with annular or nodular enhancing [77–79]. Other organs affected are the liver, spleen, and pancreas, simulating malignant neoplasia [80, 81].

The prevalence of PCM in immunocompromised patients by AIDS, is relatively low and occurs more frequently in advanced AIDS patients (CD4 counts <200 cells/mL, viral copies >100,000 mL). In the Brazilian State of São Paulo, the prevalence of this comorbid association was estimated to be 1.4% of HIV-infected patients [82]. In those patients, skin lesions are extremely common and 40% of them had received trimethoprim-sulfamethoxazole treatment for pneumocystosis and toxoplasmosis prophylaxis. This finding suggests that prevention of PCM with this medication, had proved ineffective. In these cases, treatment follows the same schemes used for the non-HIV patients, except that maintenance therapy should be required up to the time when CD4 counts are consistently greater than 200/cells/mL and viral copies undetectable through administration of antiretroviral therapy [75, 83].

The most important differential diagnosis is tuberculosis, a disease that may coexist with the mycosis in 8–13% of patients [84]. Other diseases that should be considered are cancer and neoplastic disorders (including lymphoma), histoplasmosis, leishmaniasis, leprosy, and syphilis. Only laboratory testing can establish the correct diagnosis [45, 85].

9.5 Diagnosis Trends

Current diagnostic approaches include traditional techniques, namely direct examination, culture, histology, antigen and antibody detection, and nucleic acid-based methods such as polymerase chain reaction (PCR) assays [77, 86–88]. More recently, commercial platforms have also been developed such as MALDI-TOF MS (matrix-assisted laser desorption ionization time-of-flight mass spectroscopy), which is unique in that it does not depend on analysis of genetic information, but is instead based on analyses of the spectrum of organic biomolecules/proteins present in the isolate [89, 90]. The definitive diagnosis of PCM depends on microscopic examination and positive culture. Current culture-based diagnostic tools continue to be the "gold standard" for diagnosis; however, isolation and identification of *Paracoccidioides* spp. from clinical specimens may take up to 4 weeks. The best approach is typically a use of a combination of methods for diagnosis when possible.

9.5.1 Direct Examination and Histopathology

In clinical specimens, such as sputum, bronchoalveolar lavage fluid, oral, pharyngeal, adrenal glands, pus from draining lymph nodes, or biopsy materials from any other tissues, the characteristic translucent-thick refracting wall of yeast cells of *Paracoccidioides* with multiple peripheral buds (pilot wheel configuration) plus prominent intracytoplasmatic vacuoles can be found (Fig. 9.3a). Often, yeast cells appear in chains and have single buds and some large yeast cells can also be observed. Sensitivities vary from 85 to 100% depending on specimen, clinical manifestations, and treatment status on direct examination [67, 91, 92]. Several procedures are

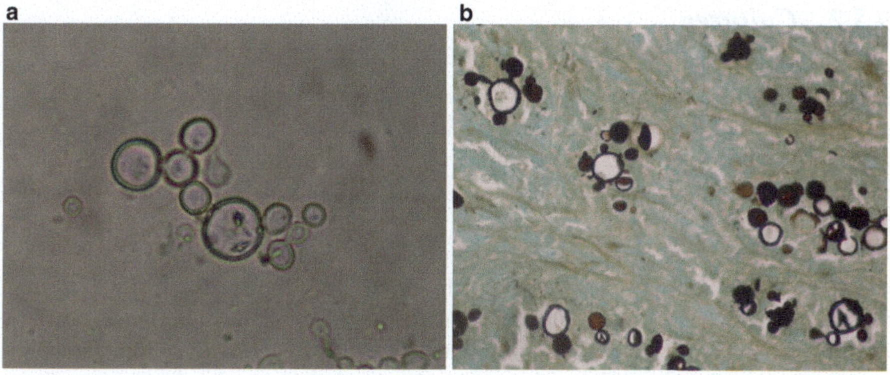

Fig. 9.3 Wet mount of (**a**) sputum in KOH showing blastoconidia of *P. brasiliensis*. Magnification ×200; (**b**) Gomori methenamine-silver stain of lung tissue showing blastoconidia of *P. brasiliensis*. Magnification ×100

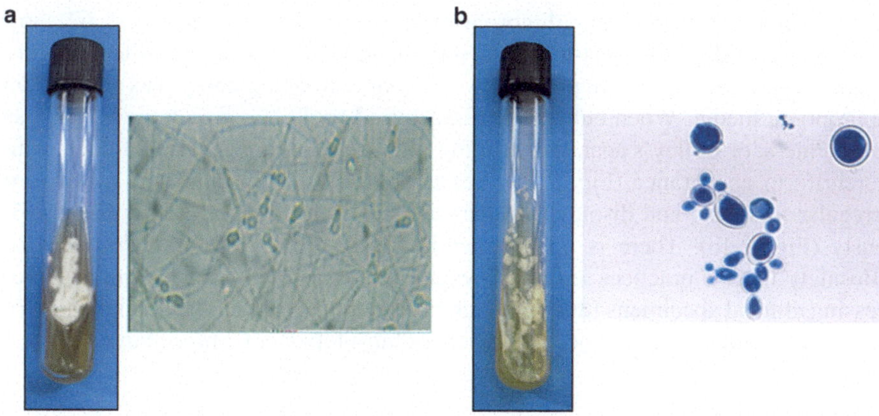

Fig. 9.4 (**a**) *P. brasiliensis* colonies of mycelial phase incubated at 25 °C and microscopic observation showing thin septate hyphae and microconidia (magnification ×100). (**b**) *P. brasiliensis* colonies of yeast phase incubated at 37 °C and microscopic observation showing round mother cell surrounded by spherical to elongated daughter cells (magnification ×40)

adequate for visualizing fungal elements, including fresh or wet KOH preparations, as well as calcofluor and immunofluorescence methods. Staining of preparations with lactophenol cotton blue and Gram staining can also be helpful. Histopathology preparations stained with hematoxylin and eosin (H&E), Gomori methenamine-silver (GMS; Fig. 9.3b), or periodic Acid-Schiff are also useful as they reveal the multiple budding yeast elements, especially within the granulomatous foci commonly observed. It is important to differentiate small yeast of *P. brasiliensis* from *Cryptococcus neoformans*, *Blastomyces dermatitidis*, and even from *Histoplasma capsulatum* [43, 67, 68, 86, 91–93].

9.5.2 Cultures

P. brasiliensis is a thermally dimorphic fungus. At room temperature (19–24 °C) it grows as a mold and requires a lengthy incubation, of up to 30 days (Fig. 9.4 a). Isolation of *P. brasiliensis* from clinical specimens requires a battery of culture media such as simple Sabouraud dextrose agar, more enriched media like brain-heart infusion (BHI) plus glucose, and a selective medium, Mycosel™, which includes antibacterial drugs (chloramphenicol or gentamicin) and mold inhibitors (cycloheximide) that provide a useful differential tool in the process of identifying the fungus [43, 67, 86]. When incubated at 25–30 °C, the isolate produces a variety of forms from flat to glabrous colonies, with a few tufts of aerial mycelium. Colonies can also look wrinkled, and a variation in color can be observed, which can go from white to beige-brownish. Microscopically, the mold shows thin septate hyphae and occasionally chlamydospores (15–30 µm) are present. Under conditions of nutritional deprivation, some isolates produce conidia, which vary in structure from arthroconidia to microconidia measuring less than 5 µm (Fig. 9.4a). The mycelial form is not distinctive and dimorphism must be demonstrated [43, 67, 86]. Conversion to the yeast phase is also very slow. Conidia respond to temperature changes, germinating into hyphae at 20–24 °C or converting into yeasts at 36 °C on appropriate media. When cultures are transferred to 37 °C on rich media such as BHI, Pine's, or Kelley's agar, the resulting yeast colonies are generally folded, with cerebriform appearance (Fig. 9.4b). Yeasts are 2–30 µm in diameter, are oval or irregular in shape, and display the characteristic "pilot's wheel" described previously (Fig. 9.4b). There is no commercial DNA probe test for *P. brasiliensis*. Biosafety level 2 practices and facilities are recommended for handling and processing clinical specimens [86], although procedures involving the manipulation of sporulating cultures should be handled under biosafety level 3 procedures.

9.5.3 Immunodiagnostic Tests

Immune-based methods for antibody and antigen detection are useful not only for diagnosis but also for monitoring the patient's course [86, 94–96]. The most popular methods are immunodiffusion (ID), double immunodiffusion (DID), complement fixation (CF), but other tests, such as ELISAs, counter-immunoelectrophoresis (CIE), dot blot, and immunoblotting have also been employed [86, 94]. The ID test is highly specific and is positive in 65–100% of cases of acute or chronic pulmonary infection or disseminated PCM. The CF test allows a more precise evaluation of the patient's response to treatment, but is less specific than the ID as cross-reactions with *H. capsulatum* occur. As it is simple to perform, the gel immunodiffusion or DID test is typically the most common method used in endemic countries. This test uses antigen preparations from either mycelial or yeast cell lysates. Cell wall antigens have proved less useful than culture filtrate antigens, largely because cell wall antigens are dominated by cross-reactive galactomannan [86, 94, 95]. While a series of immunoreactive antigens are present in the yeast cell lysates (27 kDa, 43 kDa, 60 kDa, 70 kDa and 87 kDa), the predominant antigen is a 43-kDa glycoprotein, gp43 [67, 94]. Commercial mycelial-form culture filtrate antigen can be obtained for "in-house use in ID from IMMY (*Paracoccidioides* ID Antigen), but its sensitivity has not been widely studied [86]. Complement fixation (CF) test is performed with *P. brasiliensis* yeast-form culture-filtrate antigen but this reagent is not commercially available. CF titers of ≥1:8 are considered presumptive evidence of PCM and falling CF titers are often predictive of successful treatment [86]. No commercial kits or reagents are available. Several laboratories have developed their own "in house" immunodiagnostic tests based on methods such as ELISA, inhibition ELISA, competition ELISA, Western blot, or dot blot [94–96]. Improvements include the detection of antibodies against chemically characterized and/or recombinant *P. brasiliensis* antigens, notably gp43, pb27, and the 87-kDa heat-shock protein [97–99]. A combination of two recombinant products has resulted in increased sensitivity (92%) and specificity (88%) [100]. Antigen detection tests have some important advantages over antibody detection in the diagnosis of PCM, particularly in immunocompromised patients and in those previously exposed to *P. brasiliensis* who may have preexisting antibody titers [86, 94].

Many efforts are being made and are underway to search for new serological markers for the diagnosis and monitoring of PCM, especially because of the recent advances in characterizing the taxonomy of the genus, and this is certainly a new challenge [24, 94]. It has been found by DID that the serum from patients with PCM due to *P. brasiliensis* does not recognize antigen contained in the cell-free preparations of *P. lutzii*, unlike the serum from patients with PCM due to *P. lutzii*, which recognizes both antigens (*P. lutzii* and *P. brasiliensis*), suggesting that *P. lutzii* is antigenically more complex and has species-specific antigens but also shares common antigens with *P. brasiliensis* [24]. Many laboratories in Latin America have had difficulty obtaining antigens because the method requires many steps, and efforts have been made to simplify the methodologies. However, there is no standard way of obtaining the antigens, and the diagnosis techniques used in different mycology reference centers show large inter-laboratory variability in results when assessing the same clinical sample [95, 101]. Understanding the uses and limitations of the battery of serological methods is essential for the diagnosis of PCM [94, 95]. New technologies have also started to be applied, with the use of gold-nanoparticles functionalized with specific antibodies, or antigens for detecting circulating antigens of antibodies alone or in combination with PCR for the early diagnosis of PCM [102, 103].

9.5.4 Molecular Detection

A variety of molecular methods have been designed and tested over the years, and they now play a significant and growing role in clinical mycology and offer distinct advantages over conventional methods, as they provide faster, more sensitive and more specific identification. However, for the detection of *Paracoccidioides* in human clinical samples the number of "in house" tests developed and evaluated so far is still very limited, they have no extensive validation across laboratories and they have not been made commercially available [87, 88]. Amplification and sequencing of conserved rDNA regions, especially 5.8 and 28S subunits, as well as intergenic spacer regions have been successfully developed and enabled discrimination between *Paracoccidioides* and other human pathogenic fungi by PCR using DNA extracted from cultures [87, 104]. Although in more recent studies, genes specific for *Paracoccidioides* have been chosen as PCR targets and have often had higher specificity. A real-time PCR that used as a target the ITS1 region of ribosomal DNA (rDNA) was developed to detect *P. brasiliensis* DNA in both cultures, and clinical specimens and, although the evaluation was carried out with a low number of clinical isolates ($n = 10$) and human samples ($n = 2$) the authors reported 100% sensitivity and specificity [105]. Some other attempts to develop a PCR with the same target have had very limited success.

Among the *Paracoccidioides*-specific targets, gp43 is considered the immunodominant antigen for diagnosis of PCM. A nested PCR assay to amplify the gp43 gene was evaluated using an experimental mouse model of PCM, and the test was positive in 91% of lung homogenates and did not cross-react with other pathogens [106]. The same technique was then evaluated and validated for use in a variety of clinical samples ($n = 25$ cases with proven PCM), and a 100% sensitivity and specificity was reported [107]. Another PCR assay based on the 5′ nuclease assay, using a fluorescent probe derived from the sequence of the gene coding for the gp43 antigen was developed, and the authors reported that the assay could detect at least 10 copies of this DNA sequence, providing sufficient accuracy to be useful for diagnosis of PCM [108]. A loop-mediated isothermal amplification (LAMP) assay was tested for its ability to detect the gp43 gene of *P. brasiliensis* [109], and positive results were reported for DNA extracted from formalin fixed paraffin embedded tissues from PCM patients [109]. In another study, respiratory samples were evaluated by LAMP, with reported sensitivity and specificity values of 61% and 100%, respectively [110]. LAMP methodology exhibits advantages in speed in developing the assay and in simplicity, when compared to the classic diagnostic methods, as it does not require sophisticated equipment. Molecular detection and discrimination of *P. brasiliensis* and *P. lutzii* has been reported by Single Nucleotide Polymorphism (SNP) genotyping [21]. Much work still needs to be done to reach a consensus for extraction protocols, amplification targets and best approaches and platforms for measurement. Extensive clinical validation of the techniques is also much needed.

9.5.5 MALDI-TOF Mass Spectroscopy

Isolates of *P. brasiliensis* and *P. lutzii* ($n = 22$), previously characterized by molecular techniques, including multilocus sequence typing, internal transcribed spacer (ITS) rRNA gene sequencing, and PCR of the hsp70 gene, were recently identified for the first time by MALDI-TOF mass spectrometry [90]. The use of a commercial platform does not require extensive training, could be well adapted to reference routine laboratories and could become a method of choice for species differentiation within the genus *Paracoccidioides*, providing to epidemiologists and clinicians new tools to address unresolved issues of PCM [90].

9.6 Virulence Factors

9.6.1 Dimorphism

Once conidia or hyphal fragments are inhaled into the lung alveoli, *Paracoccidioides* spp., as well as other endemic dimorphic fungi, switch to a unicellular yeast form due to temperature, oxidative stress, changes in carbon dioxide tension, and hormones. This morphological switching, known as dimorphic transition, represents an essential step for the survival and maintenance of the fungus within the host. In dimorphic fungi, three signaling pathways have been identified that induce the dimorphic switching and yeast growth at 37 °C; the two-component signaling, the heteromeric G protein and Ras signaling, and calcium signaling [111]. The two-component signaling systems are regulated through DRK1 (dimorphism-regulating histidine kinase 1) [111]. In *B. dermatitidis* and *H. capsulatum*, *DRK1* mutants are avirulent in a murine model of infection, failing to convert to the pathogenic yeast form, and growing as mycelia at 37 °C [112]. The *DKR1* orthologue in *P. brasiliensis* is highly expressed in the virulent yeast phase, and is fundamental in the mycelial to yeast transition, suggesting a potential new drug target [113, 114].

9.6.2 Cell Wall Polysaccharides Composition

The dimorphic transition promotes changes in the cell wall composition and carbohydrate polymer structure. *Paracoccidioides* yeast cell walls' most remarkable change is the occurrence of α-(1,3)-glucan and the drastic reduction of β-(1,3)-glucan that correlates with pathogenicity [115]. The initial observation in *P. brasiliensis* that α-(1,3)-glucan might act as a protective shield against host defense was demonstrated 30 years later in *H. capsulatum* [116]. The presence of α-(1,3)-glucan in the outermost layer of the cell wall of *H. capsulatum* yeast masks β-(1,3)-glucan, avoiding its recognition from pattern recognition receptors (PRR) found on host phagocytic cells. Furthermore, disturbance of the α-(1,3)-glucan synthesis by deleting the *H. capsulatum* α-(1,4)-amylase (*AMY1*), involved in priming the oligosaccharide synthesis, reduces cell wall α-(1,3)-glucan content and fungal virulence [117, 118].

9.6.3 Adhesins

Multiple studies have characterized diverse molecules synthesized by *Paracoccidioides* spp., known as surface adhesins, including: gp43 [119, 120] HAD32 [121, 122], a 30-kDa adhesin [123], enolase (ENO) [124], malate synthase (MLS) [125], triosephosphate isomerase (TPI) [126], glyceraldehyde 3-phosphate dehydrogenase (GAPDH) [127] and fructose-1,6-bisphosphate aldolase (ALD) [128]. These surface proteins interact with host cells to promote successful colonization and/or dissemination of the fungus into the host organism, by mediating fungal cell binding to host extracellular matrix components (mainly fibronectin, laminin, fibrinogen, type I and IV collagen, and plasminogen) and to epithelial lung cells. Adhesins play a critical role in the establishment of the infection; lately, it is also reported that their differential expression among *Paracoccidioides* spp. leads to different clinical manifestations [129, 130].

9.6.4 Defenses against Oxidative Stress

Macrophages produce abundant reactive oxygen species (ROS) and reactive nitrogen species (RNS), that along with the release of hydrolytic enzymes and toxic metabolites inside the phagolysosome aim to kill fungal pathogens. However, high throughput analyses [131, 132] and functional studies [133, 134] in *Paracoccidioides* revealed that upon macrophage phagocytosis, the fungus can cope with oxidative stress. In response to H_2O_2, *Paracoccidioides* spp. is able to resist highly oxidative stress not only by generating ROS scavenging enzymes, such as superoxide dismutases, thioredoxins, cytochrome C peroxidase (CCP), and catalases; but also through preventing the protein misfolding or unfolding under long term stress conditions [132]. These data correlate with the transcriptional upregulation of genes encoding peroxisomal catalase and Mn-superoxide dismutase in yeast cells derived from infected macrophages [131]. Moreover, it was shown that in *Paracoccidioides* spp. *ccp*-aRNA strains were more sensitive to mitochondrial-generated ROS stress, suggesting that CCP avoids cell damage caused by oxidative stress, thus playing a critical role in fungal survival inside macrophages. A *ccp* mutant in *C. neoformans* has been shown to reduce intracellular growth when cultured with macrophages [135] Other study have reported *Paracoccidioides* ability to reduce nitric oxide (NO) levels by secreting adhesin gp43, a mannose-rich glycoprotein, that prevents the release of NO from macrophages and stimulates release of IL-10, hence reducing the iNOS expression and its enzymatic activity [136].

9.6.5 Other Virulence Factors

In addition to the described proteins and enzymes, many other microbial determinants have been associated with *Paracoccidioides* spp. virulence, such as biofilm formation, secretion of vesicles and melanin production.

A recent study reported that *Paracoccidioides* spp. can colonize surfaces and form biofilms [137]. This biofilm consisted of a dense network of yeasts cells characterized by a tightly regulated expression of genes encoding adhesins (gp43, GAPDH) and hydrolytic enzymes (aspartyl proteinase), consistent with the adhesion, invasion, and tissue destruction also reported for *Candida albicans* biofilms [138]. Biofilm formation by *Paracoccidioides* spp. is a critical factor in the persistence of the fungal infection, since it hinders the action of antifungal drugs and may contribute to a chronic state of the disease.

Fungal extracellular vesicles, resembling mammalian exosomes have also been reported to act as "virulence bags" [139]. These extracellular compartments composed of lipid bilayers have the potential to regulate key pathogenic steps during fungal infections. In *Paracoccidioides* spp. yeast cells, pioneer studies demonstrated that extracellular vesicles carrying antigenic components bearing highly immunogenic α-linked galactosyl epitopes were recognized by both anti-α-Gal and MOA (*Marasmius oreades* agglutinin) lectin. Additionally, these secreted vesicles could stimulate cytokine expression of macrophages in vitro [140].

Melanin pigments are remarkable and unique substances present in all biological kingdoms, closely associated to virulence in fungal pathogenesis [141]. Fungal melanin is deposited in the cell wall and cytoplasm; particularly in *Paracoccidioides* spp. melanization contributes to virulence by acting as a scavenger of NO and ROS, and through the binding to antifungal drugs, thus changing their activities [142, 143].

9.7 Recognition by Immune Cells and Immunoevasion Strategies

The host cellular immunity, mediated by cells of the innate and adaptive systems, is fundamental to achieve a successful microbial clearance. Initially, dendritic cells and macrophages recognize and present fungal antigens (e.g. chitin, β-glucans, and mannans), known as pathogen-associated molecular patterns (PAMPs), to T-lymphocytes. An effective T-cell response leads the generation of Th1 cytokines, such as tumor necrosis factor (TNF-α) and interferon gamma (IFN-γ), resulting in a classic activation of macrophages to produce ROS and RNS that kill fungi or inhibit their growth [144, 145].

Paracoccidioides spp. exploit several strategies to bypass host defense mechanisms, some of which are described in the following sections.

9.7.1 Shielding of Stimulatory PAMPs

An iconic survival strategy used by dimorphic fungi to avoid recognition by host phagocytic cells' pattern recognition receptors (PRRs) is their quick cell wall turnover during infection, as a consequence of the morphological switching. One PRR,

Dectin-1, present on host phagocytic cells recognizes fungal cell wall β-(1,3)-glucan, triggering phagocytosis, respiratory burst, and release of cytokines such as TNF-α, IL-12, and other interleukins; however, due to the spatial arrangement of the yeast cell wall polysaccharides in *Paracoccidioides* spp. and *H. capsulatum*, α-(1,3)-glucan can actively interfere with these events [115, 116]. Indeed, *H. capsulatum* mutants lacking α-(1,3)-glucan, induced an increased production of proinflammatory TNF-α by macrophages [116]. When *dectin-1*-silenced macrophages were in contact with the *H. capsulatum* mutant depleted of α-(1,3)-glucan, they failed to produce proinflammatory TNF-α, suggesting that α-(1,3)-glucan effectively shields β-(1,3)-glucan from innate immune recognition by the Dectin-1 receptor [116].

9.7.2 Intracellular Survival

Aside from *Paracoccidioides* spp. yeast cell adaptation to high temperatures within the host, the fungus manages to ease its invasion to pulmonary epithelial cells and keratinocytes by affecting host cells cytoskeleton structure, a process that is also promoted by gp43, which participates in the cytokeratin [146, 147]. On the other hand, Cdc42, a fungal Rho GTPase involved in regulating morphogenesis by controlling actin-mediated polarized growth, supports the large, multibudded state of *Paracoccidioides* spp. yeast cells, a morphology that inhibits phagocytosis. RNAi *cdc42* strains were more efficiently phagocytosed by macrophages and displayed decreased pathogenicity [148]. Similarly, studies have shown that *Paracoccidioides* gp43 inhibits the phagocytic and fungicidal capacity of macrophages through binding to mannose receptors and inducing IL-18 production [136, 149].

Host cells inhibit intracellular microbial proliferation by limiting available iron through its sequestration by high-affinity iron-binding proteins such as transferrin and ferritin. Recent proteomic analyses in *Paracoccidioides* spp. demonstrated that under iron deprivation and carbon starvation, proteins involved in alternative carbon metabolism, such as enzymes involved in gluconeogenesis, β-oxidation of fatty acids and amino acids catabolism, were positively regulated, whereas proteins related to glycolysis and protein synthesis were down-regulated [134, 150, 151]. This suggests that under stress conditions similar to that found in the macrophage, *Paracoccidioides* spp. shifts to a starvation mode prioritizing iron-independent metabolic pathways in order to persist in vivo microenvironmental conditions.

Paracoccidioides spp. must also tolerate and overcome stress conditions caused by low oxygen levels. It is well established that in inflamed tissues, oxygen supply is restricted by the high volume of host phagocytic cells or the microbe itself at blood vessels. Recent characterization of *Paracoccidioides* hypoxia response, revealed that members of the genus contain homologues of the key regulator of hypoxia adaptation in fungi, SrbA, a sterol regulatory element binding protein (SREBP) [150]. Functional complementation of an *Aspergillus fumigatus srbA* null mutant by *Paracoccidioides srbA* (*PbsrbA*) restored the null mutant hyphal growth under hypoxia, suggesting that

PbsrbA promotes adaptation to hypoxic microenvironments. Moreover, this study also showed that *Paracoccidioides* SrbA is likely involved in azole drug resistance responses by regulating ergosterol biosynthesis to compensate the effects on membrane fluidity due to low oxygen levels.

The induction or prevention of apoptosis can be a critical step in the development of the infectious processes. It is also reported that *P. brasiliensis* benefits its intracellular survival and further dissemination by modulating programmed cell death of macrophages and epithelial cells. Such strategy is accomplished by the expression of caspase-2, 3 and 8, strongly influenced by the 14-3-3 and gp43 adhesins [152, 153].

9.7.3 Altering T-Cell Repertoire

Proper differentiation and maturation of T-cells occurs in the thymus, thus integrity of the thymic microenvironment is crucial for the maturation of thymocytes. Experimental data have shown that acute infection with *Paracoccidioides* yeast cells promotes thymus invasion, causing its microenvironment atrophy, as consequence of epithelial cell spatial disarrangement and increased gene expression of inflammatory mediators [154, 155]. These studies suggested that impaired differentiation of pathogen-specific T-cells leads to host immunosuppression favoring the fungal ability to thrive and multiply in the thymic microenvironment.

Data summarized in this section intend to show *Paracoccidioides* species remarkable ability to become successful pathogens, which could also be partly explained by recalling the fungi soil-related environment. The precise ecological niche of these fungi remains undefined; however, there is close association with armadillos, disturbed vegetation, and shaded mildly acidic soil, in other words a diverse and harsh microenvironment that has stimulated the fungus evolution to enhance its adaptation and survival mechanisms. The emergence of CRISPR (Clustered regularly interspaced short palindromic repeats) technology might help to increase our understanding of the pathophysiology of these fungi, which could eventually be translated into benefits for patients.

9.8 Treatment

Because of the systemic nature of the mycosis, consideration should be given to patients' general status, target organs and systems, and to the form of presentation (acute/subacute, chronic). Existence of comorbidities, such as tuberculosis, AIDS, and cancer, should be determined and treated accordingly. Treatment should include supportive general measures, such as bed rest, anemia correction, and suspension of tobacco use and alcohol intake, as well as balanced nutrition. Associated clinical complications should also be treated [156, 157].

Antifungal therapy is the mainstay for patient improvement, with three groups of medications being available: sulfonamides, amphotericin B, and azole derivatives [158].

Trimethoprim/sulfamethoxazole (160–240 mg/days / 800–1,200 mg/days) with a high rate of response, but the treatment should be prolonged for 2 or more years, and the probability of relapse is high (20%) [159].

Amphotericin B administered intravenously, is reserved for patients with serious disease, for those who do not tolerate the oral route, for women in their first 12 weeks of pregnancy, and for patients with fourfold increase in hepatic enzymes. Cumulative total dosages vary from 1 to 2 g, with 0.6–1 mg/kg/days of daily doses. Amphotericin B is not fungicidal in vivo, and all patients thus treated should also receive maintenance sulfonamide or azole therapy to avoid relapses [158, 160].

Ketoconazole for oral administration has proven effective, but its hepatotoxicity and multiple interactions with other medications argue against its use.

Among the newly developed triazoles, **itraconazole** 200–400 mg/day for 6 months, is the choice for treatment of patients with both minor and moderate clinical manifestations. It is also recommended for maintenance therapy in severe cases, once the initial amphotericin B course has been concluded. Relapse rates are lower than with other medications (3–5%) [159].

Voriconazole has been used successfully in patients with neuroparacoccidioidomycosis because it has better penetration into the central nervous system than itraconazole and other drugs [161, 162].

Alternative triazoles have been incompletely evaluated in the treatment of paracoccidioidomycosis, although **posaconazole** and **isavuconazole** have exhibited efficacy [163].

References

1. Lutz A (1908) Uma micose pseudo-coccídica localizada na boca e observada no Brasil: Contribucao ao conhecimento das hiphoblastomicoses americanas. Brasil Med 22:121–124
2. Matute DR, McEwen JG, Puccia R et al (2006) Cryptic speciation and recombination in the fungus *Paracoccidioides brasiliensis* as revealed by gene genealogies. Mol Biol Evol 23:65–73
3. Teixeira MM, Theodoro RC, Nino-Vega G et al (2014) *Paracoccidioides* species complex: ecology, phylogeny, sexual reproduction, and virulence. PLoS Pathog 10(10):e1004397. doi:10.1371/journal.ppat.1004397
4. Salazar-Salgado C, Jones LR, Restrepo A et al (2010) The human fungal pathogen *Paracoccidioides brasiliensis* (Onygenales: Ajellomycetaceae) is a complex of two species: phylogenetic evidence from five mitochondrial markers. Cladistics 26:11
5. Carrero LL, Nino-Vega G, Teixeira MM et al (2008) New *Paracoccidioides brasiliensis* isolate reveals unexpected genomic variability in this human pathogen. Fungal Genet Biol 45:605–612
6. Teixeira MM, Theodoro RC, de Carvalho MJ et al (2009) Phylogenetic analysis reveals a high level of speciation in the *Paracoccidioides* genus. Mol Phylogenet Evol 52:273–283
7. Carbonell LM (1969) Ultrastructure of dimorphic transformation in *Paracoccidioides brasiliensis*. J Bacteriol 100:1076–1082

8. Queiroz-Telles F (1994) *Paracoccidioides brasiliensis* ultrastructural findings. In: Franco M, Lacaz CS, Restrepo-Moreno A, del Negro G (eds) Paracoccidioidomycosis. CRC Press, Boca Raton, FL, pp 27–47
9. Bustamante-Simon B, McEwen JG, Tabares AM et al (1985) Characteristics of the conidia produced by the mycelial form of *Paracoccidioides brasiliensis*. Sabouraudia 23:407–414
10. Restrepo A, Salazar ME, Cano LE et al (1986) A technique to collect and dislodge conidia produced by *Paracoccidioides brasiliensis* mycelial form. J Med Vet Mycol 24:247–250
11. Montoya AE, Moreno MN, Restrepo A, McEwen JG (1997) Electrophoretic karyotype of clinical isolates of *Paracoccidioides brasiliensis*. Fungal Genet Biol 21:223–227
12. Molinari-Madlum EE, Felipe MS, Soares CM (1999) Virulence of *Paracoccidioides brasiliensis* isolates can be correlated to groups defined by random amplified polymorphic DNA analysis. Med Mycol 37:269–276
13. Montoya AE, Alvarez AL, Moreno MN, Restrepo A, McEwen JG (1999) Electrophoretic karyotype of environmental isolates of *Paracoccidioides brasiliensis*. Med Mycol 37:219–222
14. Nino-Vega GA, Calcagno AM, San-Blas G, San-Blas F, Gooday GW, Gow NA (2000) RFLP analysis reveals marked geographical isolation between strains of *Paracoccidioides brasiliensis*. Med Mycol 38:437–441
15. Hahn RC, Macedo AM, Santos NL, Resende JC, Hamdan JS (2002) Characterization of *Paracoccidioides brasiliensis* atypical isolates by random amplified polymorphic DNA analysis. Rev Iberoam Micol 19:49–51
16. Hahn RC, Macedo AM, Fontes CJ, Batista RD, Santos NL, Hamdan JS (2003) Randomly amplified polymorphic DNA as a valuable tool for epidemiological studies of *Paracoccidioides brasiliensis*. J Clin Microbiol 41:2849–2854
17. Nascimento E, Martinez R, Lopes AR et al (2004) Detection and selection of microsatellites in the genome of *Paracoccidioides brasiliensis* as molecular markers for clinical and epidemiological studies. J Clin Microbiol 42:5007–5014
18. Taylor JW, Jacobson DJ, Kroken S, Kasuga T, Geiser DM, Hibbett DS, Fisher MC (2000) Phylogenetic species recognition and species concepts in fungi. Fungal Genet Biol 31:21–32
19. Kasuga T, White TJ, Koenig G, McEwen J, Restrepo A, Castaneda E, Da Silva LC, Heins-Vaccari EM, De Freitas RS, Zancope-Oliveira RM et al (2003) Phylogeography of the fungal pathogen *Histoplasma capsulatum*. Mol Ecol 12:3383–3401
20. Koufopanou V, Burt A, Taylor JW (1997) Concordance of gene genealogies reveals reproductive isolation in the pathogenic fungus *Coccidioides immitis*. Proc Natl Acad Sci U S A 94:5478–5482
21. Theodoro RC, Teixeira MM, Felipe MS et al (2012) Genus *Paracoccidioides*: species recognition and biogeographic aspects. PLoS One 7(5):e37694. doi:10.1371/journal.pone.0037694
22. Teixeira Mde M, Theodoro RC, Oliveira FF, Machado GC, Hahn RC, Bagagli E, San-Blas G, Soares Felipe MS (2014) Paracoccidioides lutzii sp. nov.: biological and clinical implications. Med Mycol 52:19–28
23. Machado GC, Moris DV, Arantes TD, Silva LR, Theodoro RC, Mendes RP, Vicentini AP, Bagagli E (2013) Cryptic species of *Paracoccidioides brasiliensis*: impact on paracoccidioidomycosis immunodiagnosis. Mem Inst Oswaldo Cruz 108:637–643
24. Gegembauer G, Araujo LM, Pereira EF, Rodrigues AM, Paniago AM, Hahn RC, Pires de Camargo Z (2014) Serology of paracoccidioidomycosis due to *Paracoccidioides lutzii*. PLoS Negl Trop Dis 8(7):e2986
25. Batista J Jr, de Camargo ZP, Fernandes GF, Vicentini AP, Fontes CJ, Hahn RC (2010) Is the geographical origin of a *Paracoccidioides brasiliensis* isolate important for antigen production for regional diagnosis of paracoccidioidomycosis? Mycoses 53:176–180
26. Queiroz Junior Lde P, de Camargo ZP, Tadano T, Rodrigues AM, Takarara DT, Gegembauer G, Araujo LM, Hahn RC (2014) Serological and antigenic profiles of clinical isolates of *Paracoccidioides* spp. from central western Brazil. Mycoses 57:466–472
27. Andrade ALSS (1983) Lymphatic-abdominal forms of paracoccidioidomycosis. Rev Pat Trop 12:91

28. Barbosa W, Daher R, Oliveira AR (1968) Lymphatic abdominal forms of south American blastomycosis. Rev InstMed Trop Sao Paulo 10:11
29. Scorzoni L, de Paula e Silva AC, Singulani Jde L, Leite FS, de Oliveira HC, da Silva RA, Fusco-Almeida AM, Mendes-Giannini MJ (2015) Comparison of virulence between *Paracoccidioides brasiliensis* and *Paracoccidioides lutzii* using *Galleria mellonella* as a host model. Virulence 6:766–776
30. de Oliveira HC, da Silva Jde F, Scorzoni L, Marcos CM, Rossi SA, de Paulae Silva ACA, Assato PA, da Silva RAM, Fusco-Almeida AM, Mendes-Giannini MJS (2015) Importance of adhesins in virulence of *Paracoccidioides* spp. Front Microbiol 6(303). doi:10.3389/fmicb.2015.00303
31. Desjardins CA, Champion MD, Holder JW, Muszewska A, Goldberg J, Bailao AM, Brigido MM, Ferreira ME, Garcia AM, Grynberg M et al (2011) Comparative genomic analysis of human fungal pathogens causing paracoccidioidomycosis. PLoS Genet 7(10):e1002345
32. Munoz JF, Gallo JE, Misas E, Priest M, Imamovic A, Young S, Zeng Q, Clay OK, McEwen JG, Cuomo CA (2014) Genome update of the dimorphic human pathogenic fungi causing paracoccidioidomycosis. PLoS Negl Trop Dis 8(12):e3348
33. Muñoz JF, Farrer RA, Desjardins CA, Gallo JE, Sykes S, Sakthikumar S, Misas E, Whiston EA, Bagagli E, Soares CMA et al (2016) Diversity, recombination, and virulence across the major lineages of *Paracoccidioides*. mSphere 1(5):e00213–e00216. doi:10.1128/mSphere.00213-16
34. Arantes TD, Theodoro RC, Da Graca Macoris SA, Bagagli E (2013) Detection of *Paracoccidioides* spp. in environmental aerosol samples. Med Mycol 51:83–92
35. Arantes TD, Theodoro RC, Teixeira Mde M, Bosco Sde M, Bagagli E (2016) Environmental mapping of *Paracoccidioides* spp. in Brazil reveals new clues into genetic diversity, biogeography and wild host association. PLoS Negl Trop Dis 10(4):e0004606
36. Kwon-Chung KJ (1972) Sexual stage of *Histoplasma capsulatum*. Science 175:326
37. Mcdonough ES, Lewis AL (1967) Blastomyces dermatitidis: production of the Sextual stage. Science 156:528–529
38. Teixeira Mde M, Theodoro RC, Derengowski Lda S, Nicola AM, Bagagli E, Felipe MS (2013) Molecular and morphological data support the existence of a sexual cycle in species of the genus *Paracoccidioides*. Eukaryot Cell 12:380–389
39. Torres I, Garcia AM, Hernandez O, Gonzalez A, McEwen JG, Restrepo A, Arango M (2010) Presence and expression of the mating type locus in *Paracoccidioides brasiliensis* isolates. Fungal Genet Biol 47:373–380
40. Gomes-Rezende JA, Gomes-Alves AG, Menino JF, Coelho MA, Ludovico P, Goncalves P, Sturme MH, Rodrigues F (2012) Functionality of the *Paracoccidioides* mating alpha-pheromone-receptor system. PLoS One 7(10):e47033
41. Almeida AJ, Matute DR, Carmona JA, Martins M, Torres I, McEwen JG, Restrepo A, Leao C, Ludovico P, Rodrigues F (2007) Genome size and ploidy of *Paracoccidioides brasiliensis* reveals a haploid DNA content: flow cytometry and GP43 sequence analysis. Fungal Genet Biol 44:25–31
42. Colombo AL, Tobón A, Restrepo A et al (2011) Epidemiology of endemic systemic fungal infections in Latin America. Med Mycol 49:785–798
43. Restrepo A, Gómez BL, Tobón A (2012) Paracoccidioidomycosis: Latin America's own fungal disorder. Curr Fungal Infect Rep 6:303–311
44. Martinez R (2015) Epidemiology of paracoccidioidomycosis. Rev Int Med Trop Sao Paulo 57(Suppl 19):11–20
45. Kurai H, Ohmagari N, Ito K et al (2012) A case of oral paracoccidioidomycosis suspected to be pharyngeal cancer. Japanese J Med Mycol 53:49–52
46. Bocca AL, Amaral AC, Teixeira MM et al (2013) Paracoccidioidomycosis: eco-epidemiology, taxonomy and clinical and therapeutic issues. Future Microbiol 8:1177–1191
47. Siqueira IM, Fraga CL, Amaral AC et al (2016) Distinct patterns of yeast cell morphology and host responses induced by representative strains of *Paracoccidiodes brasiliensis* (Pb18) and *Paracoccidioides lutzii* (Pb01). Med Mycol 54:177–188

48. Coutinho ZF, Silva D, Lazera M et al (2002) Paracoccidioidomycosis mortality in Brazil (1980–1995). Cad Saude Publica 18:1441–1454
49. Loth EA, Castro SV, Silva JR et al (2011) Occurrence of 102 cases of paracoccidioidomycosis in 18 months in the Itaipu Lake region, western Paraná. Rev Soc Bras Med Trop 44:636–637
50. Vieira DG, Alves TC, Lima SM et al (2014) Paracoccidioidomycosis in a western Brazilian Amazon state: clinical-epidemiologic profile and spatial distribution of the disease. Rev Soc Bras Med Trop 47:63
51. Bittencourt JIM, Oliveira RM, Coutinho FZ (2005) Paracoccidioidomycosis mortality in the state of Paraná, Brazil, 1980/1998. Cad Saúde Pública 21:1856–1864
52. Coutinho ZF, Wanke B, Travassos C et al (2015) Hospital morbidity due to paracoccidioidomycosis in Brazil (1998–2006). Tropical Med Int Health 20:673–680
53. Prado M, Silva MB, Laurenti R et al (2009) Mortality due to systemic mycoses as a primary cause of death or in association with AIDS in Brazil: a review from 1996 to 2006. Mem Inst Oswaldo Cruz 104:513–521
54. Franco M (1987) Host-parasite relationships in paracoccidioidomycosis. J Med Vet Mycol 25:5–18
55. Restrepo A, Tobón AM, Agudelo CA (2008) Paracoccidioidomycosis. In: Hospenthal DR, Rinaldi MG (eds) Diagnosis and treatment of human mycoses. Humana Press, Totowa, NJ, pp 331–342
56. Marques SA (2013) Paracoccidioidomycosis: epidemiological, clinical, diagnostic and treatment up-dating. Ann Bras Dermatol 88:700–711
57. Restrepo A, Benard G, de Castro CC et al (2008) Pulmonary paracoccidioidomycosis. Semin Resp Crit Care Med 29:182–197
58. Bellissimo-Rodrigues F, Machado AA, Martinez R (2011) Paracoccidioidomycosis: epidemiological features of a 1,000-cases series from a hyperendemic area on the southeast of Brazil. Am J Trop Med Hyg 85:546–550
59. Restrepo A (2000) Morphological aspects of *Paracoccidioides brasiliensis* in lymph nodes: implications for the prolonged latency of paracoccidioidomycosis? Med Mycol 38:317–322
60. Bellissimo-Rodrigues F, Bollela VR, Fonseca BAL et al (2013) Endemic paracoccidioidomycosis: relationship between clinical presentation and patients' demographic features. Med Mycol 51:313–318
61. Martinez R (2010) Paracoccidioidomycosis: the dimension of the problem of a neglected disease. Rev Soc Bras Med Trop 43:480
62. Guglielmino CM, Ivanov ML, Verea MA et al (2012) Paracoccidioidomicosis: Presentación de la casuística de diez años y revisión de la literature. Arch Argent Dermatol 62:92–97
63. Shankar J, Restrepo A, Clemons KV et al (2010) Hormones and the resistance of women to paracoccidioidomycosis. Clinic Microbiol Rev 24:296–313
64. Restrepo A, Robledo M, Ospina S et al (1968) Distribution of paracoccidioidin sensitivity in Colombia. Am J Trop Med 17:25–37
65. Taicz M, Rosanova MT, Bes D et al (2014) Paracoccidioidomycosis in pediatric patients: a description of 4 cases. Rev Iberoam Micol 31:141–144
66. Buccheri R, Khoury Z, Barata LC et al (2016) Incubation period and early natural history events of the acute form of paracoccidioidomycosis: *Paracoccidioides* spp., exposure. Mycopathologia 181:435–439
67. Restrepo A, Tobón A, Cano L (2015) Paracoccidioides brasiliensis. In: Bennett JE, Dolin R, Blaser MJ (eds) Principles and practice of infectious diseases, 8th edn. Elsevier, Philadelphia, pp 2995–3002
68. Nucci M, Colombo AL, Queiroz-Telles F (2009) Paracoccidioidomycosis. Curr Fungal Infect Rep 3:315–320
69. Tobón A, Agudelo C, Osorio ML et al (2003) Pulmonary abnormalities in adult patients with chronic Paracoccidioidomycosis: prolonged follow-up after Itraconazole therapy. Clin Infect Dis 37:894–904
70. Barreto MM, Marchiori E, Amorim VB et al (2012) Thoracic paracoccidioidomycosis: radiographic and CT findings. Radiographics 32:71–84

71. Yamaga LY, Benard G, Hironaka FH et al (2003) The role of gallium-67 scan in defining the extent of disease in an endemic deep mycosis, paracoccidioidomycosis: a predominantly multifocal disease. Eur J Nucl Med Mol Imaging 30:888–894
72. de Abreu ESMA, Salum FG, Figueiredo MA et al (2012) Important aspects of oral paracoccidioidomycosis—a literature review. Mycoses 56:189–199
73. Bonifaz A Paracoccidioidomicosis. In: Bonifaz A (ed) Micología Médica Básica, 4th edn. Mc Graw Hill Americana, Mexico, pp 297–309
74. Amorim Pellicioli AC, Neves-Silva R, Santos-Silva AR, Vargas PA, Ajudarte Lopes M (2015) Lessons from patients with a single synchronous oral paracoccidioidomycosis and pulmonary TB in an immunocompetent patient. Mycopathologia 179:459–464
75. Morejón KM, Machado AA, Martinez R (2009) Paracoccidioidomycosis in patients infected with and not infected with human immunodeficiency virus: a case-control study. Am J Trop Med Hyg 80:359–366
76. Tobón AM, Agudelo CA, Restrepo CA et al (2010) Adrenal function status in patients with paracoccidioidomycosis after prolonged post-therapy follow-up. Am J Trop Med Hyg 83:111–114
77. Restrepo A, Tobón-Orozco AM, Gómez BL, Benard G (2015) Paracoccidioidomycosis. In: Hospenthal DR, Rinaldi M (eds) Diagnosis and treatment of fungal infections, 2nd edn. Springer International Publishing, Cham, pp 225–236
78. de Almeida SM, Queiroz-Telles F, Teivea HAG et al (2004) Central nervous system paracoccidioidomycosis: clinical features and laboratorial findings. J Infect 48:193–198
79. Reis F, Collier PP, Souza TF et al (2013) Neuroparacoccidioidomycosis (NPCM): magnetic resonance imaging (MRI) findings. Mycopathologia 175:181–186
80. Vermelho MB, Correia AS, Michailowsky TC et al (2015) Abdominal alterations in disseminated paracoccidioidomycosis: computed tomography findings. Radiol Bras 48:81–85
81. Lima TB, Custodino-Dominguez MA, Caramori CA et al (2013) Pancreatic paracoccidioidomycosis simulating malignant neoplasia: Case report. World J Gastroenterol 19:5750–5753
82. Sarti EC, de Oliveira SM, dos Santos LF et al (2012) Paracoccidioidal infection in HIV patients at an endemic area of paracoccidioidomycosis in Brazil. Mycopathologia 173:145–149
83. Correa-Riveiro L, Hahn RC, Cometi Favalessa O et al (2009) Systemic mycoses: factors associated with death among patients infected with human immunodeficiency virus, Cuiba, state of Mato Grosso, Brasil 2005-2008. Rev Soc Bras Med Trop 42:698–705
84. Quagliato-Junior R, Grangeia TA, Massucio RA et al (2007) Association between paracoccidioidomycosis and tuberculosis: reality and misdiagnosis. J Bras Pneumol 33:295–300
85. Da Silva RG, Bittencourt Severo C, de Maltos OF et al (2010) Association between paracoccidioidomycoses and cancer. J Bras Pneumol 36:356–362
86. Thomposon G, Gómez BL (2015) Histoplasma, *Blastomyces*, *Coccidioides*, and other dimorphic fungi causing systemic mycoses. In: Versalovic J, Warnock D (eds) Manually of clinical microbiology, vol Vol 2, 11th edn. ASM press, Washington DC, pp 2109–2127
87. Gómez BL (2014) Molecular diagnosis of endemic and invasive mycoses: advances and challenges. Rev Iberoam Micol 31:35–41
88. Teles FRR, Martins ML (2011) Laboratorial diagnosis of paracoccidioidomycosis and new insights for the future of fungal diagnosis. Talanta 85:2254–2264
89. Das S, Larone DH (2011) Basics of molecular methods for fungal identification. In: Larone DH (ed) Medically important fungi: a guide to identification, 5th edn. ASM Press, Washington DC, pp 317–335
90. Nobrega de Almeida J Jr, Del Negro GM, Grenfell RC, Vidal MS, Thomaz DY, de Figueiredo DS, Bagagli E, Juliano L, Benardb G (2015) Matrix-assisted laser desorption ionization–time of flight mass spectrometry for differentiation of the dimorphic fungal species *Paracoccidioides brasiliensis* and *Paracoccidioides lutzii*. J Clin Microbiol 53:1383–1386
91. Lacaz CS, Porto E, Martins JEC, Heins-Vaccari E, de Melo NT Paracoccidioidomicose (2002) In: Lacaz CS, Porto E, Martins JEC, Heins-Vaccari E, de Melo NT (eds) Tratado de Micologia médica Lacaz, 9th edn. Servier, Sao Paulo, pp. 639–729

92. Restrepo A, Gónzalez A, Agudelo CA (2011) Paracoccidioidomycosis. In: Dismukes W, Kauffman C, Pappas P, Sobel J (eds) Essentials of medical mycology, 2nd edn. Springer, New York, pp 367–385
93. San-Blas G, Niño-Vega G, Iturriaga T (2002) *Paracoccidioides brasiliensis* and paracoccidioidomycosis: molecular approaches to morphogenesis, diagnosis, epidemiology, taxonomy and genetics. Med Mycol 40:225–242
94. Da Silva Jde F, de Oliveira HC, Marcos CM, Assato PA, Fusco-Almeida AM, Mendes-Giannini MJ (2016) Advances and challenges in paracoccidioidomycosis serology caused by *Paracoccidioides* species complex: an update. Diagn Microbiol Infect Dis 84:87–94
95. de Camargo ZP (2008) Serology of paracoccidioidomycosis. Mycopathologia 165:289–302
96. Lindsley MD, Warnock DW, Morrison CJ (2006) Serological and molecular diagnosis of fungal infection. In: Rose NR, Hamilton RG, Detrick B (eds) Manual of clinical laboratory and immunology. ASM Press, Washington, pp 569–605
97. Ortiz BL, Díez S, Urán ME, Rivas JM, Romero M, Caicedo V, Restrepo A, McEwen JG (1998) Use of the 27kilodalton recombinant protein from *Paracoccidioides brasiliensis* in serodiagnosis of paracoccidioidomycosis. Clin Diagn Lab Immunol 5:826–830
98. Marques da Silva SH, Colombo AL, Blotta MH, Lopez JD, Queiroz Tellez F, Pires de Camargo Z (2003) Detection of circulating gp43 antigen in serum, cerebrospinal fluid, and bronchoalveolar lavage fluid of patients with paracoccidioidomycosis. J Clin Microbiol 41:3675–3680
99. Díez S, Gómez BL, Restrepo A, Hay RJ, Hamilton AJ (2002) *Paracoccidioides brasiliensis* 87kilodalton antigen, a heat shock protein useful in diagnosis: characterization, purification, and detection in biopsy material via immunohistochemistry. J Clin Microbiol 40:359–365
100. Díez S, Gómez BL, McEwen JG, Restrepo A, Hay RJ, Hamilton AJ (2003) Combined use of *Paracoccidioides brasiliensis* recombinant 27 kilodalton and purified 87kilodalton antigens in an enzymelinked immunosorbent assay for serodiagnosis of paracoccidioidomycosis. J Clin Microbiol 41:1536–1542
101. Vidal MS, Del Negro GM, Vicentini AP, Svidzinski TI, Mendes-Giannini MJ, Almeida AM, Martinez R, de Camargo ZP, Taborda CP, Bernadrd G (2014) Serological diagnosis of paracoccidioidomycosis: high rate of inter-laboratorial variability among medical mycology reference centers. PLoS Negl Trop Dis 8(9):e3174
102. Ferreira I, Ferreira-Strixino J, Castilho ML, Campos CBL, Tellez C, Raniero L (2016) Characterization of *Paracoccidioides brasiliensis* by FT-IR spectroscopy and nanotechnology. Spectrochim Acta A Mol Biomol Spectrosc 152(5):397–403
103. Silva Ferreira C, de Castro Ribeiro EM, Miranda Goes AD, Mello Silva BD (2016) Current strategies for diagnosis of paracoccidioidomycosis and prospects of methods based on gold nanoparticles 2016. Future Microbiol 11:973–985
104. Motoyama AB, Venancio EJ, Brandão GO, Petrofeza-Silva S, Pereira IS, Soares CM, Felipe MS (2000) Molecular identification of *Paracoccidioides brasiliensis* by PCR amplification of ribosomal DNA. J Clin Microbiol 38:3106–3109
105. Buitrago MJ, Merino P, Puente S, Gomes-Lopez A, Arribi A, Zancop'e-Oliveira RM, Gutierrez MC, Rodriguez-Tudela JM, Cuenca-Estrella M (2009) Utility of real-time PCR for the detection of *Paracoccidioides brasiliensis* DNA in the diagnosis of imported paracoccidioidomycosis. Med Mycol 47:879–882
106. Bialek R, Ibricevic A, Aepinus C, Najvar LK, Fothergill AW, Knobloch J, Graybill JR (2000) Detection of *Paracoccidioides brasiliensis* in tissue samples by a nested PCR assay. J Clin Microbiol 38:2940–2942
107. Gaviria M, Rivera V, Muñoz-Cadavid C, Cano LE, Naranjo TW (2015) Validation and clinical application of a nested PCR for paracoccidioidomycosis diagnosis in clinical samples from Colombian patients. Braz J Infect Dis 19:376–383
108. Semighini CP, Camargo ZP, Puccia R, Goldman MH, Goldman GH (2002) Molecular identification of *Paracoccidioides brasiliensis* by 5′ nuclease assay. Diagn Microbiol Infect Dis 44:383–386

109. Endo S, Komori T, Ricci G, Sano A, Yokoyama K, Ohori A, Kamei K, Franco M, Miyaji M, Nishimura K (2004) Detection of gp43 of *Paracoccidioides brasiliensis* by the loop mediated isothermal amplification (LAMP) method. FEMS Microbiol Lett 1:93–97
110. Tatibana BT, Sano A, Uno J, Kamei K, Igarashi T, Mikami Y, Nishimura K, Itano EN (2009) Detection of *Paracoccidioides brasiliensis* gp43 gene in sputa by loop-mediated isothermal amplification method. J Clin Lab Anal 23:139–143
111. Boyce KJ, Andrianopoulos A (2015) Fungal dimorphism: the switch from hyphae to yeast is a specialized morphogenetic adaptation allowing colonization of a host. FEMS Microbiol Rev 39:797–811
112. Nemecek JC, Wuthrich M, Klein BS (2006) Global control of dimorphism and virulence in fungi. Science 312:583–588
113. Bastos KP, Bailao AM, Borges CL et al (2007) The transcriptome analysis of early morphogenesis in *Paracoccidioides brasiliensis* mycelium reveals novel and induced genes potentially associated to the dimorphic process. BMC Microbiol 7:29
114. Chaves AF, Navarro MV, Castilho DG et al (2016) A conserved dimorphism-regulating histidine kinase controls the dimorphic switching in *Paracoccidioides brasiliensis*. FEMS Yeast Res 16(5):pii. doi:10.1093/femsyr/fow047
115. San-Blas G, Vernet D (1977) Induction of the synthesis of cell wall alpha-1,3-glucan in the yeastlike form of *Paracoccidioides brasiliensis* strain IVIC Pb9 by fetal calf serum. Infect Immun 15:897–902
116. Rappleye CA, Eissenberg LG, Goldman WE (2007) *Histoplasma capsulatum* alpha-(1,3)-glucan blocks innate immune recognition by the beta-glucan receptor. Proc Natl Acad Sci U S A 104:1366–1370
117. Marion CL, Rappleye CA, Engle JT, Goldman WE (2006) An alpha-(1,4)-amylase is essential for alpha-(1,3)-glucan production and virulence in *Histoplasma capsulatum*. Mol Microbiol 62:970–983
118. Camacho E, Sepulveda VE, Goldman WE et al (2012) Expression of *Paracoccidioides brasiliensis AMY1* in a *Histoplasma capsulatum amy1* mutant, relates an alpha-(1,4)-amylase to cell wall alpha-(1,3)-glucan synthesis. PLoS One 7(11):e50201. doi:10.1371/journal.pone.0050201
119. Vicentini AP, Gesztesi JL, Franco MF, De Souza W, De Moraes JZ, Travassos LR, Lopes JD (1994) Binding of *Paracoccidioides brasiliensis* to laminin through surface glycoprotein gp43 leads to enhancement of fungal pathogenesis. Infect Immun 62:1465–1469
120. Torres I, Hernandez O, Tamayo D, Munoz JF, Leitao NP Jr, Garcia AM, Restrepo A, Puccia R, Mcewen JG (2013) Inhibition of PbGP43 expression may suggest that gp43 is a virulence factor *in Paracoccidioides brasiliensis*. PLoS One 8:e68434
121. Hernandez O, Almeida AJ, Gonzalez A, Garcia AM, Tamayo D, Cano LE, Restrepo A, Mcewen JG (2010) A 32-kilodalton hydrolase plays an important role in Paracoccidioides *brasiliensis* adherence to host cells and influences pathogenicity. Infect Immun 78:5280–5286
122. Hernandez O, Almeida AJ, Tamayo D, Torres I, Garcia AM, Lopez A, Restrepo A, Mcewen JG (2012) The hydrolase PbHAD32 participates in the adherence of *Paracoccidioides brasiliensis* conidia to epithelial lung cells. Med Mycol 50:533–537
123. Da Silva JF, De Oliveira HC, Marcos CM et al (2013) *Paracoccidoides brasiliensis* 30 kDa adhesin: identification as a 14-3-3 protein, cloning and subcellular localization in infection models. PLoS One 8(4):e62533. doi:10.1371/journal.pone.0062533
124. Nogueira SV, Fonseca FL, Rodrigues ML, Mundodi V, Abi-Chacra EA, Winters MS, Alderete JF, De Almeida Soares CM (2010) *Paracoccidioides brasiliensis* enolase is a surface protein that binds plasminogen and mediates interaction of yeast forms with host cells. Infect Immun 78:4040–4050
125. Da Silva Neto BR, De Fatima Da Silva J, Mendes-Giannini MJ et al (2009) The malate synthase of *Paracoccidioides brasiliensis* is a linked surface protein that behaves as an anchorless adhesin. BMC Microbiol 9:272

126. Da Fonseca CA, Jesuino RS, Felipe MS et al (2001) Two-dimensional electrophoresis and characterization of antigens from *Paracoccidioides brasiliensis*. Microbes Infect 3:535–542
127. Barbosa MS, Bao SN, Andreotti PF et al (2006) Glyceraldehyde-3-phosphate dehydrogenase of *Paracoccidioides brasiliensis* is a cell surface protein involved in fungal adhesion to extracellular matrix proteins and interaction with cells. Infect Immun 74:382–389
128. Chaves EG, Weber SS, Bao SN et al (2015) Analysis of *Paracoccidioides* secreted proteins reveals fructose 1,6-bisphosphate aldolase as a plasminogen-binding protein. BMC Microbiol 15:53
129. Mendes-Giannini MJ, Andreotti PF, Vincenzi LR, Da Silva JL, Lenzi HL, Benard G, Zancope-Oliveira R, De Matos Guedes HL, Soares CP (2006) Binding of extracellular matrix proteins to *Paracoccidioides brasiliensis*. Microbes Infect 8:1550–1559
130. De Oliveira HC, Assato PA, Marcos CM et al (2015) *Paracoccidioides*-host interaction: an overview on recent advances in the Paracoccidioidomycosis. Front Microbiol 6:1319
131. Tavares AH, Silva SS, Dantas A, Campos EG, Andrade RV, Maranhao AQ, Brigido MM, Passos-Silva DG, Fachin AL, Teixeira SM, Passos GA, Soares CM, Bocca AL, Carvalho MJ, Silva-Pereira I, Felipe MS (2007) Early transcriptional response of *Paracoccidioides brasiliensis* upon internalization by murine macrophages. Microbes Infect 9:583–590
132. De Arruda GD, Bailao AM, Vieira Rezende TC, Borges CL et al (2013) Response to oxidative stress in *Paracoccidioides* yeast cells as determined by proteomic analysis. Microbes Infect 15:347–364
133. Parente AF, Naves PE, Pigosso LL, Casaletti L, Mcewen JG, Parente-Rocha JA, Soares CM (2015) The response of *Paracoccidioides* spp. to nitrosative stress. Microbes Infect 17:575–585
134. Parente-Rocha JA, Parente AF, Baeza LC, Bonfim SM, Hernandez O, Mcewen JG, Bailao AM, Taborda CP, Borges CL, Soares CM (2015) Macrophage interaction with *Paracoccidioides brasiliensis* yeast cells modulates fungal metabolism and generates a response to oxidative stress. PLoS One 10:e0137619
135. Giles SS, Perfect JR, Cox GM (2005) Cytochrome c peroxidase contributes to the antioxidant defense of *Cryptococcus neoformans*. Fungal Genet Biol 42:20–29
136. Flavia Popi AF, Lopes JD, Mariano M (2002) GP43 from *Paracoccidioides brasiliensis* inhibits macrophage functions. An evasion mechanism of the fungus. Cell Immunol 218:87–94
137. Sardi Jde C, Pitangui Nde S, Voltan AR, Braz JD, Machado MP, Fusco Almeida AM, Mendes Giannini MJ (2015) In vitro *Paracoccidioides brasiliensis* biofilm and gene expression of adhesins and hydrolytic enzymes. Virulence 6:642–651
138. Ramage G, Coco B, Sherry L, Bagg J, Lappin DF (2012) In vitro *Candida albicans* biofilm induced proteinase activity and *SAP8* expression correlates with in vivo denture stomatitis severity. Mycopathologia 174:11–19
139. Rodrigues ML, Nimrichter L, Oliveira DL, Nosanchuk JD, Casadevall A (2008) Vesicular trans-Cell Wall transport in fungi: a mechanism for the delivery of virulence-associated macromolecules? Lipid Insights 2:27–40
140. Vallejo MC, Matsuo AL, Ganiko L, Medeiros LC, Miranda K, Silva LS, Freymuller-Haapalainen E, Sinigaglia-Coimbra R, Almeida IC, Puccia R (2011) The pathogenic fungus *Paracoccidioides brasiliensis* exports extracellular vesicles containing highly immunogenic alpha-Galactosyl epitopes. Eukaryot Cell 10:343–351
141. Gomez BL, Nosanchuk JD (2003) Melanin and fungi. Curr Opin Infect Dis 16:91–96
142. Da Silva MB, Marques AF, Nosanchuk JD et al (2006) Melanin in the dimorphic fungal pathogen *Paracoccidioides brasiliensis*: effects on phagocytosis, intracellular resistance and drug susceptibility. Microbes Infect 8:197–205
143. Baltazar LM, Werneck SM, Soares BM et al (2015) Melanin protects *Paracoccidioides brasiliensis* from the effects of antimicrobial photodynamic inhibition and antifungal drugs. Antimicrob Agents Chemother 59:4003–4011

144. Gauthier G, Klein BS (2008) Insights into fungal morphogenesis and immune evasion: fungal conidia, when situated in mammalian lungs, may switch from mold to pathogenic yeasts or spore-forming spherules. Microbe Wash DC 3:416–423
145. Thind SK, Taborda CP, Nosanchuk JD (2015) Dendritic cell interactions with *Histoplasma* and *Paracoccidioides*. Virulence 6:424–432
146. Mendes-Giannini MJ, Hanna SA, Da Silva JL, Andreotti PF, Vincenzi LR, Benard G, Lenzi HL, Soares CP (2004) Invasion of epithelial mammalian cells by *Paracoccidioides brasiliensis* leads to cytoskeletal rearrangement and apoptosis of the host cell. Microbes Infect 6:882–891
147. Puccia R, Vallejo MC, Matsuo AL, Longo LV (2011) The *Paracoccidioides* cell wall: past and present layers toward understanding interaction with the host. Front Microbiol 2:257
148. Almeida AJ, Cunha C, Carmona JA et al (2009) Cdc42p controls yeast-cell shape and virulence of *Paracoccidioides brasiliensis*. Fungal Genet Biol 46:919–926
149. Dias-Melicio LA, Fernandes RK, Rodrigues DR et al (2015) Interleukin-18 increases TLR4 and mannose receptor expression and modulates cytokine production in human monocytes. Mediat Inflamm 2015:236839. doi:10.1155/2015/236839
150. Parente AF, Bailao AM, Borges CL, Parente JA, Magalhaes AD, Ricart CA, Soares CM (2011) Proteomic analysis reveals that iron availability alters the metabolic status of the pathogenic fungus *Paracoccidioides brasiliensis*. PLoS One 6:e22810
151. Lima Pde S, Chung D, Bailao AM, Cramer RA, Soares CM (2015) Characterization of the *Paracoccidioides* hypoxia response reveals new insights into pathogenesis mechanisms of this important human pathogenic fungus. PLoS Negl Trop Dis 9:e0004282
152. Silva Jde F, Vicentim J, Oliveira HC, Marcos CM, Assato PA, Andreotti PF, Silva JL, Soares CP, Benard G, Almeida AM, Mendes-Giannini MJ (2015) Influence of the *Paracoccidioides brasiliensis* 14-3-3 and gp43 proteins on the induction of apoptosis in A549 epithelial cells. Mem Inst Oswaldo Cruz 110:476–484
153. Vericimo MA, Franca KM, Arnholdt AC, Kipnis TL (2006) Increased apoptosis during the early phase of experimental paracoccidioidomycosis as a phenotypic marker of resistance. Microbes Infect 8:2811–2820
154. Brito VN, Souto PC, Cruz-Hofling MA et al (2003) Thymus invasion and atrophy induced by *Paracoccidioides brasiliensis* in BALB/c mice. Med Mycol 41:83–87
155. Di Gangi R, Alves Da Costa T, Thome R et al (2016) Paracoccidioides brasiliensis infection promotes thymic disarrangement and premature egress of mature lymphocytes expressing prohibitive TCRs. BMC Infect Dis 16:209
156. Shikanai-Yasuda MA, Telles Filho F d Q, Mendes RP et al (2006) Consenso em paracocidioidomicoses. Rev Soc Bras Med Trop 39:297–310
157. Froes H, Coligiorne RB (2011) Fundamentals of paracoccidioidomycosis treatment. Drug Develop Res 72:528–537
158. Lorenzetti Bocca A, Correa Amoral A, Teixeira MM et al (2013) Paracoccidioidomycosis: eco-epidemiology, taxonomy and clinical and therapeutic issues. Future Microbiol 8:1177–1191
159. Shikanai-Yasuda MA, Benard G, Higaki Y et al (2002) Randomized trial with itraconazole, ketoconazole and sulfadiazine in paracoccidioidomycosis. Med Mycol 40:411–417
160. Benard G, Campos AF, Netto LC et al (2012) Treatment of severe forms of paracoccidioidomycosis: is there a role for corticosteroids? Med Mycol 50:641–648
161. Lutzor I, Roffey S, Troke P (2003) Voriconazole concentration in the cerebrospinal fluid and brain tissue of guinea pigs and immunocompromised patients. Clin Infect Dis 37:728–732
162. Queiroz-Telles F, Goldani LZ, Schlamm HT et al (2007) An open-label comparative pilot study of oral voriconazole and itraconazole for long-term treatment of paracoccidioidomycosis. Clin Infect Dis 45:1462–1469
163. Thompson GR, Rendon A, dos Santos R et al (2016) Isavuconazole treatment of Cryptococcosis and dimorphic mycoses. Clin Infect Dis 63:356–362

Chapter 10
Sporothrix and Sporotrichosis

Leila M. Lopes-Bezerra, Héctor M. Mora-Montes, and Alexandro Bonifaz

Abstract Sporotrichosis is the mycosis due to implantation of greatest global distribution. It is caused by a complex of thermodymorphic fungi of cryptic species, of which the most frequent and important are *Sporothrix schenckii* and *Sporothrix brasiliensis*. These fungi live in nature, matter in decomposition and several plants. The route of entry usually involves traumatic inoculation, but sometimes it could be by the respiratory route. In general, they produce a condition confined to the subcutaneous cellular tissue and, to a lesser extent, affect various viscera, such as the lungs and central nervous system. Clinically, it is a polymorphic condition with various manifestations depending mainly on the patient's immune status and strain virulence. In cases associated with immunosuppression, cutaneous and extracutaneous features may occur. The main clinical form is lymphangitic cutaneous sporotrichosis. In this chapter, the authors discuss the biological aspects of the main pathogenic species of the *Sporothrix schenckii* complex, including virulence factors, clinical-epidemiological aspects, diagnosis, and treatment.

L.M. Lopes-Bezerra, Ph.D.
Laboratory of Cellular Mycology and Proteomics, Biology Institute, University of Rio de Janeiro State (UERJ), Rio de Janeiro 20550-013, Brazil
e-mail: lmlb23@globo.com

H.M. Mora-Montes, Ph.D.
Departamento de Biología, División de Ciencias Naturales y Exactas, Campus Guanajuato, Universidad de Guanajuato,
Noria Alta s/n, col. Noria Alta, C.P, 36050 Guanajuato, Gto., Mexico
e-mail: hmora@ugto.mx

A. Bonifaz, Ph.D. (✉)
Department of Mycology and Dermatology Service, Hospital General de México, "Dr. Eduardo Liceaga", Dr. Balmis 148, Col Doctores, CP 06720 Mexico, DF, Mexico
e-mail: a_bonifaz@yahoo.com.mx

H.M. Mora-Montes, L.M. Lopes-Bezerra (eds.), *Current Progress in Medical Mycology*, DOI 10.1007/978-3-319-64113-3_10

10.1 Definition

Sporotrichosis is a worldwide-distributed human and animal mycosis caused by implantation; however, cutaneous disseminated and extracutaneous forms also are reported [1]. The causative agent usually affects skin and subcutaneous tissues and less frequently the invasion of deep tissues, such as lungs and other viscera [1–4]. The disease often undergoes a subacute or chronic phase and is caused by dimorphic fungi belonging to the *Sporothrix schenckii* complex [1, 4, 5]. These organisms generally affect the skin and lymphatic ganglions and may cause problems in bones and joints or, in rare occasions, to the lungs and other deep organs. This mycosis has been traditionally named the rose picker's disease or the gardener's disease [1, 4, 6–10]. Zoonotic transmission is mainly observed in Latin America [10]. The zoonosis and sapronosis are currently related to the distinct species of *Sporothrix*.

10.2 Etiology

As mentioned, the causative agents are thermodymorphic organisms of the *Sporothrix* genus. Originally, the disease was related to infection with *S. schenckii*; however, molecular and phenotypical studies have revealed that this is in fact a group of cryptic, phylogenetically closely related cryptic species, named the *Sporothrix schenckii* complex [5]. The complex is composed of at least four pathogenic species, including *S. schenckii sensu stricto, Sporothrix globosa*, *Sporothrix brasiliensis*, and *Sporothrix lurei*. The rest of the complex members include *Sporothrix mexicana* and *Sporothrix pallida*, which are a less common cause of the disease; most of the isolates are considered a nonpathogenic environmental clade [1, 5, 6, 10–12]. No teleomorphic phase has been reported for any of the complex members, and molecular studies indicate that they are not related to the species of the genus *Ophiostoma*. The species *S. schenckii s. str.* is regarded as the most important of the complex, with a worldwide distribution and accounting for 80% of the cases reported. Infections caused by *S. brasiliensis* are restricted to Brazil, and *S. globosa* is mainly reported in Asia [1, 10, 13–15].

Members of the *S. schenckii* complex are thermo-dimorphic organisms and share most of the morphological and biochemical traits, making a complicated task the establishment of the species by traditional means [1, 4]. In *S. schenckii s. str.*, the conidium size is bigger (5–10 μm) and generates big blastoconidia composed of pigmented, pseudoparenchymatous mycelia [1, 4, 7]. Based on molecular studies conducted by Japanese groups, the analysis of mitochondrial DNA allowed to propose two *S. schenckii* groups: A and B; the former is widely distributed in America and the latter is limited to Japan and other areas of Asia [16]. The use of the DNA sequence of the encoding genes for chitin synthase, ß-tubulin and calmodulin, helped to classify the species in five clades: *S. brasiliensis* in clade I, *S. schenckii s.*

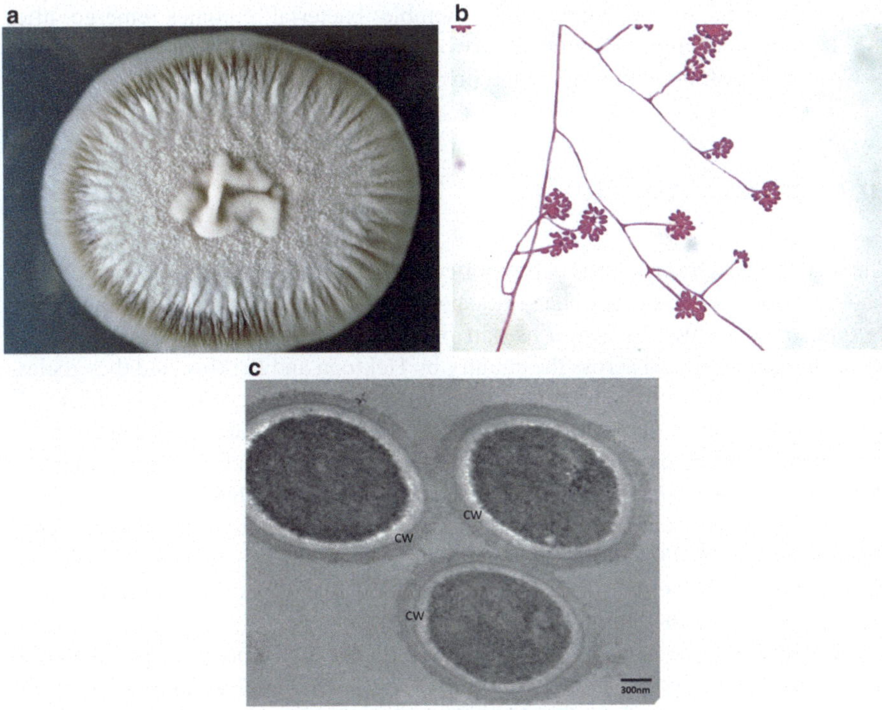

Fig. 10.1 (**a**) Culture of *S. schenckii*, on Saboraud dextrose agar, 10 days at 28 °C, (**b**) *S. schenckii*, sympodial microconidia (filamentous phase), like a "daisy flowers"; (**c**) Transmission electron microscopy showing the cell wall (CW) of *S. brasiliensis* yeast cells

str. in clade II, *S. globosa* in clade III, *S. mexicana* in clade IV, and *S. pallida* (formerly known as *S. albicans*) in clade V [1, 5, 10, 17–21].

The morphology of the different members of the complex is similar when cultured at 28 °C in Sabouraud-dextrose agar, YPD agar, or potato-dextrose agar (Fig. 10.1a). Under microscopic inspection, it is easy to observe thin, hyaline, and septate hyphae, 1–3 μm in diameter, that usually produce branches (Fig. 10.1b). The anamorphic phase generates microconidia that adopt an ovoid, round, elongated, or piriform shape (the predominant morphology may vary depending on the species of the complex). These microconidia can be generated by two ways: Ovoid or pyriform microconidia, derived from the denticle (sympudolic) emerge from a conidiophore of 10–30 μm in length, in an arrangement that resembles a "peach flower or daisy flower" (Fig. 10.1b); and sessile conidia that directly emerge from hypha, which are usually named microaleurioconidia or raduloconidia. After detachment, conidia become thicker and tend to be grouped, generating the melanic pigment of the fungal colony [1, 2, 4, 6, 22].

The dimorphism is usually triggered at 37 °C in rich media, such as blood agar, chocolate agar, and BHI agar. The generation of yeast-like cells (Fig. 10.1c) also can be stimulated by growing the fungus in presence of 5% CO_2. On plate, white-yellowish, soft colonies that resemble bacterial colonies emerge after 3–5 days of incubation. They contain ovoid or enlarged yeast cells approximately 2–4 × 3–6 μm, along with fragmented mycelia [1, 2, 4, 6, 23].

10.3 Historic Aspects

The first case of sporotrichosis was reported in 1898 in the United States by Schenck who described a classic lymphangitic sporotrichosis. The organism rescued from lesions was classified by Smith as part of the fungal genus *Sporotrichum*. Other cases were documented across the country by Hektoen and Perkins, and they reclassified this fungus within the genus *Sporothrix* and described the new species *S. schenckii*. In France, de Beurmann isolated a pigmented strain, different from the etiological agent of the American cases already reported, and it was classified by Matruchot as *Sporotrichum* variety *beurmanni*. In 1962, Michael described the differences of the two genera and confirmed *Sporothrix schenckii* as the sole species of this genus [1, 2, 7, 23]. Simon documented the first epidemic in 1947, which took place in South Africa [2, 15]. The yeast parasitic phase was described in Brazil by Lutz and Spendore in 1947 [13].

Most recently, the work of Guarro's group (2006–2008) has reshaped the etiology of this disease, proposing the *Sporothrix schenckii* complex, a group of cryptic species, whereas the work by Sasaki et al. on the chromosomal polymorphisms has helped to explain the genetic diversity of members of the *Sporothrix* complex [1–4, 11, 69].

10.4 Epidemiology

Geographic distribution. The sporotrichosis is a worldwide-distributed mycosis caused by implantation [10, 15] and has been reported in all the continents, although the cases tend to be concentrated in specific areas. Example of this is the epidemic outbreak reported in 1947 in Transvaal, South Africa, where 3,000 cases with lesions in lungs and skin were documented; most of them were related to contaminated wood inside a mine [15, 24]. Other endemic areas with high burden of cases are China (northeast, Jillin province, south of Guangdong and across the Yangtze River banks) [17, 25], Japan (Sendai and Nagasaki provinces) [26, 27], India (sub-Himalayan and Kangra regions), and Australia (the southeast coast and the coast of New South Wales) [14, 28, 29]. However, the main number of cases are found in America (in the area limited by the tropics), with particular emphasis in Southeast Brazil (where the most important zoonotic outbreak of sporotrichosis transmitted by domestic cats (*Felis catus*) has been reported [1, 9, 15, 30–33]; Peru (Andes region and Abancay) [34–36], Mexico (Jalisco and Puebla) [2, 4, 6, 37–41], Colombia (Cundinamarca and Boyaca provinces) [42], Venezuela (Aragua and Miranda provinces) [43], Uruguay [43, 44], and Guatemala [44, 45]. There are few reports of this infection in Europe, and most of them are immigrants or patients who previously visited endemic areas. France, Italy, and Spain are the countries that gather most of the reported cases of sporotrichosis [4, 14]. In United States, there also are reports of outbreaks related to working with sphagnum moss [13].

Source of infection. Sporotrichosis has been regarded as an occupational disease, in particular of farmers, housekeepers, children, people in close contact with flowers, veterinaries, hunters, miners, fishermen, etc. [1–4, 6, 22, 37].

Sporothrix spp. are usually found in Tropical and Subtropical areas with humid and template climates, temperatures between 20 °C and 30 °C, and a relative humidity over 80%; however, there are reports of strains showing thermotolerance to higher temperatures and capable of surviving in cold weathers, such as those isolated from Jilin, China (collected from regions with 2–6 °C of environmental temperature). *Sporothrix* spp. usually grows in a wide range of pH (3.5–9.6). Although the disease is found throughout the year, most of the affected population refers contact with the pathogen during the fall and winter seasons [1, 2, 4, 6, 17]. This interesting observation can be associated with the end of the rainy season (in Mexico, for example), where the optimum humidity and temperature for the fungal growth are reached [4, 6]. In addition, during this time of the year, the chances to have contact with plant biomass increases due to the harvesting of vegetables and flowers with ornamental purposes. *Sporothrix* has been even isolated from the roots and stalk of maize plant in China [17]. *Sporothrix* spp. as a saprophyte lives in soil with abundance in cellulose, associated with vegetal detritus and decaying wood. The fungus was isolated directly from soil and grass [1, 4, 46, 47]. In the United States, these organisms have been isolated from sphagnum moss used for gardening and to grow Bonsai trees [48].

The disease also has been documented after traumatic interactions with rodents, such as rats, mice, or squirrels, insects, reptiles, spiders, and bats [1, 2, 4, 6, 49]. It is important to stress that cats have a natural susceptibility to be infected by these organisms, which makes likely the establishment of disseminated cases with fatal outcomes [50–57]. The cat to human transmission of sporotrichosis is a zoonosis, and the fungus has been isolated from claws, lesions, and the animal habitat. Thus far, the biggest epidemic outbreak in animals has been reported in Rio de Janeiro and has spread to the southern regions of Brazil. Domestic cats and dogs have been affected since 1998, and it is estimated that approximately 10,000 cats and 200 dogs have been infected. As expected, this has been extended to humans, with more than 4,000 cases of sporotrichosis diagnosed only in one health center. The main etiologic agent associated with this zoonotic outbreak is *S. brasiliensis* [57]. However, all of these numbers are an underestimation of the problem, as sporotrichosis is not an infection of compulsory notification. In other regions of the globe, there also are reports of the disease transmitted by cats but associated with *S. schenckii s. str.*, which indicates that the feline is susceptible to the *Sporothrix* genus rather than to one species in particular.

Route of entry. The fungus usually penetrates tissues after a traumatic lesion, i.e., requires loss of continuity in the epithelial and mucosal tissues. Although rare, it could be an airborne disease, establishing the primary infection in lungs. The incubation period varies but tends to be 3 weeks in the cutaneous cases. When affecting the lungs, the incubation period is difficult to establish, because it depends on the size of the inoculum and the patient's immune status [1–4, 6, 7–10]. The gender distribution of the disease is similar, with a ratio 1:1 in most studies, or with a slight bias towards being more associated with males. However, data from the epidemic outbreak in Brazil [57, 58] indicate an increased number of cases associated with females and elderly people. The most likely explanation is that this population is usually in closer contact with domestic cats. There is not a specific age period to acquire the disease; it is found from kids to elderly people, but the increased number of cases often is associated with occupational age: children attending school (age 5–15 years, represent 30% of cases) and young adults (age 16–35 years, representing 50% of cases) [40, 41, 59]. There are some exceptions; for example, in the mountains region in Peru and Guerrero (Mexico), there are hyperendemic areas where more than 60% of cases are children younger than 15 years old [35, 36, 60].

Predisposing factors. Besides those already mentioned and related to the occupational risk, other factors, such as diabetes mellitus, chronic alcohol abuse, cancer, and chronic treatment with steroids are usually associated with the disseminated disease, with a fatal outcome. The real incidence and prevalence of the disease is unknown; as mentioned, this is not an infection of compulsory notification worldwide. In Mexico, there are not real figures about the incidence of the disease, but Mayortga et al. [37] has reported more than 1,000 cases in Jalisco state, whereas an incidence of 25 cases per 1,000 habitants in the highlands in Puebla state has been documented. The 53.2% of this population has a positive reactivity to sporotrichin, which is similar to the hyperendemic region for sporotrichosis in Acambay, Peru [35, 60, 61].

10.5 Pathogenesis

The cutaneous disease begins with the traumatic inoculation of the causative agent, generating in approximately 10 days a cutaneous-lymphatic complex or sporotrichotic chancre in the infection site. It is possible to have auto-limitation and spontaneous resolution of the infection, or it can spread and generate nodular and progressive lesions localized in the regional lymphatic vessels that stop in the main lymphatic ganglion. Alternatively, it could be established as a sole lesion that grows slowly with a verrucous appearance. The disseminate disease (cutaneous and deep seated) is associated with temporal or permanent immunodeficiency [1–4, 6–10].

Some groups have tried to correlate the clinical form with the patient's immunological status; however, there is a study from Zhang et al. [62] where a strain from a disseminated cutaneous disease was compared with organisms from fixed cutaneous or cutaneous-lymphangitic diseases. The authors found that the strain from the disseminate disease showed genotype variations, with a deletion of 10 pb in the NTS ribosomal region, suggesting that the virulence of the strains can account for the capability of causing the disseminate infection. Moreover, an increased virulence in *S. brasiliensis* strains isolated from chronic cases of disseminate sporotrichosis has been demonstrated [10, 63]. Also, reports indicate a relationship between the strain genotype and the clinical form, but no formal confirmation has been obtained [18]. The primary lung disease, acquired by aspirating the conidia, is asymptomatic in most cases. If the fungal load is big or the host's immune system is altered, it is likely to develop classic symptoms of pneumonia, similar to those found in tuberculosis [64, 65].

The three main pathogenic species, *S. schenckii s. str*, *S. brasiliensis*, and *S. globosa*, exhibit significant differences in virulence when assayed in animal models [66–68]. The virulence factors already known of *Sporothrix* spp. are the fungal dimorphism, thermotolerance, melanin, production of extracellular enzymes (proteases and phosphatases), adhesion to host tissues and antigenic molecules, such as the Gp60-Gp70 antigen, and also the peptidorhamnomannnan sugar epitopes [1, 2, 4, 18, 61, 64–73].

The cell-mediated immune response is the key to control sporotrichosis, participating lymphocytes T $CD4^+$, dendritic cells, macrophages, and neutrophils. These cells are responsible for establishing a granuloma, which often is suppurative and in rare occasions tuberculoid. Using immunohistochemistry, it has been possible to demonstrate that lymphocyte produce INF-γ, a classic marker of a Th1-polarized immune response that activates macrophages. Other studies suggest the recognition of the fungal cells via the receptor TLR4, stimulating the secretion of proinflammatory cytokines, such as IL-6 and TNF-α. In summary, for the establishment of a protective anti-*Sporothrix* immune response, both the Th1 and Th2 branches of the immune system are required [74–79]. Moreover, humanized antibodies raised against the Gp 60-70 antigen are protective to mice infected with either *S. schenckii* or *S. brasiliensis* [52, 53].

10.6 Clinical Aspects

Sporotrichosis is a subcutaneous infection with great clinical polymorphism. The main forms of the disease are:

10.6.1 Cutaneous-Lymphangitic Sporotrichosis

This is the most frequent form of the disease, accounting for 65–85% of all reports. It often affects upper limbs and, to a lesser extent, lower limbs, shoulders, and the face. It develops within the inoculation area, forming a cutaneous-lymphatic complex or sporotrichotic chancre (Fig. 10.2). The affected area usually shows swollen tissue, erythema, and the lesions are asymptomatic or with minimal erythema. After 2 to 3 weeks of the initial lesion, patients usually develop ulcerative nodules in

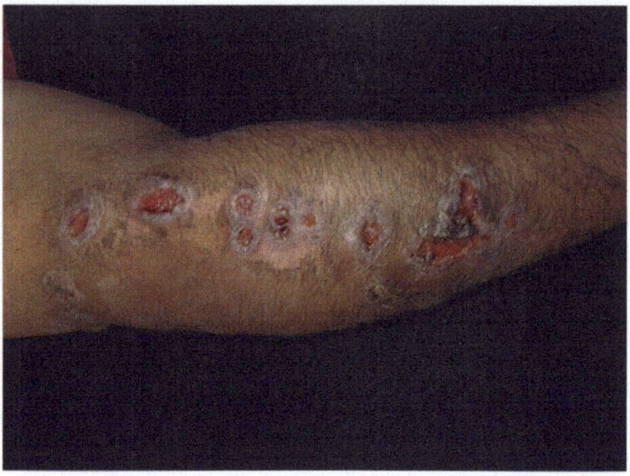

Fig. 10.2 Cutaneous lymphangitic sporotrichosis, nodular and ulcerative lesions

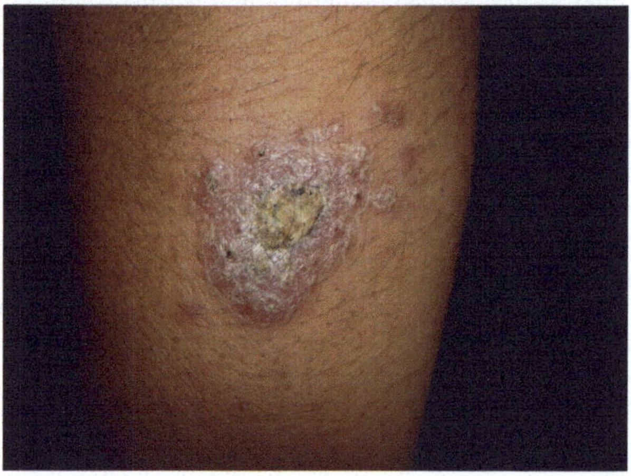

Fig. 10.3 Fixed cutaneous sporotrichosis, chronic verrucose plaque

lineal and progressive form, affecting regional lymphatic vessels and ganglions [2, 4, 6, 8, 22]. The nodules are covered with blood-containing crusts and develop an erythematosus-violaceous halo. In chronic situations, they could either progress to auto-resolution of the disease or develop more nodules. When inoculation occurs in multiple sites in the same area, it is possible to generate a mycetoma-like lesion with lymphostasis and elephantiasis [1, 2, 4, 8]. It is noteworthy that in children the facial topography of the disease occurs in approximately 40% of cases, either with unilateral or bilateral manifestations [4, 23, 31, 40, 41, 59, 79].

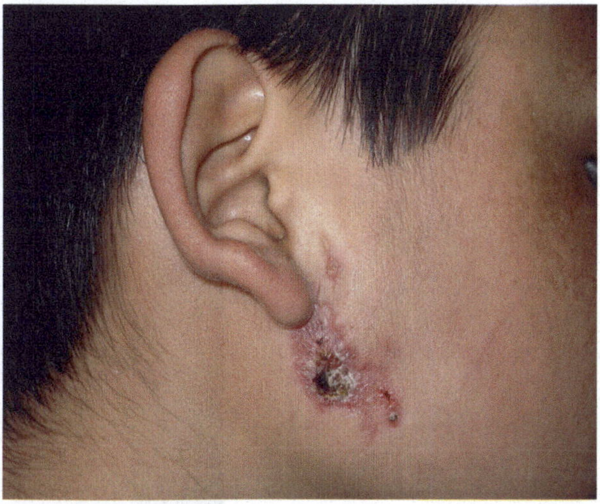

Fig. 10.4 Verrucous plaque of fixed facial cutaneous sporotrichosis (child)

10.6.2 Fixed Cutaneous Sporotrichosis

This also is named vegetative form and is the chronic manifestation of the disease. It is found in 25% of cases; however, in some countries, such as Japan, Korea, and Costa Rica, it is the most frequent manifestation of the infection and is present in approximately 60% of patients [1, 4, 6, 26, 27]. The lesion is limited to the inoculation area and looks like a sporotrichotic chancre that grows slowly, with a verrucous shape, well-delimited borders, and an erythematosus-violaceous color (Figs. 10.3 and 10.4). The lesion is usually covered with flakes and is asymptomatic. It is regarded as a disease in patients with good immunological constitution; therefore, it has good prognosis and in most cases is spontaneously resolved [4, 6, 23, 31, 41].

10.6.3 Disseminated Cutaneous Sporotrichosis

This is the third most prevalent form of the disease and is found in 1–5% of all cases. It could start as a lymphangitic form in immunosuppressed patients and is mostly reported in noncontrolled, diabetic patients, chronic alcohol abusers, patients with hematological malignances (lymphomas and leukemia), pregnancy, AIDS, and patients with prolonged exposure to systemic steroids. The lesions are ulcerated, necrotic, and tend to form verrucous plaques (Fig. 10.5). They can be found in any part of the skin and mucosal tissue, such as mouth, pharynx, nose, and genitals [2, 4, 6, 80–88].

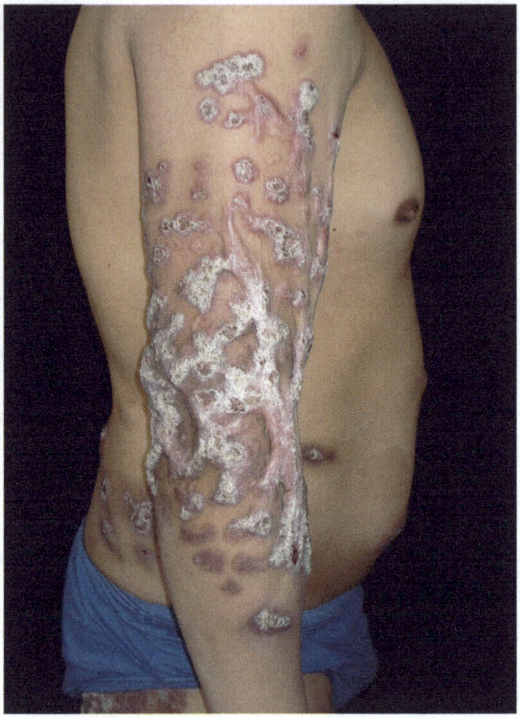

Fig. 10.5 Cutaneous disseminated sporotrichosis associated with chronic alcoholism

10.6.4 Pulmonary Sporotrichosis

This is a rare form, with only a few reports in the literature from hyperendemic areas, and is accompanied by immunosuppression. This form is divided in two groups: the most frequent is the chronic disease, which is asymptomatic (98% of cases), and similar to pulmonary tuberculosis, i.e., auto-limited with cavitary areas. Patients with the symptomatic form have pneumonia, with moderate coughing and expectoration, and the radiological inspection shows condensation areas and miliar infiltrates. The second group is the acute and progressive form that affects the hilar and tracheobronchial lymphangitic ganglions. These patients present massive adenopathy, symptomatology associated with weight loss, productive coughing, dyspnea, and cachexy. The radiological inspection is accompanied with hilar adenopathies and mediastinal widening [1, 2, 64, 65, 92, 93].

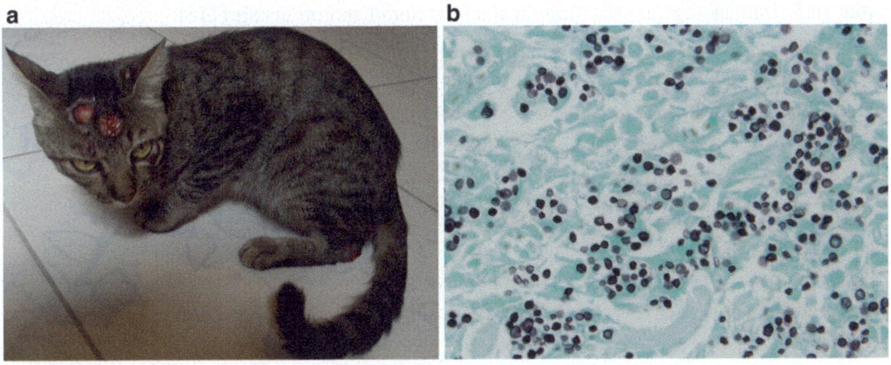

Fig. 10.6 (**a**) Feline sporotrichosis with multiple ulcers; (**b**) Multiple yeasts of *S. schenckii*, from feline sporotrichosis (Grocott, 40×) (Courtesy of Elisabete Aparecida da Silva and Hildebrando Montenegro, part of the Brazilian epidemic)

10.6.5 Extracutaneous Sporotrichosis

From these cutaneous lesions, the fungus can disseminate to other tissues or organs, which is associated with a bad prognosis. Lungs, bones, and joint are frequently affected (especially those found in the elbows and knees where osteomyelitis is generated), as well as the central nervous system, where meningitis caused by sporotrichosis is frequently fatal. The systemic form of the disease is rare and has been mainly associated with HIV infection (phase C) and chronic alcoholism [4, 8, 85, 88–91]. The disseminated or systemic cases can be found in immunosuppressed patients, which have fungaemia and fungal spreading to deep-seated organs, such as liver, spleen, testicles, and the central nervous system. The neurological infection has been exclusively associated with AIDS patients, with a fatal outcome [1, 2, 33, 85, 91–98]. In rare occasions, the infection can affect eyes and cardiac tissue [1, 99].

10.7 Veterinary Sporotrichosis

The most susceptible animals to *Sporothrix* spp. are cats (*Felis catus*) and to lesser degree dogs (ratio 30:1 for cats and dogs, respectively). Both animals are capable of transmitting the disease to humans. Most of the animal cases are caused by *S. brasiliensis* as discovered in the epidemic zoonotic outbreak in Brazil, as already commented [57]. *S. schenckii s. str.* also can infect these animals, and anecdotic cases of infection in horses, cows, goats, camels, pigs, birds, rats, mice, armadillo, and chimpanzee have been documented [1, 3, 50–58, 100, 101]. In cats and dogs, the clinical forms of the disease can be similar to those described in humans; the disseminated and systemic forms are the most frequent in cats [31]. Male felines that fight for breeding are mostly affected, because the bites and scratching can inoculate the

Table 10.1 Immunological classification and behavior of sporotrichosis [118]

	I. Hyperergic and normergic cases	II. Anergic and hypoergic cases
Clinical features	– Cutaneous lymphatic – Cutaneous fixed	– Hematogenous – Cutaneous-disseminated – Osteoarticular – Pulmonary/visceral
Yeast forms in direct examination and in tissues	Rare	Frequent
Histopathology	Suppurative granuloma	Unspecific granuloma
Sporotrichin (IDR)	Always positive	Frequently negative
Resistance	High	Low
Prognostic	Good	Bad
Spontaneous regression	Possible	Never
Treatment response	Good	Bad

fungal cells [3, 51, 55]. For this reason, most of the lesions are usually found in the head, nose, and ears and are ulcerative nodules with crust. They usually develop fistulas with production of purulent material (Fig. 10.6). Most cases have affectation of the mucosal tissue (nasal, oral, and genital), fever, anorexia, and cachexy [1, 3, 51–58, 100].

In the systemic or disseminated cases, the organs that often are affected include the liver and lungs, and to a lesser extent, the spleen, kidneys, testicles, eyes, and central nervous system. The laboratory examination of animals usually reports anemia, eosinophilia, and neutropenia [1, 3, 58, 100, 102] (Table 10.1).

10.8 Differential Diagnosis

Cutaneous-lymphangitic sporotrichosis: cutaneous tuberculosis, syphilis, mycetoma, tularemia, chromoblastomycosis, pyogenic infections, tuberculoid leprosy, and infections caused by nontuberculosis mycobacteria. The infiltrated lesions with keloid scars can be confused with lacaziosis (lobomycosis), whereas the acne-like forms can be confused with acne and cryptococcosis [2, 3, 4, 6–9].

Fixed cutaneous sporotrichosis: verrucous tuberculosis, chromoblastomycosis, leishmaniasis, squamous carcinoma, impetigo, infections caused by nontuberculosis mycobacteria, and Orf disease [4, 6, 7].

Disseminated cutaneous sporotrichosis: tuberculosis, syphilis, coccidioidomycosis, and infections caused by nontuberculosis mycobacteria [6–9, 103–106].

Pulmonary sporotrichosis: pneumonia, tuberculosis, aspergilloma, and other diseases caused by accumulation of fungal material [1, 62, 23].

Extracutaneous sporotrichosis: histoplasmosis, bacterial infections, and cryptococcosis [1, 4, 94, 95, 98].

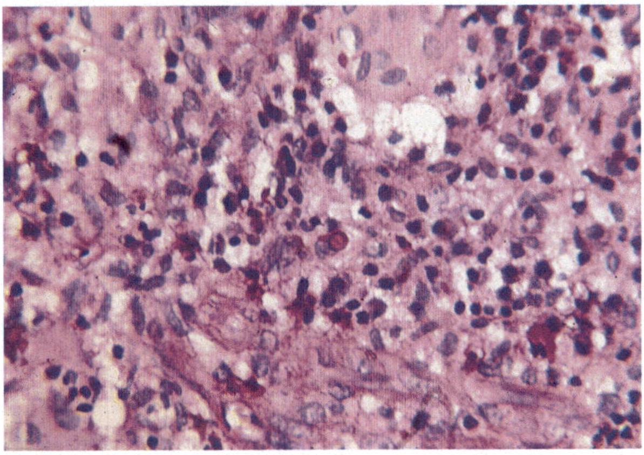

Fig. 10.7 Suppurative granuloma with *S. schenckii* yeast (PAS, 100×)

10.9 Laboratory Diagnosis

The direct examination using 10–20% KOH, the Gram, Giemsa, PAS, or Grocott straining have poor diagnostic value in cutaneous lymphangitic and fixed cutaneous sporotrichosis, because in only 5–35% of cases the yeast-like cells are observed. The staining with calcofluor white or immunofluorescence are more suitable to help in the diagnosis of the disease. The structures usually found during the inspection are elongated yeast-like cells that resemble a cigar or a ship, they are scare, and are visualized only in a handful of microscopic fields, being difficult to confirm the fungus. It is possible to observe asteroid bodies, which are clumps of yeast-like cells, forming structures similar to a crown (Splendore-Hoepli phenomenon) [1–4, 8, 9, 107]. It is of clinical value to determine IFN-γ by immunohistochemistry [108]. In cases of animal sporotrichosis or human cutaneous disseminated and systemic infections, the isolation of a significant number of yeast-like cells is common; and therefore, it is likely to confuse the infection with histoplasmosis [1, 58, 102, 105].

The current diagnostic tool available is the culturing of the organism from the lesion exudates, skin flakes, tissue samples, or sputum. The most useful culture media are Sabouraud-dextrose agar, Sabouraud-dextrose agar supplemented with antibiotics, and yeast-extract agar. When incubated at 28 °C, the fungal growth occurs after 5–8 days, with limited and membranous colonies of radial shape and whitish or beige-brown color. Then, the aerial mycelia are developed with the formation of *coremium.* Some species are unable to generate pigments, such as *S. pallida* [1–4, 7, 8]. It is important to keep in mind that the causative agents of sporotrichosis are dimorphic fungi; thus, if cultured in rich media, such as blood agar, chocolate agar, or BHI agar at 37 °C, the organisms will grow as yeast-like cells, easy to confound with bacterial colonies.

The histopathology is not specific of this disease. In lymphangitic and fixed cases, it is possible to observe asteroid bodies with budding cells, which are surrounded by eosinophilic material. In general, the fungal cells are well stained with

Grocott or PAS techniques. The histological inspection often shows a combination of a granulomatous and suppurative imagen with a pyogenic reaction, and this is composed of three zones: the central or chronic area with polymorphonuclear, histiocytes and lymphocytes microabscesses; the second zone surrounds the central one and shows a tuberculoid image composed of epithelioid cells and multinuclear giant cells similar to Langerhans cells; and the third zone contains lymphocytes, plasmocytes, and fibroblasts (Fig. 10.7). Quintella et al. [107], in a work with a significant number of cases, reported that 84% of cases showed suppurative granuloma, whereas the rest was tuberculoid, against a strange body, caseous and fibrinoid, with the confirmation of yeast-like cells or asteroid bodies only in 35% of the cases [2, 4, 109–114].

10.10 Cell Wall Antigens

The cell wall of *S. schenckii s. str.* is composed of beta-glucans, chitin, and a rare polysaccharide found only in a handful of fungal species, the peptidorhamnomannan. The latter is a potent immunogen, as infected patients have antibodies against it [13, 73]. Structural reports showed that the *N*-linked chain of rhamnomannan contains two antigenic epitopes: one found in both filament and yeast-like cells, and the second only on the surface of filament cells. These structures are a lateral chain with monorhamnosyl unit (Rha1 → 3 Man) and a lateral chain with dirhamnosyl unit (Rha 1 → 2 Rha 1 → 3 Man), respectively [13, 73]. Further analyses indicated the presence of other antigenic epitopes on the *O*-linked tetra- and pentasaccharides of peptido-rhamnomannans [73, 115]. These are in fact the main epitopes found within the *Sporothrix* cell wall and recognized by human IgG antibodies [115, 116].

Besides these antigenic polysaccharides, the cell wall also contains a peptide antigen, Gp70, a protein only found in the cell surface of pathogenic species of the *S. schenckii* complex, i.e., *S. schenckii s. str.*, *S. brasiliensis*, and *S. globosa*, but not in *S. mexicana* [73]. This antigen was sequenced and characterized using proteomic approaches [68], and interestingly, *S. brasiliensis* expresses an isoform of 60 kDa (Gp60) [117]. It is worth mentioning that a monoclonal antibody against Gp70 from *S. schenckii s. str.* is capable of conferring passive immunization in the mouse model of sporotrichosis [52, 53], which may lead to strategies to control both human and animal sporotrichosis.

10.11 New Molecular and Serological Diagnostic Tests

Most of the developed serological tests have not been efficient and are applicable only in systemic cases [2, 4, 118, 119]. However, in the past decade the use of ELISA-based methods, using specific epitopes from the *S. schenckii* cell wall have showed good specificity, sensitivity, and reliability [120, 121]. The ELISA test is a fast and sensitive diagnostic method, which can be used to diagnose all clinical forms of sporotrichosis with a good specificity. This new serological tool is being successfully applied for therapeutic follow-up of patients, including those affected with *S. brasiliensis* [120, 121].

Another attractive approach that is currently being developed is the strain identification by molecular means, using biopsy samples and the PCR technique. The gene targets include those encoding for the chitin synthetase I, 26s rRNA, and topoisomerase II [1, 4, 122–127]. The tools already developed for the molecular identification of *Sporothrix spp.* can be used in parallel with the serological diagnosis.

10.12 Treatment

Fixed and Cutaneous-lymphangitic sporotrichosis: According to the North American Guidelines for the treatment of sporotrichosis, itraconazole (A-II) is the choice for these forms of the disease [128, 129]. The suggested posology is 200 mg/day for 3–6 months. There are reports indicating similar successful rates using intermittent doses or pulses, i.e., 400 mg/day during 1 week and then no treatment for the next 3 weeks [130, 131]. The advantage of the latter is the use of a lower number of doses. Nonresponsive patients to the indicated options are candidates for an increment in the dose to 300–400 mg/day; for children, the suggested dose is 6–10 mg/kg/day [4, 89, 129, 132–136]. Another therapeutic option is terbinafine (A-II), using doses of 250–500 mg/day or even 1 g/day in special cases. This drug is a good choice for children, and the recommendation is 125–250 mg/day [1, 4, 129, 136–140]. Of importance, most of the *S. schenckii* strains have increased *in vitro* sensitivity to terbinafine, but *in vivo*, itraconazole is more active [1, 141–143, 158].

Another good alternative for cutaneous cases is potassium iodide (A-II), which is the option in low-income countries [128, 129, 144]. It could be used as normal or saturate suspension. The posology of the former is 3–6 g/day, starting with 1 g/day to assess the patient tolerance. The administration route is oral, and it is recommended to prepare a suspension of 20 g in 300 mL of water in a dark container, having 1 g in 15 mL. In children, a dose of 1–3 g/day is recommended, starting with 500 mg. When the saturated suspension is selected, the treatment starts with 5 drops three times per day, gradually increasing the number of drops until reaching 40–50 drops three times per day. This treatment has to continue for an average of at least 3–5 months. The potassium iodide is usually well tolerated by patients, but it is possible to present secondary effects, such as gastritis and, rarely, rhinitis, bronchitis, rash, and nodosum erythema. It is important to assess the use of this drug in patients with diseases involving the heart or thyroid. This is not a patented drug; therefore, it has to be prepared by specialized personnel at the drugstore. In feline sporotrichosis, the potassium iodide has shown good results and is regarded as the treatment of choice [4, 136, 144–147].

If none of the above indicated drugs is available, fluconazole at 400–800 mg/day may be used [129, 147]. Another concomitant option is hyperthermia or heating the site of the lesion with warm baths at 45 °C per 15–20 min two or three times per day. This is because the members of the *S. schenckii* complex do not grow at temperatures higher than 40 °C [4, 129, 136, 148, 149]. This treatment is recommended for a limited number of sporotrichosis cases, only as a concomitant alternative. Pregnant patients or those incapable of receiving the mentioned drugs are candidates. It is likely that this approach will not kill the fungal cells but will help to stop dissemination to other tissues. After childbirth, the systemic treatment should be installed [129, 140, 150].

Disseminated cutaneous or extracutaneous (osteoarticular and pulmonary) sporotrichosis: The treatment of choice is the lipid formulation of amphotericin B (B-III) at 3–5 mg/kg/day or deoxycholate amphotericin B (B-III) at 0.7–1 mg/kg/day. Both must be applied within the hospital to assess the presence of collateral effects. In pregnant patients, it is the treatment of choice [1, 129, 151–153]. The second option for these forms of the infection is itraconazole (B-III) at 400 mg/day; it can be the primary treatment or the option after the use of amphotericin B. Alternatively, it could be a concomitant treatment for a period of 6–12 months, in particular in patients with AIDS, hematological cancer, and other immunosuppressive conditions [129, 153].

In particular cases of cutaneous-osteoarticular sporotrichosis, potassium iodide or itraconazole plus trimethoprim/sulfamethoxazole at 400 mg and 80 mg, respectively, during 3–4 months can be used as an option to control the infection [4, 154].

Posaconazole has been used recently in combination with amphotericin B in refractive cases to the conventional therapeutic schemes [155, 156].

It is important that, during the typification of isolates of the *S. schenckii* complex, the *in vitro* susceptibility to antifungal drugs will be part of the standard protocol of identification, because the resistance to antifungal drugs is significantly increasing [1, 142, 144, 157].

References

1. Barros MB, de Almeida PR, Schubach AO (2011) Sporothrix schenckii and Sporotrichosis. Clin Microbiol Rev 24:633–654
2. Arenas R (2005) Sporotrichosis. En: Topley & Wilsongs, Merz WG and Hay R (eds) Microbiology and microbial infections, 10th ed. Hodder-Arnold, London. p. 367–384
3. Werner AH, Werner BE (1994) Sporotrichosis in man and animal. Int J Dermatol 33:692–700
4. Bonifaz A, Vázquez-González D (2010) Sporotrichosis: an update. G Ital Dermatol Venereol 145:659–673
5. Marimon R, Cano J, Gené J, Sutton DA, Kawasaki M, Guarro J (2007) Sporothrix brasiliensis, S. globosa, and S. mexicana, three new *Sporothrix* species of clinical interest. J Clin Microbiol 45:3198–3206
6. Vásquez-del-Mercado E, Arenas R, Padilla-Desgarenes C (2012) Sporotrichosis. Clin Dermatol 30:437–443
7. Lavalle P, Mariat F (1983) Sporotrichosis. Bull Inst Pasteur 81:295–322

8. Kauffman CA (1999) Sporotrichosis. Clin Infect Dis 29:231–237
9. Schechtman RC (2010) Sporotrichosis: part I. Skinmed 8:216–220
10. Rodrigues AM, de Hoog S, de Camargo ZP (2013) Emergence of pathogenicity in the Sporothrix schenckii complex. Med Mycol 51:405–412
11. Marimon R, Gene J, Cano J, Guarro J (2008) Sporothrix luriei: a rare fungus from clinical origin. Med Mycol 10:1–5
12. Dias NM, Oliveira MM, Santos C, Zancope-Oliveira RM, Lima N (2011) Sporotrichosis caused by Sporothrix Mexicana, Portugal. Emerg Infect Dis 17:1975–1976
13. López-Romero E, Reyes-Montes MR, Pérez-Torres A, Ruiz-Baca E, Villagómez-Castro JC et al (2011) Sporothrix schenckii complex and sporotrichosis, an emerging health problem. Future Microbiol 6:85–102
14. Chakrabarti A, Bonifaz A, Gutierrez-Galhardo MC, Mochizuki T, Li S (2015) Global epidemiology of sporotrichosis. Med Mycol 53:3–14
15. Hay RJ, Morris-Jones R (2008) Outbreaks of sporotrichosis. Curr Opin Infect Dis 21:119–121
16. Ishizaki I, Kawasaki M, Aoki M et al (1998) Mitocondrial DNA analysis of Sporothrix schenckii in North and South America. Mycopathologia 142:115–118
17. Yu X, Wan Z, Zhang Z, Li F, Li R, Liu X (2013) Phenotypic and molecular identification of Sporothrix isolates of clinical origin in Northeast China. Mycopathologia 176:67–74
18. Kong X, Xiao T, Lin J, Wang Y, Chen HD (2006) Relationships among genotypes, virulence and clinical forms of Sporothrix schenckii infection. Clin Microbiol Infect 12:1077–1081
19. Mesa-Arango AC, Reyes-Montes R, Pérez-Mejía A et al (2002) Phenotyping and genotyping of Sporothrix schenckii isolates according to geographic origin and clinical form of sporotrichosis. J Clin Microbiol 40:3004–3011
20. Madrid H, Cano J, Gené J, Bonifaz A, Toriello C, Guarro J (2009) Sporothrix globosa, a pathogenic fungus with widespread geographical distribution. Rev Iberoam Micol 30:218–222
21. Oliveira MM, Almeida-Paes R, Gutierrez-Galhardo MC, Zancope-Oliveira RM (2013) Molecular identification of the Sporothrix schenckii complex. Rev Iberoam Micol S1130-1406(13):00099–00095
22. Schechtman RC (2010) Sporotrichosis: part II. Skinmed 8:275–280
23. Bonifaz A, Rojas-Padilla R, Tirado-Sánchez A, Ponce RM (2015) Sporotrichosis: State of the art (Chapter 9). In: Razzaghi-Abyaneh M, Shams-Ghahfarokhi M, Rai M (eds) Medical Mycology: Current Trends and Future Prospects. CRC Press, London, pp 234–253
24. Findlay GH (1970) The epidemiology of the sporotrichosis in the Transvaal. Sabouraudia 7:231–242
25. Song Y, Li SS, Zhong SX, Liu YY, Yao L, Huo SS (2013) Report of 457 sporotrichosis cases from Jilin province, Northeast China, a serious endemic region. J Eur Acad Dermatol Venereol 27:313–318
26. Takenaka M, Yoshizaki A, Utani A, Nishimoto K (2014) A survey of 165 sporotrichosis cases examined in Nagasaki prefecture from 1951 to 2012. Mycoses 57:294–298
27. Fujii H, Tanioka M, Yonezawa M, Arakawa A, Matsumura Y, Kore-eda S, Miyachi Y, Tanaka S, Mochizuki T (2008) A case of atypical sporotrichosis with multifocal cutaneous ulcers. Clin Exp Dermatol 33:135–138
28. Ghosh A, Chakrabarti A, Sharma VK, Singh K, Singh A (1999) Sporotrichosis in Himachal Pradesh (north India). Trans R Soc Trop Med Hyg 93:41–45
29. McGuinness SL, Boyd R, Kidd S, McLeod C, Krause VL, Ralph AP (2016) Epidemiological investigation of an outbreak of cutaneous sporotrichosis, Northern Territory. Australia BMC Infect Dis 16:16. doi:10.1186/s12879-016-1338-0
30. Silva MB, Costa MM, Torres CC, Galhardo MC, Valle AC, Magalhães M de A, Sabroza PC, Oliveira RM. Urban sporotrichosis: a neglected epidemic in Rio de Janeiro, Brazil. Cad Saude Publica 2012; 28:1867–1880
31. Barros MB, Costa DL, Schubach TM et al (2008) Endemic of zoonotic sporotrichosis: profile of cases in children. Pediatr Infect Dis J 2:246–250

32. Schubach A, Barros MB, Wanke B (2008) Epidemic sporotrichosis. Curr Opin Infect Dis 21:129–133
33. Freitas DF, Valle AC, da Silva MB, Campos DP, Lyra MR, de Souza RV, Veloso VG, Zancopé-Oliveira RM, Bastos FI, Galhardo MC (2014) Sporotrichosis: an emerging neglected opportunistic infection in HIV-infected patients in Rio de Janeiro, Brazil. PLoS Negl Trop Dis 8(8):e3110
34. Bustamante B, Campos PE (2001) Endemic sporotrichosis. Curr Opin Infect Dis 14:145–149
35. Pappas PG, Téllez I, Deep AE, Nolasco D, Holgado W, Bustamante B (2000) Sporotrichosis in Peru: description of an area of hyperendemicity. Clin Infect Dis 30:65–70
36. Ramírez-Soto MC, Andagua-Castro J, Lizárraga-Trujillo J, Aguilar-Ancori EG, Pezo-Ochoa JD (2011) Esporotricosis en pacientes que acuden a un centro de referencia en Abancay, Perú. Rev Peru Med Exp Salud Publica 28:508–512
37. Mayorga J, Tarango MV, Barba-Rubio J (1999) Esporotricosis 100 años después (1898-1998). Dermatologia Rev Mex 43:S22–S29
38. Campos P, Arenas R, Coronado H (1994) Epidemic cutaneous sporotrichosis. Int J Dermatol 33:38–41
39. Espinoza-Texis A, Hernández-Hernández F, López-Martínez R et al (2001) Estudio de 50 pacientes con esporotricosis. Evaluación clínica y de laboratorio. Gac Med Mex 137:111–116
40. García-Vargas A, Mayorga J, Soto Ortiz A, Barba Gómez JF (2008) Esporotricosis en niños. Estudio de 133 casos en el Instituto Dermatológico de Jalisco "Dr. José Barba Rubio". Med Cutan Iber Lat Am 36:18–22
41. Padilla-Desgarenes MC, Medina-Castillo DE, Cortés-Lozano N (2004) Esporotricosis en edad pediátrica: experiencia del Centro Dermatológico Pascua. Piel 19:350–363
42. Rubio G, Sánchez G, Porras L, Alvarado Z (2010) Esporotricosis: prevalencia, perfil clínico y epidemiológico en un centro de referencia en Colombia. Rev Iberoam Micol 27:75–79
43. Mata-Essayag S, Delgado A, Colella MT, Landaeta-Nezer ME, Rosello A, Perez de Salazar C et al (2013) Epidemiology of sporotrichosis in Venezuela. Int J Dermatol 52:974–980
44. Conti Díaz IA (1989) Epidemiology of sporotrichosis in Latin America. Mycopathologia 108:113–116
45. Mayorga R, Cáceres A, Toriello C et al (1978) Etude d'une zone d'endemie sporotrichosi que au Guatemala. Sabouraudia 16:185–198
46. Mayorga J, Martínez LD, Méndez GA (1999) Aislamiento de Sporothrix schenckii en la naturaleza (suelo y plantas). Med Cut ILA 27:25–28
47. Sánchez-Alemán MA, Araiza J, Bonifaz A (2004) Aislamiento y caracterización de cepas silvestres de Sporothrix schenckii e investigación de reactores a la esporotricina. Gac Med Mex 140:507–512
48. Coles FB, Schuchat A, Hibbs JR, Kondracki SF, Salkin IF, Dixon DM et al (1992) A multistate outbreak of sporotrichosis associated with sphagnum moss. Am J Epidemiol 136:475–487
49. Martínez-Rodríguez C, Enríquez-Guerra MA, Arce-Romero R, Vallejos-Médic C, Sánchez-Hernández JA et al (2013) Esporotricosis cutánea linfangítica por mordedura de murciélago vampiro (Desmodus rotundus). Dermatol Rev Mex 57:468–472
50. Alencar-Marques S, Pires de Camargo RM, Haddad-Junior V (1998) Human sporotrichosis: transmitted by feline. An Bras Dermatol 73:559–562
51. Schubach A, Schubach TM, Barros MB, Wanke B (2005) Cat-transmitted sporotrichosis, Rio de Janeiro, Brazil. Emerg Infect Dis 11:1952–1954
52. Almeida HL Jr, Lettnin CB, Barbosa JL, Dias MC (2009) Spontaneous resolution of zoonotic sporotrichosis during pregnancy. Rev Inst Med Trop São Paulo 51:237–238
53. Almeida SR (2012) Therapeutic monoclonal antibody for sporotrichosis. Front Microbiol 28:409
54. Kano R, Okubo M, Siew HH, Kamata H, Hasegawa A (2015) Molecular typing of Sporothrix schenckii isolates from cats in Malaysia. Mycoses 58(4):220–224
55. Cordeiro FN, Bruno CB, Paula CD, Motta JO (2011) Familial occurrence of zoonotic sporotrichosis. An Bras Dermatol 86(4 suppl. 1):S121–S124

56. Borges TS, Rossi CN, Fedullo JD, Taborda CP, Larsson CE (2013) Isolation of Sporothrix schenci from the claws of domestic cats (indoor and outdoor) and in captivity in São Paulo (Brazil). Mycopathologia 176:129–137
57. Rodrigues AM, de Melo TM, de Hoog GS, Schubach TM, Pereira SA, Fernandes GF et al (2013) Phylogenetic analysis reveals a high prevalence of Sporothrix brasiliensis in feline sporotrichosis outbreaks. PLoS Negl Trop Dis 7(6):e2281
58. Freitas DF, do Valle AC, de Almeida Paes R, Bastos FI, Galhardo MC (2010) Zoonotic sporotrichosis in Rio de Janeiro, Brazil: a protracted epidemic yet to be curbed. Clin Infect Dis 50:453
59. Bonifaz A, Saúl A, Paredes-Solís V et al (2007) Sporotrichosis in childhood. Clinical and therapeutic experience in 25 cases. Pediatr Dermatol 24:369–372
60. Ramírez Soto MC. Sporotrichosis: The Story of an Endemic Region in Peru over 28 Years (1985 to 2012). PLoS One. 2015;10(6):e0127924. doi: 10.1371
61. Bonifaz A, Araiza J, Pérez-Mejía A, Ochoa LA, Toriello C (2013) Prueba intradérmica con esporotricina en una comunidad de la Sierra Norte de Puebla. Dermatol Rev Mex 57:428–432
62. Zhang Z, Liu X, Lv X, Lin J (2011) Variation in genotype and higher virulence of a strain of Sporothrix schenckii causing disseminated cutaneous sporotrichosis. Mycopathologia 172:439–446
63. Freitas DF, Santos SS, Almeida-Paes R, de Oliveira MM, do Valle AC, Gutierrez-Galhardo MC, Zancopé-Oliveira RM, Nosanchuk JD (2015) Increase in virulence of Sporothrix brasiliensis over five years in a patient with chronic disseminated sporotrichosis. Virulence 6:112–120
64. Bearson SD, Brandt FA (1977) Primary pulmonary sporotrichosis with unusual morphology. Thorax 32:505–508
65. Aung AK, Teh BM, McGrath C, Thompson PJ (2013) Pulmonary sporotrichosis: case series and systematic analysis of literature on clinico-radiological patterns and management outcomes. Med Mycol 51:534–544
66. Fernandes GF, dos Santos PO, Rodrigues AM, Sasaki AA, Burger E, de Camargo ZP (2013) Characterization of virulence profile, protein secretion and immunogenicity of different *Sporothrix schenckii sensu stricto* isolates compared with *S. globosa* and S. *brasiliensis* species. Virulence 4:241–249
67. Clavijo-Giraldo DM, Matínez-Alvarez JA, Lopes-Bezerra LM, Ponce-Noyola P, Franco B, Almeida RS, Mora-Montes HM (2016) Analysis of *Sporothrix schenckii sensu stricto* and *Sporothrix brasiliensis* virulence in *Galleria mellonella*. J Microbiol Methods 122:73–77
68. Castro RA, Kubitschek-Barreira PH, Teixeira PA, Sanches GF, Teixeira MM, Quintella LP et al (2013) Differences in cell morphometry, cell wall topography and gp70 expression correlate with the virulence of Sporothrix brasiliensis clinical isolates. PLoS One 8(10):e75656
69. Romeo O, Criseo G (2013) What lies beyond genetic diversity in Sporothrix schenckii species complex?: New insights into virulence profiles, immunogenicity and protein secretion in S. schenckii sensu stricto isolates. Virulence 4:203–206
70. Alba-Fierro CA, Pérez-Torres A, López-Romero E, Cuéllar-Cruz M, Ruiz-Baca E (2013) Cell wall proteins of Sporothrix schenckii as immunoprotective agents. Rev Iberoam Micol S1130-1406(13):00114–00119
71. Almeida-Paes R, Frases S, Araújo Gde S, de Oliveira MM, Gerfen GJ, Nosanchuk JD, Zancopé-Oliveira RM (2012) Biosynthesis and functions of a melanoid pigment produced by species of the sporothrix complex in the presence of L-tyrosine. Appl Environ Microbiol 78:8623–8630
72. Sasaki AA, Fernandes GF, Rodrigues AM, Lima FM, Marini MM, Dos S, Feitosa L et al (2014) Chromosomal polymorphism in the Sporothrix schenckii complex. PLoS One 9(1):e86819
73. Lopes-Bezerra LM (2011) Sporothrix schenckii cell wall peptidorhamnomannans. Front Microbiol 21:243

74. Kajiwara H, Saito M, Ohga S, Uenotsuchi T, Yoshida S (2004) Impaired host defense against Sporothrix schenckii in mice with chronic granulomatous disease. Infect Immun 72:5073–5079
75. Uenotsuchi T, Takeuchi S, Matsuda T et al (2006) Differential induction of Th1-prone immunity by human dendritic cells activated with Sporothrix schenckii of cutaneous and visceral origins to determine their different virulence. Int Immunol 18:1637–1646
76. Sassá M, Ferreira L, Ribeiro L, Carlos I (2012) Immune response against Sporothrix schenckii in TLR-4-deficient mice. Mycopathologia 174:21–30
77. Verdan FF, Faleiros JC, Ferreira LS, Monnazzi LG, Maia DC, Tansine A et al (2012) Dendritic cell are able to differentially recognize Sporothrix schenckii antigens and promote Th1/Th17 response in vitro. Immunobiology 217:788–794
78. Zhang X, Zhang J, Huang H, Xue R, Hu X, Li M, Zhong Y, Yuan L (2012) Taenia taeniae formis in rat favors protracted skin lesions caused by Sporothrix schenckii infection: Dectin-1 and IL-17 are dispensable for clearance of this fungus. PLoS One 7(12):e52514
79. Tirado-Sánchez A, Bonifaz A (2016) Sporotrichosis in children: an update. Cur Fungal Infect Rep. doi:10.1007/s12281-016-0259-0
80. Bibler MR, Luber HJ, Glueck HI et al (1986) Disseminated sporotrichosis in a patient with HIV infection after treatment for acquired factor VIII inhibitor. JAMA 256:3125–3126
81. Bonifaz A, Saúl A, Montes de Oca G, Mercadillo P (1999) Superficial cutaneous sporotrichosis in specific anergic case. Int J Dermatol 38:700–703
82. Bonifaz A, Peniche A, Saúl A et al (2001) Successful treatment of AIDS related disseminated cutaneous sporotrichosis with itraconazole. AIDS Patient Care STDs 15:603–606
83. Chang S, Hersh AM, Naughton G, Mullins K, Fung MA, Sharon VR (2013) Disseminated cutaneous sporotrichosis. Dermatol Online J 19:20401
84. Eustace KE, Sampaio FM, Lyra MR, Quintella L, do Valle AC. Cutaneous disseminated sporotrichosis complicated by osteomyelitis. Acta Derm Venereol 2013; 93:192–193
85. Castrejón OV, Robles M, Zubieta Arroyo OE (1995) Fatal fungaemia due to Sporothrix schenckii. Mycoses 38:373–376
86. Espinoza-Hernández CJ, Jesús-Silva A, Toussaint-Caire S, Arenas R (2013) Disseminated sporotrichosis with cutaneous and testicular involvement. Actas Dermosifiliogr 105:204–206
87. Freitas DF, de Siqueira HB, do Valle AC, Fraga BB, de Barros MB, de Oliveira Schubach A et al (2012) Sporotrichosis in HIV-infected patients: report of 21 cases of endemic sporotrichosis in Rio de Janeiro, Brazil. Med Mycol 50:170–178
88. Baker JH, Goodpasture HC, Kuhns HR Jr, Rinaldi MG (1989) Fungemia caused by an amphotericin B-resistant isolate o Sporothrix schenckii. Successful treatment with itraconazole. Arch Pathol Lab Med 113:1279–1281
89. Galhardo MC, Silva MT, Lima MA, Nunes EP, Schettini LE, de Freitas RF et al (2010) Sporothrix schenckii meningitis in AIDS during immune reconstitution syndrome. J Neurol Neurosurg Psychiatry 81:696–699
90. Kosinski RM, Axelrod P, Rex JH, Burday M, Sivaprasad R, Wreiole A (1992) Sporothrix schenckii fungemia without disseminated sporotrichosis. J Clin Microbiol 30:501–503
91. Ewing GE, Bosl GJ, Peterson PK (1980) Sporothrix schenckii meningitis in a farmer with Hodgkin's disease. Am J Med 68:455–457
92. Orofino-Costa R, Unterstell N, Carlos Gripp A, de Macedo PM, Brota A, Dias E, de Melo TM et al (2013) Pulmonary cavitation and skin lesions mimicking tuberculosis in a HIV negative patient caused by Sporothrix brasiliensis. Med Mycol Case Rep 16:65–71
93. Rufino R, Marques BL, Costa CH (2013) Rose Gardener's disease. Thorax 68:602
94. Aronson EN (1992) Disseminated sporotrichosis. JAMA 268:2021
95. Yelverton CB, Stetson CL, Bang RH et al (2006) Fatal sporotrichosis. Cutis 78:253–256
96. Severo LC, Festugato M, Bernardi C, Londero AT (1999) Widespread cutaneous lesions due to Sporothrix schenckii in a patient under a long-term steroids therapy. Rev Inst Med Trop Sao Paulo 41:59–62

97. Yamaguchi T, Ito S, Takano Y, Umeda N, Goto M, Horikoshi M et al (2012) A case of disseminated sporotrichosis treated with prednisolone, immunosuppressants, and tocilizumab under the diagnosis of rheumatoid arthritis. Intern Med 51:2035–2039
98. Moreira JA, Freitas DF, Lamas CC (2015) The impact of sporotrichosis in HIV-infected patients: a systematic review. Infection 43:267–276
99. Silva-Vergara ML, de Camargo ZP, Silva PF, Abdalla MR, Sgarbieri RN, Rodrigues AM et al (2012) Disseminated Sporothrix brasiliensis infection with endocardial and ocular involvement in an HIV-infected patient. Am J Trop Med Hyg 86:477–480
100. Madrid IM, Mattei AS, Fernandes CG, Nobre Mde O, Meireles MC (2012) Epidemiological findings and laboratory evaluation of sporotrichosis: a description of 103 cases in cats and dogs in Southern Brazil. Mycopathologia 173:265–273
101. Pereira SA, Gremião ID, Kitada AA, Boechat JS, Viana PG, Schubach TM (2014) The epidemiological scenario of feline sporotrichosis in Rio de Janeiro, State of Rio de Janeiro, Brazil. Rev Soc Bras Med Trop 47:392–393
102. Lloret A, Hartmann K, Pennisi MG, Ferrer L, Addie D, Belák S et al (2013) Sporotrichosis in cats: ABCD guidelines on prevention and management. J Feline Med Surg 15:619–623
103. Morgan MA, Cockerill FR 3rd, Cortese DA, Roberts GD (1984) Disseminated sporotrichosis with Sporothrix schenckii fungemia. Diagn Microbiol Infect Dis 2:151–155
104. Nassif PW, Granado IR, Ferraz JS, Souza R, Nassif AE (2012) Atypical presentation of cutaneous sporotrichosis in an alcoholic patient. Dermatol Online J 18:12
105. Romero-Cabello R, Bonifaz A, Romero-Feregrino R, Sánchez CJ, Linares Y, Tay-Zavala J et al (2011) Disseminated sprotrichosis. BMJ Case Rep. doi:10.1136/bcr.10.2010.3404
106. Ware AJ, Cockerell CJ, Skiest DJ et al (1999) Disseminated sporotrichosis with extensive cutaneous involvement in a patient with AIDS. J Am Acad Dermatol 40:350–355
107. Quintella LP, Passos SR, do Vale AC, Galhardo MC, Barros MB, Cuzzi T, Reis Rdos S et al (2011) Histopathology of cutaneous sporotrichosis in Rio de Janeiro: a series of 119 consecutive cases. J Cutan Pathol 38:25–32
108. Koga T, Duan H, Furue M (2002) Immunohistochemical detection of interferon-γ-producing cells in granuloma formation of sporotrichosis. Med Mycol 40:111
109. Gezuele E, Da Rosa D (2005) Importance of the sporotrichosis asteroid body for the rapid diagnosis of sporotrichosis. Rev Iberoam Micol 22:147–150
110. Lurie HL (1963) Histophatology of sporotrichosis. Arch Pathol 75:121–137
111. Novales SJ, Navarrete FG, Ramos GA (1995) Esporotricosis. Aspectos histológicos de 50 casos. Rev Centr Dermatol Pascua 4:149–152
112. Quintella LP, Passos SR, de Miranda LH, Cuzzi T, Barros MB, Francesconi-do-Vale AC et al (2012) Proposal of a histopathological predictive rule for the differential diagnosis between American tegumentary leishmaniasis and sporotrichosis skin lesions. Br J Dermatol 167:837–846
113. Miranda LH, Conceição-Silva F, Quintella LP, Kuraiem BP, Pereira SA, Schubach TM (2013) Feline sporotrichosis: histopathological profile of cutaneous lesions and their correlation with clinical presentation. Comp Immunol Microbiol Infect Dis 36:425–432
114. Miyaji M, Nishimura K (1982) Defensive role of granuloma against Sporothrix schenckii infection. Mycopathologia 80:117–124
115. Lopes Alves L, Travassos LR, Previato JO, Mendonça-Previato L (1994) Novel antigenic determinants from peptidorhamnomannans of *Sporothrix schenckii*. Glycobiology 4(3):281–288
116. Penha CV, Lopes-Bezerra LM (2000) Concanavalin A-binding cell wall antigens of *Sporothrix schenckii*: a serological study. Med Mycol 38(1):1–7
117. Rodrigues AM, Kubitschek-Barreira PH, Fernandes GF, de Almeida SR, Lopes-Bezerra LM, de Camargo ZP (2015) Immunoproteomic analysis reveals a convergent humoral response signature in the *Sporothrix schenckii* complex. J Proteome 115:8–22
118. Saúl A, Bonifaz A (2011) Clasificación de la esporotricosis. Una propuesta con base en el comportamiento inmunológico. Dermatologia Rev Mex 55:200–208

119. Toriello C, Reyes-Montes R, Taylor ML (1997) Production of fungal antigens from local strains for the immunodiagnosis of mycoses in Mexico. Rev Investig Clin 49:501–505
120. Bernardes-Engemann AR, de Lima BM, Zeitune T, Russi DC, Orofino-Costa R, Lopes-Bezerra LM (2015) Validation of a serodiagnostic test for sporotrichosis: a follow-up study of patients related to the Rio de Janeiro zoonotic outbreak. Med Mycol 53(1):28–33
121. Bernardes-Engemann AR, Costa RC, Miguens BR, Penha CV, Neves E, Pereira BA, Dias CM, Mattos M, Gutierrez MC, Schubach A, Oliveira Neto MP, Lazéra M, Lopes-Bezerra LM (2005) Development of an enzyme-linked immunosorbent assay for the serodiagnosis of several clinical forms of sporotrichosis. Med Mycol 43(6):487–493
122. Hu S, Chung WH, Hung SI (2003) Detection of Sporothrix schenckii in clinical samples by a nested PCR assay. J Clin Microbiol 41:1414–1418
123. Kawasaki M, Anzawa K, Mochizuki T, Ishizaki H (2012) New strain typing method with Sporothrix schenckii using mitochondrial DNA and polymerase chain reaction restriction fragment length polymorphism (PCR-RFLP) technique. J Dermatol 39:362–365
124. Liu X, Zhang Z, Hou B, Wang D, Sun T, Li F, Wang H, Han S (2013) Rapid identification of Sporothrix schenckii in biopsy tissue by PCR. J Eur Acad Dermatol Venereol 27:1491–1497
125. Rodrigues AM, de Hoog GS, de Camargo ZP (2015) Molecular Diagnosis of Pathogenic Sporothrix Species. PLoS Negl Trop Dis 9(12):e0004190. doi:10.1371/journal.pntd.0004190
126. Xu TH, Lin JP, Gao XH, Wei H, Liao W, Chen HD (2010) Identification of Sporothix schenckii of various mtDNA types by nested PCR assay. Med Mycol 48:161–165
127. Oliveira MM, Maifrede SB, Ribeiro MA, Zancope-Oliveira RM (2013) Molecular identification of Sporothrix species involved in the first familial outbreak of sporotrichosis in the state of Espírito Santo, Southeastern Brazil. Mem Inst Oswaldo Cruz 108:936–938
128. Bustamante B, Campos PE (2004) Sporotrichosis: a forgotten disease in the drug research agenda. Expert Rev Anti-Infect Ther 2:85–94
129. Kauffman CA, Bustamante B, Chapman SW, Pappas PG, Infectious Diseases Society of America (2007) Clinical practice guidelines for the management of sporotrichosis: 2007 update by the Infectious Diseases Society of America. Clin Infect Dis 45:1255–1265
130. Chaves AR, de Campos MP, Barros MB, do Carmo CN, Gremião ID, Pereira SA, Schubach TM. Treatment abandonment in feline sporotrichosis - study of 147 cases. Zoonoses Public Health 2013; 60:149–153
131. Song Y, Zhong SX, Yao L, Cai Q, Zhou JF, Liu YY, Huo SS, Li SS (2011) Efficacy and safety of itraconazole pulses vs. continuous regimen in cutaneous sporotrichosis. J Eur Acad Dermatol Venereol 25:302–305
132. Sharkey-Mahis PK, Kauffman CA, Graybill JR et al (1993) Treatment of sporotrichosis with itraconazole. Am J Med 95:279–285
133. Bolao F, Podzanczer D, Ventin M et al (1994) Efficacy of acute phase and maintenance therapy with itraconazole in an AIDS patient with sporotrichosis. Eur J Clin Microbiol Infect Dis 13:609–612
134. Stalkup JR, Bell K, Rosen T (2002) Disseminated cutaneous sporotrichosis treated with itraconazole. Cutis 69:371–374
135. De Lima-Barros MB, Schubach AO, de Vasconcellos C, de Oliveira R, Martins EB, Teixeira JL, Wanke B (2011) Treatment of cutaneous sporotrichosis with itraconazole: study of 645 patients. Clin Infect Dis 52:200
136. Bonifaz A, Vázquez-González D (2013) Diagnosis and treatment of sporotrichosis lymphocutaneous: What are the options. Curr Fungal Infect Rev 7:252–259
137. Hull PR, Vismer HF (1992) Treatment of cutaneous sporotrichosis with terbinafine. Br J Dermatol 126(suppl. 39):51–55
138. Bonifaz A, Saúl A, Paredes-Solis V, Araiza J, Fierro-Arias L (2005) Treatment of chromoblastomycosis with terbinafine: experience with four cases. J Dermatolog Treat 16:47–51
139. Francesconi G, Valle AC, Passos S, Reis R, Galhardo MC (2009) Terbinafine (250 mg/day): an effective and safe treatment of cutaneous sporotrichosis. J Eur Acad Dermatol Venereol 23:1273–1276

140. Orfino-Costa R, Bernardes-Engemann AR, Azulay-Abulafia L, Benvenuto F, Neves Mde L, Lopes-Bezerra LM (2011) Sporotrichosis in pregnancy: case reports of 5 patients in a zoonotic epidemic in Rio de Janeiro, Brazil. An Bras Dermatol 86:995–998
141. Marimon R, Serena C, Gené J, Cano J, Guarro J (2008) In vitro antifungal susceptibilities of five species of Sporothrix. Antimicrob Agents Chemother 52:732–734
142. Zhang X, Huang H, Feng P, Zhang J, Zhong Y, Xue R et al (2011) In vitro activity of itraconazole in combination with terbinafine against clinical strains of itraconazole-insensitive Sporothrix schenckii. Eur J Dermatol 21:573–576
143. Ottonelli Stopiglia CD, Magagnin CM, Castrillón MR, Mendes SD, Heidrich D, Valente P, Scroferneker ML (2014) Antifungal susceptibilities and identification of species of the Sporothrix schenckii complex isolated in Brazil. Med Mycol 52(1):56–64
144. Macedo PM, Lopes-Bezerra LM, Bernardes-Engemann AR, Orofino-Costa R (2015) New posology of potassium iodide for the treatment of cutaneous sporotrichosis: study of efficacy and safety in 102 patients. J Eur Acad Dermatol Venereol 29(4):719–724
145. Ramírez-Soto MC (2014) Effects of potassium iodide in concentrations of TSH, tT3 and tT4 in serum of subjects with sporotrichosis. Mycoses 57:460–465
146. Reis EG, Gremião ID, Kitada AA, Rocha RF, Castro VS, Barros MB et al (2012) Potassium iodide capsule treatment of feline sporotrichosis. J Feline Med Surg 14:399–404
147. Castro LG, Belda W, Cucé LC et al (1993) Successful treatment of sporotrichosis with oral fluconazole: a report of three cases. Br J Dermatol 128:352–356
148. Pavón N, Bonifaz A, Ponce RM (2007) Termoterapia en esporotricosis cutánea infantil. Comunicación de dos casos. Dermatologia Rev Mex 51:68–72
149. Watanabe S (2014) Local heat treatment of sporotrichosis by pocket warmer “Kairo”. Gilaberte Y, Aspiroz C, Alejandre MC, Andrés-Ciriano E, Fortuño B, Charlez L et al. Cutaneous sporotrichosis treated with photodynamic therapy: an in vitro and in vivo study. Photomed Laser Surg 32:54–77
150. Ferreira CP, do Valle AC, Freitas DF, Reis R, Galhardo MC (2012) Pregnancy during a sporotrichosis epidemic in Rio de Janeiro, Brazil. Int J Gynaecol Obstet 117:294–295
151. Sharon VR, Kim J, Sudhakar S, Fung MA, Maniar A (2013) Disseminated cutaneous sporotrichosis. Lancet Infect Dis 13:95
152. Yap FB (2011) Disseminated cutaneous sporotrichosis in an immunocompetent individual. Int J Infect Dis 15:727–729
153. Paixão AG, Galhardo MC, Almeida-Paes R, Nunes EP, Gonçalves ML, Chequer GL, Lamas CC (2015) The difficult management of disseminated Sporothrix brasiliensis in a patient with advanced AIDS. AIDS Res Ther 7:12–16
154. Solorzano S, Ramirez R, Cabada MM, Montoya M, Cazorla E (2015) Esporotricosis cutánea diseminada con compromiso articular en una mujer con diabetes tipo 2. Rev Peru Med Exp Salud Publica 32:187–190
155. Fernández-Silva F, Capilla J, Mayayo E, Guarro J (2012) Efficacy of posaconazole in murine experimental sporotrichosis. Antimicrob Agents Chemother 56:2273–2277
156. Bunce PE, Yang L, Chun S, Zhang SX, Trinkaus MA, Matukas LM (2012) Disseminated sporotrichosis in a patient with hairy cell leukemia treated with amphotericin B and posaconazole. Med Mycol 50:197–201
157. Stopiglia CD, Heidrich D, Sorrentino JM, Vieira FJ, Landell MF, Valente P, Scroferneker ML (2013) Susceptibility of species within the Sporothrix schenckii complex to a panel of killer yeasts. J Basic Microbiol. doi:10.1002/jobm.201200516
158. Martínez-Álvarez JA, Pérez-García LA, Flores-Carreón A, Mora-Montes HM (2014) The immune response against Candida spp. and Sporothrix schenckii. Rev Iberoam Micol 31:62–66

Chapter 11
Fonsecaea and Chromoblastomycosis

Peiying Feng and G.S. de Hoog

Abstract Chromoblastomycosis is a chronic fungal infection of cutaneous and subcutaneous tissues caused by traumatic inoculation of a specific group of melanized fungi, with species of *Fonsecaea* and *Cladophialophora* as prevalent etiological agents. Chromoblastomycosis has a global distribution, particularly in tropical and subtropical rural areas. *Fonsecaea* spp. are prevalent in humid tropical climates, whereas *Cladophialophora* is found under arid conditions. The disease is difficult to treat due to its recalcitrant nature, which may lead to severe clinical forms with high morbidity, even leading to neoplastic transformation. In this review, we summarize current knowledge on *Fonsecaea* and chromoblastomycosis, including the taxonomy of *Fonsecaea*, pathogenic potentials of species, their epidemiology, and clinical manifestations. Notes on diagnostics and therapeutic options are provided.

11.1 Introduction

Chromoblastomycosis (CBM) is defined as a chronic cutaneous and subcutaneous infection caused by melanized fungi and characterized by verrucose lesions and dark-colored, thick-walled muriform cells in infected tissue; the latter provide a histopathological criterion for definite diagnosis. The disease has a global distribution but is prevalent in tropical and subtropical rural areas, such as Madagascar, Brazil, and China. The etiologic agents are hypothesized to gain entrance through the skin by traumatic implantation of contaminated materials, as the majority of lesions are observed on extremities of outdoor workers [1, 2]. Agents of CBM are members of the family *Herpotrichiellaceae* of the ascomycete order *Chaetothyriales*. By far three genera with five species have been proven as recurrent causative agents of the disease, i.e., *Cladopholophora carrionii, Fonsecaea monophora, F. nubica, F.*

P. Feng, Ph.D. (✉)
Department of Dermatology, The Third Affiliated Hospital of Sun Yat-Sen University, No. 600, Tianhe Road, Guangzhou, China
e-mail: fengpeiying@medmail.com.cn

G.S. de Hoog, Ph.D.
Westerdijk Fungal Biodiversity Institute, Uppsalalaan 8, Utrecht, The Netherlands
e-mail: s.hoog@westerdijkinstitute.nl

H.M. Mora-Montes, L.M. Lopes-Bezerra (eds.), *Current Progress in Medical Mycology*, DOI 10.1007/978-3-319-64113-3_11

pedrosoi, and *Rhinocladiella aquaspersa*. Several *Exophiala, Cladophialophora,* and *Phialophora* species have been reported as occasional agents of the disease [3].

11.2 History

CBM was described for the first time by Dr. Maximilliano Willibaldo Rudolph, a German physician living in Brazil, who published the manuscript "Über die brasiliensische Figueira" (Brazilian fig tree) in a German Journal in 1914 [4, 5]. Rudolph noticed six patients with warty lesions on the lower limbs popularly known as fig tree. He was able to isolate black and velvety cultures in four patients, which today we notice that their microscopic features are quite similar to those of *F. pedrosoi*, and he successfully inoculated the disease in four white rats and two monkeys. Surprisingly, he did not describe the histologic aspects of the disease or the pathognomonic sclerotic cells, which Lane and Medlar described 1 year later in separated publications. Lane and Medlar [6, 7] reported the first CBM North American case observed in an Italian patient living in Boston. The patient presented a warty violet plaque lesion on the right buttock simulating verrucous tuberculosis. They found a number of spherical, pigmented, parasitic cells (Medlar bodies) in the tissue. Previously, Bruas in 1903 and Fontoynont 1909 in Madagascar, and Guiteras in 1904 in Cuba, respectively, referred to some possible CBM cases, but the isolation of the fungus by these authors to verify identification was not possible, neither the cases were published. In 1911, Alexandre Pedroso and Jose Maria Gomes in São Paulo, Brazil, began scientific research on four cases with verrucous dermatitis in Brazilian patients. They revealed the presence of spherical, brownish cells in skin biopsies, corresponding to the current muriform cell, the hallmark of CBM. The cultivation of the patient's skin lesions yielded dark fungal colonies and was classified as *Phialophora verrucosa*. They named the disease *blastomycose negra* at that time. However, Pedroso published their observations late only in 1920 [8]. Two years later, Brumpt [9] concluded that the agents isolated by Pedroso and Gomes could not be classified as *Phialophora* species, and he coined the denomination *Hormodendrum pedrosoi*, later renamed *Fonsecaea pedrosoi* by Negroni [10].

The term chromoblastomycosis was introduced by Terra et al. [11] to differentiate a cutaneous fungal disease that they were studying in Brazil from the confusing clinical entity known as "verrucous dermatitis". The name chromoblastomycosis suggests the disease is a fungal infection caused by pigmented "*Blastomyces*" fungus (Greek *chroma*, ~atos = complexion, skin color, + *Blastomyces* Greek *blastós* = sprout, bud + *mykes* = fungus + *osis* Greek suffix denoting the state of something/somebody). However, as the etiologic agents of the disease produce muriform cells but not budding yeast forms in tissue, Moore and Almeida [12] proposed a new term "chromomycosis" as a replacement of "chromoblastomycosis". With time, the name "chromomycosis" was used as an umbrella to encompass a heterogenic and diverse group of fungal infections caused by a wide spectrum of dark melanized fungi. With the introduction of the concept of phaeohyphomycosis by Ajello [13] and McGinnis [14], differentiation among these diseases became more obvious. Nowadays, the term chromoblastomycosis is restricted to the cases in which muriform cells are present in tissue. Muriform cells, also known as sclerotic cells or Medlar bodies, are globe-shaped, cigar-colored, thick-walled structures that are 4–12 μm in diameter. In 1992, the International Society for Human and Animal Mycology (ISHAM) recommended that the best name to define the disease was chromoblastomycosis [15], which Terra et al. coined in 1922. Currently the disease is defined in the International Classification of Disease (ICD) as follows: ICD-9 no. 117.2, ICD 10-B43.

11.3 Taxonomy of *Fonsecaea*

Fonsecaea is an anamorph genus of black yeast and relatives in the family *Herpotrichiellaceae* (order *Chaetothyriales*). Colonies are restricted, dark olivaceous, velvety to cottony in texture and have an olivaceous-black reverse. Mophologically the genus *Fonsecaea* is characterized by absence of budding cells, and the presence of sympodial condidiogenesis with blunt, scattered denticles bearing conidia singly or in short chains (Figs. 11.1). *Cladophialophora* species such as the neurotropic fungus *C. bantiana* are located in the 'bantiana-clade' [16] together with *Fonsecaea.* These species are different by having very long conidial chains, but some species show intermediate morphology and are difficult to attribute to either one of the genera on a morphological basis. Note that *Cladophialophora carrionii*, frequent agent of CBM in arid climates, is located in another clade and to only distantly related to *C. bantiana* and *Fonsecaea*. A phialidic synanamorph may be produced on nutritionally poor media. All strains grow at 37 °C but not at 40 °C. Recent molecular investigation and characterization of environmental siblings of black agents of human CBM by multilocus sequence analysis has shown that the genus *Fonsecaea* contains five clinically relevant cryptic species i.e. *Fonsecaea pedrosoi*, *F. monophora*, *F. nubica*, *F. multimorphosa,* and *F. pugnacius* [17–19], and three environmental fungi i.e. *Fonsecaea minima*, *F. erecta*, and *F. brasiliensis* [20] (Fig. 11.2).

Fonsecaea pedrosoi was first isolated in 1914 as an etiological agent of CBM by Pedroso and later described and named by Brumpt [9] and Negroni [10]. Traditionally *F. compacta* was another species in the genus with reduced denticles and barrel-shaped conidia, which was described by Carrión from a CBM case in Puerto Rico. This taxon is now known to be a morphological variant of *F. pedrosoi*. In 2004, a new species, *F. monophora* was described using ITS rDNA and RAPD data. The type stain of *F. monophora* CBS 269.37 was isolated from a human CBM case in South America [22]. Morphologically, *F. pedrosoi* and *F. monophora* are very similar. *Fonsecaea monophora* on average has slightly longer conidial chains and slightly shorter denticles than that of *F. pedrosoi*. In 2010, *Fonsecaea nubica* was firstly described based on AFLP profiles and on sequences of ITS, *CDC42*, *BT2,* and *ACT1* genes in a retrospective study [23, 24]. However, detailed clinical

Fig. 11.1 *Fonsecaea monophora*, CBS 117238, (**a**) Colony on MEA after 4 weeks of incubation at 30 °C; (**b**, **c**) Conidiophores and conidia; (**d**) conidia. *Fonsecaea nubica*, CBS 269.64, (**e**) Colony on MEA after 4 weeks of incubation at 30 °C; (**f**, **g**) Conidiophores and conidia; (**h**) conidia. *Fonsecaea pedrosoi*, CBS 273.66. (**i**) Colony on MEA after 4 weeks of incubation at 30 °C; (**j**, **k**) Conidiophores and conidia; (**l**) conidia

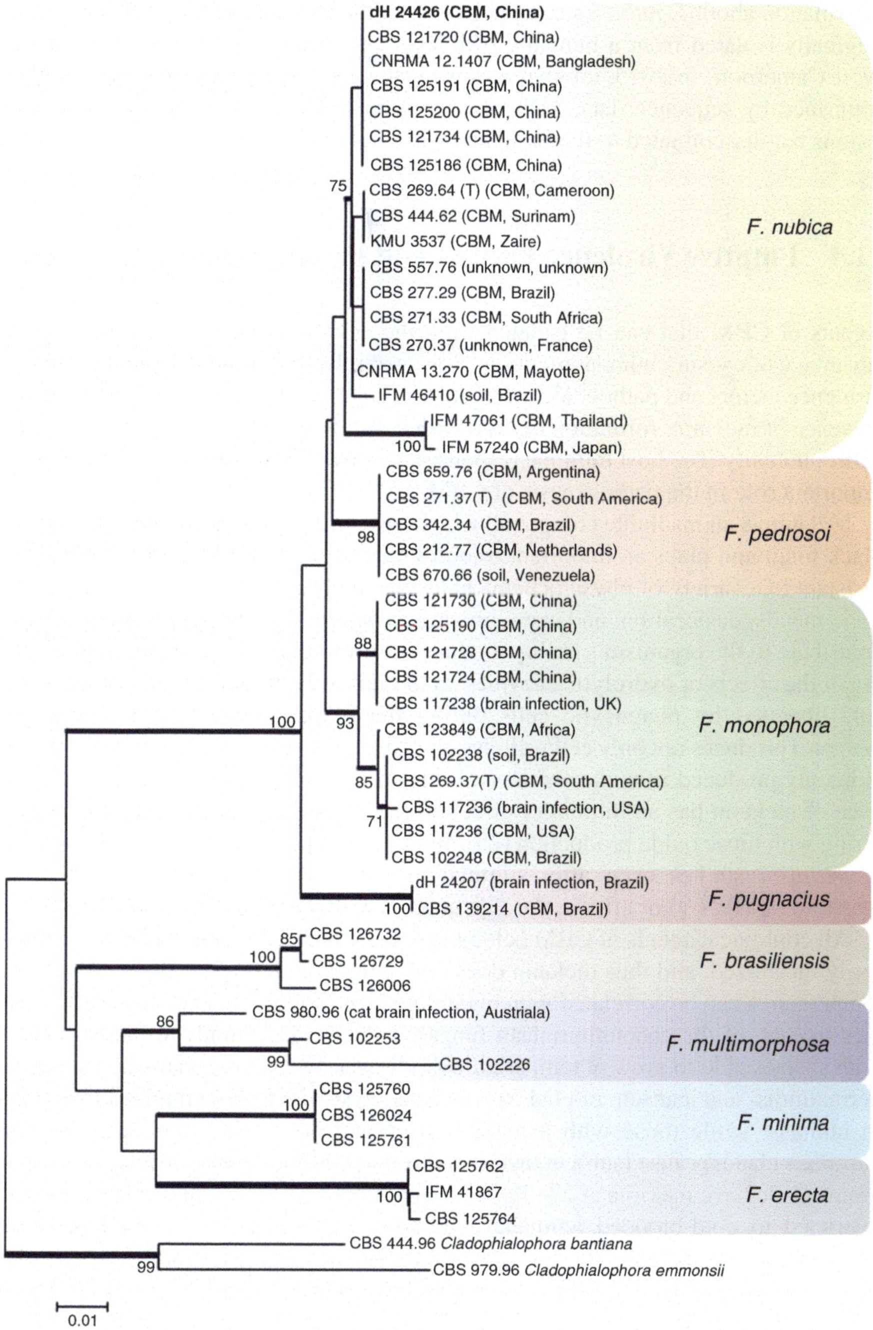

Fig. 11.2 Neighbor Joining tree based on ITS of the genus Fonsecaea. (Reprint from reference [21])

information about *F. nubica* scant, including on the type strain, CBS 269.64. It was originally isolated from a human CBM case with a dull appearance of lesions in West Cameroon. In 2015, another species was added to *Fonsecaea*, *F. pugnacius*, confirmed by sequence data. The case, from central Brazil, showed chronic skin lesions but disseminated to the brain and was fatal [17].

11.4 Putative Virulence Factors and Pathogenesis

Agents of CBM that can be isolated from the environment, but seem to have an advantage of causing human infection. However, relatively little is known regarding virulence factors and pathogenic mechanisms. Putative virulence factors include the presence of melanin, formation of muriform cells, thermotolerance, adherence, and hydrophobicity. The host immune responses to above virulence factors may play an important role in the pathogenesis of CBM.

Melanin is an insoluble compound which is mainly deposited in the cell walls of black fungi and plays an important role in virulence and pathogenicity. Melanin is resistant to a variety of physicochemical agents, including free radical compounds, toxic metals, desiccation, and even ionizing radiation [25, 26], and it is believed to contribute to the organism's ability to elude host immune responses through blocking of the effects of hydrolytic enzymes on the cell wall and scavenging of free radicals liberated by phagocytic cells during the oxidative burst [27]. *Fonsecaea pedrosoi* produces not only cell wall-associated but also extracellular melanin which is mainly produced in melanosomes associated with Fe^{2+} and Ca^{2+} [28]. The presence of melanin has an inhibitory effect on receptor-mediated phagocytosis, interfering with nitric oxide production [29]. Inhibition of phagocytosis was observed in *F. pedrosoi* studies even after stimulation with lipopolysaccharide (LPS) and interferon-γ (IFN-γ) or after in vitro incubation with the whole melanized cell wall.

All etiological agents of CBM belong to chaetothyrialean fungi which are obligatorily melanized, and thus melanin does not explain these predilections. Host body temperatures can be correlated with maximum growth temperatures of invasive species in each of the chaetothyrialean fungal clades [16]. Roughly speaking, clades with species able to grow at temperatures well over 37 °C or higher (the bantiana-, dermatitidis- and jeanselmei-clades) may cause systemic or disseminated infections in humans, while those with a maximum around 36–37 °C (the carrionii- and europaea-clades) cause (sub)cutaneous and superficial infections. Species of the salmonis clade have maxima at 27–33 °C, exceptionally at 36 °C, and cause infection restricted to cold-blooded animals, and rarely in humans, and then mostly on extremities with decreased blood circulation. The clinically relevant *Fonsecaea* species are located in the bantiana-clade with relatively well growth at 36–37 °C which could partly explain they cause not only CBM but also brain phaeophyphomycosis.

It is not known how and under what conditions muriform cells are formed in tissue. This special cellular plasticity is regarded as an important virulence factor for black yeasts and relatives. The presence of the meristematic growth forms in general is thought to be an adaptation to harsh environmental conditions, such as extreme temperatures, low water availability, acidity, nutrient deficiency, high UV exposure, or high salt concentrations. In addition, microbial adherence and hydrophobicity are two of the most important determinants of fungal pathogenesis [30]. For CBM, infectious propagules adhere to epithelial cells inside the host leading to differentiation of muriform cells that effectively resist destruction by host effector cells and allow the establishment of chronic disease. This phenomenon may be enhanced by relative cellular hydrophobicity due to the presence of hydrophilic extracellular polysaccharides. Other compounds such as peptidases, glycosphingolipids, and sialidase may be involved in pathogenesis as well [31].

The immunological mechanisms involved in prevention and control of CBM are still not very well understood. However, they probably include innate and adaptive immunity consisting of cellular humoral responses, which has been well reviewed by Seyedmousavi et al. [32]. The innate immune response functions as the first line of defense against invading pathogens. Macrophages are crucial cell in the immunological response in CBM. Rozental et al. [33] showed that activated macrophages were fungistatic to *F. pedrosoi*, delaying germ tube and hypha formation. Bocca et al. [34] showed that during infection with *F. pedrosoi*, macrophages were unable to produce NO, even after stimulation with lipopolysaccharide (LPS) and IFN-γ. Dendritic cells (DCs) are an important line of defense against black fungi when the infection starts with cutaneous or subcutaneous inoculation into tissues. Sousa et al. [35] reported that DCs are the first line of defense against inoculation of black yeasts and relatives into subcutaneous tissues. In their study, the interaction between *F. pedrosoi* and DCs obtained from patients with CBM was investigated. Monocyte-derived DCs from patients with severe forms of CBM induced $CD4^+$ T-cell activation and increased the expression of human leukocyte antigen D-related (HLA-DR) and co-stimulatory molecules (such as CD86, TNF-α, IL-10, and IL-12) in vitro. Notably, in the presence of conidia, the expression of HLA-DR and CD86 was upregulated by DCs from both patients and controls.

In cell-mediated immunity, it was demonstrated that $CD4^+$ T-helper cells and B-lymphocytes have a major role in defense against black yeasts [36]. The severity of diseases caused by black fungi is dependent on Th1/Th2 activation [32, 37]. Patients with chronic forms of infection exhibit increased IL-10 and low levels of IFN-γ. The T-helper 2 profile has been linked to extensive, severe verrucous forms. However, patients with erythematous atrophic plaques presented with a T-helper 1 profile. Recent studies showed that $CD4^+$ lymphocytes are potential key cells for the control of chromoblastomycosis. The role of Th17 lymphocytes in the host defense against CBM remains to be established. Recently, meristematic cells of *F. pedrosoi* were demonstrated to have a chitin-like component that inhibits dectin-1 mediated Th17 development, impairing immune response and contributing for CBM chronicity [38–40]. The host defense mechanism against CBM has not been fully understood yet and the role of humoral response in controlling the disease still remains unknown. In serum of patients with CBM antimicrobial antibodies are detected, however, these antibodies do not seem to play a role in CBM pathogenesis [41].

11.5 Update on Epidemiology

Chromoblastomycosis is not a reportable disease in most countries, thus the incidence of this disorder is not known, and epidemiological estimates are derived from surveys and case series. Furthermore, most of the species were identified by morphology; only during the last decade part of these reports have been well confirmed with molecular methods.

Chromoblastomycosis occurs worldwide, but the incidence is higher in the tropical and subtropical climates within a zone between 30° latitude North and 30° latitude South. Most cases are reported from Latin America, Africa, Asia and Australia countries like Brazil, Mexico, Venezuela, Madagascar, India and China account for the highest frequencies [2, 4]. In Brazil the estimate incidence rate of CBM is 3/100,000 [42]. Silva et al. [43] reported 325 cases in the Amazon region of Brazil, 98% of which had *Fonsecaea* spp. as the etiologic agent. In Mexico, more than 600 cases had been reported during the last seven decades, 90% of which was also caused by *Fonsecaea* spp. [44]. In Africa, Esterre et al. [45] reported 1343 cases of CBM from Madagascar, of which 98.5% were confirmed by histopathology, and *Fonsecaea* spp. were identified from 61.8% of the positive cultivation. Kombila et al. [46] reported 64 cases in Gabon, all caused by *Fonsecaea* spp. In China, Jiajun You first reported chromoblastomycosis in 1951. Since that time, more than 600 cases have been described in the Chinese and English literature, with the highest prevalence being reported in Guangdong (84/196) and Shandong Provinces (38/196). In this review, *Cladophialophora carrionii* is shown to be the predominant species in northern China, and *Fonsecaea* spp. in southern China [47]. In Japan, Kano et al. [48] described the first case of CBM. Later investigation showed Japan had the highest incidence of CBM among populations in Asia (1/416,000). Fukushiro [49] described approximately 296 cases of CBM reported in Japan from 1955 to 1982, and Kondo et al. [50] analyzed 212 additional cases reported from 1982 to 2001, using the database of the "Japan Centra Revuo Medicina". Recently molecular studies confirm *F. monophora* as the most prevalent species in Japan [51]. In Sri Lanka, 94% of 71 CBM cases were caused by *Fonsecaea* spp. [52]. Chromoblastomycosis is rarely seen in temperate Europe. A recent molecular confirmed case and review of the literature revealed a total of 31 cases in Europe since the first European case in Leningrad in 1929 [53]. In Australia, to date approximately 200 cases were published which is mainly due to *C. carrionii*, occurring in the dry bush county area [54]. In general, *Fonsecaea* spp. are prevalent in humid tropical climates, whereas *C. carrionii* infection is normally found in dry countries and desert regions.

In many published CBM case reports, etiologic agents were identified with histopathologic features in the tissue or phenotypic features of the culture pending precise molecular verification. More recently, Najafzadeh et al. [18, 19] assessed the geographic distribution of 81 *Fonsecaea* strains using amplified fragment length polymorphism (AFLP) technology and multilocus sequence typing. Judging from AFLP fingerprinting, *F. pedrosoi* is restricted to Central and South America, whereas

F. monophora and *F. nubica* have global geographical distributions and have been recovered from clinical specimens in Asia (Southern China, Japan), Western Hemisphere (Brazil, Mexico), Africa and Europe (France, England). It is noteworthy that both *F. monophora* and *F. nubica* clusters could be subdivided into two groups, and population 4 of *F. monophora* together with population 1 of *F. nubica* were prevalently confined to Southern China. Considering that the taxonomy of *Fonsecaea* species has been reevaluated, it becomes necessary to identify clinical isolates at the species level in order to study their epidemiology distribution, and to determine if different clinical presentation and response to antifungal therapy are associated with each of the taxa, which would be clinically critical for appropriate patient management.

The majority of the patients with CBM are between 30 and 50 years of age, with a male-to-female ratio of 4:1 to 9:1 [4, 44, 47, 55]. Most lesions are observed on the extremities of farmers and outdoor rural workers. The preponderance of males in most reported cases classically has been attributed to their higher exposure risk than gender susceptibility, but given the male preponderance in the neurotrope *Cladophialophora bantiana* and many other fungi, a hormonal cause seems more likely.

The ecological niches of *Fonsecaea* spp. and the mode of transmission of CBM remain unclear, though traumatic implantation is regarded as the main mode of infection of this disorder. For example, a Bangladesh CBM case recalled a trauma with a knife, while another remembered an injury with scrap wood [56, 57]. Thus far all known *F. nubica* and *F. pugnacius* strains were clinical CBM strains, however, a few *F. pedrosoi* and *F. monophora* strains were isolated from plant debris [20]. Occasional cases of CBM were acquired through different route, such as animal bite and insect sting [4]. Recently Slesak et al. [58] reported a CBM case associated with a leech bite. Other CBM cases associated with insect bites were reported in Thailand, due to *F. pedrosoi* [59] and in Venezuela, by a possibly undescribed species [60].

Numerous reports of epizootics of CBM-like infections in animals have been published [16, 22, 61]. In the animal model using immunosuppressed mice, *F. pedrosoi* and *F. monophora* replicate the infection with intracutaneous or intercutaneous lesions, but as yet there is no confirmed case of animal CBM.

11.6 Clinical Manifestations

Chromoblastomycosis is a chronic fungal infection of the skin and subcutaneous tissue. The infection usually appears after minor trauma preceding the lesions, however, only few patients recall any history of trauma as it is usually innocuous, occurs months or decades earlier, and is mostly forgotten. The exposed body parts, particularly the extremities, are involved most frequently, especially feet and shins [1, 2, 4]. Cases of auricle, face, trunk, and buttock infection have also been reported [21, 53,

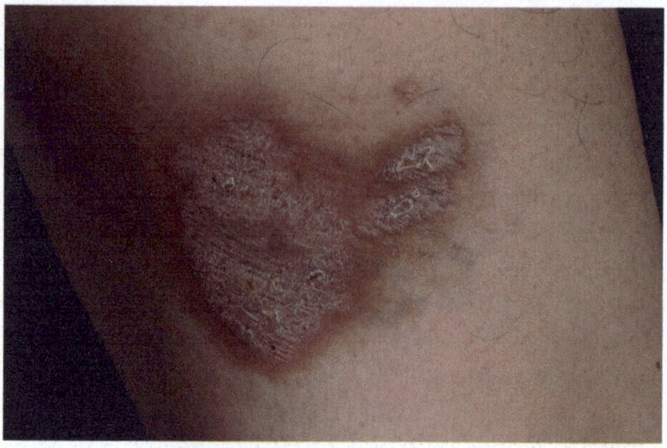

Fig. 11.3 Different types of lesions of chromoblastomycosis

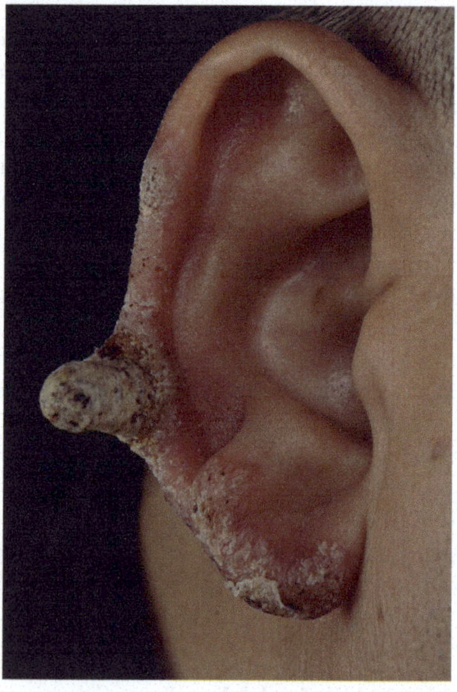

Fig. 11.4 Different types of lesions of chromoblastomycosis

62–64]. The primary lesion may start as a solitary erythematous macule, then progresses to a papular shape with pink smooth-surface [1] which gradually enlarges from the site of infection to become polymorphic over extended periods of time. In general, there are five different types of lesions. Nodular, tumoral (cauliflower-like), verrucous, plaque and cicatricial [1, 2, 4]. Recently pseudo-vacuolar, eczematous and horn-like lesions were also observed [21, 47, 65]. In advanced and severe cases,

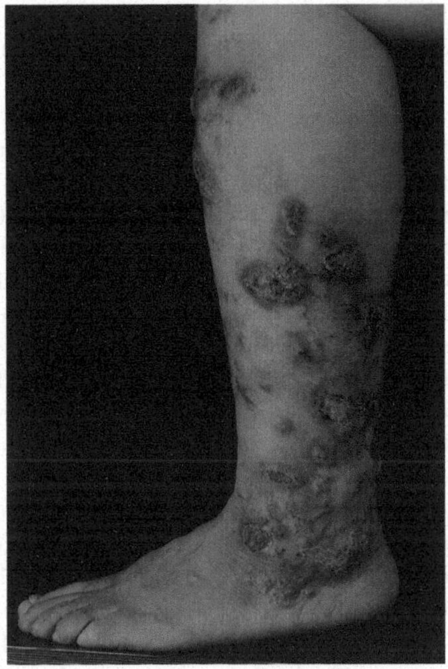

Fig. 11.5 Different types of lesions of chromoblastomycosis

Table 11.1 Differential diagnosis of chromoblastomycosis lesions

Infectious diseases	Fungi	Blastomycosis
	Fungi	Coccidiodomycosis
	Fungi	Phaeohyphomycosis
	Fungi	Sporotrichosis
	Protozoa	Leishmaniasis
	Protozoa	Rhinosporidiosis
	Bacteria	Ecthyma
	Bacteria	Nocardiosis
	Bacteria	Tuberculosis
	Virus	Warts
Non-infectious diseases		Carcinoma
		Lupus
		Psoriasis
		Sarcoidosis

more than one type of lesion can be observed in the same patient (Figs 11.3, 11.4, and 11.5; Table 11.1).

Apart from the division presented above, the lesions in CBM may be classified according to severity [2, 65]. Three levels are distinguished: (1) Mild form – a solitary plaque or nodules measuring <5 cm in diameter; (2) Moderate form – solitary or multiple lesions of plaque, verrucous or nodular type, isolated or conjoined, covering one or two adjacent body areas measuring <15 cm in diameter; (3) Severe form – any type of single or multiple lesions covering extensive skin areas, whether adjacent or non-adjacent. Severe lesions tend to poorly respond to treatment or become treatment-resistant.

The initial and mild skin lesions are usually asymptomatic, may pruritic or become painful. Satellite lesions can develop from autoinoculation through scratching and from lymphatic dissemination. Common complications of CBM include ulceration and secondary bacterial infection. In severe cases, chronic lymphedema and ankyloses may develop, rarely, malignant transformation into squamous cell carcinoma may arise, which may lead to long term disability of the patient. Extracutaneous infection due to contiguous spread, hematogenic or lymphatic dissemination involving bone, lymph nodes, lungs and brain have been rarely reported [50, 66–70], but should be noted that these results are tentative pending precise identification of the etiologic agents by molecular methodology. Recently confirmed cases of central nervous system and invasive tissue infections show that the etiological agents are mostly attributed to *F. monophora* [70–76] and once to *F. pugnacius* [17]. An experimental murine model of disseminated infection with *F. monophora* was also documented [77]. In brain tissue histopathology is with pigmented hyphae rather than muriform cells, probably due to lower levels of immunity in the brain.

Clinical presentations of human CBM may vary according to the immunological status of the host, the time of evolution, the load of the inoculum, involved site, patient's hygienic habits, compliance to therapy and the pathogenicity of the strain, among other factors. In order to better plan dose and duration of therapy as well as the patient's prognosis, the CBM lesions must be also classified according to the predominant clinical type and severity grade.

11.7 Classical and Novel Diagnostics

Chromoblastomycosis is diagnosed through a combination of clinical and laboratory findings, i.e. direct examination of skin scrapings or biopsy tissue and/or histopathology. Direct examination of specimens is conducted with 10–40% potassium hydroxide (KOH) which displays muriform cells unambiguously. Thick-walled, chestnut brown, rounded cells (4–12 μm in diam) with transversal and longitudinal septation are readily seen, also known as sclerotic cells, Medlar bodies, or copper pennies; these are a pathognomonic criterion for diagnosis. Muriform cells are easily found in black dots at the lesion's surface which are small hematic crusts with cellular debris and fungal structures resulting from transepithelial elimination. Occasionally, pigmented hyphae can be also observed, especially when the patient has decreased immunity, muriform cells undergoing morphological transformation for germination (Fig. 11.6.).

Muriform cells are not informative for species identification of the agent. As CBM could be caused by a diverse group of melanized fungi, and the knowledge of fungal identification down to species level contributes to our understanding of their

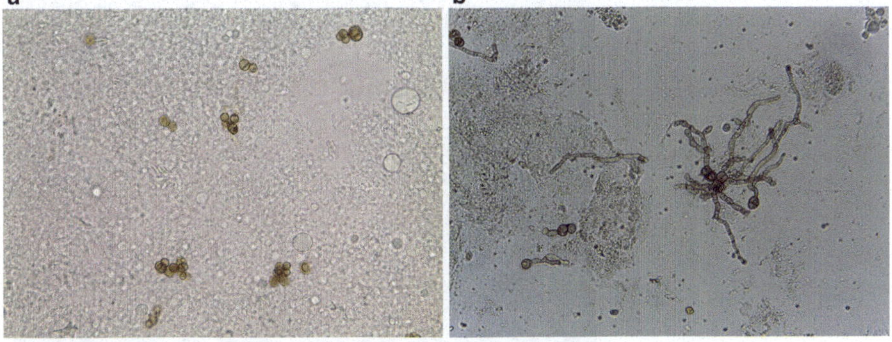

Fig. 11.6. Muriform cells (**a**) and germinating muriform cells (**b**)

epidemiological, biodiversity and clinical features. In addition, the causative agents of CBM could also cause other clinical syndromes, such as phaeohyphomycosis. For definitive CBM diagnosis isolation and identification of the etiological agent it is recommended to use the term "chromoblastomycosis" [27, 78]. Culturing is done on selective media supplemented with cycloheximide and chloramphenicol avoiding fast growing fungi and bacteria. Most agents of chromoblastomycosis have a maximum growth temperature around 36–37 °C [16] and require 2–3 weeks before small colonies are obtained. Colonies are flat to heaped and folded, velvety to cottony, olivaceous to black with black reverse. Conidiophores are short, branched, with short, sympodial, chains of conidia on denticles. Budding cells are absent. Phialides with collarettes may be present in low abundance on poor media. rDNA ITS is recommend as a genetic barcoding marker for *Fonsecaea* species. Shared traits between species are absence of growth with inulin as sole carbon source, absence of fermentation, tolerance of cycloheximide and intolerance of 10% NaCl. Members of the order *Dothideales*, including *Aureobasidium* and *Hortaea*, mostly grow with 10% NaCl. Urease production varies with the strain [22].

Muriform cells seen in CBM are frequently detected by direct microscopy, however, tissue invasion is best visualized and documented by permanent histopathological preparations. Although muriform cells may be easily seen in tissue with the routinely used hematoxylin and eosin (H&E) stain, other special stains such as Gomori methenamine silver (GMS) stain and periodic acid-Schiff (PAS) stain could be employed to enhance fungal detection, especially where fungal burden are low or with secondary infections. In skin, a characteristic pseudo-epitheliomatous hyperplasia with hyperkeratosis and parakeratosis occurs in the epidermis, and mixed granulomatous suppuration with fungal elements with irregular acanthosis alternating with areas of atrophy are seen in the dermis. Muriform cells are found in the stratum corneum and the following layers of the epidermis, and in the granulomatous processes and in giant cells (Fig. 11.7.).

Several serological tests have been developed to aid CBM diagnosis, though they are not routinely used because of relatively low sensitivity and specificity, and only

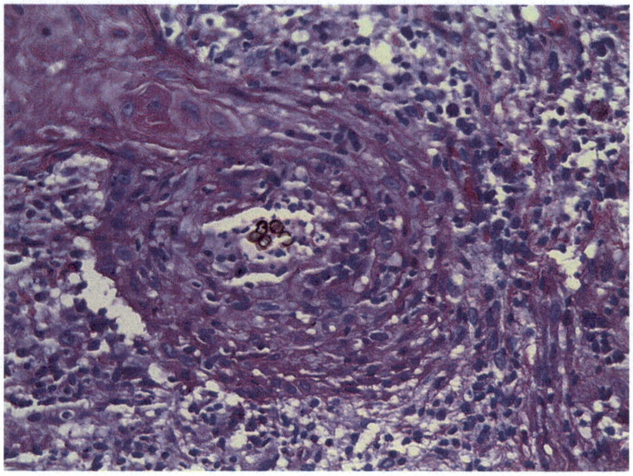

Fig. 11.7. Muriform cells in a granulomatous process

small numbers of patients have been studied. *Fonsecaea pedrosoi* was analyzed using immunoblotting, a 54-KD antigen with high sensitivity and specificity was found [79]. Double immunodiffusion (DID), counterimmunoelectrophoresis (CIE) and immunoenzymatic test (ELISA) have also been employed to evaluate humoral immune responses. Metabolic antigen (chromomycin) was used for immunoprecipitation tests (DID and CIE), while somatic antigens for ELISA were found. Results show variable sensitivity for all techniques, with relatively high specificity ranging from 83 to 90% [80]. Chromomycin was effective in detecting delayed hypersensitivity in patients with *F. pedrosoi* infection, with sensitivity and specificity of 90.0% and 98.8%, respectively. Recently, Bruña-Romero and Marques [81] evaluated the association between IgG antibody levels and the severity of CBM and therapeutic response of patients to itraconazole. The levels of antibodies in 55 CBM patients expressed in optical density and titration showed that, before treatment, patients with severe disease had higher levels of IgG, IgG1, IgG2, and IgG3 when compared with moderate or mild disease. Main antibody titers of IgG, IgG1, and IgG2 were reduced after treatment. However, the anti-*F. pedrosoi* antibody did not exhibit a protective role against infection.

Currently, nonculture methods are in use to improve sensitivity and specificity of mycological diagnosis. Molecular identification of individual *Fonsecaea* species is done with the rDNA ITS region, which is currently a universal DNA barcode marker for fungi. For distinction of closely related *Fonsecaea* species causing CBM, an additional gene such as partial β-tubulin (*BT2*) or translation elongation factor 1-α (*TEF1*) may be recommended. In addition, a quality-controlled reference databases should be used in the comparison. For example, Slesak et al. [58] recently reported a CBM case associated with a leech bite, identified by ITS rDNA sequencing, yielding 100% identity with *F. pedrosoi*, *F. monophora*, and *F. nubica* in GenBank. However, comparing HQ616145 with the CBS database (http://www.westerdijkinstitute.nl/) 100% similarity with *F. monophora* and 99.8% with *F. pedrosoi* was obtained, leading to final identification as *F. monophora*. The ITS database of the

International Society of Human and Animal Mycology (http://www.isham.org/ and http://its.mycologylab.org/) provides a quality-controlled standard tool for routine identification [82, 83].

Molecular identification may be hampered by sequence variability in the ITS domain caused by difficult-to-sequence homopolymeric regions and by poor taxonomic attribution of sequences deposited in GenBank. Heinrichs et al. [84] proposed short barcode identifiers (27 to 50 bp) based on sufficiently conserved ITS2 for the common clinical *Fonsecaea* species. Using the proposed barcode identifiers, a 4.1-fold increase of 100% matches in GenBank was achieved.

Isothermal amplification methods, such as Loop-mediated isothermal amplification (LAMP) and rolling circle amplification (RCA) have been developed for rapid detection and identification of *Fonsecaea* stains [18, 19, 85]. LAMP proved to be a fast and sensitive method for direct amplification of fungal DNA from environmental samples but could not distinguish individual *Fonsecaea* species, whereas for RCA ITS amplicons are needed but is species-specific.

11.8 Treatment

Despite in vitro susceptibility of agents to most antifungals, CBM is often recalcitrant and difficulty to cure. Except for the initial lesions, which can be surgically removed, CBM lesions constitute a true therapeutic challenge. A clinical guideline for diagnosis and management of infections by melanized fungi was published by ESCMID/ECMM [86], in which CBM is treated with itraconazole (ITZ) plus surgery (recommendation AII) or terbinafine (TBF) monotherapy plus surgery (recommendation BIII). Recommendations for refractory cases are combined antifungal therapy and cryotherapy or surgery (recommendation BIII). Laser, heat and potassium iodide therapies have also been used with successful outcome (recommendation BIII). Based on experimental and in vitro studies the new triazole drug posaconazole (PCZ) is promising and could be useful when other therapies have failed (recommendation BIII). However, this guideline did not distinguish between etiologic agents. In general, *Fonsecaea* infections are more difficult to treat than those by *C. carrionii* [65].

Up to now, breakpoints have not been established for melanized fungi, including *Fonsecaea* species. Several studies have documented that *F. pedrosoi* isolates exhibit high in vitro susceptibility to triazoles such as ITZ, voriconazole (VCZ), PCZ and isavuconazole (ISA), but limited in vitro susceptibility to amphotericin B (AMB), fluconazole (FCZ) and 5-flucytosine (5-FC). Najafzadeh et al. [23, 24] tested a total of 55 strains of *Fonsecaea* species using the CLSI document M38-A2. All isolates exhibited low MIC values for ITZ, VCZ, PCZ, ITZ and ISA. In contrast, higher MIC values were documented against AMB, FCZ, and echinocandins including caspofungin (CAS) and anidulafungin (AND). Although there were no significant differences in the in vitro activities of the new triazoles against *F. pedrosoi*, *F. monophora* and *F. nubica*, it was observed that the MICs to VCZ and ISA were 1 or 2 dilutions higher in strains of *F. pedrosoi* compared to *F. monophora* and/or *F. nubica*.

Itraconazole is a fungistatic first generation triazole which is metabolized by the cytochrome P450. It is the most widely used drug in the treatment of CBM with satisfactory results and good tolerance, even with long-term treatment. The dose varies from 200–400 mg/day, and the duration of therapy ranges from months to years, with most of the cases showing improvement in 8–10 months. In an early study, safety and efficacy of ITZ were evaluated in 15 CBM patients caused by *F. pedrosoi* [87]. Patients with mild to severe disease were cured with ITZ after a mean duration of 7.2 months and 25.1 months, respectively. No significant side effects nor biochemical alteration during this trial were important enough to interrupt the treatment. However, it has been suggested that ITZ resistance might be acquired in patients under chronic use of this drug. Andrade et al. [88] found that sequential isolates from four patients presented higher ITZ MIC values than initial strains; for two of these patients, no response to therapy was observed.

Terbinafine (TBF) is an allylamine derivative causing a fungicidal and fungistatic effect through lanosterol synthesis. It has less drug-drug interactions than ITZ since TBF does not interfere with P450 cytochrome. Several reports have demonstrated efficacy and tolerability of TBF 250–500 mg/day obtaining cure CBM due to *Fonsecaea* species [89–91]. In an open trial, long courses of terbinafine at a dosage of 500 mg/day were administered orally to 43 CBM patients. Mycological cure was observed in 82.5% of patients infected with *F. pedrosoi* after 12 months of therapy. The efficacy of terbinafine in *C. carrionii*-infected patients seemed higher than in *Fonsecaea*-infected patients [90].

Combination therapy is an option for refractory or severe CBM cases, and its clinical effect in vivo has been clearly demonstrated. Zhang et al. [92] evaluated the in vitro effects of a combination of TBF plus ITZ on 18 clinical *F. monophora* isolates. The fractional inhibitory concentration index (FICI) analysis demonstrated that 12 (67%) were synergistic, 4 (22%) were additive, and 2 (11%) were indifferent, with no antagonism being observed. The minimal inhibitory concentrations (MICs) obtained with the terbinafine-itraconazole combination were within levels that can be achieved in plasma at clinically relevant doses. Gupta et al. [93] used ITZ and TBF in an alternate week or combination therapy on four patients with longstanding CBM (8–23 years) caused by *F. pedrosoi* with poor response to standard oral antifungal monotherapy. The combination therapy was considered effective in all four patients.

Posaconazole (PCZ) is a new, extended-spectrum triazole antifungal agent that has demonstrated in vitro activity against many melanized fungi including *Fonsecaea* species. Two clinical reports suggested that PCZ 800 mg/day was associated with success rates of more than 80% in refractory CBM patients [94, 95]. The efficacy of PCZ was also evaluated in a murine model of disseminated infection by *F. monophora*. The results suggested that PCZ may have a clinical role in the treatment of disseminated infections by *F. monophora*, because it prolonged survival significantly and reduced the fungal load in most of the organs tested [77].

In most cases, several modalities of physical methods may be used but as an adjuvant therapy, always in combination with antifungal drugs. Physical therapies that can be applied include surgery, thermotherapy, laser and photodynamic therapy. Surgical excision can be used for initial lesions only; unfortunately CBM initial lesions are scarcely observed in clinic. Thermotherapy encompasses heat and cold therapy. The use of pocket warmers or electric bed warmer has proven successful in the treatment of a limited number of cases, because the increase in skin temperature somehow impairs fungal development [96]. However, heat therapy as single treatment shows insufficient result, combination with other treatment modalities being required [97]. Cryotherapy is convenient and cost-effective, especially for localized lesions. Dissemination of lesions to adjacent areas after cryotherapy has been noted and therefore it is important to combine cryotherapy with ITZ or TBF. Photodynamic therapy (PDT) using 5-aminolevulinic acid (ALA) irradiation may be adjunctive in combination with antifungal medication [98, 99]. Hu et al. [100] combined oral terbinafine with weekly ALA-PDT in a case of CBM and with this combination, apparent clinical improvement was achieved within less than a year and no recurrence was observed.

Acknowledgments Sarah Abdallah Ahmed is thanked for preparing the photoplates of *Fonsecaea*.

References

1. Queiroz-Telles F, Esterre P, Perez-Blanco M et al (2009) Chromoblastomycosis: an overview of clinical manifestations, diagnosis and treatment [J]. Med Mycol 47(1):3–15
2. Queiroz-Telles F, de Hoog GS, Santos DWCL et al (2017) Chromoblastomycosis [J]. Clin Microbiol Rev 30(1):233–276
3. de Hoog GS, Guarro J, Gené J, Figueras MJ (2016) Atlas of clinical fungi, 4th edn. Centraalbureau voor Schimmelcultures/Universitat Rovira i Virgili, Utrecht
4. Queiroz-Telles F (2015) Chromoblastomycosis: a neglected tropical disease [J]. Rev Inst Med Trop Sao Paulo 57:46–50
5. Rudolph M (1914) Über die brasilianische "Figueira" (Vorläufige Mitteilung) [J]. Archiev Schiffs und Tropen-Hyg 18:498–499
6. Lane CG (1915) A cutaneous disease caused by a new fungus (Phialophora verrucosa) [J]. J Cutan Dis 33:840–846
7. Medlar EM (1915) A cutaneous infection caused by a new fungus, Phialophora verrucosa, with a study of the fungus [J]. J Med Res 32(3):507
8. Pedroso A, Gomes JM (1920) Sôbre quatro casos de dermatite verrucosa produzida pela Phialophora verrucosa [J]. Ann Paulistas de Medicina e Cirurgia 11:53–61
9. Brumpt E (1922) Prés de Parasitologie, 3rd edn. Masson, Paris, France, p 1105
10. Negroni P (1936) Estudio del 1 caso argentino de cromomicosis, Fonsecaea (ng) pedrosoi (Brumpt 1921) [J]. Rev Inst Bact Dpto Nat de Higiene 7:419–426
11. Terra F, Torres M, Da Fonseca O et al (1922) Novo typo de dermatite verrucosa mycose por Acrotheca com associacao de leishmaniosa [J]. Brasil Medico 2:363–368
12. Moore M, De Almeida F (1935) Etiologic agents of chromomycosis (chromoblastomycosis of Terra, Torres, Fonseca and Leao, 1922) of north and South America [J]. Revista de Bilologia e Hygiene 6:94–97

13. Ajello L (1975) Phaeohyphomycosis: definition and etiology [J]. PAHO Sci Publicatios 304:126–130
14. McGinnis MR (1983) Chromoblastomycosis and phaeohyphomycosis: new concepts, diagnosis, and mycology [J]. J Am Acad Dermatol 8(1):1–16
15. Odds FC, Arai T, Disalvo AF et al (1992) Nomenclature of fungal diseases: a report and recommendations from a sub-Committee of the International Society for human and animal mycology (ISHAM) [J]. J Med Vet Mycol 30(1):1–10
16. de Hoog GS, Vicente VA, Najafzadeh MJ et al (2011) Waterborne Exophiala species causing disease in cold-blooded animals [J]. Persoonia 27(1):46–72
17. de Azevedo CMPS, Gomes RR, Vicente VA et al (2015) Fonsecaea pugnacius, a novel agent of disseminated chromoblastomycosis [J]. J Clin Microbiol 53(8):2674–2685
18. Najafzadeh MJ, Sun J, Vicente VA et al (2011a) Molecular epidemiology of Fonsecaea species [J]. Emerg Infect Dis 17(3):464–469
19. Najafzadeh MJ, Sun J, Vicente VA et al (2011b) Rapid identification of fungal pathogens by rolling circle amplification using Fonsecaea as a model [J]. Mycoses 54(5):e577–e582
20. Vicente VA, Najafzadeh MJ, Sun J et al (2014) Environmental siblings of black agents of human chromoblastomycosis [J]. Fungal Divers 65(1):47–63
21. Chen Y, Yin S, Li M et al (2016) A case of chromoblastomycosis by Fonsecaea nubica indicating a possible insect route of transmission [J]. Mycoses 59(10):662–667
22. de Hoog GS, Attili-Angelis D, Vicente VA et al (2004) Molecular ecology and pathogenic potential of Fonsecaea species [J]. Med Mycol 42(5):405–416
23. Najafzadeh MJ, Sun J, Vicente V et al (2010a) Fonsecaea nubica sp. nov, a new agent of human chromoblastomycosis revealed using molecular data [J]. Med Mycol 48(6):800–806
24. Najafzadeh MJ, Badali H, Illnait-Zaragozi MT et al (2010b) In vitro activities of eight antifungal drugs against 55 clinical isolates of Fonsecaea spp. [J]. Antimicrob Agents Chemother 54(4):1636–1638
25. Nosanchuk JD, Casadevall A (2003) The contribution of melanin to microbial pathogenesis [J]. Cell Microbiol 5(4):203–223
26. Wheeler MH, Bell AA (1988) Melanins and their importance in pathogenic fungi [M]//current topics in medical mycology. Springer, New York, pp 338–387
27. Revankar SG, Sutton DA (2010) Melanized fungi in human disease [J]. Clin Microbiol Rev 23(4):884–928
28. Franzen AJ, Cunha MML, Miranda K et al (2008) Ultrastructural characterization of melanosomes of the human pathogenic fungus Fonsecaea pedrosoi [J]. J Struct Biol 162(1):75–84
29. Zhang J, Wang L, Xi L et al (2013) Melanin in a meristematic mutant of Fonsecaea monophora inhibits the production of nitric oxide and Th1 cytokines of murine macrophages [J]. Mycopathologia 175(5–6):515–522
30. Cunha MML, Franzen AJ, Seabra SH et al (2010) Melanin in Fonsecaea pedrosoi: a trap for oxidative radicals [J]. BMC Microbiol 10(1):1
31. Santos ALS, Palmeira VF, Rozental S et al (2007) Biology and pathogenesis of Fonsecaea pedrosoi, the major etiologic agent of chromoblastomycosis [J]. FEMS Microbiol Rev 31(5):570–591
32. Seyedmousavi S, Netea MG, Mouton JW et al (2014) Black yeasts and their filamentous relatives: principles of pathogenesis and host defense [J]. Clin Microbiol Rev 27(3):527–542
33. Rozental S, Alviano CS, De Souza W (1994) The in vitro susceptibility of Fonsecaea pedrosoi to activated macrophages [J]. Mycopathologia 126(2):85–91
34. Bocca AL, Brito PPMS, Figueiredo F et al (2006) Inhibition of nitric oxide production by macrophages in chromoblastomycosis: a role for Fonsecaea pedrosoi melanin [J]. Mycopathologia 161(4):195–203
35. Sousa MG, Ghosn E, Nascimento RC et al (2009) Monocyte-derived dendritic cells from patients with severe forms of chromoblastomycosis induce CD4+ T cell activation in vitro [J]. Clin Exp Immunol 156(1):117–125

36. d'Ávila SCGP, Pagliari C, Duarte MIS (2003) The cell-mediated immune reaction in the cutaneous lesion of chromoblastomycosis and their correlation with different clinical forms of the disease [J]. Mycopathologia 156(2):51–60
37. Gimenes VMF, de Souza MDG, Ferreira KS et al (2005) Cytokines and lymphocyte proliferation in patients with different clinical forms of chromoblastomycosis [J]. Microbes Infect 7(4):708–713
38. de Lima SAA, Criado PR, Nunes RS et al (2014) In situ immune response in human chromoblastomycosis – a possible role for regulatory and Th17 T cells [J]. PLoS Negl Trop Dis 8(9):e3162
39. Dong B, Li D, Li R et al (2014) A chitin-like component on sclerotic cells of Fonsecaea pedrosoi inhibits Dectin-1-mediated murine Th17 development by masking β-glucans [J]. PLoS One 9(12):e114113
40. Wüthrich M, Wang H, Li M et al (2015) Fonsecaea pedrosoi-induced Th17-cell differentiation in mice is fostered by Dectin-2 and suppressed by Mincle recognition [J]. Eur J Immunol 45(9):2542–2552
41. Esterre P, Jahevitra M, Andriantsimahavandy A (2000) Humoral immune response in chromoblastomycosis during and after therapy [J]. Clin Diagn Lab Immunol 7(3):497–500
42. Queiroz-Telles F, Nucci M, Colombo AL et al (2011) Mycoses of implantation in Latin America: an overview of epidemiology, clinical manifestations, diagnosis and treatment [J]. Med Mycol 49(3):225–236
43. Silva JP, de Souza W, Rozental S (1998) Chromoblastomycosis: a retrospective study of 325 cases on Amazonic region (Brazil) [J]. Mycopathologia 143(3):171–175
44. Navarrete MR, Arenas R, Estrada VFM et al (2014) Cromoblastomicosis en México: revisión de 603 casos en siete décadas [J]. Dermatología Cosmética, Médica y Quirúrgica 12(2):87–93
45. Esterre P, Andriantsimahavandy A, Ramarcel ER et al (1996) Forty years of chromoblastomycosis in Madagascar: a review [J]. Am J Trop Med Hygiene 55(1):45–47
46. Kombila M, de Diaz MG, Richard-Lenoble D et al (1995) La chromoblastomycose au Gabon Étude de 64 cas [J]. Cahiers d'études et de recherches francophones/Santé 5(4):235–244
47. Lu S, Lu C, Zhang J et al (2013) Chromoblastomycosis in mainland China: a systematic review on clinical characteristics [J]. Mycopathologia 175(5–6):489–495
48. Kano K (1937) Über die Chromoblastomykose durch einen noch nicht als pathogen beschriebenen Pilz: Hormiscium dermatitidis n. sp [J]. Arch Dermatol Res 176(3):282–294
49. Fukushiro R (1983) Chromomycosis in Japan. International Journal of Dermatology 22(4):221–229
50. Kondo M, Hiruma M, Nishioka Y et al (2005) A case of chromomycosis caused by Fonsecaea pedrosoi and a review of reported cases of dematiaceous fungal infection in Japan [J]. Mycoses 48(3):221–225
51. Yaguchi T, Tanaka R, Nishimura K et al (2007) Molecular phylogenetics of strains morphologically identified as Fonsecaea pedrosoi from clinical specimens [J]. Mycoses 50(4):255–260
52. Attapattu MC (1997) Chromoblastomycosis–a clinical and mycological study of 71 cases from Sri Lanka [J]. Mycopathologia 137(3):145–151
53. Pindycka-Piaszczyńska M, Krzyściak P, Piaszczyński M et al (2014) Chromoblastomycosis as an endemic disease in temperate Europe: first confirmed case and review of the literature [J]. Eur J Clin Microbiol Infect Dis 33(3):391–398
54. Weedon D, van Deurse M, Allison S et al (2013) Chromoblastomycosis in Australia: an historical perspective [J]. Pathology 45(5):489–491
55. Minotto R, Bernardi CDV, Mallmann LF et al (2001) Chromoblastomycosis: a review of 100 cases in the state of Rio Grande do Sul, Brazil [J]. J Am Acad Dermatol 44(4):585–592
56. Sophie B, Coralie Z, Ba HM et al (2015) First case of chromoblastomycosis from Bangladesh [J]. Med Mycol Case Rep 10:1–3
57. Yanagihara S, Kobayashi H, Kamo R et al (2015) Chromoblastomycosis caused by Fonsecaea nubica: first report from Japan [J]. J Dermatol 42(8):833–834

58. Slesak G, Inthalad S, Strobel M et al (2011) Chromoblastomycosis after a leech bite complicated by myiasis: a case report [J]. BMC Infect Dis 11(1):14
59. Ungpakorn R, Reangchainam S (2006) Pulse itraconazole 400 mg daily in the treatment of chromoblastomycosis [J]. Clin Exp Dermatol 31(2):245–247
60. Sauerteig E, Hernández R, Salfelder K et al (1998) Acute chromoblastomycosis provoked by an insect bite in an immunosuppressed patient [J]. Mycoses 41(5–6):191–194
61. Seyedmousavi S, Guillot J, de Hoog GS (2013) Phaeohyphomycoses, emerging opportunistic diseases in animals [J]. Clin Microbiol Rev 26(1):19–35
62. Sharma MM, Misra RN, Gandham NR et al (2012) Chromoblastomycosis of the face: a rare case report from the district of western Maharashtra, India [J]. J Clin Diagn Res 6(5):899–901
63. Silva CM, Da Rocha RM, Moreno JS et al (1994) [the coconut babacu (Orbignya Phalerata martins) as a probable risk of human infection by the agent of chromoblastomycosis in the state of Maranhao, Brazil] [J]. Rev Soc Bras Med Trop 28(1):49–52
64. Verma S, Verma GK, Singh G et al (2012) Facial chromoblastomycosis in sub-Himalayan region misdiagnosed as cutaneous leishmaniasis: brief report and review of Indian literature [J]. Dermatol Online J 18(10):3
65. Queiroz-Telles F, de Santos DW (2013) Challenges in the therapy of chromoblastomycosis [J]. Mycopathologia 175(5–6):477–488
66. de Guzman L, Perlman DC, Hubbard CE (2012) Septic arthritis and osteomyelitis due to the chromoblastomycosis agent Fonsecaea pedrosoi [J]. Am J Orthop 41(7):328–331
67. Morris A, Schell WA, McDonagh D et al (1995) Pneumonia due to Fonsecaea pedrosoi and cerebral abscesses due to Emericella nidulans in a bone marrow transplant recipient [J]. Clin Infect Dis 21(5):1346–1348
68. Nóbrega JPS, Rosemberg S, Adami AM et al (2003) Fonsecaea pedrosoi cerebral phaeohyphomycosis (" chromoblastomycosis"): first human culture-proven case reported in Brazil [J]. Rev Inst Med Trop Sao Paulo 45(4):217–220
69. Sharma NL, Sharma VC, Mahajan V et al (2007) Chromoblastomycosis with underlying osteolytic lesion [J]. Mycoses 50(6):517–519
70. Surash S, Tyagi A, de Hoog GS et al (2005a) Cerebral phaeohyphomycosis caused by Fonsecaea monophora [J]. Med Mycol 43(5):465–472
71. Bagla P, Loeffelholz M, Blanton LS (2016) Cerebral phaeohyphomycosis by Fonsecaea monophora: report in a patient with AIDS and a ring enhancing lesion [J]. Medical Mycology Case Reports 12:4–7
72. Doymaz MZ, Seyithanoglu MF, Hakyemez I et al (2015) A case of cerebral phaeohyphomycosis caused by Fonsecaea monophora, a neurotropic dematiaceous fungus, and a review of the literature [J]. Mycoses 58(3):187–192
73. Koo S, Klompas M, Marty FM (2010) Fonsecaea monophora cerebral phaeohyphomycosis: case report of successful surgical excision and voriconazole treatment and review [J]. Med Mycol 48(5):769–774
74. Madhugiri VS, Bhagavatula ID, Mahadevan A et al (2011) An unusual infection, an unusual outcome-Fonsecaea pedrosoi cerebral granuloma: case report [J]. J Neurosurg Pediatr 8(2):229–232
75. Takei H, Goodman JC, Powell SZ (2006) Cerebral phaeohyphomycosis caused by Cladophialophora bantiana and Fonsecaea monophora: report of three cases [J]. Clin Neuropathol 26(1):21–27
76. Varghese P, Jalal MJA, Ahmad S et al (2016) Cerebral phaeohyphomycosis caused by Fonsecaea monophora: first report from India [J]. Int J Surg Med 2(1):44–49
77. Calvo E, Pastor FJ, Rodríguez MM et al (2010) Murine model of a disseminated infection by the novel fungus Fonsecaea monophora and successful treatment with posaconazole [J]. Antimicrob Agents Chemother 54(2):919–923
78. Barton K, Miller D, Pflugfelder SC (1997) Corneal chromoblastomycosis [J]. Cornea 16(2):235–239
79. Vidal MSM, Castro LGM, Cavalcante SC et al (2004) Highly specific and sensitive, immunoblot-detected 54 kDa antigen from Fonsecaea pedrosoi [J]. Med Mycol 42(6):511–515

80. da Azevedo S, da Silva AAM, Marques SG et al (2013) Detection of delayed hypersensitivity to Fonsecaea pedrosoi metabolic antigen (Chromomycin) in healthy people in an endemic area [J]. Journal of Life Sciences 7(3):267
81. Bruña-Romero O, Marques SG, Nascimento FR et al (2014) Association of IgG immunoglobulin and subclasses level with the severity of chromoblastomycosis due to Fonsecaea pedrosoi and therapeutic response to itraconazole [J]. Eur J Clin Microbiol Infect Dis 33(10):1791–1797
82. Irinyi L, Serena C, Garcia-Hermoso D et al (2015) International Society of Human and Animal Mycology (ISHAM)-ITS reference DNA barcoding database—the quality controlled standard tool for routine identification of human and animal pathogenic fungi [J]. Med Mycol 53(4):313–337
83. Schoch CL, Seifert KA, Huhndorf S et al (2012) Nuclear ribosomal internal transcribed spacer (ITS) region as a universal DNA barcode marker for fungi [J]. Proc Natl Acad Sci 109(16):6241–6246
84. Heinrichs G, de Hoog GS, Haase G (2012) Barcode identifier-a practical tool for reliable species assignment of medically important black yeast species [J]. J Clin Microbiol 50(9):3023–3030
85. Sun J, Najafzadeh MJ, Vicente V et al (2010) Rapid detection of pathogenic fungi using loop-mediated isothermal amplification, exemplified by Fonsecaea agents of chromoblastomycosis [J]. J Microbiol Methods 80(1):19–24
86. Chowdhary A, Meis JF, Guarro J et al (2014) ESCMID and ECMM joint clinical guidelines for the diagnosis and management of systemic phaeohyphomycosis: diseases caused by black fungi [J]. Clin Microbiol Infect 20(s3):47–75
87. Queiroz-Telles F, Purim K, Fillus J et al (1992) Itraconazole in the treatment of chromoblastomycosis due to Fonsecaea pedrosoi [J]. Int J Dermatol 31(11):805–812
88. Andrade TS, Castro LGM, Nunes RS et al (2004) Susceptibility of sequential Fonsecaea pedrosoi isolates from chromoblastomycosis patients to antifungal agents [J]. Mycoses 47(5–6):216–221
89. Bonifaz A, Saul A, Paredes-Solis V et al (2005) Treatment of chromoblastomycosis with terbinafine: experience with four cases [J]. J Dermatol Treat 16(1):47–51
90. Esterre P, Inzan CK, Ratsioharana M et al (1998) A multicentre trial of terbinafine in patients with chromoblastomycosis: effect on clinical and biological criteria [J]. J Dermatol Treat 9(sup1):S29–S34
91. Sevigny GM, Ramos-Caro FA (2000) Treatment of chromoblastomycosis due to Fonsecaea pedrosoi with low-dose terbinafine [J]. Cutis 66(1):45
92. Zhang J, Xi L, Zhang H et al (2009) Synergistic effects of terbinafine and itraconazole on clinical isolates of Fonsecaea monophora [J]. Eur J Dermatol 19(5):451–455
93. Gupta AK, Taborda PR, Sanzovo AD (2002) Alternate week and combination itraconazole and terbinafine therapy for chromoblastomycosis caused by Fonsecaea pedrosoi in Brazil [J]. Med Mycol 40(5):529–534
94. Keating GM (2005) Posaconazole [J]. Drugs 65(11):1553–1567
95. Negroni R, Tobón A, Bustamante B et al (2005) Posaconazole treatment of refractory eumycetoma and chromoblastomycosis [J]. Rev Inst Med Trop Sao Paulo 47(6):339–346
96. Tagami H, Ginoza M, Imaizumi S et al (1984) Successful treatment of chromoblastomycosis with topical heat therapy [J]. J Am Acad Dermatol 10(4):615–619
97. Ameen M (2009) Chromoblastomycosis: clinical presentation and management [J]. Clin Exp Dermatol 34(8):849–854
98. Lyon JP, de Maria Pedroso e Silva Azevedo C, Moreira LM et al (2011) Photodynamic antifungal therapy against chromoblastomycosis [J]. Mycopathologia 172(4):293–297
99. Yang Y, Hu Y, Zhang J et al (2012) A refractory case of chromoblastomycosis due to Fonsecaea monophora with improvement by photodynamic therapy [J]. Med Mycol 50(6):649–653
100. Hu Y, Huang X, Lu S et al (2015) Photodynamic therapy combined with terbinafine against chromoblastomycosis and the effect of PDT on Fonsecaea monophora in vitro [J]. Mycopathologia 179(1–2):103–109

Chapter 12
Mycetoma

Ahmed Hassan Fahal

Abstract Mycetoma is a unique neglected disease, endemic in many tropical and subtropical regions, characterized by devastating deformities, disability, high morbidity and serious negative socioeconomic impacts on patients, families and community. It enjoys meager attention by health and social sectors across the world, as it is a non-glorious disease. Mycetoma commonly affects young adults and children of low socioeconomic status from rote rural areas in poor countries and it is an excellent example of a neglected tropical disease.

To date, its actual incidence, prevalence, and route of infection are not well characterized, as is its susceptibility, resistance, and response to medical treatment. This data scarcity is reflected on the available diagnostic and treatment modalities, which proved to be ineffective, have serious side effects, and are expensive for patients and health authorities in endemic areas. Early-stage detection of mycetoma and subsequent medical intervention lead to more favorable prognosis and minimization of disease recurrence, its sequelae, and economic burden.

12.1 Background

Mycetoma is a common health problem, endemic in many tropical and subtropical regions characterized by devastating deformities, disability and high morbidity (Figs. 12.1 and 12.2) [1, 2]. It is a debilitating disease, which progresses relatively silently. Mycetoma has serious negative medical and socioeconomic impacts on patients, families, communities, and health authorities. Yet, it enjoys meager attention across the globe which is translated in massive knowledge gaps in various aspects of mycetoma [1, 2].

Mycetoma is a chronic granulomatous subcutaneous inflammatory disease caused by either true fungi (eumycetoma) or certain bacteria (actinomycetoma). It affects the poorest in poor populations in the poor and most remote areas. Typically, young adult male farmers, laborers, and students between 15 and 30 years old of low socioeconomic status are mostly affected [3–6].

A.H. Fahal, MBBS, FRCS, FRCSI, FRCS (Glasg) (✉)
Department of Surgery, University of Khartoum, EL Qasr Avenue, Khartoum 11111, Sudan
e-mail: ahfahal@hotmail.com; ahfahal@uofk.edu

H.M. Mora-Montes, L.M. Lopes-Bezerra (eds.), *Current Progress in Medical Mycology*, DOI 10.1007/978-3-319-64113-3_12

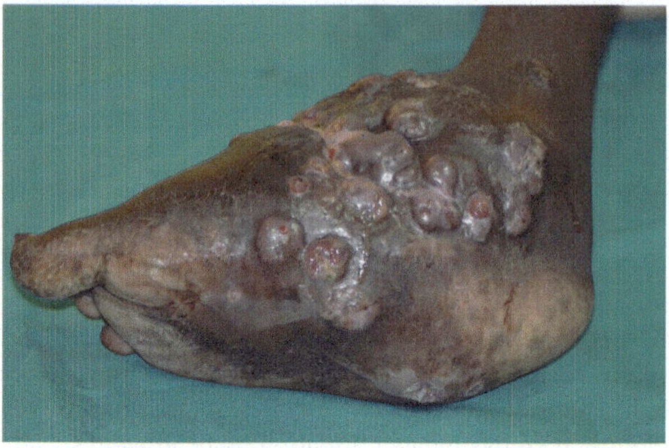

Fig. 12.1 Showing massive foot actinomycetoma

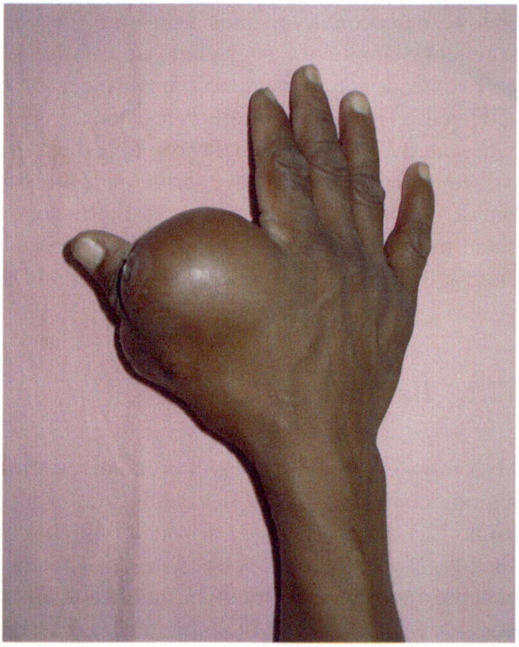

Fig. 12.2 Showing massive eumycetoma with deformity

12.2 Historical Note

The first reports on mycetoma appeared in 1694 by the German physician Engelbert Kaempfer. In 1842, Gill reported the disease in Madurai, Southern India, hence the name Madura foot, a term that is no longer used. Van Dyke Carter (1860) named the disease mycetoma (mykes = fungus and oma denotes tumor) after isolating a fungus for the first time. Although the first reports on mycetoma originated from India, by the beginning of the twentieth century, interest in mycetoma research had shifted to Africa. Gemy and Vincent reported the first African mycetoma case from Tunisia in

1894. Since then, many reports on the disease were published and have formed the foundation of the current knowledge and research on mycetoma [7].

12.3 Epidemiology

12.3.1 Incidence and Prevalence

For various understandable reasons, the true incidence and prevalence of mycetoma throughout the world are not exactly known. These reasons include the nature of the disease, which is usually painless, slowly progressive, and the late presentation of the majority of patients due to the lack of health education and financial constraints. In many centers, amputation is still the sole treatment for advanced cases, which may contribute to the fear of many mycetoma patients to seek medical advice. The medical records and statistical information, in most of the mycetoma endemic health facilities are deficient, which contributes to the inadequate data on mycetoma incidence. In many rural communities, mycetoma is a social stigma, particularly among females [8, 9].

In mycetoma endemic areas, due to several reasons, many individuals pursue medical help from local native healers and that is another cause for the late admission of patients in clinics or hospitals. Therefore, most of the data on mycetoma are related to hospital cases with advanced disease [1, 2, 9].

Although Sudan is considered the mycetoma homeland, data on its prevalence and incidence are meager and scanty. Since no real prevalence studies are performed, prevalence can only be roughly estimated based on epidemiological studies. Two large epidemiological studies have been performed on which an artificial and largely underestimated prevalence can be calculated. These are the studies performed by Abbott in Sudan in 1952–1955, and that reported by Lopez Martinez and colleagues from Mexico between 1956 and 1985 [7, 8, 10].

In Abbott's study, 1231 mycetoma cases were admitted to hospitals throughout the country in a period of 2.5 years. By dividing this number of cases seen in hospitals by the total population of Sudan in those years, a prevalence of 4.6 cases per 100,000 inhabitants would be obtained. No incidence data can be calculated [10].

Lopez Martinez reported on 2105 mycetoma cases from 14 dermatological centers throughout Mexico in a period of 30 years. Again, by applying the same formula an average prevalence of 0.6 cases per 100,000 inhabitants would be obtained. These grossly underestimated incidences are comparable to those of other neglected tropical infections such as Buruli ulcer, African trypanosomiasis, dracunculiasis, and leprosy [10].

More recent data were reported from the Mycetoma Research Centre, WHO Collaborating Centre on mycetoma, Khartoum, Sudan, the only one of its kind in the country. This center reported on 6792 patients with mycetoma and an incidence of 370 new cases per year in the period 1991–2015 [10].

12.3.2 Geographical Distribution

Mycetoma has a worldwide distribution but this is extremely uneven. It is endemic in many tropical and subtropical regions. The African continent seems to be the area of the highest prevalence. It prevails in the mycetoma belt, which includes Sudan, Somalia, Senegal, India, Yemen, Mexico, Venezuela, Columbia, Argentina, and others [2, 3, 7].

In Africa, mycetoma is most frequently seen in Sudan, Senegal, Mauritania, Kenya, Niger, Nigeria, Ethiopia, Chad, Cameroon, Djibouti, and Somalia. In Asia, it has been extensively reported from India and Yemen [2, 3, 7]. However, mycetoma has been reported in many temperate regions as well. Cases were reported from the United Kingdom, most of those afflicted were immigrants, who probably contracted the infection overseas, with the pale grain eumycetoma as the most common type. There are few reports on mycetoma from the United States, Ceylon, Germany, Egypt, Turkey, Philippines, Japan, Lebanon, Thailand, Saudi Arabia, Tunisia, and Iran [2, 3, 7].

Areas where mycetoma prevails are relatively arid zones with a short rainy season of 4–6 months, with the rainfall of 50–1000 mm per year, a relative humidity of 60–80% and constant temperatures of 30–37 °C day and night. This is followed by a dry season of 6–8 months with a relative humidity of 12–18%, day temperatures of 45–60 °C and night temperatures of 15–18 °C [2, 3, 7].

The geographical distribution of the individual mycetoma organism shows considerable variations, which can be convincingly explained by climatic factors, including the temperature, humidity and the rainfall; having the latter as the most relevant factor [1, 7].

Madurella mycetomatis, which is the commonest cause of eumycetoma, predominates in the tropical areas of Africa and India with a rainfall of 250–500 mm per year. Whereas, *Streptomyces somaliensis* is seen more often in the Middle East, Central, and West Africa and the arid region adjacent to the Sahara Desert, with a rainfall of 50–100 mm per year. However, it is sometimes seen in areas with higher rainfall as in Mexico [1, 7].

Actinomadura pelletierii is more prevalent in the relatively humid areas where the rainfall ranges from 250 to 1000 mm per year. *Nocardiae* are usually the causative organisms of mycetoma in temperate regions [1, 7].

The mycetoma belt encases an area of forest trees and savannah. The dominant plants usually are various species of Acacia and other thorny bushes and trees, like *Balanytesa egyptica,* which are armed with strong thorns.

There is an association between mycetoma pathogenesis and thorns. The penetrating thorns facilitate the entry of the organism into the subcutaneous tissue as it lives saprophytically on or inside the thorn. This was documented in serial sections of thorns removed from two mycetoma patients. In these patients, the thorns cells contained many fungal elements in different stages of development were reported. From one of them, *Leptosphaeria senegaliensis* was isolated and *Pyrenochaetaromeroifrom* in the other [7]. The presence of thorns in the vicinity of the mycetoma granuloma had been observed histologically and during surgery. Thorns may have the ability to produce the granuloma, and probably they provide an appropriate nucleus for the development of the mycetoma organism. The fact that the disease is common among rural people who go barefooted is in line with these observations. However, in many patients there is no history of local trauma at the mycetoma site.

12.3.3 Susceptibility and Resistance

Mycetoma is reported in both males and females of all age groups, but some groups are affected most. It occurred more frequently in the young adult males, with ages ranging from 15 to 30 years, but no age is exempted, especially in endemic areas. Patients at this age group represent the most active and the earning members of the society, particularly in developing societies where the life expectancy is short [8, 11].

Not all occupations are equally encountered in the patient population. Most of the mycetoma patients appear to earn their living by working on the land, peasant and herdsmen and the farmers are affected more often. However, in a recent study, 30% of reported patients appeared to be young students. This may due to the fact that, mycetoma is commonly seen in this age group, increased educational facilities in the endemic areas and the children commonly share the various outdoor activities with their families. In endemic areas, other professionals such as workers, clerks and health workers may also be affected [2, 8, 11].

In all previous reports, male predominance is a constant finding in mycetoma. In general, the gender ratio in eumycetoma is (3:1) which is higher than that encountered in actinomycetoma (2.7:1) and the explanation behind this observation is unclear. The male predominance is commonly attributed to the greater risk of exposure to organisms in the soil during the outdoor activities. However, in areas where mycetoma is prevalent in the Sudan, both sexes go barefooted, and in Western Sudan women are more committed to outdoor activities than males and yet, the incidence of the disease is the same as that observed in other parts of the country. These findings are in line with reports from Kenya, Congo, and others [2, 8, 11].

It may be that women are inherently less susceptible to infection with mycetoma due to differences in sex-related hormone levels, but this seems not to be the case. Women have much higher levels of 17 β-estradiol than men, but in male mycetoma patients higher 17 β-estradiol levels were reported than in non-infected males [12]. However, it is interesting to note, during pregnancy the mycetoma lesions become more aggressive, active with formation of new discharging sinuses (Personal observation), and the depressed immunity during pregnancy may partially explain this observation; however, other causes must be considered.

Another difference between men and women, which could be of influence in developing mycetoma, is the make-up of the skin. Men have a thicker dermal compartment of the skin, less subcutaneous fat, higher basal blood flow in the skin, generate more sebum and sweat, have larger pores, and most importantly, have a slower wound healing. All these differences can be important in a subcutaneous infection [10].

In general, mycetoma patients are considered to have a normal and responsive immune system, although there are some reports, which showed partial impairment of the cell-mediated immune (CMI) response in patients severely infected or not responding to medical treatment [7]. This finding was supported by animal studies, since mycetoma was more successfully induced in athymic mice than in immunocompetent animals [13]. In another report, some impairment in the innate immune response was noted. Overall, still no convincing evidence exists that showed mycetoma patients have particular immune defects.

12.4 Mycetoma Entry Route

Many believe that, the causative organisms may be present in the soil in the form of grains. After they are moistened by rain, they form conidia or other fungal forms capable of infecting the host. This infecting agent is then implanted into the host tissue through a breach in the skin produced by local trauma caused by sharp objects such as thorn pricks, stone, or splinters [2, 3, 7].

In areas where mycetoma is frequent, the habit of going barefooted is common and thorns are plentiful. As a result, natural infection is expected to be more frequent than it actually is, if this hypothesis of route of infection is true. Many workers in the field believe that, there is an intermediate host for the infection to develop but it is not known [2, 3, 7].

It is interesting to note, that it was not possible to isolate *Madurella mycetomatis*, the main causative agent of human eumycetoma in Sudan from different soils and thorn tree samples from different endemic areas in the country. However, PCR-mediated detection followed by restriction fragment length polymorphism (RFLP) analysis for the identification of *Madurella mycetomatis* DNA from the environmental samples was done, and it was positive in 23% of soil and in 5% of the thorn samples. These observations support the hypothesis that eumycetoma is primarily environmentally acquired, and suggest that *Madurella mycetomatis* needs special conditions for growth, as direct isolation from the environment seems to be impossible [14].

The Central Sudan is endemic area for eumycetoma due to *M. mycetomatis*. The majority of villagers in that area are farmers and live on animal husbandry. They have many cattle, goats, sheep, dogs, chickens, and donkeys, and they are kept in cages surrounded by walls made of mud or dry thorny bushes. The floors of the cages are covered with dry animal dung, thorns, and trash. The villagers' homes are in close proximity to these cages. As the frequency of *M. mycetomatis* on thorns is low as documented previously, and the common habit of going barefooted, contaminated animal dung and its role as an adjuvant may have important roles in contracting mycetoma [15].

The disease can occur naturally in several animals including goats, horses, donkeys, dogs, and cats. However, the disease is not contagious from animal to human or from one person to another and there is no report on hospital cross- infection [2, 3, 7].

12.5 The Incubation Period

The incubation period in mycetoma is unknown due to the difficulty in establishing the time of initial infection; however, in experimental animals the formation of the mycetoma lesion was noted after a period of 3 weeks from the inoculation of the organism [2, 7].

12.5.1 Mycetoma Pathogenesis

Unlike the life-threatening systemic fungal infections, there is no clear immune defect or deficiency known that could explain why people develop mycetoma. Serological studies with the *M. mycetomatis* specific antigens showed that Translationally Controlled Tumour Protein, fructose bisphosphate aldolase, and pyruvate kinase antibodies are elevated in both mycetoma patients and endemic controls. However, only the patients developed mycetoma. This suggested that all people from the endemic area have contacted this fungus at some stage but only a small portion developed the disease [16].

Mahgoub and his colleagues in 1973 studied the cell-mediated immunity of a group of mycetoma patients by means of the tuberculin test, 2,4-dinitrochlorobenzene sensitization and lymphocyte proliferation induced by phytohemagglutinin, and found defective T cell-mediated responses especially in severely infected patients and in those patients who did not respond well to treatment [17]. These findings were supported by animal studies, since mycetoma was more successfully induced in athymic mice than in immunocompetent mice [13]. However, on the other hand, Bendl and his colleagues had studied 31 patients with mycetoma but no immune defects were detected in that cohort.

Since it was observed in endemic areas that mycetoma is seen more frequent in certain families, either environmental or genetic factors seem to be important in the development of mycetoma. Van de Sande and colleagues therefore addressed the genetic susceptibility towards mycetoma by determining differences in allele frequencies for several single nucleotide polymorphisms (SNPs) in patients and healthy endemic controls. Two of those genetic studies addressed the role of single nucleotide polymorphisms (SNPs) in genes involved in the function of the immune system [18]. In the first study, the role of SNPs in genes neutrophil function was addressed. In that study, 11 SNPs in eight genes involved in neutrophil function were studied. Significant differences were found in genes encoding for interleukin

8 (CXCL8), its receptor CXCR2, thrombospondin-4 (TSP4), nitric oxide synthase 2 (NOS2) and complement receptor 1. The genotypes more commonly found in mycetoma patients for CXCL8, its receptor CXCR2 and TSP4 all were associated with a higher CXCL8 production in other studies. Therefore, this finding was confirmed by the measurement of high levels of CXCL8 in serum of mycetoma patients. In contrast, the NOS2 genotype obtained from the patients was associated with a lower Nitric oxide production. This finding was also confirmed by lower nitrite and nitrate levels in mycetoma patients.

Mhmoud and colleagues determined the role of interleukin 10 (IL-10) and CC chemokine ligand 5 (CCL5) in the granuloma formation of mycetoma [19]. Two SNPS in the promoter region of IL-10 and three SNPs in the promoter region of CCL5 were determined. Significant differences in allele distribution were demonstrated for one of the SNPs in the IL-10 promoter and two of the SNPs in the CCL5 promoter between mycetoma patients and healthy controls. Both IL-10 and CCL5 were found to be present in the mycetoma granuloma and secreted in the serum of patients. Since both, CCL5 and IL-10, play important roles in granuloma formation in general, it looks like the granuloma formation itself also plays an important role in the pathology of mycetoma.

Based on the results of these three genetic association studies performed in mycetoma patients, it can be concluded that, there are indeed subtle genetic differences between the Sudanese people who develop mycetoma and the people who don't. The exact role of these genetic differences should be further studied.

The role of sex-related hormone synthesis was studied because a clear male predominance was seen in mycetoma. In one study, 5 SNPs in 5 genes involved in sex hormone synthesis were studied. Significant differences in allele frequencies were found for catechol-O-methyltransferase (COMT) and cytochrome p450 subfamily 19 (CYP19). The genotypes more often obtained in mycetoma patients for COMT and CYP19 were previously described as influencing 17β-estradiol production. Also in the male mycetoma patients, significantly elevated serum levels of 17β-estradiol were found. In contrast, in these patients, also lowered levels of dehydroepiandrosteron (DHEA) were found. The differences in hormone levels noted between mycetoma patients and healthy controls did not directly affect the fungus itself. Indirect effects on the patients' hormone regulated immune states are the more likely explanations for mycetoma susceptibility [12].

As described above, SNPs in genes involved in the pathology of mycetoma can contribute to the risk of developing mycetoma once the causative agent is introduced into the subcutaneous tissue, but next to that it can also contribute to the extent of the disease. In the mentioned studies, it was noted that two of the SNPs found to be associated with the size of the mycetoma lesion, namely NOS2 and COMT.

12.6 Clinical Presentation

Male predominance is a constant finding in mycetoma with a sex ratio of 3:1. This is commonly attributed to the greater risk of exposure to organisms in the soil during the outdoor activities but many reports are questioning this. No age is exempted but mycetoma commonly affects adults between 15 and 30 years of age and these are the earning members of the society, especially in under developed countries. However, in endemic regions children and elderly people may also be affected [2, 4].

The clinical presentation of mycetoma is almost identical irrespective of the causal organism. However, the rate of progress is more rapid with actinomycetoma than with eumycetoma. In eumycetoma, the lesion grows slowly with clear defined margins and remains encapsulated for a long period, whereas, in actinomycetoma the lesion is not encapsulated, more inflammatory, more destructive and invades the bone at an earlier period [2, 4].

The characteristic triad, of a painless subcutaneous mass, sinuses and the presence of discharge containing grains are pathognomic of mycetoma. It presents as a slowly progressive painless subcutaneous swelling sometimes at the site of previous trauma. The swelling is usually firm and rounded but it may be soft, lobulated, rarely cystic and it is often mobile.

Multiple secondary nodules then evolve as well, the nodules may suppurate and drain through multiple sinuses tracts and these sinuses may close transiently after discharge during the active phase of the disease. Fresh adjacent sinuses may open while some of the old ones may heal completely. They are connected with each other, with deep abscesses and with the skin surface [2, 4].

The discharge is usually serous, serosanguinous or purulent. During the active phase of the disease, the sinuses discharge grains, the color of which depends on the causative organism. The grains can be black, yellow, white or red and they are of variable size and consistency. The black grains are usually due to *M. mycetomatis,* the red ones are due to *A. pelletierii*, the yellow are due to *Streptomyces somaliensis* and the white grains can be due to *A. madurae*. Pus, exudate, the dressing gauze and biopsy material should be examined for the presence of the grains.

Mycetoma is usually painless in nature, and it was suggested that produces substances that have an anesthetic action. At a late stage of the disease, the pain may become negligible due to nerve damage by the tense fibrous tissue reaction. Pain may be produced due to a secondary bacterial infection.

As the mycetoma granuloma increases in size, the skin over it becomes attached and stretched. The skin may become smooth, shiny and areas of hypo or hyperpigmentation may develop.

Local hyperhidrosis in the skin overlying the lesion is commonly seen in mycetoma. This corresponds with the hyperplasia and hypertrophy of sweat glands found in surgical biopsies. This can be explained by the increase in local temperature that is the result of increased blood flow to the area induced by the chronic inflammatory process. The increased local blood flow in mycetoma was confirmed by an angiographic study, which showed dilated and tortuous terminal arterial branches, vascular blush and dilated veins proximal to the lesion.

Mycetoma eventually invades the subcutaneous tissue, fat, muscles and bone. This is usually gradual and delayed in eumycetoma while in actinomycetoma it is earlier and extensive especially in infections induced by *A. pelletierii.*

For unknown reasons, the tendons and the nerves are curiously spared until very late in the disease process, this may explain the rarity of neurological and trophic changes even in patients with long-standing mycetoma. The absence of trophic changes may also be explained by the adequate blood supply in the mycetoma area [2, 4].

In most patients, the regional lymph nodes are small and shotty. An enlarged regional lymph node is not uncommon and this may be due to secondary bacterial infection, genuine lymphatic spread of mycetoma or it may be due to immune complex deposition as part of a local immune response to mycetoma infection.

The infection remains localized and the constitutional disturbances are rare but when they do occur, they are generally due to secondary bacterial infection of the open sinuses tracts, fistula formation in some patients or generalized immunosuppression [20–35].

Cachexia and anemia may be seen in late mycetoma. This is often due to malnutrition, sepsis, and mental depression. Mycetoma can produce many disabilities, distortion and deformity. It can be fatal especially with cranial mycetoma [20–35].

Secondary bacterial infection commonly occurs in mycetoma. In material obtained by deep swabs and fine needle aspiration, 65% of patients have concomitant superficial bacterial infection; of these 56% are attributable to *Staphylococcus aureus*, 34% to *Streptococcus pyogenes* and 10% to *Proteus mirabilis.* The coexisting bacterial infections disturb the local environment and reduce the responses to various antifungal and antimicrobial agents. Elimination of coexisting infection results in better clinical response. [36].

It is interesting to note that in a recent study from a mycetoma endemic village in the Sudan, the villagers' knowledge of the disease was poor in 96.3% of them, 70% had appropriate attitudes and beliefs towards interaction with mycetoma patients and treatment methods, and 49% used satisfactory or good practices in the management of mycetoma. This indicates the need for objective and effective health education to improve awareness among affected communities [15].

12.7 Mycetoma Spread

In mycetoma, spread occurs locally and through the lymphatics thereby resembling sporotrichosis. Metastatic lesions may occur at various lymph node stations in which new satellites may develop. This phenomenon is more common in actinomycetoma, especially in cases with repeated surgery.

Hematological spread has also been described; *M. mycetomatis* and *A. pelletierii* grains have been detected in an intact blood vessel. In many spinal mycetoma, these lesions occurred without involvement of the skin or surrounding tissues suggesting hematogenous spread [20–35].

12.7.1 Site of Mycetoma

The commonest site for mycetoma is the foot (79.2%), most of the lesions are seen on the dorsal aspect of the forefoot. Both feet are equally affected. The hand ranks as the second commonest site (6.6%), the right hand is more affected. This may imply a traumatic basis of the infection in this site [20–35].

In endemic areas, other parts of the body may be involved but less frequently and these include the knee, arm, leg, head and neck, thigh, and the perineum. Rare sites such as the chest and abdominal walls, fascial bones, mandible, paranasal sinuses, eyelid, vulva, orbit, scrotum and surgical incisions may be affected [20–35].

12.7.2 Differential Diagnosis

In endemic areas, subcutaneous masses should be considered as mycetoma until it proved otherwise. The clinical presentation, the etiological agent, and the prevalence of infectious diseases that can mimic mycetoma influence the differential diagnosis of mycetoma.

The differential diagnosis of mycetoma includes many of the soft tissue tumors such as Kaposi's sarcoma, fibroma, malignant melanoma, or fibrolipoma, as well as thorn granuloma. The radiological features of advanced mycetoma may be comparable to osteogenic sarcoma and bone tuberculosis. Primary osseous mycetoma is to be differentiated from chronic osteomyelitis, osteoclastoma, bone cysts, and from syphilitic osteitis [2, 3].

12.8 Mycetoma Diagnosis

In many endemic areas, due to diagnostic facilities scarcity, the disease diagnosis is often made clinically only, a practice, which may lead to the administration of inappropriate management.

The proper diagnosis of mycetoma needs the accurate identification of the causative organism and the disease extent. Various diagnostic tools are available for the precise identification of the organism and that includes the classical grain culture, histopathological and cytopathological techniques, serodiagnosis, as ELISA and CIE, and molecular techniques such as PCR. For the disease, extent numerous imaging techniques such as radiography, ultrasonography, CT, MRI are in use. It is

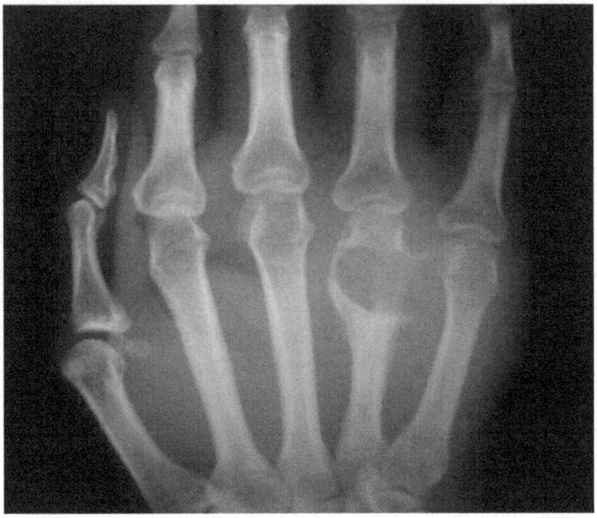

Fig. 12.3 X-ray of right hand showing massive soft tissue mass, periosteal reaction, and bone cavities in line with eumycetoma appearance

interesting to note most of these techniques are not available in majority of mycetoma endemic regions [37–42].

Presently, in endemic areas, it is recommended for diagnosis to start with lesion ultrasound examination and fine needle aspiration for cytology and cell blocks, to identify the causative organism; in cases where this is not successful, deep surgical biopsy is carried out for histological and mycological identification. For massive lesions, radiography and MRI are mandatory to determine the disease extent. Molecular and phenotyping assays are only performed in highly specialized laboratories.

12.9 Mycetoma Imaging

12.9.1 Radiology

A series of radiological changes are seen in mycetoma. This is due to the fact that all mycetoma agents are osteophilic and it may be due to the effect of the granuloma on both, the affected bone and its blood supply (Fig. 12.3).

In early disease, the radiography is essentially normal. Eventually with the development of the granuloma in size, a soft tissue mass with obliteration of fascial planes may be seen. The granuloma, is shown as a dense shadow or as scattered multiple soft tissue shadows. Calcification and obliteration of the fascial planes may sometimes be the disease progresses, the cortex may be compressed from outside by the granuloma leading to bone scalloping, this is followed by a variable amount of periosteal reaction. Periosteal new bone spicules are laid down at right angle to the cortex to create a sun-ray appearance and Codman triangle, an appearance that may be undistinguishable from that due to osteogenic sarcoma [37–39].

Late in the disease, there may be multiple punched out cavities through the normal density of the bone. These cavities are large in size, few in number with well-defined margins in eumycetoma. Whereas, the bone cavities in actinomycetoma are usually smaller in size, numerous and have no definite margins. The cavities are produced by the replacement of the osseous tissue by the grains. Their size is due to the size of the grains of the causative organism. The cavities are usually filled with solid masses of grains and fibrous tissue, which provides bone support. This may explain the rarity of pathological fractures in mycetoma. The bony changes in the skull are unique: they are purely sclerotic with dense bone formation and loss of trabeculation [2, 3, 37–39].

Osteoporosis at, and distal to, the affected part is well observed in mycetoma and this may be due to disuse atrophy or due to the compression of the bone and its blood supply by the mycetoma granuloma.

Chemotherapy causes radiological improvement consisting of remoulding, absorption of the sclerotic bone and reappearance of the normal trabecular pattern.

In a retrospective study of 516 patients seen at the Mycetoma Research Centre, Sudan, a tertiary center, only 3% had a normal radiograph. The most common abnormalities were soft tissue swelling (93%), bone sclerosis (56%), and bone invasion (46%). Other abnormalities included bone cavities (32%) and osteoporosis (19%) [39].

12.9.2 Ultrasonic Imaging of Mycetoma

The mycetoma lesions, its capsule and the accompanying inflammatory granuloma have characteristic ultrasonic appearances (Fig. 12.4). Ultrasound imaging can differentiate between eumycetoma and actinomycetoma and between mycetoma and other non-mycetoma lesions. In eumycetoma lesions, the grains produce numerous sharp bright hyperreflective echoes, which are consistent with the black grains. The grain cement substance is most probably the origin of these sharp echoes. In addition, there are multiple thick-walled cavities with absent acoustic enhancement. In actinomycetoma lesions, the findings are similar but the grains are less distinct. This may be due to their smaller size and consistency, individual embedding of the grains or the absence of the cement substances in few of them [40].

The ultrasonic diagnosis of mycetoma is more precise and accurate in lesions with no sinuses. The extent of the lesion can be accurately determined ultrasonically and this is useful in planning surgical incisions and procedures.

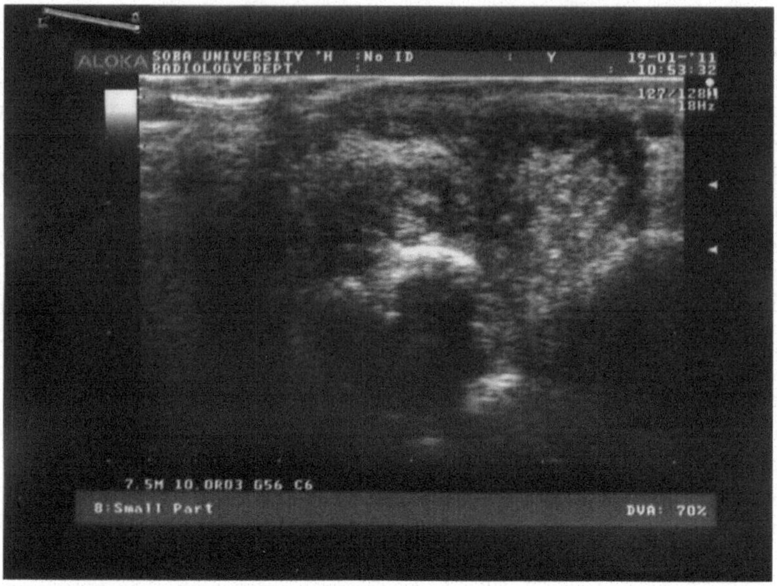

Fig. 12.4 Ultrasound showing typical eumycetoma lesion; multiple thick-walled cavities with sharp hyper-reflective echoes denoting grains without acoustic enhancement

12.9.3 Magnetic Resonance Imaging

Magnetic Resonance Imaging (MRI) is useful for determining the extent of the lesions and the invasion of structures (Fig. 12.5). [41] It has greater sensitivity than radiographs, ultrasound, and CT.

MRI usually shows multiple 2–5 mm lesions of high signal intensity that indicates the granuloma, interspersed within a low-intensity matrix denoting the fibrous tissue. The 'dot in circle' sign which indicates the presence of grains, is highly characteristic.

A MRI grading system; the Mycetoma Skin, Muscle, Bone Grading System (MSMBS), was recently proposed, it grades disease severity, compares patients and helps to manage them. The MRI appearance should be differentiated from chronic osteomyelitis, granulomas, soft tissue tumors, bone tuberculosis, and cold abscesses. This procedure, however, is only possible in well-equipped hospitals.

In general, for small early lesions, ultrasound examination is recommended, meanwhile, radiography and MRI are recommended for large lesions [41].

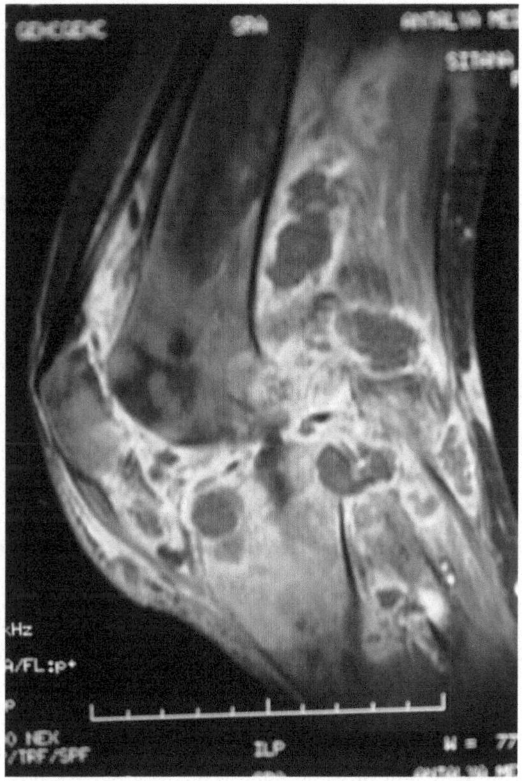

Fig. 12.5 MRI showing involvement of soft tissue and bone by mycetoma and the *dot-in-circle* sign which is characteristic of mycetoma

12.9.4 CT Scan

CT findings in mycetoma are not specific but are helpful to detect early bone involvement. Helical computed tomography is superior to plain CT scanning as it allows three-dimensional reconstruction and more precise information on the degree of organ involvement. It also allows vascular involvement to be visualized.

12.9.5 Organism Identification

Identification of the agent is important to guide treatment. Goodfellow and colleagues have described a stepwise procedure from the collection of specimens through to the isolation and identification of established causal agents of actinomycetoma [44].

To identify the causative agent, grains need to be isolated from the lesion. Grain material can be obtained by surgical excision or by extraction with a cotton swab from the sinuses or by fine needle aspiration (FNA). Deep-seated grains obtained by surgical excision are preferred over those extruded through sinuses because the latter are often inviable and contaminated [44].

A first indication of the causative agent is obtained by looking at the grain itself: the size, shape, color and consistency help to identify the organism; however, causal agents must be isolated for definitive identification. In direct examination, grains are mounted on a slide and crushed under a cover glass. The size of the filaments, septation, morphological characteristics and pigment formation can be used to differentiate between actinomycetoma and eumycetoma.

In actinomycetoma, fine filaments are seen and can be stained with Gram staining. In eumycetoma, the filaments stain with Periodic Acid Schiff (PAS). Grains are plated onto appropriate culture media and incubated for at least 6 weeks. Identification based on colony morphology can be difficult, due to a large variety of colony types, and to the fact that many species resemble each other. Lack of conidiation can further complicate matters.

12.9.6 Histopathological Identification

Stained sections usually show the grain morphology and the tissue reaction to the organisms. Since it does not require aseptic procedure or the rigid time schedule required for culture, this technique is attractive; however, it lacks the precision of culture [45–49]. Histopathological identification requires a deep biopsy with the risk of disease spread. It is less reliable than culture and does not allow differentiation between most species. This tool has been used as the sole identification tool in many centers in the past decades.

Various staining procedures are in use and that include H&E, PAS and Gram. Still, there are many difficulties in identifying causative agents based on histology alone, particularly for *Nocardia* spp. It is known that, it is practically impossible to define *Nocardiae* to the species level due to their similar appearance [50].

In eumycetoma, differentiation between *Acremoniumspp, Fusariumspp* and *Pseudallescheriaboydii,* and between *Exophialajeanselmei, Leptosphaeria senegalensis, Leptosphaeriatompkinsii, Nigrogranaromeroi,* and *Madurella griseais* is also difficult. Furthermore, other fungi, such as *M. mycetomatis*, can present multiple grain types in histological slides, complicating the identification process even more.

The inflammation reaction around the grain is variable. Three types of tissue reaction can be seen. In type I tissue reaction, there is a zone of neutrophils in the vicinity of the grain. Some histiocytes may also be seen among the neutrophils, but they are more numerous outside the neutrophil zone [45–49].

In type II tissue reaction, the neutrophil zone is absent, instead a layer of histiocytes and multinucleated giant cells is seen. Some of these giant cells contain fragments of grain or pigmented cement substance. At this stage, the grain is usually small and fragmented [45–49].

The most uncommon reaction is the type III tissue reaction, in which the grain material has largely or completely disappeared. This leaves a compact epithelioid granuloma with or without Langerhans giant cells.

12.9.7 Fine-Needle Aspiration Cytology of Mycetoma

Mycetoma can be accurately diagnosed in smears obtained by fine-needle aspiration (FNA) [42, 43]. Mycetoma lesion has a distinct appearance in a cytology smear characterized by the presence of polymorphous inflammatory cells consisting of an admixture of neutrophils, lymphocytes, plasma cells, histiocytes, macrophages, and foreign body giant cells and grains.

In sections, the grain is closely surrounded by, and occasionally infiltrated by, neutrophils causing its fragmentation. Outside the neutrophil zone, monocytic cells and giant cells are seen. This is surrounded by granulation tissue rich in fibroblasts and blood vessels [42, 43].

The technique allows morphological identification of mycetoma and its classification into eumycetoma and actinomycetoma, this is important as the treatment depends mainly on the etiological agents.

The technique is simple, cheap, rapid, sensitive, and can be tolerated by patients. It can be used not only in routine diagnosis, but can be used as an effective mean in collection of material for culture and immunological studies. Due to the simplicity of the technique, it can be used in epidemiological survey of mycetoma and for detection of early cases, in which the radiological and serological techniques may not be helpful [42, 43].

12.10 Culture

A large variety of microorganisms are capable of producing mycetoma. They can be identified by their textural description, morphological and biological activities in pure culture [44, 50].

The grains are the source of the culture and they should be alive and free of contaminants. For successful culture, a deep-seated surgical biopsy specimen is preferred to collect grains discharged with pus through the sinuses. The specimen obtained is divided into two portions. One portion is used for direct microscopy, the other for culture [44, 50].

Direct microscopy can be performed by examination of crushed grains in 10% KOH. Black grains usually are composed of hyphae of approximately 15 μm in width.

For culture, grains are washed in saline buffer containing chloramphenicol, grown on blood agar and Sabouraud dextrose agar and incubated at 37 °C and 26 °C for 6–8 weeks. Identification of isolates is achieved by observing the rate of growth, colony morphology, production of conidia, and assimilation patterns.

Culturing of fungi remains troublesome, often no growth is obtained or cultures are contaminated with bacteria. Identification of the various fungi responsible for black-grain mycetoma remains difficult with standard mycological procedures, since these fungi rarely produce conidia. Misidentification, therefore, often occurs. Furthermore, most of these fungi have low growth rates, which may delay the identification until up to 12 weeks.

Many culture media are in use, e.g., Sabouraud, blood agar, and Malt extract agar.

The culture technique is cumbersome, time-consuming, and chance contamination may give a false positive result. It also requires experience to identify the causative organisms [44, 50].

12.11 Serodiagnosis in Mycetoma

In the absence of the classical triad of mycetoma, the demonstration of significant antibodies titers against the causative organism may be of diagnostic value. Serodiagnosis is of a great help in identification and classification of the various organisms, which is an essential prerequisite for medical treatment, and is mandatory for the follow-up of these patients [51–54].

It has many advantages over the culture and histopathological techniques, as both require surgical biopsy, which may enhance the spread of the organism. The common serodiagnostic tests for mycetoma are the counter-immuno-electrophoresis (CIE) and Enzyme-linked immunosorbent assay (ELISA). CIE distinguishes between eumycetoma and actinomycetoma, and between Actinomycetes themselves; but cross reactivity between *A. madurae* and *A. pelletierii* is very common. However, it is time-consuming and the preparation of the antigens take a considerable time [51–54].

ELISA was used in the diagnosis of mycetoma. It appeared to be a sensitive test for the detection of antibodies; however, the high sensitivity of the test makes the cross reactivity unavoidable. ELISA may be a useful tool in community studies, as sero-epidemiological surveys could give valuable information on the distribution and prevalence of exposure to mycetoma.

Recombinant based ELISA performed with pure antigens of *M. mycetomatis,* had detected translationally controlled tumor protein (TCTP) and luminex assays based on TCTP, fructose-bisphosphatealdolase (FBA) and pyruvate kinase (PK). Although patients had higher antibody levels, antibodies were detected in healthy controls, making these techniques unsuitable as a diagnostic tool.

It is clear that these serodiagnostic tests have many limitations that include tedious and lengthy preparation of antigens, the antigens are crude and not standardized hence cross reactivity between different mycetoma causative organisms is found.

12.12 Organism Biological Activity Identification

The causative organism biological activity may include acid fastness, optimal temperature, proteolytic activity, utilization of sugars, and nitrogenous compounds.

These biological activities, while effective in distinguishing between genera of actinomycetes, are laborious and time-consuming and are being complemented and replaced by molecular systematic procedures, notably by 16S rRNA gene sequencing studies [44, 55, 56].

12.12.1 Molecular Diagnosis

Polymerase Chain Reaction (PCR) can be used for molecular typing of the causative agents. Different techniques are in use, which include restriction endonuclease analyses (REA), random amplified polymorphic DNA (RAPD), and amplification fragment length polymorphism (AFLP) [57–60].

For *M. mycetomatis*, a species-specific PCR procedure was developed. This PCR-RFLP analysis showed strict homogeneity between *M. mycetomatis* isolates and can be used to identify the causative agent in clinical material and also from soil and thorn samples [57].

Although results with RAPD are variable, REA and AFLP were able to differentiate between *M. mycetomatis* isolates from different countries, or even within a country.

Certain AFLP types were associated with the origin of the strain or the size of the lesion.

For eumycetoma, various molecular techniques have been used to identify causative agents and all of them are based on the identification of the internal transcribed spacer (ITS). To identify all fungal mycetoma causative agents, the ITS regions are usually amplified with pan-fungal primers and sequenced. Identification is based on comparing the resulting sequence with sequences previously submitted to Genbank [57, 61].

Using this approach, multiple studies have shown that a number of causal agents for eumycetoma are underspeciated, as exemplified by the identification of three new *Madurella* species; *Madurella fahalii, Madurella pseudomycetomatis* and *Madurella tropicana*, and *Pleurostomophoraochracea* [62, 63].

All of these tests are expensive, are not field friendly, nor are available in endemic areas.

12.13 Management

The management of this distressing and devastating disease is disappointing. The treatment depends mainly on its etiological agent, the site and extent of the disease. Until recently, the only available treatment for mycetoma was amputation or multiple mutilating, disfiguring surgical excisions. Combined medical treatment in the form of antifungals for the eumycetoma, and antibiotics and antimicrobial agents for actinomycetoma, and various surgical excisions is the gold standard in mycetoma [64–70].

Reports on medical treatment in eumycetoma are scarce and inadequate. Over the years and till now, the treatment of eumycetoma was based on personal clinical experience, and on the results of sporadic case reports, rather than controlled clinical trials. In general, massive surgical excisions or amputation of affected part is the treatment of choice.

Various antifungal agents have been tried with little success. This is perhaps surprising, as the eumycetoma causative agents are low-grade infective organisms and their eradication should be readily achieved by the administration of safe systemically antifungal drugs. Local administration of mycetoma chemotherapeutic agents was used, but the results were not impressive with a high rate of failure and complications.

For eumycetoma, surgery is always needed. The aim of surgery is complete excision of the lesion. This is only possible when the mycetoma lesion is small and well encapsulated. In larger lesions, only reduction of the amount of infected tissue is possible, and occasionally, multiple surgeries are needed to excise most of the lesion. Especially in large lesions, effective surgery needs aggressive excision or debridement under general anesthesia, which usually cripples the limb or leads to permanent disability due to mandatory amputation. To prevent recurrent infections, surgery is always combined with antifungal treatment, both before and after the procedure. Antifungal therapy for eumycetoma still depends mainly on the azole antifungals, and itraconazole is the drug of choice. The duration of treatment varies with the severity of the infection and the general health status of individual patients. Treatment may need to be continued for 18–24 months or more, and the liver function of the patients' needs regular monitoring, especially with long-term itraconazole treatment in cases of advanced mycetoma. However, long-term treatment may lead to antifungal resistance, which may in turn complicate patient management. Itraconazole is implicated in encapsulating the mycetoma lesion. This renders final surgical treatment a likely option, but in advanced lesions, especially when the bone tissue is involved, the response to chemotherapeutic treatment still remains very poor. In the search for new, alternative antifungal strategies in the treatment of mycetoma, terbinafine has been tried in the treatment of eumycetoma. Recently, the effect of posaconazole was studied and it showed some success [64–67].

Generally, actinomycetoma is amenable to medical treatment with antibiotics and other chemotherapeutic agents. Combined drug therapy is always preferred to a single drug, to avoid drug resistance and for disease eradication. Cure is possible, although a prolonged period of treatment is needed. Combined medical and surgical treatments facilitate surgery, accelerate healing and reduce the chance of relapse. However, a good number of patients respond to medical treatment alone [67–70].

There is more information on the treatment of actinomycetoma, although well-controlled comparative studies are lacking. In the 1960s, trimethoprim-sulfamethoxazole (TMP-SMX) became the gold standard for treatment, replacing sulphonamides. It was used as therapy in early lesions and has a cure rate of 60%. In the early 70s, for extensive lesions, streptomycin was added as a second drug, achieving a cure rate of 63% and a varying degree of improvement in the remaining patients [71].

Streptomycin was later replaced by amikacin, an aminoglycoside that is effective in vitro, in vivo, and clinically against *N. brasiliensis* mycetoma. In extensive and unresponsive cases, TMP-SMX 8/40 mg/kg/day in cycles for 5 weeks and amikacin 15 mg/kg/day in a divided dose every 12 h for 3 weeks are administered. The two-week interval of amikacin in the five-week cycle is used for renal and audiometric monitoring. This is currently the first line of treatment in many centers [71]. This regime has been proven effective (90%) in the cure of severe cases, if adverse effects do not occur. It is [71] important to maintain treatment if there are no adverse effects until cure or remission of the infection is achieved, which is generally obtained after 10–25 weeks. In case of resistance or allergy to TMP-SMX and amikacin, amoxicillin-clavulanic acid and netilmicin, respectively, may be used as alternatives. Amoxicillin-clavulanic acid has been used as rescue therapy with good results in some cases [71]. It can also be used alone during pregnancy; however, acquired resistance may develop, and it is generally not effective against *A. madurae*. Amikacin combined with a carbapenem, such as imipenem or meropenem, may also be used in refractory cases. Surgery is rarely required. It is recommended to determine the MICs to carbapenems because of resistant strains and the cost of these drugs [71].

12.13.1 Surgery for Mycetoma

Surgery in mycetoma is indicated when the disease process is limited and localized, cases resistant to medical treatment, or as a lifesaving procedure for advanced disease complicated by secondary bacterial infection, massive bone involvement, and poor general condition.

The goal of surgery is the complete removal of the lesion or the reduction of size, followed by medical treatment. Eumycetoma is well encapsulated and great care must be exercised not to rupture the capsule, which may lead to recurrence by transferring the fungal element into other parts of the operative field [72].

Actinomycetoma has an ill-defined border; therefore, a margin of healthy tissue should always be excised with the lesion. Simple bone curettage and soft tissue excision is recommended for localized bony lesions.

A bloodless operative field using a tourniquet is mandatory to identify margins of the lesion to avoid rupturing it. It is advisable to flood the wound at the end of surgery with tincture of iodine and hydrogen peroxide to clean the surgical field from left out grains, especially if there is doubt about the completeness of the excision or contamination of the field [72].

The wound can be closed primarily or by delayed primary sutures and in many cases skin grafts may be required. The open postoperative wounds are usually dressed with iodine and hydrogen peroxide.

In general, in mycetoma, advanced cases of mycetoma not responding to medical treatment for a prolonged period nothing short of amputation is likely to succeed. The amputation rate ranges from 25 to 50%. Extensive repeated excisions of the diseased tissue, including bone may be carried out several times to avoid the social consequences of amputation. This debulking procedure must be coupled with chemotherapy. In less advanced cases, less mutilating surgery is advised, for example, toe, mid tarsal, or Syme's amputation. However, in many cases of inadequate surgery, recurrence is inevitable and radical amputation is the only treatment.

Recurrence is more common after an incomplete or irregular course of medical treatment. With drug incompliance, there is a good chance for the organism to develop drug resistance [72].

Medical treatment for both types of mycetoma must continue until the patient is clinically, radiologically, ultrasonically, and cytologically cured. Cure is considered when the skin becomes normal, the mass disappears, the sinuses heal and the organisms are eliminated from the tissue. Clinical improvement is judged by reduction in the size of the mass and healing of most of the sinuses.

Radiological examination is an essential tool for follow-up of patients on medical treatment. It usually shows reappearance of normal bone pattern and the disappearance of the soft-tissue mass. Absent grains cytologically with type III tissue reaction and the disappearance of the grains and cavities ultrasonically are reliable evidences for cure.

Mycetoma has many serious medical and socioeconomic impacts on patients, community and health authorities. The postoperative recurrence rate varies from 25 to 50%. The available treatment although not very effective, has many serious side effects, they are expensive and it may amount to more than US$3000 per year per patient with eumycetoma. Most of the patients cannot afford to have that and the drugs are not always available in endemic regions. The treatment is of a variable duration and may lapse for 2–3 years with a mean of 18 months [72].

References

1. Hay RJ, Fahal AH (2015) Mycetoma: an old and still neglected tropical disease. Trans R Soc Trop Med Hyg 109(3):169–170. doi:10.1093/trstmh/trv003
2. Zijlstra EE, van de Sande WW, Welsh O, Mahgoub ES, Goodfellow M, Fahal AH (2016) Mycetoma. Lancet Infect Dis 16(1):100–112. doi:10.1016/S1473-3099(15)00359-X

3. Fahal AH, Hassan MA (1992) Mycetoma. Br J Surg 79(11):1138–1141
4. Fahal AH (2004) Mycetoma thorn on the flesh review article. Trans R Soc Trop Med Hyg 98(1):3–11
5. Fahal AH (2011) Mycetoma review article. Khartoum Med J 4(1):514–523
6. Nenoff P, van de Sande WW, Fahal AH, Reinel D, Schöfer H (2015) Eumycetoma and actinomycetoma - an update on causative agents, epidemiology, pathogenesis, diagnostics and therapy. J Eur Acad Dermatol Venereol 29(10):1873–1883. doi:10.1111/jdv.13008
7. Mahgoub ES, Murray IG (1973) Mycetoma. William Heinemann, London, pp 2–25
8. Fahal AH, EL Hassan AM, Mahgoub ES, Rahman ME (2015) Mycetoma in the Sudan: the Mycetoma research Centre update. PLoS Negl Trop Dis 9(3):e0003679. doi:10.1371/journal.pntd.0003679. eCollection 2015 Mar
9. Ezaldeen EA, Fahal AH, Osman A (2013) Mycetoma herbal treatment: the mycetoma research centre, Sudan experience. PLoS Negl Trop Dis 7(8):e2400
10. van de Sande WW (2013) Global burden of human mycetoma: a systematic review and meta-analysis. PLoS Negl Trop Dis 7(11):e2550. doi:10.1371/journal.pntd.0002550. eCollection 2013 Nov
11. Fahal AH (2013) Mycetoma. In: Williams N, O'Connell PR (eds) Bailey and Love's short practice of surgery, 26th edn. CRC Press, Oxford
12. van de Sande WWJ, Fahal AH, Tavakol M, van Belkum A (2010) Polymorphisms in catechol-O-methyltransferase and cytochrome p450 subfamily 19 genes predispose towards Madurella mycetomatis-induced mycetoma susceptibility. Med Mycol 48(7):959–968
13. Mahgoub ES (1978) Experimental infection of athymic nude New Zealand mice, nu nu strain with mycetoma agents. Sabouraudia 16(3):211–216
14. Ahmed AOA, Adelmann D, Fahal AH, Verbrugh HA, Van Belkum A, De Hoog S (2002) Environmental occurrence of Madurella mycetomatis, major agent of human eumycetoma in Sudan. J Clin Microbiol 40(3):1031–1036
15. Fahal A, Mahgoub ES, EL Hassan AM, Abdel-Rahman ME, Alshambaty Y, Zijlstra EE (2014) A new model for Management of Mycetoma in the Sudan. PLoS Negl Trop Dis 8(10):e3271. doi:10.1371/journal.pntd.0003271
16. de Klerk N, de Vogel C, Fahal A, van Belkum A, van de Sande WW (2012) Fructose-bisphosphate aldolase and pyruvate kinase, two novel immunogens in Madurella mycetomatis. Med Mycol 50(2):143–151
17. Mahgoub ES, Gumaa SA (1977) El Hassan AM immunological status of mycetoma patients. Bull Soc Pathol Exot Filiales 70(1):48–54
18. van de Sande WW, Fahal A, Verbrugh H, van Belkum A (2007) Polymorphisms in genes involved in innate immunity predispose toward mycetoma susceptibility. J Immunol 179(5):3065–3074
19. Mhmoud NA, Fahal AH, van de Sande WW (2012) The association between the interleukin-10 cytokine and CC chemokine ligand 5 polymorphisms and mycetoma granuloma formation. Med Mycol 51(5):527–533
20. Omer RF, Seif EL Din N, Abdel Rahim FA, Fahal AH (2016) Hand Mycetoma: the Mycetoma research Centre experience and literature review. PLoS Negl Trop Dis 10(8):e0004886. doi:10.1371/journal.pntd.0004886
21. Ezaldeen EA, Raif Mohamed Ahmed, EL Sammani Wadella, Nadia EL Dawi, Ahmed Hassan Fahal (2015) Actinomycetoma induced cervical spinal cord compression: a rare and serious complication. JMM Case Reports. doi: 10.1099/jmmcr.0.000074
22. Scolding PS, Abbas MAQ, Omer RF, Fahal AH (2016) Madurella mycetomatis-induced massive shoulder joint destruction: a management challenge. PLoS Negl Trop Dis 10(8):e0004849. doi:10.1371/journal.pntd.0004849
23. Mohamed ESW, Seif El Din N, Fahal AH (2016) Multiple mycetoma lung secondaries from knee eumycetoma: an unusual complication. PLoS Negl Trop Dis 10(7):e0004735. doi:10.1371/journal.pntd.0004735. eCollection 2016

24. Fahal AH (2011) Mycetoma. In: Guerrant RL, Walker DH, Weller PF (eds) Tropical infectious diseases: principles, pathogens and practice, 3rd edn. Elsevier, Boston, pp 565–568
25. Taha H, Fahal A, van de Sande WWJ (2015) Mycetoma: epidemiology, treatment challenges and progress. Mycetoma: epidemiology, treatment challenges and progress. Res Reports Trop Med 6:31–36
26. Fahal AH, Abu Sabaa AH (2010) Mycetoma in children. Trans R Soc Trop Med Hyg 104:117–112
27. Fahal AH, Suliman SH (1994) Clinical presentation of mycetoma. Sudan Med J. 32:46–66
28. Fahal AH, Sharfy ARA (1998) Vulval mycetoma: a rare cause of bladder outlet obstruction. Trans R Soc Trop Med Hyg 92:652–653
29. Fahal AH, Sharfi AR, Sheikh HE, EL Hassan AM (1996) Mycetoma: Uncommon complication. Trans R Soc Trop Med Hyg 89:550–552
30. Fahal AH, El Hassan AM, Abdelalla AO, Sheik HE (1998) Cystic mycetoma: an unusual clinical presentation of Madurella mycetomatis. Trans R Soc Trop Med Hyg 92:6–67
31. Fahal AH, Suliman SH, Gadir AFA, EL Hag IA, EL Amin FI, Gumaa SA, Mahgoub ES (1994) Abdominal wall mycetoma: an unusual presentation. Trans R Soc Trop Med Hyg 88(1):78–80
32. Mahgoub ES (1968) Clinical aspects of mycetoma. Systemic mycoses. A Ciba foundation symposium. J & A. Churchill Ltd, London, pp 125–127
33. Fahal AH, Yagi HI, EL Hassan AM (1996) Mycetoma induced palatal deficiency and pharyngeal plexus dysfunction. Trans R Soc Trop Med Hyg 90:676–677
34. Gumaa SA, Mahgoub ES, EL Sid MA (1986) Mycetoma of the head and neck. Trans R Soc Trop Med Hyg 35(3):596–600
35. Fahal A, Mahgoub ES, EL Hassan AM, Jacoub AO, Hassan D (2015) Head and neck Mycetoma: the Mycetoma research Centre experience. PLoS Negl Trop Dis 9(3):e0003587. doi:10.1371/journal.pntd.0003587
36. Ahmed AOA, Fahal AH, Abugroun EAM, Zijlstra E, van Belkum A, Verbrugh HA (1998) Unexpected high prevelance of secondary bacterial infection in mycetoma. J Clin Microbiol 36(3):850–851
37. Fahal AH, Sheikh HE, EL Hassan AM (1996) Pathological fracture in mycetoma. Trans R Soc Trop Med Hyg 90:675–676
38. Abd Bagi ME, Fahal AH, Sheik HE, Abdul Wahab O, Taifoor MK, Osmanr EM (2003) High incidence of pathological fracture in mycetoma. Trans R Soc Trop Med Hyg 97(5):582–584
39. Abd El-Bagi MEB, Fahal AH (2009) Mycetoma revisited. Incidence of various radiographic signs. Saudi Med J 30(4):529–533
40. Fahal AH, Sheikh HE, EL Lider MA, Homeida MA, EL Arabi YE, Mahgoub ES (1997) Ultrasonic imaging in mycetoma. Br J Surg 78:765–766
41. El Shamy ME, Fahal AH, Shakir MY, Homeida MM (2012) New MRI grading system for the diagnosis and Management of Mycetoma. Trans R Soc Trop Med Hyg 106(12):738–742
42. EL Hag IA, Fahal AH, Khalil EAG (1996) Fine needle aspiration cytology of mycetoma. Acta Cytol 40(3):461–446
43. Yousif BM, Fahal AH, Yahia MY (2010) The cell block technique: a new diagnostic tool for Mycetoma. Trans R Soc Trop Med Hyg 104(1):6–9
44. Wendy WJ van de Sande, Ahmed H. Fahal, Michael Goodfellow, El Sheikh Mahgoub, Oliverio Welsh, Ed E. Zijlstra (2014) The merits and pitfalls of the currently used diagnostic tools in mycetoma. PLoS Negl Trop Dis. doi: 10.1371/journal.pntd.0002918
45. EL Hassan AM, Fahal AH, EL Hag IA, Khalil EAG (1994) The pathology of mycetoma: light microscopic and ultrastructural features. Sudan Med J 32:23–45
46. Fahal AH, EL Toum EA, EL Hassan AM, Gumaa SA, Mahgoub ES (1995) Host tissue reaction to Madurella mycetomatis: new classification. J Med Vet Mycol 33:15–17
47. Ibrahim AI, El Hassan AM, Fahal A, van de Sande WW (2013) A histopathological exploration of the Madurella mycetomatis grain. PLoS One 8(3):e57774. doi:10.1371/journal.pone.0057774. Epub 2013 Mar 6

48. Fahal AH, EL Toum EA, EL Hassan AM, Gumaa SA, Mahgoub ES (1994) A preliminary study on the ultrastructure of Actinomadura pelletieri and its host tissue reaction. J Med Vet Mycol 32:343–348
49. EL Hassan AM, Fahal AH, Ahmed AO, Ismail A, Veress B (2001) The immunopathology of actinomycetoma lesions caused by Streptomyces Somaliensis. Trans R Soc Trop Med Hyg 95(1):89–92
50. Ludwig L, Euzeby J, Schumann P (2012) Road map of the phylum Actinobacteria. In: Goodfellow M, Kämpfer P, Busse HJ, Trujillo ME, Suzuki KI, Ludwig W, Whitman WB (eds) Bergey's manual of systematic bacteriology, The Actinobacteria, part a, vol 5, 2nd edn. Springer, New York, pp 1–28
51. Gumaa SA, Mahgoub ES (1975) Counterimmunoelectrophoresis in the diagnosis of mycetoma and its sensitivity as compared to immunodiffusion. Sabouraudia 13(3):309–315
52. Salinas-Carmona MC, Welsh O, Casillas SM (1993) Enzyme-linked immunosorbent assay for serological diagnosis of Nocardia Brasiliensis and clinical correlation with mycetoma infections. J Clin Microbiol 31(11):2901–2906
53. van de Sande WW, Janse DJ, Hira V et al (2006) Translationally controlled tumor protein from Madurella mycetomatis, a marker for tumorous mycetoma progression. J Immunol 177(3):1997–2005
54. McLaren ML, Mahgoub ES, Georgakopoulos E (1978) Preliminary investigation of the use of the enzyme linked immunosorbent assay (ELISA) in the serodiagnosis of mycetoma. Sabouraudia 16(3):225–228
55. de Hoog GS, Adelmann D, Ahmed AO, van Belkum A (2004) Phylogeny and typification of Madurella mycetomatis, with a comparison of other agents of eumycetoma. Mycoses 47(3–4):121–130
56. de Hoog GS, Guarro J, Gene J, Figueras M (2000) Atlas of clinical fungi, 2nd edn. Centraalbureau voor Schimmelcultures, Utrecht, The Netherlands
57. Ahmed AO, Mukhtar MM, Kools-Sijmons M, Fahal AH, de Hoog S, van den Ende BG, Zijstara ED, Verburgh H, Abugroun ESA, EL Hassan AM, van Beklkum A (1999) Development of a species-specific PCR RFLP procedure for the identification of Madurella mycetomatis. J Clin Microbiol 37(10):3175–3178
58. Borman AM, Desnos-Ollivier M, Miles S-J, et al (2009) Molecular characterisation of the Madurella grisea complex reveals at least three new taxa associated with human mycetomas. ISHAM
59. Rodriguez-Nava VCA, Devulder G, Flandrois JP, Boiron P, Laurent F (2006) Use of PCR-restriction enzyme pattern analysis and sequencing database for hsp65 gene-based identification of Nocardia species. J Clin Microbiol 44:536–546
60. Desnos-Ollivier M, Bretagne S, Dromer F, Lortholary O, Dannaoui E (2006) Molecular identification of black-grain mycetoma agents. J Clin Microbiol 44(10):3517–3523
61. Zhi X, Li W, Stackebrandt E (2009) An update of the structure and 16S rRNA gene sequenced-based definition of higher categories of the class Actinobaceria, with the proposal of two new suborders and four new families and amended descriptions of the existing higher taxa. Int J Syst Evol Microbiol 59:589–608
62. Mhmoud NA, Ahmed SA, Fahal AH, de Hoog GS, van den Ende AH G, van de Sande WW (2012) Pleurostomophora ochracea, a novel agent of human eumycetoma with yellow grains. J Clin Microbiol 50(9):2987–2994
63. de Hoog GS, van Diepeningen AD, Mahgoub el S, van de Sande WW (2012) New species of Madurella, causative agents of black-grain mycetoma. J Clin Microbiol 50(3):988–994
64. Fahal AH, Rahman IA, El-Hassan AM, Rahman ME, Zijlstra EE (2011) The safety and efficacy of itraconazole for the treatment of patients with eumycetoma due to Madurella mycetomatis. Trans R Soc Trop Med Hyg 105(3):127–132
65. Mahgoub ES, Gumaa SA (1984) Ketoconazole in the treatment of eumycetoma due to Madurella mycetomii. Trans R Soc Trop Med Hyg 78(3):376–379

66. Hay RJ, Mahgoub ES, Leon G, al-Sogair S, Mycetoma WO (1992) J Med Vet Mycol 30(Suppl 1):41–49
67. Venugopal PV, Venugopal TV (1993) Treatment of eumycetoma with ketoconazole. Australas J Dermatol 34(1):27–29
68. Mahgoub ES (1972) Treatment of actinomycetoma with sulphamethoxazole plus trimethoprim. Am J Trop Med Hyg 21(3):332–335
69. Welsh O, Sauceda E, Gonzalez J, Ocampo J (1987) Amikacin alone and in combination with trimethoprim-sulfamethoxazole in the treatment of actinomycotic mycetoma. J Am Acad Dermatol 17(3):443–448
70. Welsh O, Vera-Cabrera L, Welsh E, Salinas MC (2012) Actinomycetoma and advances in its treatment. Clin Dermatol 30(4):372–381
71. Welsh O, Al-Abdely HM, Salinas-Carmona MC, Fahal AH (2014) Mycetoma medical therapy. PLoS Negl Trop Dis 8(10):e3218. doi:10.1371/journal.pntd.0003218
72. Suleiman SH, Wadaella el S, Fahal AH (2016) The surgical treatment of Mycetoma. PLoS Negl Trop Dis 10(6):e0004690. doi:10.1371/journal.pntd.0004690. eCollection 2016

Chapter 13
Dermatophytes and Dermatophytosis

Roberto Arenas, María del Rocío Reyes-Montes, Esperanza Duarte-Escalante, María Guadalupe Frías-De-León, and Erick Martínez-Herrera

Abstract Dermatophytosis, are the most common fungal infection worldwide. Transmission is mostly by direct contact with infected animals, humans or contact with fomites. Clinical features vary according to the etiological agent. Dermatophytes belong to the genera Microsporum, Trichophyton, Epidermophyton (anamorphic state), and Arthroderma (teleomorphic state). The main etiological agents in humans are T. rubrum, T. tonsurans, T. mentagrophytes complex, and M. canis, M. gypseum, and E. floccosum. Recent phylogenetic studies indicate the existence of a fourth genus, Chrysosporium. In this chapter we analyze epidemiology, biology, clinical forms, pathogenesis, diagnosis, detection methods and treatment. Also molecular taxonomy, molecular epidemiology, molecular techniques, molecular identification and treatment are widely analyzed.

R. Arenas, M.D. (✉)
Sección de Micología, Hospital General "Manuel Gea González",
Av. Calzada de Tlalpan 4800, Tlalpan 4502B Unidad Investigadores. edif B 301. CP, 14080 México, Cd.Mx, Mexico
e-mail: rarenas98@hotmail.com

M. del Rocío Reyes-Montes, Ph.D. • E. Duarte-Escalante, Ph.D.
Departamento de Microbiología y Parasitología, Facultad de Medicina, Universidad Nacional Autónoma de México (UNAM), Ciudad Universitaria No. 3000, Coyoacán, 04510 México, Cd.Mx, Mexico
e-mail: remoa@unam.mx; dupe@unam.mx

M.G. Frías-De-León, Ph.D.
División de Investigación, Hospital Juárez de México, Edificio E. Av. Instituto Politécnico Nacional 5160, Col. Magdalena de las Salinas, 07760 México, Cd.Mx, Mexico

Dirección de Investigación, Hospital Regional de Alta Especialidad de Ixtapaluca, Carretera Federal México – Puebla Km. 34.5, Ixtapaluca, Estado de México 56530, Mexico
e-mail: magpefrias@gmail.com

E. Martínez-Herrera, Ph.D.
Dirección de Investigación, Hospital Regional de Alta Especialidad de Ixtapaluca, Carretera Federal México – Puebla Km. 34.5, Ixtapaluca, Estado de México 56530, Mexico
e-mail: martinezerickh@gmail.com; erickmartinez_69@hotmail.com

H.M. Mora-Montes, L.M. Lopes-Bezerra (eds.), *Current Progress in Medical Mycology*, DOI 10.1007/978-3-319-64113-3_13

13.1 Introduction

Dermatophytosis, also known as *tinea* or ringworm, has become the most common fungal infection worldwide. It is caused by a group of keratinophilic fungi that affect the skin, hair, and nails, and is of both public and veterinary health concern [1]. These fungal infections are rarely associated with a fatal outcome; regardless, they are associated with high morbidity [2]. The transmission of dermatophytosis is mostly by direct contact with infected animals or humans or by indirect contact with fomites. Clinical cases vary according to the etiological agent (species) and anatomical location. Symptoms can be mild or severe based on the immunological status of the host; subcutaneous or internal organ infections are usually rare. The typical lesions resulting from skin infection are presented in the form of an erythematous, flaky, itchy, circular plaque as a direct result of fungal infection, indicating a hypersensitive reaction to the fungi per se or to their metabolites [3]. This condition affects a disproportionate number of infants, young adults, and people of low economic status; likewise, the distribution of these fungi can vary considerably, depending on the geographical area of origin and other epidemiological factors such as age, gender, and seasonality [4, 5].

Dermatophytes belong mostly to the genera *Microsporum*, *Trichophyton*, *Epidermophyton* (in the anamorphic state), and *Arthroderma* (in the teleomorphic state), and so far they include 40 known species responsible of infections in humans and other mammals [6]. The fungal species *T. rubrum*, *T. tonsurans*, and the *T. mentagrophytes* complex, as well as *M. canis*, *M. gypseum*, and *E. floccosum*, are considered the main etiological agents of dermatophytosis in humans [5]. Based on their ecology, dermatophytes have been classified into three groups: anthropophilic, zoophilic, and geophilic. Anthropophilic dermatophytes are associated with humans, mainly causing *tinea capitis*, *tinea corporis*, *tinea pedis,* and *tinea unguium,* and they rarely infect animals. Zoophilic species are common pathogens in animals [5, 7, 8], although they can occasionally infect humans too. Geophilic dermatophytes are associated with keratinized materials (hair, feathers, nails, and horns) scattered throughout the environment [5]. These fungi, which are able to degrade the keratin from hair, nails, and feathers, can also produce superficial infections (dermatophytosis) in the skin of human and animal hosts and penetrate into deeper tissues in immunocompromised hosts [9].

13.2 Classification

The sexual stages of these organisms are the basis of the taxonomy and nomenclature of these fungi. To date, the anamorphic states of dermatophytes (*genera Epidermophyton, Microsporum,* and *Trichophyton*) pertain to the *class* Hyphomycetes, *phylum* Deuteromycota, and teleomorphs (most zoophilic and geophilic species of *Microsporum* and *Trichophyton*) are classified within the teleomorphic *genus Arthroderma*, *order* Onygenales, *phylum* Ascomycota [10].

Based on their ecology, dermatophytes have been divided into three groups: anthropophilic, zoophilic, and geophilic; their members are included within the *genera Trichophyton*, *Microsporum*, *Epidermophyton*, and *Arthroderma*. Based on the morphology of anamorphic states, Simpanya [11] considers two species within the *genus Epidermophyton*, approximately 18 species in *Microsporum*, and 25 species in *Trichophyton*. Dermatophytes included in these *genera* are closely related phylogenetically and taxonomically, and therefore, their phylogeny is not yet clear. Since the morphological characteristics of these fungi are rather limited to facilitate their identification, Gräser et al. [8] recommend a polyphasic approach involving, in addition to phenotypic identification, other characteristics such as physiological and biochemical traits and the production of secondary metabolites, among others, along with universally applied molecular methods, to achieve an accurate identification.

Thus, a report by Molina de Diego [10] considers two species within the *genus Epidermophyton*: *E. floccosum* and *E. stockdaleae*, among which only the former is pathogenic for humans; 16 species within the *genus Microsporum* (*M. amazonicum, M. audouinii, M. boullardii, M. canis* complex, *M. cookei, M. ferrugineum, M. equinum, M. fulvum, M. gypseum, M. gallinae, M. nanum, M. praecox, M. persicolor, M. racemosum, M. vanbreyseghemii*, and *M. ripariae*), of which only 10 are pathogenic for men; and 31 within the *genus Trichophyton* (*T. ajelloi, T. concentricum, T. equinum, T. eboreum, T. erinacei, T. fischeri, T. flavescens, T. fluviomuniense, T. gallinae, T. gloriae, T. gourvilii, T. interdigitale, T. kanei, T. krajdenii, T. longifusum, T. megninii, T. mentagrophytes* complex, *T. phaseoliforme, T. quinckeanum, T. raubitschekii, T. rubrum, T. sarkisorii, T. schoenleinii, T. simii, T. soudanense, T. terrestre, T. tonsurans* complex, *T. vanbreuseghemii, T. verrucosum, T. violaceum*, and *T. yaoundei*), with 10 species that are pathogenic for humans.

However, molecular methods have provided a large contribution toward improving knowledge regarding the biological diversity of dermatophytes, although it has also raised challenges in terms of their taxonomic classification. Cafarchia et al. [12] mention that most phylogenetic reconstructions from DNA sequences support the separation of geophilic species from the remaining members of the *family* arthrodermataceae, known as the “true” dermatophytes. Moreover, the scarce available information on the mating of some species have underscored the importance of phylogenetic studies for defining *species*, which have led to a reduction in the number of species or varieties of dermatophytes. Based on the classification proposed by Gräser et al. [8], Cafarchia et al. [12] consider five species in the *Arthroderma otae* complex (*M. canis, M. equinum, M. distortum, M. audouinii*, and *M. ferrugineum*); 10 species in the *A. vanbreuseghemii* complex (*T. tonsurans, T. equinum, T. interdigitale, T. mentagrophytes* var. *mentagrophytes, T. mentagrophytes* var. *granulosum, T. verrucosum* var. *autotrophicum, T. mentagrophytes* var. *goetzii, T. mentagrophytes* var. *interdigitale* and var. *nodulare*, and *T. krajdenii*); six species in the *A. simii* complex (*T. simii, T. mentagrophytes, T. mentagrophytes* var. *quinckeanum, T. langeronii, T. sarkisovii*, and *T. schoenleinii*); six species in the *A. benhamiae* complex (*T. erinacei, T. mentagrophytes* var. *granulosum, T. verrucosum, T. concentricum, T. bullous*, and *T. eriotrephon*); and 11 species in the *T. rubrum* complex

(*T. rubrum, T. raubitschekii, T. soudanense, T. gourvilii, T. megninii, T. fischeri, T. kanei, T. raubitschekii, T. violaceum, T. yaoundei*, and *T. glabrum*).

However, recent phylogenetic studies indicate the existence of a fourth *genus*, *Chrysosporium*, which is congruent with the ecological groups based on studies of mitochondrial DNA or ribosomal RNA-coding DNA (mtDNA or rDNA) [13].

13.3 Epidemiology

Simpanya [11] mentions that some dermatophyte species evolved from their natural habitat in soil and developed an affinity toward a specific host, with the latter mostly attributed to differences in the keratin of each host.

Likewise, the severity of an infection caused by different species of the *genera Epidermophyton, Microsporum*, and *Trichophyton* depends largely on the pathogenic potential of the invading species, socioeconomic factors, and the local environmental conditions. Nevertheless, it is generally acknowledged that between 15 and 20% of the global population is infected [14].

The epidemiology of dermatophytes has been modified during the last 70 years; a review by Seebacher et al. [15] mentions that before the Second World War, in Germany, *M. audouinii* and *E. floccosum* were the main causative agents of dermatophytosis, whereas from the 1950s onwards, *T. rubrum* has become the most common dermatophyte, representing 80–90% of the cases of infection, followed by *T. mentagrophytes*. This switch also occurred in Central America and Northern Europe in association with the incidence of *tinea pedis*. In contrast, in Southern Europe and Arab countries, zoophilic dermatophytes such as *M. canis* or *T. verrucosum* are more frequent. Moreover, this review also mentions that the infection incidence of *M. canis* in Mediterranean countries has increased dramatically during the past few years; this dermatophyte is commonly associated with *tinea capitis* in infants. Table 13.1 shows the most recent epidemiological data (2013–2016) on dermatophytosis worldwide.

In addition, an extensive review by Sigurgeirsson and Baran [41] on the global incidence of onychomycosis reveals that the average prevalence of this ailment in Europe and North America was 4.3% in studies based on the general population and 8.9% in studies based on nosocomial data. Both studies demonstrated that onychomycosis is more common in toenails and more frequent in males, and the main causative agent was *T. rubrum* (65.0%).

Furthermore, *tinea pedis*, which can be a chronic or recurrent infection that involves interdigital spaces or other regions of the feet, is often caused by anthropophilic etiological agents, including *T. rubrum sensu stricto*, the most common agent, followed by *T. interdigitale* and *E. floccosum*. Notwithstanding, Ilkit and Durdu [42] consider the current situation and the changing patterns of *tinea pedis* to necessitate revision beginning from the second half of the twentieth century, during which a global increment in this mycosis and clonal propagation of *T. rubrum* were observed. They mention that these phenomena are probably due to increased urbanization, the use of sport and physical conditioning facilities, the growing prevalence of obesity, and aging.

Table 13.1 Epidemiological data on dermatophytosis worldwide

Continent	Clinical form Frequency (%)	Etiological agents Frequency (%)	Age	Sex Frequency (%)	References
America					
Brazil	*Tinea corporis* (31.81) *Tinea capitis* (31.81) *Tinea unguium* (15.15) *Tinea pedis* (12.12) *Tinea cruris* (9.09)	*T. mentagrophytes* (40.0) *T. rubrum* (30.0) *T. tonsurans* (10.0) *T. verrucosum* (5.0) *M. gypseum* (5.0) *M. canis* (5.0) *E. floccosum* (5.0)	1–20	Male (42.42) Female (57.57)	Silveira-Gomes et al. [16]
Brazil	*Tinea unguium* (62.26) *Tinea pedis* (25.42) *Tinea corporis* (6.26) *Tinea cruris* (3.42) *Tinea manuum* (1.65) *Tinea capitis* (0.99)	*T. rubrum* (96.9) *T. mentagrophytes* (3.0) *M. gypseum* (0.9) *T. tonsurans* (0.6) *E. floccosum* (0.5) *M. canis* (0.4)	0 > 60	Male (29.0) Female (71.0)	Di Chiacchio et al. [17]
Brazil	*Tinea unguium* (100)	*T. rubrum* (68.6) *T. mentagrophytes* (6.8) *Trichophyton* spp. (3.2) *T. tonsurans* (2.5) *T. raubistscheckii* (0.1) *M. gypseum* (0.2) *E. floccosum* (0.1)	1–100	Male (34.0) Female (66.0)	de Carvalho Ribeiro et al. [18]
Chile	*Tinea unguium* (95.24) *Tinea pedis* (4.76)	*T. rubrum* (59.45) *T. mentagrophytes* (40.54)	ND	Male (40.32) Female (59.68)	Díaz Jarabrán et al. [19]
Brazil	*Tinea unguium* (48.5) *Tinea pedis* (33.1)	*T. rubrum* (59.6) *T. interdigitale* (34) *T. tonsurans* (0.9) *M. canis* (2.6) *E. floccosum* (2.5) *M. gypseum* (1.3)	1–98	Male (79.3) Female (53.9)	Heidrich et al. [20]

(continued)

Table 13.1 (continued)

Continent	Clinical form Frequency (%)	Etiological agents Frequency (%)	Age	Sex Frequency (%)	References
Europe					
Greece	*Tinea unguium pedum* (42.2) *Tinea pedis* (41.4) *Tinea corporis* (5.6) *Tinea manuum* (4.1) *Tinea unguium manuum* (3.0) *Tinea capitis* (2.4) *Tinea cruris* (1.3)	*T. rubrum* (54.2) *T. mentagrophytes* (37.9) *E. floccosum* (3.1) *M. canis* (3.0)	20–78	Male (54.91) Female (45.08)	Budak et al. [21]
Italy	*Tinea capitis* (67.92) *Tinea corporis* (28.30 *Tinea unguium* (3.77)	*T. violaceum* (43.39) *T. soudanense* (56.69)	2–43	Male (44.23) Female (55.76)	Farina et al. [22]
Italy	*Tinea unguium* (14.2)	Dermatophytes	1–60	Male (41.09) Female (53.83)	Papini et al. [23]
France	*Tinea unguium* (37.1) *Tinea pedis* (21.6) *Tinea corporis* (27) *Tinea manuum* (2.4) *Tinea unguium manuum* (7.36)	*T. rubrum* (78.6) *T. interdigitale* (9.5) *T. tonsurans* (1.8) *T. soudanense* (1.4) *M. audouinii* (0.8) *T. violaceum* (0.6) *E. floccosum* (0.4)	15–70	Male (52.7) Female (47.3)	Faure-Cognet et al. [24]
Greece	*Tinea corporis* (16.1) *Tinea capitis* (3.6) *Tinea unguium manuum* (16.7) *Tinea unguium pedum* (63.6)	*T. rubrum* (53.9) *T. mentagrophytes* (6.9) *T. interdigitale* (10.8) *M. canis* (22.5)	2–85	ND	Nasr et al. [25]

Africa					
Cameroon	*Tinea capitis* (3.6)	*T. soudannense* (56.8) *T. schoenleinii* (2.7) *T. violaceum* (2.7) *T. tonsurans* (2.7) *M. canis* (2.7)	4–15	Male (63.7) Female (36.3)	Kechia et al. [26]
Senegal	*Tinea unguium* (37.92)	*T. rubrum* (53.6) *T. interdigitale* (26.1)	0–82	Male (28.2) Female (71.8)	Seck et al. [27]
Senegal	*Tinea capitis* (34.51)	*T. soudanense* (56.18) *T. rubrum* (18.37) *M. langeronii* (12.72) *M. canis* (6.36) *T. mentagrophytes* (4.60).	1–83	Male (17.13) Female (82.86)	Ndiaye et al. [28]
Kenya	*Tinea capitis* (81.3)	*Trichophyton* spp. (61.3) *Microsporum* spp. (13.3) *Epidermophyton* spp. (7.3)	3–14	Male (59.3) Female (40.7)	Ndunge Moto et al. [29]
Mali	*Tinea capitis* (39.3)	*M. audouinii* (ND) *T. soudanense* (ND)	6–15	Male (48.47) Female (51.52)	Coulibaly et al. [30]
Morocco	*Tinea capitis* (64.67)	*M. canis* (63.26) *T. violaceum* (26.51) *T. mentographytes* (4.81) *M. langeronii* (3.01) *T. verrucosum* (1.8) *T. schoenleinii* (0.61)	0–32	Male (63.88) Female (36.12)	El Mezouari et al. [31]
Asia					
Iran	*Tinea unguium* (44.2)	*T. mentagrophytes* (17.7) *T. rubrum* (1.7) *E. floccosum* (0.7) *T. violaceum* (0.2) *T. verrucosum* (0.2) *T. tonsourans* (0.2) *M. gypseum* (0.2%)	0–70	Male (37.1) Female (62.9)	Afshar et al. [32]

(continued)

Table 13.1 (continued)

Continent	Clinical form Frequency (%)	Etiological agents Frequency (%)	Age	Sex Frequency (%)	References
Israel	*Tinea unguium pedum* (89.6) *Tinea unguium manuum* (33.4)	*T. rubrum* (38.2) *T. mentagrophytes* (5) *T. tonsourans* (1) *T. violaceum* (ND) *T. terrestre* (ND)	0–60	Toenails: Male (57.8) Female (42.2) Fingernails: Male (74.0) Female (26.0)	Segal et al. [33]
Iran	*Tinea unguium* (56.4)	*T. mentagrophytes* var. *interdigitale* (75) *T. rubrum* (25)	0–60	Male (29.3) Female (70.7)	Soltani et al. [34]
India	*Tinea corporis* (78.0) *Tinea cruris* (10.0) *Tinea manuum* (2.5) *Tinea faciei* (1.8) *Tinea pedis*(0.7) *Tinea unguium* (6.7)	*T. rubrum* (79.0) *T. mentagrophytes* (14.5) *M. canis* (3.2) *M. gypseum* (3.2)	22–45	Male (56.0) Female (443.0)	Lakshmanan et al. [35]
China	*Tinea unguium* (28.55) *Tinea capitis* (15.64) *Tinea pedis* (15.06) *Tinea cruris* (10.33) *Tinea corporis* (8.75) *Tinea manuum* (5)	*T. rubrum* (56.24) *T. mentagrophytes* (13.35) *M. canis* (10.19) *T. violaceum* (ND) *T. tonsurans* (ND) *M. gypseum* (ND) *E. floccosum* (ND)		*Tinea unguium* Male (23.8) Female (34.73) *Tinea capitis* Male (15.58) Female(15.76) *Tinea pedis* Male (14.54) Female (15.76) *Tinea cruris* Male (16.88) Female (2.25) *Tinea corporis* Male (10.65) Female (6.43) *Tinea manuum* Male (2.08) Female (3.86)	Cai et al. [36]

Iran	*Tinea corporis* (32.0) *Tinea cruris* (22.0) *Tinea capitis* (12.0) *Tinea manuum* (11.2) *Tinea pedis* (2.9) *Tinea unguium* (6.9) *Tinea faciei* (5.2) *Tinea barbae* (0.4)	*T. interdigitale* (58.7) *E. floccosum* (35.4) *M. canis* (3) *T. rubrum* (1.5) *A. benhamiae* (0.5) *T. tonsurans* (0.3) *T. violaceum* (0.3)	1–89	ND	Rezaei-Matehkolaei et al. [37]
Palestine	*Tinea pedis* (ND) *Tinea capitis* (ND) *Tinea cruris* (ND) *Tinea corporis* (ND) *Tinea unguium* (ND)	*M. canis* (75.4) *T. rubrum* (22.5) *T. mentagrophytes* (2.2)	2–70	Male (61.36) Female (38.63)	Ali-Shtayeh et al. [38]
Oceania					
New Zealand	*Tinea unguium* (ND) *Tinea corporis* (ND) *Tinea capitis* (ND)	*T. rubrum* (69.0) *T. mentagrophytes* (19.0) *M canis* (6.0) *E. floccosum* (3.0) *T. tonsurans* (1.0) *T. violaceum* (1.0)	0–99	ND	Singh et al. [39]
New Zealand	*Tinea capitis*	*T. soudanense* (47.82) *M. audouinii* (26.08) *T. violaceum* (17.39)	6–12	Male (56.0) Female (44.0)	McPherson et al. [40]

ND Non-determined

13.4 Biology

In general, each *genus* of dermatophytes is described based on the colony morphology, conidia, and its formation; the following paragraphs describe the distinctive morphology of each pathogenic *genus*.

Epidermophyton **spp.** usually shows grainy, folded colonies with a suede-like texture and a yellowish-green color; these colonies are easily bleached and become floccose and sterile. It forms abundant, typically smooth, clavate macroconidia with thin or moderately thick walls, 8–59 septa, and a size of 20–60 by 4–13 μm. Microconidia are typically absent [10].

Microsporum **spp.** forms dusty, grainy colonies with a cottony surface and yellowish-orange color. It generates macroconidia with a fusiform, pyriform, or cylindro-fusiform shape; with echinulate or verrucous walls, and containing 1–15 septa, moderately thick to thick, and have a size of 6–160 by 6–25 μm. It presents sessile or clavate microconidia that are borne laterally directly on the hyphae or in clusters [10].

Trichophyton **spp.** has variable macroscopic features, presenting differences between species; the colonies can be grainy, cottony, cerebriform, and hairy, among others. The underside of the colonies can have reddish or brownish pigmentation. When present, macroconidia have smooth, usually thin, walls with 1–12 septa, either individual or clustered, clavate to fusiform. They have a variable size ranging from 8 to 86 by 4 to 14 μm. They also present microconidia, which might be more abundant than macroconidia, with a spherical, clavate, or fusiform shape, sessile, and borne laterally directly on the hyphae or in pedicels [10].

13.5 Clinical forms of Dermatophytosis

Dermatophytoses, also known as *tineas* or ringworm, result in the development of clinical characteristics such as inflammation, erythematous flaky skin, vesicles, and pustules, although there are also specific clinical characteristics for each case of *tinea* or dermatophytosis. The characteristics are described individually in the following paragraphs [43–46].

Tinea capitis. An infection that occurs mainly in infants and teenagers, affecting the hair and scalp. It has a broad spectrum of clinical manifestations that range from mild peeling lesions to inflammatory lesions such as vesicles and pustules. According to the type of lesion, it is classified as either dry or inflammatory; however, some authors have reported the existence of a less frequent form of *tinea capitis* in adults older than 70 years that presents a loss of sebum production in perimenopausal women, as well as in patients presenting some degree of immunosuppresion (leukemia, diabetes, treatment with corticosteroids, transplants, among others) [47–51]. In agreement with this classification, dry *tinea capitis* is divided into two varieties: microsporic and trichophytic. The microsporic variety is associated with the *genus Microsporum,* displaying one or several big plaques with short hair that can become confluent and form a single, extended plaque; major apparent symptoms include an erythematous area with irregular alopecia and dry scales. If the infection is not treated soon and persists without spontaneous healing, it may result in severe peeling of the scalp, rupture of the cuticle, and fragile and brittle hair. *M. canis* is the main etiological agent of this clinical form, followed by *M. gypseum* and *M. audouinii* [52, 53]. The trichophytic variety of dry *tinea*, associated with the *genus Trichophyton*, is commonly observed in African-American infants. The lesion starts as an erythematous flaky plaque in the scalp, and although multiple lesions are often present, single lesions are seen sporadically. Hair is usually broken at the shaft, and detritus can be observed within the follicle aperture, showing the appearance of black grains similar to gunpowder. The main causal agent of this condition is *T. tonsuras*, followed by *T. mentagrophytes* and *T. rubrum* [52, 53]. Inflammatory *tinea*, also known as Celsus' kerion, is associated with immunological mechanisms of the host, and lesions can be secondary to bacterial infections [54]. This *tinea* has evident clinical characteristics, such as an exudative plaque with pustules, adjacent edematous and erythematous areas, extensive scabbed areas, and alopecia. It is usually accompanied by fever, clearly delimited and localized adenopathies, malaise, dull and gray tinted mangy hair, and a mucopurulent exudate. The most common agents are *M. canis* and *T. tonsurans* [55, 5]. Adults suffering from *tinea capitis* have infections in the scalp similar to seborrheic dermatitis and erythematosus discoid lupus, as well as large, slightly erythematous plaques with alopecia. This type of ringworm can also appear in either dry or inflammatory forms, and the causal agents vary according to the geographical location. Several authors advise against minimizing the importance of this infection based on reports of adult patients with nonhealing scalp affectations [56, 49, 50, 57, 58].

Tinea favosa. Also known as favid, *favus* is characterized as a chronic infection, mainly of the scalp; however, it can also be observed sporadically in glabrous and hairy skin, and in nail roots. Scalp infection represents the most severe form of *tinea capitis*, presenting fibrillar clusters with the aspect of honey called scutulum, and a mousy odor. Severe alopecia and scabbing are frequent [59–61].

Some varieties can be identified using Sabouraud: (a) favus pityroides, with large, peeling plates are predominant; the hair takes on a gray coloration, and scutulae are visible after scraping, creating contusions with seborrheic dermatitis; (b) favus papyroides, observed in outbreaks and characterized by pliable, adherent, and erythematous scabs with a humid consistency; erosion is observed after scab removal; (c) favus impetigoides, characterized by suppurating scabs; lesions with scabs resembling impetigo are usually rare; no scutulae are present [61]. It must be mentioned that although *tinea favosa* is rare in adults, it can affect immunosuppressed or immunocompetent people. Common etiological agents, from high to low relevance, are *T. schoenleinii*, *M. gypseum*, *T. violaceum*, and *T. mentagrophytes* var. *quinckeanun* [60–62]. It is important to consider that in reports from industrialized countries, the causal agents are mostly related to the *genus Trichophyton*, whereas in developing countries, they are associated with species of the *genus Microsporum*; *E. floccosum*, in contrast, is rare [48, 52].

Tinea barbae. This condition is mostly exclusive to adolescent or adult males; it is limited to the beard, chin, and neck areas [63–65], and is caused by dermatophytes that infect and colonize keratinized tissues, including *T. metagrophytes*, *T. interdigitale*, *T. verrucosum*, *T. rubrum*, and more rarely *T. violaceum*, *T. erinacei*, *M. canis*, and *A. benhamiae*, the latter being a zoophilic dermatophyte of the *T. mentagrophytes* complex [63, 66–68, 65, 69]. The clinical manifestations can vary according to the pathogenic agent responsible for the disease, and clinical variants can be diagnosed as inflammatory and noninflammatory [63, 67].

The inflammatory variety is mostly related to zoophilic dermatophytes such as *T. mentagrophytes*, *T. verrucosum*, *M. canis*, and *T. erinacei*, and it is capable of inducing severe inflammatory damage ranging from inflamed nodules to multiple pustules. Broken and loose hair can also be found, as well as vesiculopustular lesions and exudate. This variety is also observed in regional adenopathies, such as fever and malaise [63, 65, 66, 68, 69].

The noninflammatory variety has a diverse range of manifestations, from flaky plates with papules, pustules, or scabs, to erythematous annular eruptions and facial edema. Broken and loose hair near the skin potentially obstructs the follicle, resulting in folliculitis. It is common to find patients lacking regional adenopathies who are not feverish. The dermatophytes associated with this variety are mostly anthropophilic, including *T. rubrum*, *T. violaceum*, and *T. tonsurans* Pruritus is present in both varieties [63, 64, 67].

Tinea pedis. Tinea pedis is also known as foot tinea or athlete's foot, with the latter name related to its high prevalence in athletes, consisting of mainly adult males. The main causal agents are *T. rubrum*, *T. mentagrophytes* var. *mentagrophytes* or *interdigitale* and, at a lower frequency, *E. floccosum* and *M. gypseum.* These fungi can be isolated from the floors of pools and showers, as well as from closed footwear, explaining the occurrence of this condition in females and infants, although only sporadically [70–73]. The causal agents of this infection often invade the plantar aspect, interdigital spaces, and lateral foot, which are commonly affected in at-risk populations (immunocompetent or immunosuppressed). It may present recurrence and resistance to treatment. Although it is not caused by them directly, bacteria are often implicated in the infection onset, and the condition is in fact exacerbated and worsened by their presence, augmenting the bad odor, maceration, inflammation, and other symptoms [74, 75, 72].

To date, *tinea pedis* has been classified in four varieties according to their clinical characteristics: (a) chronic or interdigital intertrigo, the most frequent form of the infection, onset is characterized by erosion and peeling, followed by maceration, particularly between the fourth and fifth interdigital space, which may extend to the plantar aspect and in turn develop into a bacterial infection. It may or may not present erythema and fissures between the fourth and fifth toes, and also presents hyperhidrosis, pruritus, and bad odor. Studies investigating this variety have reported that the most common causal agents are of anthropophilic origin: *T. rubrum* and *E. floccosum*

[72, 74–76]; (b) vesiculose, generally subacute, presents small vesicles or vesiculopustules over an erythematous background in the lateral foot, with or without fissures; it extends from the interdigital space toward the midfoot and plantar aspect, although the lateral foot is more commonly affected. It can be painful and exhibit pruritus; it evolves quickly and is exclusively associated with zoophilic dermatophytes, such as species of the *T. mentagrophytes* complex [72, 74–77]; (c) moccasin, also called chronic hyperkeratosis; at the onset of infection, peeling and hyperkeratosis can be observed in the plantar aspect, heel, and lateral foot, which can be extended in a superior direction along the frontiers (moccasin distribution). It also presents fissures over one or both feet. This variety is observed in patients with atopic dermatitis, and although *tinea manuum* is frequently found in these infections, *T. rubrum*, a pathogen of anthropophilic origin, is most commonly associated with this disease (two feet-one hand syndrome) [74–76] (Porche 2015); (d) acute ulcerated, is the rarest of all four varieties, and can present ulcers, pustules, erosions, and rapid extension to the interdigital spaces; it is an exacerbated form of the interdigital or chronic intertrigo varieties. Patients may acquire a secondary bacterial infection that can be severe, developing cellulite, lymphangitis, and fever. This *tinea pedis* variety is common in patients with immunosuppression, such as AIDS, diabetes, or hypertension (HT). It is associated with the zoophilic dermatophytes *T. interdigitale*, *T. mentagrophytes*, and *M. canis* [74–76] (Porche 2015).

Tinea manuum. Also known as *hand tinea*, it is not a common affliction and usually affects the interdigital regions, back, and palms of the hands; often, only one hand is affected, although rarely both hands can be infected. It is more common in adult males and rare in infants, and it can present pruritus with variable and recalcitrant lesions [78–80]. A hyperkeratosis variant has also been classified that produces eruptions with dry margins in a white or yellow color and flaky erythematous plaques; annular manifestations occasionally occur, and verrucous lesions are rare. This *tinea* is associated with the pathogen *T. rubrum* [81, 82]. The inflammatory form is manifested as an acute inflammatory plaque with suppurative folliculitis, dyshidrosis, and vesicle eruptions. The pathogenic agent associated with this type of *tinea*, *T. mentagrophytes*, is a zoophilic dermatophyte. Although some studies have reported the involvement of *T. erinacei*, *M. canis*, *T. rubrum*, and *E. floccosum*, the latter two are of anthropophilic origin [82, 83]. This infection has been observed along with the presence of two feet-one hand syndrome. Physical examination reveals a papular erythematous affection, flaky skin and superficial erosion, erythematous margins between fingers and toes, and onychomycosis [84, 78, 82].

Tinea cruris. Also known as inguinal *tinea*, *tinea cruris* is a superficial infection caused by *T. rubrum*, *T. mentagrophytes*, and *E. floccosum* that mainly affects adult males due to the occlusive function of the scrotum that promotes a humid and warm environment favoring fungal growth; although it can also affect females and infants, it is most prevalent in male teenagers [85, 78, 86–89]. The inguinal, perineum, and gluteus areas are most commonly affected, and, rarely, the scrotum or penis [85, 86]. At onset, the infection is usually, and briefly, unilateral due to predisposing factors such as contaminated underwear [85, 78, 90, 88, 89]. The acute forms present one or several well-delimited erythematous and flaky plaques that range in color from red to intense brown. These plaques present maceration and are exudative and eczematiform, which is why they have been previously known as thread-like margin eczema [85, 3, 78, 86, 87, 89]. Pruritus is common and can present inflammation, vesicles, and pustules when the causal agent is *T. mentagrophytes* [85].

Tinea imbricata. Tinea imbricata is a type of chronic *tinea* characterized as the most superficial and dry tinea. It has a limited distribution and is usually endemic to isolated rural communities where poverty and lack of hygiene are the norm; some of the most affected locations are in Oceania (Polynesia and Melanesia), where it is known as *tokelau*, and with less frequency, India, Central, and South America, where it is known as *chimbere* in Brazil [91, 92]. The onset of this *tinea* occurs during infancy, but it can be observed in all age ranges and affects both genders equally. It is thought that a genetic predisposition could play an important role in this condition and could be of autosomal recessive or dominant heredity [93, 91]. Lesions can be manifested throughout the body, although the torso and extremities are usually affected. Nails, palms, and plantar aspects are rarely infected, and hair is never affected. Large, flaky concentric or annular plaques are observed at onset, with or without erythema. These plaques are overlapping and exhibit the appearance of lamellar structures that might be hypo- or hyperpigmented with posterior detachment [93–95]. It is common to observe macules or papules in recent lesions, which are often itchy. The responsible pathogen is *T. concentricum* [93, 91, 94, 95].

Tinea corporis. This infection is present in exposed glabrous skin, often in the neck, torso, shoulders, and extremities [77, 86, 87, 96], sporadically in legs [78, 96, 89], and never in the scalp, beard, feet, or hands, although examination of the latter could reveal the primary source of infection [86]. The responsible agents are *M. canis, T. rubrum, T. mentagrophytes, T. tonsurans*, and *E. floccosum,* although others agents have been sporadically reported, such as *M. audouinii, T. verrucosum, M. aenigmaticum, M. praecox,* and a new species of the *M. gypseum* complex [97, 86, 87, 98]. The species within the *genus Microsporum* cause a microsporic *tinea corporis*, in which a large number of small plaques are observed. They can become confluent and form large plaques, whereas species of the *genus Trichophyton* cause a trichophytic *tinea corporis* in which a smaller number of large plaques are formed and become confluent, resulting in an even larger plaque with an irregular size that affects a large extent of the body surface [99, 3]. It is manifested as flaky, erythematous, annular, and localized plaques, with papules, pustules, or vesicles and an active margin. It can be observed either as a single or multiple lesions. Immunosuppression is one of the predisposing factors to this disease (diabetes, use of topical corticosteroids, AIDS, among others) [77, 3, 79]. Other varieties are present less frequently, such as gluteus dermatophytosis, or dermatophytosis of the diaper area, which affects infants younger than 3 years of age in the diaper area and neighboring zones such as the perineum, legs, and abdomen, among others. The main causal agent is *E. floccosum*, although sporadic cases caused by *T. rubrum* and *T. verrucosum* have also been observed *verrucosum* [100, 101]. It is manifested as multiple circular, flaky, and erythematous lesions, which can be circinate and serpiginous; macules, papules, and vesicles can also be observed [100, 102]. A less frequent form of this variety is *corporis gladiatorum*, which is observed among wrestlers, in which the most affected areas are skin-to-skin contact areas in wrestling sports, such as the face, neck, and extremities. In the last few years, the incidence of the disease has increased alarmingly, thus gaining the attention of researchers. Pruritus is common, provoking malaise and impacting work performance during training and competitions [103, 3], moreover, an erythematous, annular plaque with an active peripheral margin is commonly observed (vesicles) [78].

Trichophytic granuloma. Also known as Majocchi granuloma, the infection is caused by *T. rubrum, T. tonsurans*, and *T. mentagrophytes var. mentagrophytes*, and it can be manifested in the dermis as well as in subcutaneous tissue. Some predisposition has been observed in immunocompetent or immunosuppressed patients, although the latter are more severely affected. Frequently affected body areas are the scalp, beard, and the upper and lower extremities, with a few instances in either the face or neck and rare occurrences in the vulva or ankles [104–106]. Females are commonly afflicted, which has been attributed mostly to leg shaving, during which the follicle is ruptured and subsequently infected by fungi carrying keratin and necrotic material from the dermis. Immunosuppressed persons are also commonly affected. This immunosuppression can be in the form of diabetes, AIDS, the use of topical corticosteroids, autoimmune disease, or physical trauma [105, 107, 106]. This disease is characterized by the presence of dark, inflammatory nodular lesions, which can be observed in the active margin of erythematous plaques. Papules are discreet or clustered, and pustules can also be observed. It is important to highlight the rarity of cheloid or verrucous lesions [108, 105, 107, 106].

Mycetoma (Pseudomycetoma). The term pseudomycetoma is applied to mycetomas caused by fungi such as *M. canis*, *M. gypseum*, *M. audouinii*, *M. ferrugineum*, *T. rubrum*, *T. mentagrophytes*, *T. tonsurans*, *T. schoenleinii*, and *T. mentagrophytes* var. *interdigitale* [109], which present the characteristic triad of eumycetic mycetoma: (1) tumefaction of the affected tissue; (2) the appearance of microcolonies (grains); and (3) the formation of fistules for the external draining of grains. Unlike mycetoma, which is initiated by external inoculation, this disease is initiated as tinea.

Dermatophytic disease. Described by Hadida [110], this is a rare, chronic infection (Hadida disease); the few known cases have mostly been reported in Algeria in association with the high percentage of consanguineous matrimonies, which would explain the recessive autosomal heredity predisposition to this disease [111]. Immunological studies have also reported a poor cellular response against the causal agents, which may be *T. rubrum*, *T. violaceum*, *T. mentagrophytes*, and *T. schoenleinii* [111, 112]. Affecting mostly males, with an onset at infancy, this condition is usually manifested as a superficial infection that later evolves and invades the dermis, lymph nodes, and potentially even the lungs, bones, liver, intestines, and central nervous system. Lesions are manifested as polymorphic, granulomatous, and flaky plaques. Granulomas can also be found in lymph nodes [113, 111, 112].

Tinea unguium. Also known as onychomycosis, this condition is manifested in fingernails or toenails. It has an estimated incidence of 2–13% worldwide, and it is strongly associated with the presence of *tinea pedis*. The major causal agents are fungi of the *genera Trichophyton*, *Microsporum*, and *Epidermophyton* [114], which produce keratinase and degrade the keratin of nails [115], although damage to the nail plaque by other filamentous, non-dermatophyte fungi has also been observed (non-dermatophyte mold), such as dematiaceous and hyaline fungi. These fungi do not express keratinase, however, and instead use residues from hyperkeratosis that are left behind by dermatophytes. Yeast are also known to frequently affect fingernails (70–80%). According to the International Society for Human and Animal Mycology, the infection is defined according to the etiological agent, i.e., *tinea unguium* when caused by dermatophytes, onyxis when caused by yeasts, nail candidiasis when the yeast in question is from the *genus Candida*, and nail mycosis when the causal agent is a filamentous, non-dermatophyte fungi [116].

Certain clinical criteria must be considered in evaluations of fungal infection in nails, such as (1) spots or stripes with a white–yellow or orange–brown tint; (2) onycholysis; (3) subungual hyperkeratosis; (4) thickening of the nail plaque; and (5) mycological criteria such as a positive dermatophyte microscopy test and culture in vitro.

The most recent classification of *tinea unguium*, or onychomycosis, was described by Hay and Baran [117] and Baran and Hay [118], who propose the classification according to clinical manifestations during the course of infection: (1) distal and lateral subungual onychomycosis, in which fungi invade the subungual region distal to the matrix and lateral to the nail, and some alterations in tint (opaque, white, yellow, orange, grey, brown, or black) can be observed. Other alterations are observed, such as onycholysis, paronychia, or complications of dermatophytoma, which does not allow the proper function of antifungal treatments. The causal agents are *T. rubrum*, *T. mentagrophytes* var. *interdigitale*, *T. mentagrophytes* var. *nodulare*, and opportunistic fungi such as *Scopulariopsis brevicaulis, Paecilomyces,* and *Neoscytalidium* spp.; (2) superficial onychomycosis, manifested in two modalities, white or black. The white modality is also called trichophytic leukonychia, which is characterized as white, rugged patches at onset that later extend over the surface of the nail. The main etiological agents are *T. rubrum*, *T. mentagrophytes* var. *interdigitale* and the keratinase-expressing opportunistic fungi *Aspergillus*, *Fusarium*, and *Acremonium*. The black variety, melanonychia, can be caused by dermatophytes such as *T. rubrum*, opportunistic fungi such as *N. dimidiatum* and *Scytalidium* (*Neoscytalidium*) *hyalinum*, and yeasts such as *Candida albicans* and *C. parasilopsis*; (3) proximal subungual white onychomycosis, in which fungi invade the subungual area of the nail, directly on the cuticle, and progressively advance toward the lunula and the remaining the nail. The last two varieties have been associated with immunosuppression; (4) endonyx onychomycosis, which presents distal and superficial damage, without inflammation or hyperkeratosis. The associated causal agents are *T. soudanense* and *T. violaceum;* (5) mixed pattern onychomycosis, which can be manifested in several combinations of any of the previously mentioned onychomycoses; (6) total dystrophic onychomycosis, which manifests as the final stage of any variety initiated by subungual infection. The nail bed is thickened and obliterated, and the nail surface becomes disintegrated. The etiological agents are usually *T. rubrum* and filamentous, non-dermatophytic fungi; (7) secondary forms of associated diseases, in which fungi tend to invade nails that have been previously afflicted by another type of primary pathology, such as psoriasis and dystrophia caused by trauma; (8) paronychia is a secondary onyxis caused by *Candida*. Females are commonly affected due to continuous contact with predisposing factors such as humidity and carbohydrates, although individuals with metabolic disorders, such as diabetes mellitus and other hormonal disorders, are also predisposed to this infection [117, 118].

13.6 Pathogenesis

The process of dermatophyte infection must first overcome a series of obstacles, including the innate immune response of the host, the physical structure and chemistry of the skin, constant UV exposition, temperature, lack of humidity, normal microbiota, sphingosomes produced by keratinocytes, and fatty acids produced by sebaceous glands. The infectious process requires the initial inoculation and adherence of fungal elements to the corneal epithelium [119]. Arthroconidia adhere through mannose and galactose ligands to the corneal epithelium of the host; however, the host organism has several defense mechanisms against the adherence of infectious agents, consisting of renewal of the corneal layer leading to epithelial shedding and possible fungal elimination. Therefore, epidermal invasion is only possible if tissue adherence is achieved before elimination by epithelial cell shedding. Moreover, during the initial interaction between arthroconidia and the corneal layer, within 3–4 h after initial contact, elongated fibrillar structures are produced that connect the arthroconidia to the keratinocytes, thus attaching to the corneal epithelium. After germination of the arthroconidia, hyphal growth is radial and multidirectional, stabilizing the permanence of the fungi within the host; this process takes 24 h [120]. During the invasion of deeper epidermal layers, recently formed arthroconidia covering the surface become finer and shorter, taking on a flat morphology that increases the surface area in contact with the invaded tissue and resulting in greater adhesion and the capacity for nutrient acquisition. In addition to adhesion to the corneal layer, these fibrillar structures enable the connection between adjacent arthroconidia, suggesting the formation of a stable complex similar to a biofilm or even an intercellular communication mechanism [121, 122]. Once the fungal elements have adhered to the host and germinated, the invasion of hyphae into the stratum corneum occurs rapidly, followed by the formation of additional hyphae that grow and spread in a circular pattern, particularly within the inferior layers of the stratum corneum. Once established, dermatophytes seek growth-enabling nutrients and respond to the environmental pH by activating the expression of proteins and enzymes that are favored by an acidic environment, i.e., adhesins, lipases, phosphatases, DNases, and keratinolytic proteases (serine-subtilisin and fungalisin), allowing the degradation and processing of keratin and other proteins, as well as lipids and DNA, in order to thrive [119, 123]. Moreover, once within the keratinized tissue, dermatophytes must catalyze the degradation of the surrounding keratin into oligopeptides or amino acids that can be assimilated by the fungi. Regardless, these enzymes cannot function until disulfide bonds reduction (sulfitolysis) within the compact net of keratin has been achieved. It is suggested that this process is managed by a sulfite efflux pump encoded by the gene *TruSsu1,* which dissociates the disulfide bonds of keratin and releases cysteine and S-sulfocysteine. Disulfide bonds are responsible for the hardness and difficult degradation of keratin; therefore, the sulfite produced by dermatophytes from environmentally available cysteine reduces and directly dissociates these bonds, facilitating digestion by exo- and endo-peptidases, thus providing an adequate nutrition source for these pathogens [124]. An additional source of nutrition is obtained from surrounding host macromolecules, which can be degraded into basic carbon, nitrogen, phosphorus, and sulfur components. Nonetheless, the high molecular weight of proteins, starch, cellulose, and lipids complicates their transport through the cytoplasmic membrane, making their degradation a necessary process. For this purpose, dermatophytes secrete proteases, lipases, elastases, collagenases, phosphatases, and esterases that hydrolyze the structural components of the epidermal tissue, in turn providing the invasive characteristics of these pathogens. These enzymatic mechanisms are well-characterized virulence factors of dermatophytes [120, 125, 126]. In addition, the process of invasion, production, and secretion of proteases contributes to the degradation of inner epithelial layers in immunocompromised hosts [120, 127]. Furthermore, it is thought that deeper penetration is inhibited by the limited availability of iron due to the ferritin activity present within the dermal subjacent layer of the host [128].

As previously mentioned, the secretion of proteases is an important virulence factor of dermatophytes, supporting the colonization of these fungi within keratinized structures of the host [124]. The resulting proteolytic damage in the invaded tissues varies markedly according to species [129, 130]. Therefore, the pattern of excreted proteases likely determines the survival of the fungi and evolution of the infection in the host tissue. Thus, the production level of proteases, as well as the level of adaptation of a dermatophyte to its host, contributes to the intensity of the inflammatory response. Therefore, it can be stated that increased adaptation results in a less intense inflammatory response. In contrast to deep infection, superficial infections are less likely to promote inflammation, providing a superior survival strategy for dermatophytes [131]. The clinical presentation of dermatophytosis is diverse and depends on both the host and fungal species. The more severe lesions produced by dermatophytes (acute infection) present faster and more efficient resolution through a specific innate immune response; however, more adapted species (e.g., *T. rubrum* and *T. interdigitale*) cause chronic infections with few or absent symptoms [120]. In the case of *tinea unguium*, nail localization is ideal for long-term survival because the dermatophyte is inaccessible to a host immune response. Likewise, the immune response of the host against dermatophyte infection depends on several factors, including defense against fungal metabolites, strain virulence, and the anatomical site of infection. These factors, as well as those related to the host (entry site, nonspecific defense, immune response, and synchronic microbial colonization), play an important role in the inflammatory process [130, 132]. Furthermore, several factors participate in both the host and fungi, resulting in a predisposition to the development of dermatophytosis and affecting the degree of the associated inflammation [129, 130].

Dermatophytes also have the capacity to manipulate the immune response of the host, i.e., the localization within the stratum corneum, preventing the development of a strong cellular immune response. The manifestation of an immune response depends on the degree of inflammation, which varies depending on the dermatophyte species, host, and physiopathological state [120]. Zoophilic species cause more inflammatory infections that may heal spontaneously, providing resistance against reinfection. Anthropophilic species, however, usually cause chronic infections that are less likely to provide resistance against future incidents [133]. Some species promote increased production of IL-8 and TNFα, explaining the occurrence of an acute inflammatory response. Mannan-oligosaccharides, a component of the cell wall of dermatophytes, seems to be involved in suppressing the inflammatory response [134]. Furthermore, it has the capacity to suppress the expression of Toll-like receptors on the surface of keratinocytes, compromising the Th1 cellular response. Thus, infection by these pathogens becomes chronic [130].

13.7 Molecular Taxonomy

As mentioned previously, to distinguish between different dermatophyte species, certain morphological criteria are used, including macromorphology, size and pigmentation of the colonies, shape of macro- and microconidia, growth rate, physiological properties (e.g., urease production), in vitro tests for hair perforation [135], growth in rice, alkaline production in bromocresol purple medium [136], sorbitol assimilation [137], vitamin or amino acid requirement, and certain biochemical tests [138]. Regardless, these markers are limited in number (there is a finite number of measurable characteristics) and must be assessed at predetermined stages of fungal growth. Moreover, these microorganisms tend to have a variable appearance and some go through degenerative and irreversible changes associated with pleomorphism. This phenomenon, which may occur shortly after isolation and following culture reseedings, results in colonies that turn white in color and lose their sporulating capacity. Furthermore, the culture media and other growth conditions influenced by the environment can affect the macro- and microscopic morphology of dermatophytes. Therefore, the utilized criteria lack definition and objectivity in some instances. The advent of molecular biology and the development of molecular markers has become a useful tool for the elimination of inconvenient means of identification based on the exclusive analysis of phenotype, enabling the rigorous and reproducible identification of species and varieties. The different types of molecular markers used to differentiate species have a remarkable capacity to detect polymorphisms in unique or multiple *loci*. In this regard, the available techniques include restriction fragment length polymorphism (RFLP) from mtDNA, restriction enzyme analysis (REA), and markers based on DNA amplification such as polymerase chain reaction (PCR), random amplified polymorphic DNA (RAPD), arbitrary primer-PCR (AP-PCR), simple sequence repeats (SSRs), amplified fragment length polymorphism (AFLP), single nucleotide polymorphisms (SNPs), inter single sequence repeat (ISSR-PCR), microsatellites, and multilocus sequence typing (MLST) [12].

The use of these markers has made possible numerous studies of dermatophyte taxonomy. The first of such studies, by Symoens et al. [139], employed isoelectric focusing to analyze four isolates of *A. benhamiae, A. vanbreuseghemii, T. mentagrophytes*, and T. *interdigitale* to differentiate the species of the *T. mentagrophytes* complex and those in close proximity to *T. interdigitale*. Their results revealed two groups, the first of which fit the profile of *A. vanbreuseghemii*, and grouped the isolates of *T. interdigitale* and most isolates of *T. mentagrophytes*. The second fit the profiles of *A. benhamiae* and *T. mentagrophytes*. Therefore, they suggested that *T. interdigitale* could be an anamorph species of the teleomorph *A. vanbreuseghemii* and that *T. mentagrophytes* could correspond to either *A. vanbreuseghemii* or *A. benhamiae* teleomorph species.

Mochizuki et al. [140], using mtDNA restriction analysis, reported the relationship between *T. interdigitale* (*T. mentagrophytes* var. *interdigitale*) and other members of the *T. mentagrophytes* complex. The restriction profiles were compared with those of *A. simii*, *A. benhamiae*, and *A. vanbreuseghemii*, and they revealed that *T. interdigitale* is identical to *A. vanbreuseghemii*. Therefore, the authors consider these two species to be closely related.

Aiming to define the divergence between *Nannizzia incurvata*, *N. gypsea*, *N. fulva*, and *N. otae*, Kawasaki et al. [141] compared these isolates through mtDNA restriction profiling, and they observed that only *N. fulva* showed two restriction profiles. The phylogenetic analysis suggested that *N. gypsea* is more closely related to *N. fulva* than *N. incurvata* and that there is a larger phylogenetic distance between *N. otae* and the remaining species.

Kawasaki et al. [142] researched the phylogenetic relationship between the *genera Arthroderma* and *Nannizzia* using mtDNA RFLP. The phylogenetic tree was constructed with 10 species: *A. benhamiae, A. insingulare, A. quadrifidum, A. simii, A. vanbreuseghemii, N. fulva, N. grubyia, N. gypsea, N. incurvata*, and *N. otae*. Their results did not reveal any differences between the *genera Arthroderma* and *Nannizzia*, confirming that the *genera Arthroderma* and *Nannizzia* are congeneric, as previously reported by Weitzman et al. [143].

Another study, using mtDNA analysis to determine the phylogeny of the *genus Trichophyton*, showed the division of *T. rubrum* isolates into two groups, Type I and II, suggesting that the *genus* is composed of a species complex. *A. benhamiae* was closely related to *T. mentagrophytes* var. *erinacei* and *T. rubrum* Type II, whereas *T. tonsurans* and *A. vanbreuseghemii* showed a similar restriction profile, which suggested that they are identical. *T. quinckeanum* and *T. schoenlenii* also showed an identical restriction profile, differing slightly from *A. vanbreuseghemii* [144]. These results are consistent with those reported by de Bievre et al. [145].

Harmsen et al. [146], studied the phylogenetic relationship of isolates representative of the *genera Trichophyton*, *Microsporum*, and *Epidermophyton* by analyzing the 18S rDNA sequences, and they demonstrated a monophyletic origin of dermatophytes, which were further classified as a subgroup within the *order* Onygenales.

Gräser et al. [147], with the goal of identifying 26 species of the *genera Trichophyton*, *Microsporum*, and *Epidermophyton,* utilized an RAPD approach with the primers $(AC)_{10}$, $(GTG)_5$, M13, and AP3. From the obtained polymorphic patterns, 17 species showed characteristic patterns, suggesting that dermatophyte species can be identified by the characteristic amplicon patterns of each species.

Another report, utilizing rDNA RFLP to differentiate isolates of *T. rubrum* and other dermatophytes, found that RFLP analysis together with nontranscribed spacer (NTS) and ITS intergenic regions could differentiate isolates of *T. rubrum.* They further identified fourteen types (A-N) among clinical isolates, the most predominant of which were A and B. The authors observed that the NTS region was highly polymorphic in *T. rubrum* compared with the ITS region; regardless, some closely related taxa such as *T. rubrum-T. soudanense* and *T. quinkeanum-T. schoenlenii* could not be distinguished from each other [148].

Makimura et al. [149], constructed the phylogeny of a group of dermatophytes, including the *genera Trichophyton, Microsporum*, and *Epidermophyton*, identifying the species through sequence comparison of the rDNA region ITS1 with the database. The study demonstrated that species of pathogenic dermatophytes for either humans or animals, *T. tonsurans*, *T. schoenleinii*, and *T. verrucosum*, are members of group *A. vanbreuseghemii-A. simii* or group *A. benhamiae*.

Furthermore, Gräser et al. [150] delimited the molecular taxonomy of *T. mentagrophytes* and *T. tonsurans* through morphological and physiological characterizations in addition to molecular techniques (i.e., rDNA ITS regions, AFLP, and PCR fingerprinting), using the primers T3B, M13, and $(AC)_{10}$. The 24 species and/or varieties analyzed in that study were reclassified into five species: *T. tonsurans*, *T. interdigitale*, *T. mentagrophytes*, *T. simii*, and *T. erinacei.*

Using the rDNA ITS region (ITS1, ITS2, and 5.8S), Gräser et al. [151] evaluated the phylogenetic relationships within the *family* Arthrodermataceae. The authors concluded that *Trichophyton* and *Microsporum* are conserved as separate *genera*, whereas within the *Trichophyton*, there is a separation of human pathogenic and primary geophilic species. Moreover, the two species of *Epidermophyton* were not related, suggesting that *E. stockdaleae* should be reclassified. Likewise, the authors proposed three species for *T. mentagrophytes* and a reduction of the number of varieties of *T. mentagrophytes*; however, they also suggested that the *genus Keratinomyces* should be restored. They further demonstrated that the taxa *T. rubrum*, *T. megninii*, *M. gallinae*, *A. grubyi*, *M. audouinii* var. *rivalieri*, *M. audouinii* var. *langeronii*, *M. audouinii*, *M. canis*, and *M. equinum* are so closely related and likely conspecific.

Summerbell et al. [152] analyzed the shared characteristics of dermatophytes with *T. rubrum* using representative isolates of recently described species such as *T. krajdenii*, a *T. mentagrophytes* isolate that is related to the yellowish variants of *T. rubrum.* These isolates were analyzed by sequencing the ITS1 and ITS2 regions. The results showed that the isolates of *T. raubitschekii*, *T. fischeri*, and *T. kanei* share identical sequences in the ITS region with *T. rubrum, T. soudanense*, and *T. megninii*, with only a small number of differences between *T. rubrum* compared with *T. soudanense* and *T. megninii. T. krajdenii* carries a sequence that is similar to that in isolates of *T. mentagrophytes* and others related to the *A. vanbreuseghemii* teleomorph.

Weitzman et al. [143] classified dermatophytes in the *genera Epidermophyton, Keratinomyces, Microsporum*, and *Trichophyton*. The anamorphic, nonpathogenic fungi were assigned to the *genus Chrysosporium*. Teleomorphic fungi, however, were classified within the *family* Arthrodermataceae. Teleomorphic fungi of the *genus Microsporum* were classified in the *Nannizzia*, and those from the *genus Chrysosporium, Keratinomyces*, and *Trichophyton* were classified in the *genus Arthroderma*.

Gräser et al. [153] evaluated the taxonomy of *T. rubrum* and *M. canis* using phenotypic characteristics (macro- and micromorphology, urease production, hair perforation) and genotyping (PCR fingerprinting, rDNA ITS region, and AFLP) techniques. The species that were closely related to *T. rubrum*, including *T. circonvolutum, T. fischeri, T. fluviomuniense, T. glabrum, T. gourvilii, T. kanei, T. kuryangei, T. megninii, T. pedis, T. raubitschekii, T. rodhaini, T. rubrum* var. *nigricans, T. soudanense, T. violaceum* var. *indicum*, and *T. yaoundei*, were reclassified as *M. canis* (teleomorph: *A. otae*), *M. ferrugineum*, and *M. audouinii*.

Hirai et al. [154] studied dermatophytes and fungi related to the *genera Chrysosporium, Epidermophyton, Microsporum*, and *Trichophyton* by analyzing the sequence of the chitin synthase 1 (*CHS1*) gene. The phylogenetic tree revealed that the *genus Microsporum* was genetically different from *Chrysosporium* and *Trichophyton*. Furthermore, the *Trichophyton* species were divided in two groups, one consisting of pathogenic geophilic fungi including *A. curreyi*, and the other consisting solely of geophilic fungi.

Years later, Nenoff et al. [155], based on previous studies, modified the taxonomic classification of the *T. mentagrophytes* complex to include the relationship with teleomorph species.

Species classified as *T. mentagrophytes* var. *interdigitale* (anthropophilic), *T. mentagrophytes* var. *nodulare* (anthropophilic) (also *T. krajdenii*), *T. mentagrophytes* var. *goetzii* (anthropophilic), and *T. mentagrophytes* var. *granulosum* (zoophilic) were reclassified as *T. interdigitale* (anthropophilic and zoophilic) (anamorph) and its teleomorph *A. vanbreuseghemii* [156]. Species classified as *T. mentagrophytes* var. *quinckeanum* (zoophilic) (also *T. sarkisovii* and *T. langeronii*) were reclassified as the anamorphic species *T. mentagrophytes* and the teleomorph *A. simii*. Species classified as *T. mentagrophytes* var. *granulosum* (zoophilic) were reclassified as *Trichophyton* sp. (zoophilic) and *T. mentagrophytes* var. *erinacei* (zoophilic) were reclassified as *T. erinacei* (zoophilic) and the teleomorph *A. benhamiae*.

To define the biological species in the *T. mentagrophytes* complex, several authors performed a sexual stimulation between fungi, seeking to reveal the sexual phase of the dermatophytes. To achieve this goal, Anzawa et al. [157] performed crosses between *A. simii* with asexual species of *Trichophyton*, including *T. rubrum, T. concentricum*, and *T. interdigitale*. The cross between *T. rubrum* with *A. simii* produced several ascomata, suggesting that *T. rubrum* is neither an asexual nor a clonal species Symoens et al. [158] also tried to improve the definition of teleomorph species in the *T. mentagrophytes* complex by performing a crossing analysis with isolates from humans, animals, and reference strains of *A. benhamiae* and *A. vanbreuseghemii*, comparing the mating competence with the sequencing of the ITS and 28S rDNA regions; this was the first study to report the taxonomy of dermatophytes associated with a mating analysis. Based on the ITS sequence and mating capacity with *A. vanbreuseghemii* strains, the authors identified all of the *tinea pedis* and *tinea unguium* isolates as *T. interdigitale*. Therefore, *T. interdigitale* should be considered a humanized species derived from the cross with *A. vanbreuseghemii*.

Kawasaki [159] verified the taxonomy of dermatophytes based on a phylogenetic analysis and mating capacity through genealogical concordance phylogenetic species recognition (GCPSR). They utilized several DNA regions: 5.8S rDNA, ITS regions 1 and 2, a partial sequence of the actin gene including two introns (ACT), a partial sequence of the DNA topoisomerase gene including one intron (TOP), and a partial sequence of the GAPDH gene including one intron (GPD). The authors found that *A. simii, A. vanbreuseghemii, T. mentagrophytes* var. *interdigitale*, and *T. tonsurans*, as well as *A. benhamiae, T. concentricum, T verrucosum*, and *T mentagrophytes* var. *erinacei*, can be considered phylogenetic species.

Tartabinia et al. [160] utilized a set of phenotypic methods to identify fungal isolates within the *genus Trichophyton*, as well as PCR fingerprinting using the primer $(GACA)_4$. The phenotypic tests allowed the differentiation of *T. rubrum* from the isolates of the *T. mentagrophytes* complex but not from the species within the complex; regardless, the molecular technique allowed the characterization of both isolates from *T. rubrum* as well as from the *T. mentagrophytes* complex, which were identified as *T. interdigitale* and *Trichophyton* sp., anamorphs of *A. benhamiae*.

Ahmadi et al. [161] performed phylogenetic and taxonomic analyses based on inter- and intraspecific variations using a partial region of the calmodulin gene. Some intraspecific differences were found in the isolates from *T. interdigitale, A. simii, T. rubrum*, and *A. vanbreuseghemii*, whereas *T. tonsurans, T. violaceum, E. floccosum, M. canis, M. audouinii, M. cookei, M. racemosum, M. gypseum, T. mentagrophytes, T schoenleinii*, and *A. benhamiae* were conserved. The isolates from *E. floccosum/M. racemosum/M. cookei, A. obtosum/A. gertleri, T. tonsurans/T. equinum*, and *T. interdigitale* presented identical sequences for the calmodulin gene. The topology of the phylogenetic tree for this gene concurred with the results obtained for the ITS, β-tubulin (BT2), and Tef-1α molecular markers.

13.8 Molecular Epidemiology

The molecular epidemiology study of infectious diseases evaluates the dynamics of such diseases within a population. These dynamics are determined by the transmission of a parasite to its host and by how this transmission affects the dispersion of the microorganism within the host population. Moreover, molecular epidemiology also identifies the existing clonal relationship between several isolates of the same species, particularly during outbreaks caused by multiresistant strains. Applying molecular techniques permits the calculation of the number of circulating clones and the identification of the infection source or reservoir as well as the vehicle for disease transmission, which are important for the control of infectious diseases [162].

The following table (Table 13.2) shows, in chronological order, relevant reports examining the molecular epidemiology of dermatophytosis based on different molecular techniques to provide a better understanding of this disease.

13.9 Diagnosis of Dermatophytosis

The diagnosis of dermatophytosis is determined based on clinical and epidemiological data; however, diagnosis must be confirmed using a laboratory test because there are non-mycotic dermatophytosis that cannot be clinically differentiated from a mycotic dermatophytosis, resulting in poor handling of the affected patient. Prior knowledge regarding the etiology of dermatophytosis is essential for the election of an effective antifungal treatment, as well as in facilitating the determination of the origin of the causal agent (zoophilic or anthropophilic) and preventive measures to treat the infection [180]. To achieve these goals, a broad number of conventional and molecular laboratory techniques are currently available.

Conventional techniques. Before suspecting a dermatophyte skin infection, either in glabrous or hairy skin, a preliminary test can be conducted in the affected skin area through a Wood lamp examination (UV light) in a dark room. Dermatophytes typically show a greenish fluorescence, even when the fungi are no longer viable, although some dermatophytes, such as *T. tonsurans* and *T. violaceum*, lack fluorescent properties [10]. However, a KOH test for the microscopic analyses of the clinical specimen (nail, skin, or hair) and cultures in vitro are the gold standard for the diagnosis of dermatophytes because these methods allow the detection of fungal elements and, in some instances, their morphology based identification. The KOH test (10–20%) is a fast and economic tool for the diagnosis of fungal infection; however, it does not allow the differentiation between a dermatophyte and a non-dermatophyte fungal infection, especially when the analyst lacks proper experience [181, 182].

The KOH test can present false positive results, which are observed in 15% of cases, partly due to the presence of artifacts constituted by lipid vesicles, air bubbles, and textile fibers, which may be mistaken for dermatophytes [181–183]. Instead, chlorazol black and DMSO have greater sensitivity because they stain fungal structures in a black-blueish color, facilitating their identification. The use of specific dyes targeting the cell wall of fungi has long been implemented, including Periodic acid-Schiff (PAS), Parker blue, and Congo red, as well as fluorescent dyes such as Calcofluor white, rhodamine, and diaminostilbene acid [180, 182, 184]. Despite facilitating the analysis, the detection of fungal elements by fluorescence microscopy is limited in its routine application due to the high cost of reagents and equipment [185].

Table 13.2 Application of molecular epidemiology in dermatophytosis

Objective	Molecular method	Conclusion	References
Identify *T. tonsurans* and determine if there are genetic differences in the phenotypic function or the geographical region of isolation	RAPD with three random primers	*T. tonsurans* is a genetically homogenous species, independent of geographical localization	Kim et al. [163]
Detect genetic variability in *T. rubrum* isolates from Europe, North America, Africa, and Asia	RAMD, SSCP, and RFLP	No genotypic variability was revealed, indicating clonal reproduction	Gräser et al. [151]
Analyze the genetic variability between isolates of *T. mentagrophytes* and *T. rubrum*	REA and RAPD	Little variability was observed between *T. mentagrophytes*, and none was detected in *T. rubrum*	Howell et al. [164]
Analyze the genetic variability between isolates of *T. rubrum* and *T. mentagrophytes*	NTS region amplification, 18S probe, and RFLP	Few differences were found among *T. rubrum*, *T. mentagrophytes*, and *T. tonsurans* by observing intraspecific isolates of *T. rubrum* and *T. mentagrophytes*	Gupta et al. [165]
Determine if nail infection caused by *T. rubrum* is due to one or more fungal isolates	NTS region amplification	Two or more types of *T. rubrum* were identified	Yazdanparast et al. [166]
Investigate the genetic variants of *T. tonsurans*	NTS and ITS region amplification	12 strains were identified, evidencing the genetic heterogeneity of *T. tonsurans*	Gaedigk et al. [167]
Study of a scalp tinea outbreak in a school population due to *M. canis* and environmental isolates	RAPD, NTS, and ITS region amplification	Clinical and environmental isolates showed identical patterns	Yu et al. [168]
Investigate the phenotypic and genotypic correlation of *M. canis* isolates from cat, dog, and human	ITS1 and ITS2 region amplification, REA, and RAPD	Microscopic and culture characteristics revealed the variability of *M. canis*. Genotypic characteristics showed that isolates of *M. canis* from cat, dog, and human were genetically identical and can therefore be considered clonal	Brilhante et al. [169]
Typify the isolates of *M. canis* obtained from humans and animals (dogs and cats)	ISSR-PCR	The isolates of *M. canis* obtained from animals were closely related to those isolated from humans	Cano et al. [170]

(continued)

Table 13.2 (continued)

Objective	Molecular method	Conclusion	References
Analyze the genetic variability of *T. tonsurans* isolates from Japan	VIR region in IGS of rDNA	All isolates had the same genotype, suggesting that a specific genotype predominated in the studied area	Sugita et al. [171]
Study the genetic variability of *T. rubrum* isolates obtained from patients with onychomycosis	RAPD, NTS region amplification	The isolates showed great diversity in the genome of *T. rubrum* and inter-strain variations	Baeza et al. [172]
Confirm that the distribution of genotypes in *T. rubrum* is associated with the geographic origin of the human host	MLMT	A subdivision of the species into two populations was revealed, showing clonal reproduction and differences in human host predilection (scalp vs. feet), as well as a geographical differentiation	Gräser et al. [173]
Study the genetic variability of isolates of *M. persicolor* from India	Microsatellites	The analysis revealed a high intraspecific variability in isolates from India, indicating that this species was not recently introduced into the country	Sharma et al. [174]
Characterize the isolates of *M. canis*, *M. fulvum*, *M. gypseum*, *T. mentagrophytes*, and *T. terrestre* from animals (dogs, cats, rabbits) and humans	PCR and sequentiation of CHS-1 and ITS	Two genotypes of *M. fulvum* were identified and grouped more closely with *M. gypseum* than with *M. canis;* in addition, two genotypes of *T. mentagrophytes* were detected; likewise, *M. fulvum* and *M. gypseum* were grouped with *T. mentagrophytes*, and *M. canis* and *T. terrestre* were in separate groups	Cafarchia et al. [175]
Differentiate isolates pertaining to *T. mentagrophytes sensu lato*	ITS region amplification and PCR-RFLP	The sequences of *T. interdigitale/A. vanbreuseghemii* isolates revealed a unique polymorphic pattern. This differentiation was essential for the selection of an antimycotic therapy	Heidemann et al. [176]
Identify dermatophytes isolated from workers in rabbit farms with a clinical history of skin lesions	Amplification of CHS-1, ITS1, 5.8S, and ITS2 regions	The close phylogenetic relationship between strains *T. interdigitale* (zoophilic) and isolates from dogs, pigs, and mice could indicate that these animals represented an infection reservoir for dermatophytes in rabbit farms	Cafarchia et al. [1]

Evaluate the inter- and intraspecific variation in dermatophytes	BT2 PCR	*T. tonsurans, T. equinum, T. concentricum, T. verrucosum, T. rubrum, T. violaceum, T. eriotrephon, E. floccosum, M. canis, M. ferrugineum,* and *M. audouinii* were not variable. The more closely related species were *A. benhamiae* and *T. concentricum* and *T. tonsurans* and *T. equinum*; more distant species were *M. persicolor* and *M. amazonicum*	Rezaei-Matehkolaei et al. [177]
Evaluate the inter- and intraspecific variation in dermatophytes	Tef-1α PCR	*T. rubrum, T. tonsurans, T. schoenleinii, T. concentricum, T. violaceum, E. floccosum, M. ferrugineum, M. canis, M. audouinii, T. equinum, T. eriotrephon,* and *T. erinacei* were not variable	Mirhendi et al. [178]
Genotype isolates of *T. mentagrophytes sensu lato* obtained from humans and diverse animal species with clinical symptoms of dermatophytosis	ITS-PCR-RFLP sequentiation of ITS	Identified *T. interdigitale*, *A. benhamiae*, and *A. vanbreuseghemii*	Ziółkowska et al. [179]

MLMT multilocus microsatellite typing, *RAPD* randomly amplified polymorphic DNA, *NTS* nontranscribed spacer, *BT2* β-tubulin, *RFLP* restriction fragment length polymorphisms, *REA* restriction enzyme analysis, *ISSR-PCR* inter single sequence repeat, *ITS* internal transcribed spacer, *rDNA* ribosomal DNA, *RAMD* random amplified monomorphilic DNA, *SSCP* single-strand conformation polymorphism, *IGS* intergenic spacer, *Tef-1α* translation elongation factor 1-α, *CHS-1* gen chitin synthase-1

In vitro culture has some advantages over microscopy analyses because it can provide higher sensitivity thresholds due to the greater quantity of available material for analysis, and in some cases, the fungi can be identified based on their morphology alone [186]. Sabouraud dextrose agar (SDA) is the most commonly used medium for dermatophyte isolation, and it is responsible for enabling the description of several morphological characteristics of these fungi. The development of these colonies in vitro is slow, lasting between 7 days and 6 weeks [186], although Rezusta et al. [187] suggested that an incubation period of 17 days was sufficient for proper interpretation of results.

The addition of antibiotics (gentamicin, chloramphenicol, and cycloheximide) to SDA culture medium confers selectivity for the development of dermatophytes by preventing the growth of nonpathogenic fungi or bacteria. Once these colonies have developed, dermatophytes can be identified based on their macro- and micromorphology [182], although sometimes it is necessary to perform subcultures in specific media (lactrimel, potato dextrose, malt extract, Candi-Select(TM) 4, SDA, 1% Tween-80), hair perforation tests in vitro, or to evaluate the capacity to degrade or assimilate compounds such as keratin, urea, and special amino acids or vitamins, especially when trying to identify *T. rubrum* or members of the *T. mentagrophytes* complex [180–182, 188, 189].

Another commercially available medium, the selective culture medium BBL DTM (Becton, Dickinson and Company, USA), can be used for the detection and presumptive identification of dermatophytes from clinical samples, providing the advantage of direct inoculation in the clinician's office. Despite the multiple options for in vitro culture, the obtained results are often false negative (40% of analyzed samples) compared with microscopic analyses, which may be due to the inviability of fungal elements obtained from medical samples [190]. Therefore, several authors concur that the success of a culture in vitro lies in proper sampling and transport to avoid potential contamination with other microorganisms that limit detection and identification, especially when nails are concerned [181, 191].

Therefore, serial repeats with intervals of 2–5 days between samplings of in vitro microscopy and culture tests are highly recommended to improve the results [192].

Furthermore, when the results are negative despite strong clinical suspicion, a histological test provides the best alternative, possessing a higher sensitivity than in vitro culture or a KOH test [193, 194]. These histological analyses commonly employ staining methods such as silver methenamine, PAS, hematoxylin and eosin, and Gomori to highlight hyphal structures against the stratum corneum [182, 195]. Such analyses assist clinicians in deciding whether to continue or terminate treatment [196]. Regardless, histological analyses are unable to determine the viability of the analyzed fungi, nor are they able to identify the pathogenic species [180, 186].

It is important to mention that regardless of the modifications implemented to improve conventional methods, a series of factors still impede the identification of dermatophytes using these tests, including intraspecific morphological variation, atypical or similar interspecific morphology, and pleomorphism. Moreover, taxonomic changes in the classification of dermatophytes have motivated the development of new methods for the identification of these species, including molecular techniques [197].

13.10 Molecular Techniques

Among the molecular techniques most commonly used to identify dermatophyte species are different formats of PCR: endpoint, multiplex, nested, RT-PCR, qPCR, PCR-ELISA, and PCR-electrospray ionization mass spectrometry (PCR-ESI/MS).

These PCR-based methods have enabled a more precise identification of dermatophytes at both inter- and intraspecific levels, depending on the amplification target. The main targets for amplification are the genes: topoisomerase II, chitin synthase, actin, regions D1/D2 and ITS, and ribosomal subunits 28S and 18S [198–200]. PCR is sensitive enough to detect 1 fg DNA or 1 colony forming unit (CFU), particularly with modalities of nested PCR, qPCR [198, 201, 202], or PCR-ELISA, [203]. Likewise, RAPD [204], amplification of ribosomal subunits followed by restriction enzyme digestion [205, 206], and the amplification of microsatellite regions (GACA)4, (AC)10, and (GTG)5, have also been utilized for the identification of dermatophytes by comparing the migration patterns of amplicons with reference strains [147, 207, 208]. Nonetheless, these techniques are recommended for epidemiological studies rather than diagnosis.

PCR tests enable a swift identification of the etiological agent in question; therefore, these assays contribute to the selection of a proper therapeutic course [209, 210], which is important because of the growing resistance to antimycotics. Most PCR tests have been developed using clinical samples (either freshly collected or mounted in paraffin blocks), and their efficiency in comparison to conventional methods reveals that the results obtained with PCR are, in some instances, consistent with those obtained by conventional methods [211] and, in some other cases, show a higher sensitivity and specificity [198–200, 211–213]. Regardless, PCR is not exempt from false positive/negative results, even when employing commercial kits such as Dermatophyte PCR (Statens Serum Institut; Copenhagen, Denmark) [214, 215]. qPCR, in contrast, greatly reduces the incidence of false results, although it depends on the adequate collection and handling of the samples [216–218], thus recommending that negative samples are cultured in vitro and re-tested if a positive microscopy result is obtained [216]. Another important advantage of qPCR for the diagnosis of dermatophytosis is the capacity to evaluate the viability of these fungi by quantifying actin mRNA or by amplifying ITS and D1/D2 regions. Moreover, several samples from skin scales or nails have been analyzed before and after treatment with terbinafine; as therapy progressed, the results revealed a reduction in the number of actin mRNA copies per mg of analyzed sample. These results were consistent with the in vitro culture and KOH test [212, 219]. Therefore, qPCR is useful for determining whether to continue treatment [220]. Furthermore, Worek et al. [221] developed a genome in situ hybridization method (GISH) using specific probes for the successful identification of *T. interdigitale*, *T. rubrum*, and *M. canis*. The authors propose the use of this method when proper identification of a dermatophyte remains elusive due to ambiguous PCR-RFLP patterns.

Another technique that has been developed for the diagnosis of dermatophytes is soft matrix-assisted laser desorption ionization time-of-flight (MALDI-TOF) spectrometry. This technique enables the identification of dermatophytes by comparing peptide patterns from proteolytic degradation with a database containing the peptide spectral library of different dermatophytes. L'Ollivier et al. [222] reported a database including *T. mentagrophytes*, *T. interdigitale*, *T. erinacei*, *T. simii*, *A. otae*, *M. canis*, *M. audouinii*, and *M. langeronii*, as well as *T. terrestre* and *T. ajelloi*, two nonpathogenic dermatophytes. This is a time-saving procedure that allows the simultaneous and precise identification of up to 64 dermatophyte strains in 24 h from clinical samples or in 3–6 days when analyzing cultures in vitro [182]. In addition to the identification of species, MALDI-TOF MS technology offers additional possibilities, such as the detection and prediction of resistance phenotypes in fungi and clinically relevant subtypes [223]. These tests are based on pure in vitro cultures of dermatophytes as well as on skin and nail samples, achieving the correct identification of dermatophytes of the *genera Trichophyton*, *Microsporum*, *Epidermophyton*, and *Arthroderma* [224, 225].

Nenoff et al. [225] compared the findings for 285 clinical isolates obtained by mass spectrometry against conventional methods and ITS sequencing and discovered consistent results in 68.9% and 98.8%, respectively. Therefore, MALDI-TOF MS has been recommended as a routine tool for the precise and reliable identification of dermatophytes.

Regardless, MALDI-TOF MS has limitations. For example, the peptide spectra can be variable, although in some cases a sufficient number of signals are conserved and are still useful for the unequivocal identification of dermatophyte species. In other cases, however, the spectra can be uninformative, resulting in an impossible or erroneous identification [182, 222].

As previously observed, both PCR and MALDI-TOF MS also have some limitations, and thus, they do not provide a complete substitute for conventional methods; however, these techniques can still be used as complementary methods to obtain a precise diagnosis [226].

New techniques are currently utilized, such as confocal reflection microscopy, which provides in vivo imaging of the epidermis and dermis at cellular resolution, revealing the presence of dermatophyte fungi and other parasites. The advantage of this test relies on its noninvasiveness and high sensitivity, which has been rated at 100%. Moreover, it is comparable to other conventional diagnostic methods with good results [227]. Another technique, immunochromatography, is currently being tested for the detection of dermatophytes in skin and nail samples with good results [228, 229]. In addition, Raman spectroscopy, a photonic test that allows the identification of organic and inorganic materials based on the scattering of monochromatic light, has also been employed for the detection and identification of *T. rubrum*, *T. tonsurans,* and *T. mentagrophytes* in nail samples; it is also capable of identifying non-dermatophyte species such as *S. brevicaulis* and *C. albicans* [230]. The results obtained using confocal reflection microscopy, immunochromatography, and Raman spectroscopy support their potential use as diagnostic tools. Regardless, further research is needed to validate these results.

13.11 Detection Methods

Wood lamp skin examination (UV light). Utilized for the direct diagnosis of *tinea* capitis caused by a species of *Microsporum* of zoophilic origin, this test is performed in a dark room, in which UV light is projected onto the infected hair and scalp, resulting in blue-green fluorescence emitted by these organisms.

KOH test (10–40%)/Chlorazol Black E (Delasco, Council Bluffs, LA). Hair or hair scales are transferred onto a slide, and KOH is added with or without DMSO. The slides are then sealed with a coverslip and analyzed under a microscope to visualize the hyphae. Regarding the hair itself, two types of microscopic invasions can be observed: the hair shaft and medulla are uniformly invaded with spores (ectothrix), or the spores are contained within the hair structure (endothrix).

Sample collection. This process is dependent upon the anatomical area of the infection. In the case of *tinea* in glabrous skin, a No. 15 scalpel blade is used to scrape the active edge of the lesion. When the scalp is sampled, hairs must be torn out, and the scalp must also be scraped using a surgical blade or a curette. When sampling a potentially infected nail, either the nail plaque or the nail bed, depending on the localization of the infection, is carefully scraped to avoid damage or bleeding in the patient. The obtained samples are placed in a slide or petri dish and sent to the laboratory for further analysis.

Calcofluor white. This is a direct test substituting KOH with a fluorescent stain (Calcofluor white M2R 1 g/L and Evans blue 0.5 g/L). The sample is placed on a slide and observed under a fluorescence microscope (380–440 nm wavelength, 100x), revealing a bright green or blue signal.

Fungal culture. The following culture media are used for the isolation of dermatophytes: Sabouraud Chloramphenicol and Actidione agar and SDA. Although fungal colonies grow slowly, taking up to 10–15 days, these media are useful for confirming a diagnosis of *tinea*. The samples used in this process are obtained from skin, scales, nails, and hair, which are transported to the mycology laboratory in sterile containers for later inoculation in SDA with or without antibiotics. For a culture to be considered negative, it must be incubated for 21 days.

13.12 Molecular Identification

ITS PCR. The primers used were as follows: ITS1-F (5′-TCCGTAGG TGAACCTGCGG-3) and ITS4-R (5′-TCCTCCGCTTATTGATATGC-3′), as reported by White et al. [231]. PCR was performed in a final volume of 25 μL as follows: 10 ng gDNA, 2.0 mM $MgCl_2$, 200 μM dNTPs, 1.0 U *Taq* polymerase, and 0.5 μM of each primer. Amplification was conducted under the following conditions: 96 °C/6 min; 35 cycles at 94 °C/1 min, 56 °C/1 min, 72 °C/45 s; and a final extension at 72 °C/10 min. The amplicons were analyzed by electrophoresis in a 1.5% agarose gel stained with gel red (30X, Biotium) (100 V/60 min, TBE 0.5X), and a 100-bp molecular weight ladder was used to determine band sizes. The results were visualized under UV light (254/365 nm) and documented electronically (e.g., GelDoc™XR). Amplification and sequencing of other genes are recommended to aid in the molecular identification of species that form complexes.

BT2 gene PCR. The primers used for PCR amplification were as follows: T1-F (5′-AACATGCGTGAGATTGTAAGT-3′) and Bt2b-R (5′-ACCCTCAGTGTAG TGACCC-TTGGC-3′), as reported by Rezaei-Matehkolaei et al. [177]. The PCR was performed in a final volume of 25 μL as follows: 10 ng gDNA, 2.0 mM $MgCl_2$, 200 μM dNTPs, 1.0 U *Taq* polymerase and 0.5 μM of each primer. Amplification was conducted under the following conditions: 96 °C/6 min; 35 cycles at 94 °C/1 min, 56 °C/1 min, 72 °C/45 s, and a final extension at 72 °C/10 min. The amplicons are analyzed by electrophoresis in a 1.5% agarose gel stained with gel red (30X, Biotium) (100 V/60 min, TBE 0.5X). A 100-bp molecular weight ladder was used. The results will be visualized under UV light (254/365 nm) and documented electronically (e.g., GelDoc™XR). Amplification and sequencing of other genes are recommended to aid in the molecular identification of species that form complexes.

Amplicon sequencing. Amplicons were sequenced, and the electropherogram is edited using analysis software such as BioEdit (http://www.mbio.ncsu.edu/BioEdit/bioedit.html) or Geneious (http://www.geneious.com) [232, 233]. The edition criteria are based on consensus sequences.

Sequence analysis. The obtained sequences from the analyzed gene(s) were compared against the GenBank (URL4) database using the Basic Local Alignment Search Tool (BLASTn) algorithm [234]. The aligned sequences are analyzed for similar percentage, expectation and identity values.

Phylogenetic analysis. After editing and analysis, the sequences are aligned using ClustalΩ, MAFFT, KALIGN, MUSCLE, and MultiAlin (URL5-9), seeking the best alignment through pair sum and minimum entropy criteria, followed by election of the best model of evolution using the software Jmodeltest [235]and considering the parameters of likelihood, Bayesian information criterion, Akaike information criterion, and decision theory.

The phylogenetic analysis can be performed using several methods, including maximum verisimilitude, Bayesian inference, and maximum parsimony with software programs such as MEGA 7.0, Mr. Bayes 3.2.2, and TNT [236–238].

13.13 Treatment

Scalp ringworm is effectively treated with griseofulvin in daily doses of 10–20 mg/kg for 8–12 weeks [239]. Selenium disulfide (2.5%) or azole shampoos can be added to eliminate viable spores from the scalp surface. Concerning kerion, some clinicians recommend prednisone at daily doses of 2 mg/kg during the first 2 weeks along with antimycotics to decrease scabs and alopecia. Terbinafine can be used with equal efficacy, at a dose in infants of 3–6 mg/kg/day and in adults of 250 mg/day. Another alternative treatment of *Mycrosporum* tinea is itraconazole at 100 mg/day for 4 weeks or 3.3–6.6 mg/kg/day in infants. It is possible to use fluconazole at weekly doses of 6–8 mg/kg/day for 8 weeks [240].

In other anatomical locations, such as the body, groin, hands, and feet, only systemic antimycotics are recommended for the treatment of disseminated infection that is resistant to local treatment, recurrent infection or infection with severe inflammation. In adults, the oral doses administered are griseofulvin 500 mg/day and ketoconazole 200 mg; the latter is recommended only for short term treatments given the risk of hepatitis; itraconazole 100 mg/day, terbinafine 250 mg/day for 2–4 weeks, and fluconazole 150 mg in a single weekly dose.

In frequent, noncomplex manifestations, the use of topical drugs is usually sufficient. Iodine creams (0.5 and 1%), Whitfield ointment (Vaseline with 3% salicylic acid and 6% benzoic acid), tolnaftate, tolciclate, tolindate, pyrrolnitrin and undecylenic acid, topical imidazoles, naftifine or terbinafine, cyclopyroxolamine, and butenafine, are used once or twice daily.

Onychomycosis are generally resistant to topical treatment, and drug penetrance is frequently increased by occlusion and with transport coadjuvants. It is usually convenient to eliminate the infected keratin by partial surgical removal or via chemical means such as 40% urea [241]. The suggested systemic drugs are itraconazole 200 mg/day for 3 months or 400 mg/day for 1 week a month in intermittent therapy (4 months); terbinafine 250 mg/day for 3–4 months or 500 mg/day for 1 week a month in intermittent therapy (3–4 months); or fluconazole 150 mg/week for 8–24 months or at a 300 mg dose for a shorter time period. Other local treatments are 28% tioconazole, 5% amorolfine, 8% ciclopirox (alone or combined with hidroxypropilchitosan), or 1% bifonazole combined with 40% urea [242]. Additional treatments currently in trials include CO_2 laser, Q-switch, and photodynamic therapy. A combined systemic and topical treatment is more efficient because it allows the use of itraconazole, terbinafine, or fluconazole for shorter periods of time, thus avoiding secondary effects.

13.14 Conclusions

Dermatophytoses are the most common fungal infections in the world, with both public and veterinary health relevance; however, many features of their taxonomy, epidemiology, pathogeny, and diagnosis remain unknown.

In the taxonomy of dermatophytes, no consensus has been achieved regarding the number of species that make up each complex, supporting the suggestions of Guarro et al. [243] who suggest that some of the different concepts related to species being utilized are difficult to grasp. Therefore, mycologists should clearly define the criteria they are using to identify species.

Epidemiology is a subject in which new data are often made available, sufficiently altering established views regarding the epidemiological status of these fungi. Seeking to correct this situation, advances in molecular biology have provided invaluable knowledge on the identification, taxonomy, and epidemiology of dermatophyte species and have greatly contributed to unraveling the role played by dermatophytes in their associated diseases. The taxonomy of these fungi has been reviewed by phylogenetic analyses of the ITS region in particular. Although, to date, additional markers with good resolution have been used, no specific markers are yet suitable for the classification of dermatophytes. Furthermore, it must be highlighted that a single marker is insufficient to determine the relationship between isolates from different geographical areas or to place the dermatophytes within the appropriate taxon [244]. Therefore, the use of several markers is recommended to obtain optimal and reliable results for the taxonomic and epidemiological classification of dermatophytes.

During pathogenesis, dermatophytes can produce diverse clinical manifestations that develop in a manner that is dependent on a series of virulence factors. These factors act in a sequential manner to confer the capacity to adhere to the keratinized epithelium and degrade keratin. It must be mentioned that such mechanisms are potential targets for disease control concerning fungal infection, and studies are currently underway with the goal of developing an effective vaccine. Profound knowledge of these pathogenic mechanisms will provide greater insight into the development of future treatments against these microorganisms.

It is of utmost importance, when concerning dermatophytosis, to be aware of their localization and major clinical characteristics, as both shall undoubtedly assist clinicians in proper identification amidst the broad variety of skin maladies with similar symptoms. The clinician should also rely on laboratory tests to act accordingly.

Concerning diagnosis, previous knowledge regarding the etiology of dermatophytoses is imperative before the election of an adequate antifungal therapy. Moreover, it will aid in assessing both origin and prevention measures. Therefore, clinical suspicion must be promptly followed by proper confirmation with laboratory tests to determine the specific etiology of the infection in question, as fungi are notably diverse in their response to the available antifungal treatments. Electing a unique method for diagnosis confirmation in the laboratory is a complex task due to the different advantages and disadvantages of either conventional or molecular techniques. Correct identification of the pathogen will unquestionably lead to proper handling of the patient.

References

1. Cafarchia C, Weigl S, Figueredo LA et al (2012) Molecular identification and phylogenesis of dermatophytes isolated from rabbit farms and rabbit farm workers. Vet Microbiol 154(3–4):395–402
2. White TC, Oliver BG, Gräser Y et al (2008) Generating and testing molecular hypotheses in the dermatophytes. Eukaryot Cell 7(8):1238–1245

3. Degreef H (2008) Clinical forms of dermatophytosis (ringworm infection). Mycopathologia 166(5–6):257–265
4. Iorio R, Cafarchia C, Capelli G et al (2007) Dermatophytoses in cats and humans in central Italy: epidemiological aspects. Mycoses 50(6):491–495
5. Weitzman I, Summerbell RC (1995) The dermatophytes. Clin Microbiol Rev 8(2):240–259
6. Kanbe T (2008) Molecular approaches in the diagnosis of dermatophytosis. Mycopathologia 166(5–6):307–317
7. Cafarchia C, Romito D, Capelli G et al (2006) Isolation of *Microsporum canis* from the hair coat of pet dogs and cats belonging to owners diagnosed with *M. canis tinea corporis*. Vet Dermatol 17(5):327–331
8. Gräser Y, Scott J, Summerbell R (2008) The new species concept in dermatophytes-a polyphasic approach. Mycopathologia 166(5–6):239–256
9. Squeo RF, Beer R, Silvers D et al (1998) Invasive *Trichophyton rubrum* resembling blastomycosis infection in the immunocompromised host. J Am Acad Dermatol 39:379–380
10. Molina de Diego A (2011) Aspectos clínicos, diagnósticos y terapéuticos de las dermatofitosis. Enferm Infecc Microbiol Clin 29(Supl 3):33–39
11. Simpanya MF (2000) Dermatophytes: their taxonomy, ecology and pathogenicity. In: Kushwaha RKS, Guarro J, editores. Biology of dermatophytes and other keratinophilic fungi. Rev Iberoam Micol Spain: Bilbao; 1–12
12. Cafarchia C, Iatta R, Latrofa MS et al (2013) Molecular epidemiology, phylogeny and evolution of dermatophytes. Infect Genet Evol 20:336–351
13. Garcia Garces H, Hrycyk MF, Giacobino J et al (2016) Molecular identification and phylogenetical analysis of dermatophyte fungi from Latin America. Mycoses. doi:10.1111/myc.12532
14. Nweze EI, Eke I (2016) Dermatophytosis in Northern Africa. Mycoses 59(3):137–144
15. Seebacher C, Bouchara JP, Mignon B (2008) Updates on the epidemiology of dermatophyte infections. Mycopathologia 166(5–6):335–352
16. Silveira-Gomes F, de Oliveira EF, Nepomuceno LB et al (2013) Dermatophytosis diagnosed at the Evandro Chagas Institute, Pará, Brazil. Braz J Microbiol 44(2):443–446
17. Di Chiacchio N, Madeira CL, Humaire CR et al (2014) Superficial mycoses at the Hospital do servidor público municipal de São Paulo between 2005 and 2011. An Bras Dermatol 89(1):67–71
18. de Carvalho Ribeiro CS, Zaitz C, de Souza Framil VM et al (2015) Descriptive study of onychomycosis in a hospital in São Paulo. Braz J Microbiol 46(2):485–492
19. Díaz Jarabrán MC, Díaz González P, Espinoza Rodríguez J et al (2015) Evaluación del perfil de sensibilidad *in vitro* de aislamientos clínicos de *Trichophyton mentagrophytes* y *Trichophyton rubrum* en Santiago, Chile. Rev Iberoam Micol 32(2):83–87
20. Heidrich D, Garcia MR, Stopiglia CD et al (2015) Dermatophytosis: a 16-year retrospective study in a metropolitan area in Southern Brazil. Infect Dev Ctries 9(8):865–871
21. Budak A, Bogusz BZ, Tokarczyk M et al (2013) Dermatophytes isolated from superficial fungal infections in Krakow, Poland, between 1995 and 2010. Mycoses 56:422–428
22. Farina C, Fazii P, Imberti G et al (2015) *Trichophyton violaceum* and *T. soudanense*: re-emerging pathogens in Italy, 2005-2013. New Microbiol 38:409–415
23. Papini M, Bianca BM, Difonzo E et al (2015) Epidemiology of onychomycosis in Italy: prevalence data and risk factor identification. Mycoses 58(11):659–664
24. Faure-Cognet O, Fricker-Hidalgo H, Pelloux H et al (2016) Superficial fungal infections in a french teaching hospital in Grenoble area: retrospective study on 5470 samples from 2001 to 2011. Mycopathologia 181(1–2):59–66
25. Nasr A, Vyzantiadis TA, Patsatsi A et al (2016) Epidemiology of superficial mycoses in Northern Greece: a 4-year study. J Eur Acad Dermatol Venereol 30(5):837–839
26. Kechia FA, Kouoto EA, Nkoa T et al (2014) Epidemiology of *tinea capitis* among school-age children in Meiganga, Cameroon. J Mycol Med 24(2):129–134
27. Seck MC, Ndiaye D, Diongue K et al (2014) Mycological profile of onychomycosis in Dakar (Senegal). J Mycol Med 24(2):124–128

28. Ndiaye M, Diongue K, Seck MC et al (2015) Epidemiological profile of *tinea capitis* in Dakar (Senegal). A 6-year retrospective study (2008-2013). J Mycol Med 25(2):169–176
29. Ndunge Moto J, Muthini Maingi J, Kebira Nyamache A (2015) Prevalence of *tinea capitis* in school going children from Mathare, informal settlement in Nairobi, Kenya. BMC Res Notes 8:274
30. Coulibaly O, Kone AK, Niaré-Doumbo S et al (2016) Dermatophytosis among schoolchildren in three eco-climatic zones of Mali. PLoS Negl Trop Dis 10(4):e0004675
31. El Mezouari E, Hocar O, Atarguine H et al (2016) *Tinea capitis* in the military hospital Avicenna (Morocco): review of 8 years (2006-2013). J Mycol Med 26(1):e1–e5
32. Afshar P, Khodavaisy S, Kalhori S et al (2014) Onychomycosis in North-East of Iran. Iran J Microbiol 6(2):98–103
33. Segal R, Shemer A, Hochberg M et al (2015) Onychomycosis in Israel: epidemiological aspects. Mycoses 58(3):133–139
34. Soltani M, Khosravi AR, Shokri H et al (2015) A study of onychomycosis in patients attending a dermatology center in Tehran, Iran. J Mycol Med 25(2):e81–e87
35. Lakshmanan A, Ganeshkumar P, Mohan SR et al (2015) Epidemiological and clinical pattern of dermatomycoses in rural India. Indian J Med Microbiol 33(5):134–136
36. Cai W, Lu C, Li X et al (2016) Epidemiology of superficial fungal infections in Guangdong, Southern China: a retrospective study from 2004 to 2014. Mycopathologia 181(5):387–395
37. Rezaei-Matehkolaei A, Rafiei A, Makimura K et al (2016) Epidemiological aspects of dermatophytosis in Khuzestan, Southwestern Iran, an update. Mycopathologia 181(7–8):547–553
38. Ali-Shtayeh MS, Yaish S, Jamous RM et al (2015) Updating the epidemiology of dermatophyte infections in Palestine with special reference to concomitant dermatophytosis. J Mycol Med 25(2):116–122
39. Singh D, Patel DC, Rogers K et al (2003) Epidemiology of dermatophyte infection in Auckland, New Zealand. Australas J Dermatol 44(4):263–266
40. McPherson ME, Woodgyer AJ, Simpson K et al (2008) High prevalence of *tinea capitis* in newly arrived migrants at an English-language school, Melbourne, 2005. Med J Aust 189(19):13–16
41. Sigurgeirsson B, Baran R (2014) The prevalence of onychomycosis in the global population: a literature study. J Eur Acad Dermatol Venereol 28(11):1480–1491
42. Ilkit M, Durdu M (2015) *Tinea pedis*: the etiology and global epidemiology of a common fungal infection. Crit Rev Microbiol 41(3):374–388
43. AL-Janabi AAHS (2014) Dermatophytosis: causes, clinic features, signs and treatment. J Symptoms Signs 3(3):200–203
44. Diongue K, Diallo MA, Ndiaye M et al (2016a) Champignons agents de mycoses superficielles isolés à Dakar (Sénégal): une étude rétrospective de 2011 à 2015. J Mycol Med. doi:10.1016/j.mycmed.2016.08.0036
45. Feuilhade de Chauvin M (2000) Fungal and bacterial infection. Rev Prat 50(20):2223–2230
46. Hainer BI (2003) Dermatophyte infections. Am Fam Physician 67(1):101–108
47. Arrazola Guerrero J, Isa Isa R, Torres Guerrero E et al (2015) Tiña de la cabeza: descripción de los hallazgos dermatoscópicos en 37 pacientes. Rev Iberoam Micol 32(4):242–246
48. del Boz-González J (2012) *Tinea capitis*: trends in Spain. Actas Dermosifiliogr 103(4):288–293
49. Khosravi R, Shokri H, Vahedi G (2016) Factors in etiology and predisposition of adult *tinea capitis* and review of published literature. Mycopathologia 181(5–6):371–378
50. Lova-Navarro M, Gómez-Moyano E, Martínez Pilar L et al (2016) *Tinea capitis* in adults in Southern Spain. A 17-year epidemiological study. Rev Iberoam Micol 33(2):110–113
51. Zuluaga A, Cáceres DH, Arango K et al (2016) Epidemiología de la *tinea capitis*: 19 años de experiencia en un laboratorio clínico especializado en Colombia. Infectio 20(4):225–230
52. Rebollo N, López-Barcenas AP, Arenas R (2008) Tiña de la cabeza. Actas Dermosifiliog 99(2):91–100
53. Sarabi K, Khachemoune A (2007) *Tinea capitis*: a review. Dermatol Nursing 19(6):525–529

54. Torres-Guerrero E, Matínez-Herrera E, Arroyo-Camarena S et al (2015) Kerion Celsi: a report of two cases due to *Microsporum gypseum* and *Trichophyton tonsurans*. Our Dermatol Online 6(4):424–427
55. Fernandes S, Amaro C, Martins ML et al (2013) Kerion caused by *Microsporum audouinii* in a child. Med Mycol Case Rep 2:52–54
56. Bargman H (2000) Trichophyton rubrum tinea capitis in a 85-year-old woman. J Cutan Med Surg 4(3):153–154
57. Moberg S (1984) *Tinea capitis* in the eldely. A report on two cases caused by *Trichophyton tonsurans*. Dermatologica 169(1):36–40
58. Offidani A, Simoncini C, Arzeni D et al (1998) *Tinea capitis* due to *Micosporum gypseum* in an adult. Mycoses 41(5–6):239–241
59. Anane S, Chtourou O (2013) *Tinea capitis* favosa misdiagnosed as *tinea amiantacea*. Med Mycol Case Rep 2:29–31
60. Hassan I, Rather PA, Sajad P (2013) Favus in an elderly female: a rare ocurrence. Indian J Dermatol 58(5):411
61. Vallarelli AFA (2014) Goya and tinea favosa. An Bras Dermatol 89(6):992–994
62. Sanusi T, Gong J, Wang X et al (2016) Disseminated favus caused by *Microsporum gypseum* in a patient with systemic lupus erythematosus. Acta Derm Venereol 96(2):270–271
63. Baran W, Szepietowski JC, Schawarts RA (2004) Tinea barbae. Acta Dermatoven APA 13(3):91–94
64. Bonifaz A, Ramírez-Tamayo T, Saúl A (2003) Tinea barbae (tinea sycosis): experience with nine cases. J Dermatol 30:898–903
65. Trotha R, Gräser Y, Platt J et al (2003) *Tinea barbae* caused by a zoophilic strain of *Trichophyton interdigitale*. Mycoses 46(1–2):60–63
66. Braun SA, Jahn K, Westermann A et al (2013) *Tinea barbae* profunda durch *Arthroderma benhamiae*. Hautarzt 64(10):720–722
67. Sabota J, Brodell R, Rutecki GW et al (1993) Severe *tinea barbae* due to *Trichophyton verrucosum* infection in dairy farmers. CID 23(6):1308–1310
68. Sidwell RU, Chan I, Francis N et al (2014) *Trichophyton erinacei* kerion barbae from a hedgehog with direct osculatory transfer to another person. Clin Experim Dermatol 39(1):38–40
69. Yin X, Du X, Zhang H (2011) A case of *tinea barbae* due to *Trichophyton rubrum* infection by autoinoculation from the infected fingernails. Mycoses 54(6):e864–e866
70. Diongue K, Ndiaye M, Diallo MA et al (2016b) Fungal interdigital *tinea pedis* in Dakar (Senegal). J Mycol Med. doi:10.1016/j.mycmed.2016.04.002
71. Kamishama T, Kmura T, Hosokawa JI et al (1997) *Tinea pedis* outbreak in swimming pools in Japan. Public Health 111:249–253
72. Pérez-Bruzón M, Falcón-Lincheta L, Miranda-Gómez O et al (2015) Índice de severidad y afección en la tiña pedis. R Cubana Med Militar 44(2):161–169
73. Porche DJ (2006) *Tinea pedis*: a common male foot problem. J Nurse Practitioners 2(3):152–153
74. Canavan TN, Elewski BE (2015) Identifying signs of *tinea pedis*: a key to understanding clinical variable. J Drugs Dermatol 14(10):S42–S47
75. Metin A, Dilek N, Demirseven DD (2015) Fungal infections of the folds (intertriginous areas). Clin Dermatol 33(4):437–447
76. Masri-Flidling GD (1996) Dermatophytosis of the feet. Dermatol Clin 14(1):33–40
77. Davis JD (1995) Superficial fungal infections of the skin: *tinea corporis*, *tinea pedis* and *Candida* intertrigo. Infect Dis Update 2(5):157–161
78. Hawkins DM, Smidt AC (2014) Superficial fungal infection in children. Pediatr Clin 61:443–455
79. Vander-Straten MR, Hossain MA, Ghannoum MA (2003) Cutaneous infections dermatophytosis, onychomycosis, and *tinea versicolor*. Infect Dis Clin 17(1):87–112
80. Zhan P, Geng C, Li Z et al (2013) The epidemiology of *tinea mannum* in Nanchang area, South China. Mycopathologia 176(1–2):83–88

81. Drira I, Neji S, Hadrick I et al (2015) *Tinea manuum* due to *Trichophyton erinacei* from Tunisia. J Mycol Med 25:200–203
82. Jayaraman AG, Gray JP, Robinson-Bostom L (2003) Cutaneous conditions of the hands: part 1. J Am Societ Surg Hand 3(1):4–13
83. Rubio-Flores C, Martín-Díaz MA, Prats Caelles I et al (2004) *Tinea manuum* inflamatoria. An Pediatr (Barc) 61(4):344–352
84. Bjekic M (2015) Two feet-one hand syndrome: a case report. Acta facultatis medicae Naissensis 32(3):215–219
85. Charles AJ (2009) Superficial cutaneous fungal infections in tropical countries. Dermatol Therapy 22:550–559
86. Loo DS (2004) Cutaneous fungal infections in the elderly. Dermatol Clin 22(1):33–50
87. Martin ES, Elewski BE (2002) Cutaneous fungal infections in the elderly. Geriatr Dermatol 18(1):59–75
88. Otero L, Palacio V, Vásquez F (2001) *Tinea cruris* in female prostitutes. Mycopathologia 153(1):29–31
89. Varade RS, Burkemper NM (2013) Cutaneous fungal infections in the Elderly. Clin Geriatr Med 29(2):461–478
90. Naafs B, Padovese V (2009) Rural dermatology in the tropics. Clin Dermatol 27(3):252–270
91. Mason D, Marks M (2015) Images in clinical tropical medicina. Bakua: *Tinea imbricata* in the Solomon Islands. Am J Trop Med Hyg 92(5):883
92. Pihet M, Bourgeois H, Mazière JY et al (2008) Isolation of *Trichophyton concentricum* from chronic cutaneous lesions in patients from the Solomon Islands. Trans R Soc Trop Med Hyg 102(4):389–393
93. Bonifaz A, Archer-Dubon C, Saúl A (2004) Tropical medicine rounds. *Tinea imbricata* or Tokelau. Int J Dermatol 43:506–510
94. Veraldi S, Giorgi R, Pontini P et al (2015a) *Tinea imbricata* in an italian child and review of the literature. Mycopathologia 180(5–6):353–357
95. Veraldi S, Pontini P, Nazzaro G (2015b) A case of *tinea imbricata* in an italian woman. Acta Derm Venereol 95(2):235–237
96. Ran X, Zhuang K, Ran Y (2015) *Tinea corporis* on the stump leg with *Trichophyton rubrum*. Med Mycol Cas Rep 9:31–33
97. Hubka V, Dobiâsova S, Dobiâs R et al (2014) *Microsporum aenigmaticum* sp. nov. from *M. gypseum* complex, isolated as a cause of *tinea corporis*. Med Mycol 52:387–396
98. Nenoff R, Overbeck C, UhrlaB S et al (2016) *Tinea corporis* due to the rare geophilic dermatophyte *Microsporum* praecox. Kasuistiken. doi:10.1007/s00105-016-3867-2
99. Branscomb R (2005) The dermatophytoses. Lab Medicine 36(8):496–500
100. Cavanaugh RM, Greeson JD (1982) *Trichophyton rubrum*. Infection of the diaper area. Arch Dermatol 118(6):446
101. Ortíz de Da Silva D, Flores de Lacarrubba L, Guzmán A (2005) Dermatofitosis infantiles en Asunción, Paraguay. Dermatología CMQ 3(1):22–26
102. Hayden GF (1985) Dermatophyte infection in the diaper area: report of two cases. Pediatr Infect Dis J 4(3):289–291
103. Aghamiriam MR, Ghiasian SA (2009) A clinico-epidemiological study on *tinea gladiatorum* in Iranian wrestlers and contamination by dermatophytes. Mycoses 54:248–253
104. Chen H-H, Chiu H-C (2002) Facial Majocchi's granuloma caused by *Trichophyton tonsurans* in an immunocompetent patient. Acta Derm Venereol 83:65–66
105. Ilkit M, Durdu M, Karakas M (2012) Majocchi's granuloma: a symptom complex caused by fungal pathogens. Med Mycol 50:449–457
106. Ríos-Reyes R, Rodríguez-Vázquez M, Comunión-Artieda A et al (2001) Foliculitis por *Trichophyton rubrum* (granuloma de Majocchi). Actas Dermosifiliogr 92:409–411
107. Nir-Paz R, Elinav H, Pierard GE et al (2003) Deep infection by *Trichophyton rubrum* in an immunocompromised patient. J Clin Microbiol 41(11):5298–5301

108. Bonifaz A, Tirado-Sánchez A, Ponce RM (2008) Granuloma de Majocchi. Gac Méd Méx 144(5):427–433
109. Tirado-González M, Ball E, Ruíz A et al (2012) Disseminated dermatophytic pseudomycetoma caused by *Microsporum* species. Int J Dermatol 51:1478–1482
110. Hadida E, Schousboe A (1957) Generalized trichophytosis with subcutaneous and ganglionic localisation caused by Trichophyton faviforme. Bull Soc Fr Dermatol Syphiligr 39:388–391
111. Cheikhrouhou F, Makni F, Ayadi A (2010) La maladie dermatophytique: revue de la littérature. J Mycol Med 20:61–69
112. Marill FG, Liautaud B, Hamra Krouha MS (1975) Evolution mortelle d'une maladie dermatophytique a *Trichophyton schoenleinii*. Clin Dermatol 68(5):450–456
113. Aboutou CNE, Hali F, Chihab S (2016) Maladie dermatophytique de revelation tardive. Pan African Med J 24:194
114. Martínez E, Ammen M, Tejada D et al (2014) *Microsporum* spp. onychomycosis: disease presentation, risk factors and treatment resonses in an urban population. Braz J Infect Dis 18(2):181–186
115. Scher RK, Tavakkol A, Sigurgeirsson B et al (2007) Onychomycosis: diagnosis and definition of cure. J Am Acad Dermatol 56(6):939–944
116. Martínez-Herrera EO, Arroyo-Camarena S, Tejada-García DL et al (2015) Onychomycosis due to opportunistic molds. An Bras Dermatol 90(3):334–337
117. Hay RJ, Baran R (2011) Onychomycosis: a proposed revision of the clinical classification. J Am Acad Dermatol 65(6):1219–1227
118. Baran R, Hay RJ (2014) New clinical classification for onychomycoses. J Mycol Med 24(4):247–260
119. Brasch J (2010) Pathogenesis of *tinea*. J Dtsch Dermatol Ges 8(10):780–786
120. Vermout S, Tabart J, Baldo A et al (2008) Pathogenesis of dermatophytosis. Mycopathologia 166(5–6):267–275
121. Burkhart CN, Burkhart CG, Gupta AK (2002) Dermatophytoma: recalcitrance to treatment because of existence of fungal biofilm. J Am Acad Dermatol 47(4):629–631
122. Vila TV, Rozental S, de Sa Guimaraes CM (2015) A new model of in vitro fungal biofilms formed on human nail fragments allows reliable testing of laser and light therapies against onychomycosis. Lasers Med Sci 30(3):1031–1039
123. Martinez-Rossi NM, Peres NT, Rossi A (2016) Pathogenesis of dermatophytosis: sensing the host tissue. Mycopathologia. doi:10.1007/s11046-016-0057-9
124. Grumbt M, Monod M, Yamada T et al (2013) Keratin degradation by dermatophytes relies on cysteine dioxygenase and a sulfite efflux pump. J Invest Dermatol 133(6):1550–1555
125. Brouta F, Descamps F, Monod M et al (2002) Secreted metalloprotease gene family of *Microsporum canis*. Infect Immun 70(10):5676–5683
126. Leng W, Liu T, Wang J et al (2009) Expression dynamics of secreted protease genes in *Trichophyton rubrum* induced by key host's proteinaceous components. Med Mycol 47(7):759–765
127. Baldo A, Tabart J, Vermout S et al (2008) Secreted subtilisins of *Microsporum canis* are involved in adherence of arthroconidia to feline corneocytes. J Med Microbiol 57(Pt9):1152–1156
128. King RD, Khan HA, Foye JC et al (1975) Transferrin, iron, and dermatophytes. I. Serum dermatophyte inhibitory component definitively identified as unsaturated transferrin. J Lab Clin Med 86(2):204–212
129. Achterman RR, White TC (2012) Dermatophyte virulence factors: identifying and analyzing genes that may contribute to chronic or acute skin infections. Int J Microbiol 2012:35830
130. Criado PR, Oliveira CB, Dantas KC et al (2011) Superficial mycosis and the immune response elements. An Bras Dermatol 86(4):726–731
131. Guarner J, Brandt ME (2011) Histopathologic diagnosis of fungal infections in the 21st century. Clin Microbiol Rev 24(2):247–280
132. Almeida SR (2008) Immunology of dermatophytosis. Mycopathologia 166(5–6):277–283

133. Grappel SF, Bishop CT, Blank F (1974) Immunology of dermatophytes and dermatophytosis. Bacteriol Rev 38(2):222–250
134. Peres NT, Maranhão FC, Rossi A et al (2010) Dermatophytes: host-pathogen interaction and antifungal resistance. An Bras Dermatol 85(5):657–667
135. Padhye AA, Young CN, Ajello L (1980) Hair perforation as a diagnostic criterion in the identification of *Epidermophyton, Microsporon* and *Trichophyton* species. PAHO Sci Publ 396:115–120
136. Summerbell RC, Rosenthal SA, Kane J (1988) Rapid method for differentiation of *Trichophyton rubrum*, *Trichophyton mentagrophytes*, and related dermatophyte species. J Clin Microbiol 26(11):2279–2282
137. Rezusta A, Rubio MC, Alejandre MC (1991) Differentiation between *Trichophyton mentagrophytes* and by sorbitol assimilation. J Clin Microbiol 29(1):219–220
138. Rebell G, Taplin D (1978) Dermatophytes: their recognition and identification, revised edn. University of Miami Press, Coral Gables
139. Symoens F, Willenz P, Rouma V et al (1989) Isoelectric focusing applied to taxonomic differentiation of the *Trichophyton mentagrophytes* complex and the related *Trichophyton interdigitale*. Mycoses 32(12):652–663
140. Mochizuki T, Takada K, Watanabe S et al (1990) Taxonomy of *Trichophyton interdigitale* (*Trichophyton mentagrophytes* var. *interdigitale*) by restriction enzyme analysis of mitochondrial DNA. J Med Vet Mycol 28(3):191–196
141. Kawasaki M, Ishizaki H, Aoki M et al (1990) Phylogeny of *Nannizzia incurvata, N. gypsea, N. fulva* and *N. otae* by restriction enzyme analysis of mitochondrial DNA. Mycopathologia 112(3):173–177
142. Kawasaki M, Aoki M, Ishizaki H et al (1992) Phylogenetie relationships of the genera *Arthroderma* and *Nannizzia* inferred from mitochondrial DNA analysis. Mycopathologia 118:95–102
143. Weitzman I, McGinnis MR, Padhye AA et al (1986) The genus *Arthroderma* and its synonym *Nannizzia*. Mycotaxon 25(2):505–518
144. Nishio K, Kawasaki M, Ishizaki H (1992) Phylogeny of the genera *Trichophyton* using mitochondrial DNA analysis. Mycopathologia 117(3):127–132
145. de Bievre C, Dauguet C, Nguyen VH et al (1987) Polymorphism in mitocondrial DNA of several *Trichphyton rubrum* isolates from clinical specimens. Ann Insti Pasteur/Microbiol 138(6):719–727
146. Harmsen D, Schwinn A, Weig M et al (1995) Phylogeny and dating of some pathogenic keratinophilic fungi using small subunit ribosomal RNA. J Med Vet Mycol 33(5):299–303
147. Gräser Y, El Fari M, Presber W et al (1998) Identification of common dermatophytes (*Trichophyton, Microsporum, Epidermophyton*) using polymerase chain reactions. Br J Dermatol 138(4):576–582
148. Jackson CJ, Barton RC, Kelly SL et al (2000) Strain identification of *Trichophyton rubrum* by specific amplification of subrepeat elements in the ribosomal DNA nontranscribed spacer. J Clin Microbiol 38(12):4527–4534
149. Makimura K, Tamura Y, Mochizuki T et al (1999) Phylogenetic classification and species identification of dermatophyte strains based on DNA sequences of nuclear ribosomal internal transcribed spacer 1 regions. J Clin Microbiol 37(4):920–924
150. Gräser Y, El Fari M, Vilgalys R et al (1999a) Phylogeny and taxonomy of the family *Arthrodermataceae* (dermatophytes) using sequence analysis of the ribosomal ITS region. Med Mycol 37(2):105–114
151. Gräser Y, Kuhnisch J, Presber W (1999b) Molecular markers reveal exclusively clonal reproduction in *Trichophyton rubrum*. J Clin Microbiol 37(11):3713–3717
152. Summerbell RC, Haugland RA, Li A et al (1999) rRNA gene internal transcribed spacer 1 and 2 sequences of asexual, anthropophilic dermatophytes related to *Trichophyton rubrum*. J Clin Microbiol 37(12):4005–4011

153. Gräser Y, Kuijpers AFA, El Fari M et al (2000) Molecular and conventional taxonomy of the *Microsporum canis* complex. Med Mycol 38(2):143–153
154. Hirai A, Kano R, Nakamura Y et al (2003) Molecular taxonomy of dermatophytes and related fungi by chitin synthase 1 (CHS1) gene sequences. Antonie Van Leeuwenhoek 83(1):11–20
155. Nenoff P, Herrmann J, Gräser Y (2007) *Trichophyton mentagrophytes* sive interdigitale? A dermatophyte in the course of time. J Dtsch Dermatol Ges 5(3):198–202
156. Takashio M (1972) Is *Arthroderma benhamiae* the perfect state of *Trichophyton mentagrophytes*? Sabouraudia 10(2):122–127
157. Anzawa K, Kawasaki M, Mochizuki T et al (2010) Successful mating of *Trichophyton rubrum* with *Arthroderma simii*. Med Mycol 48(4):629–634
158. Symoens F, Jousson O, Planard C et al (2011) Molecular analysis and mating behaviour of the *Trichophyton mentagrophytes* species complex. Int J Med Microbiol 301(3):260–266
159. Kawasaki M (2011) Verification of a taxonomy of dermatophytes based on mating results and phylogenetic analyses. Med Mycol J 52(4):291–295
160. Tartabinia ML, Boninoa GS, Raccab L et al (2013) Estudio taxonómico de aislamientos clínicos de *Trichophyton* en Rosario, Argentina. Rev Argent Microbiol 45(4):248–253
161. Ahmadi B, Mirhendi H, Makimura K et al (2016) Phylogenetic analysis of dermatophyte species using DNA sequence polymorphism in calmodulin gene. Med Mycol 54(5):500–514
162. Lymbery AJ, Thompson RCA (2012) The molecular epidemiology of parasite infections: tools and applications. Mol Biochem Parasitol 181(2):102–116
163. Kim JA, Kayoko T, Kazutaka F et al (1999) Identification and genetic homogeneity of *Trichophyton tonsurans* isolated from several regions by random amplified polymorphic DNA. Mycopathologia 145:1–6
164. Howell SA, Barnard RJ, Humphreyst F (1999) Application of molecular typing methods to dermatophyte species that cause skin and nail infections. J Med Microbiol 48(1):33–40
165. Gupta AK, Kohli Y, Summerbbell RC (2001) Variation in restriction fragment length polymorphisms among serial isolates from patient with *Trichophyton rubrum* infection. J Clin Microbiol 39(9):3260–3266
166. Yazdanparast A, Jackson CJ, Barton RC et al (2003) Molecular strain typing of *Trichophyton rubrum* indicates multiple strain involvement in onychomycosis. Br J Dermatol 148(1):51–54
167. Gaedigk A, Gaedigk R, Abdel-Rahman SM (2003) Genetic heterogeneity in the rRNA gene locus of *Trichophyton tonsurans*. J Clin Microbiol 41(12):5478–5487
168. Yu J, Wan Z, Chen W et al (2004) Molecular typing study of the *Microsporum canis* strains isolated from an outbreak of *tinea capitis* in a school. Mycopathologia 157(1):37–41
169. Brilhante RSN, Rocha MFG, Cordeiro RA et al (2005) Phenotypical and molecular characterization of *Microsporum canis* in North-east Brazil. J Appl Microbiol 99(4):776–782
170. Cano J, Rezusta A, Guarro J (2005) Inter-single-sequencerepeat-PCR typing as a new tool for identification of *Microsporum canis* strains. J Dermatol Sci 39(1):17–21
171. Sugita T, Shiraki Y, Hiruma M (2006) Genotype analysis of the variable internal repeat region in the rRNA gene of *Trichophyton tonsurans* isolated from Japanese Judo practitioners. Microbiol Immunol 50(1):57–60
172. Baeza LC, Teruyuki-Matsumoto M, Fusco-Almeida AM et al (2006) Strain differentiation of *Trichophyton rubrum* by randomly amplified polymorphic DNA and analysis of rDNA nontranscribed space. J Med Microbiol 55(Pt 4):429–436
173. Gräser Y, Fröhlich J, Presber W et al (2007) Microsatellite markers reveal geographic population differentiation in *Trichophyton rubrum*. J Med Microbiol 56(Pt 8):1058–1065
174. Sharma R, Presber W, Rajak RC et al (2008) Molecular detection of *Microsporum persicolor* in soil suggesting widespread dispersal in central India. Med Micol 46(1):67–73
175. Cafarchia C, Otranto D, Weigl S et al (2009) Molecular characterization of selected dermatophytes and their identification by electrophoretic mutation scanning. Electrophoresis 30(20):3555–3564
176. Heidemann S, Monod M, Gräser Y (2010) Signature polymorphisms in the internal transcribed spacer region relevant for the differentiation of zoophilic and anthropophilic strains of

Trichophyton interdigitale and other species of *T. mentagrophytes sensu lato*. Br J Dermatol 162(2):282–295
177. Rezaei-Matehkolaei A, Mirhendi H, Makimura K et al (2014) Nucleotide sequence analysis of beta tubulin gene in a wide range of dermatophytes. Med Mycol 52(7):674–688
178. Mirhendi H, Makimura K, de Hoog GS et al (2015) Translation elongation factor 1-α gene as a potential taxonomic and identification marker in dermatophytes. Med Mycol 53(3):215–224
179. Ziółkowska G, Nowakiewicz A, Gnat S et al (2015) Molecular identification and classification of *Trichophyton mentagrophytes* complex strains isolated from humans and selected animal species. Mycoses 58(3):119–126
180. Robert R, Pihet M (2008) Conventional methods for the diagnosis of dermatophytosis. Mycopathologia 166(5):295–306
181. Brillowska-Da-browska A, Saunte DM, Cavling Arendrup M (2007) Five-hour diagnosis of dermatophyte nail infections with specific detection of *Trichophyton rubrum*. J Clin Microbiol 45(4):1200–1204
182. Sahoo AK, Mahajan R (2016) Management of *tinea corporis*, *tinea cruris*, and *tinea pedis*: a comprehensive review. Indian Dermatol Online J 7(2):77–86
183. Karan A, Alikhan A, Feldman SR (2009) Microscopically differentiating dermatophytes from sock fibers. J Am Acad Dermatol 61(6):1024–1027
184. Pihet M, Clément N, Kauffmann-Lacroix C et al (2015) Diagnosis of dermatophytosis: an evaluation of direct examination using MycetColor® and MycetFluo®. Diagn Microbiol Infect Dis 83(2):170–174
185. Bonifaz A, Rios-Yuil JM, Arenas R et al (2013) Comparison of direct microscopy, culture and calcofluor white for the diagnosis of onychomycosis. Rev Iberoam Micol 30(2):109–111
186. Kozielewicz D, Wernik J, Mikucka A et al (2015) Problems in the diagnosis of profound trichophytosis barbae. Indian J Med Microbiol 33(3):444–447
187. Rezusta A, de la Fuente S, Gilaberte Y et al (2016) Evaluation of incubation time for dermatophytes cultures. Mycoses 59(7):416–418
188. Ates A, Ozcan K, Ilkit M (2008) Diagnostic value of morphological, physiological and biochemical tests in distinguishing *Trichophyton rubrum* from *Trichophyton mentagrophytes* complex. Med Mycol 46(8):811–822
189. Ergın Ç, İlkit M, Gök Y et al (2014) The effect of Tween-80 on the differentiation of *Trichophyton mentagrophytes* and *Trichophyton rubrum* strains with FT-IR spectroscopy. Mikrobiyol Bul 48(3):449–460
190. Carballo GM, Ambasch G, Peralta NB et al (2002) Rojo neutro: nuestra experiencia de diagnóstico en micosis superficiales en cinco pacientes. Rev Chil Infect 19(4):267–270
191. Rich P, Elewski B, Scher RK et al (2013) Diagnosis, clinical implications, and complications of onychomycosis. Semin Cutan Med Surg 32(2 Suppl 1):S5–S8
192. Fernandes-Meireles TE, Gadelha-Rocha MF, Nogueira-Brilhante RS et al (2008) Successive mycological nail tests for onychomycosis: a strategy to improve diagnosis efficiency. Braz J Infect Dis 12(4):333–337
193. Nagar R, Nayak CS, Deshpande S et al (2012) Subungual hyperkeratosis nail biopsy: a better diagnostic tool for onychomycosis. Indian J Dermatol Venereol Leprol 78(5):620–624
194. Wilsmann-Theis D, Sareika F, Bieber T et al (2011) New reasons for histopathological nail-clipping examination in the diagnosis of onychomycosis. J Eur Acad Dermatol Venereol 25(2):235–237
195. Idriss MH, Khalil A, Elston D (2013) The diagnostic value of fungal fluorescence in onychomycosis. J Cutan Pathol 40(4):385–390
196. Gianni C, Morelli V, Cerri A et al (2001) Usefulness of histological examination for the diagnosis of onychomycosis. Dermatol 202(4):283–288
197. Guarro J (2012) Taxonomía y biología de los hongos causantes de infección en humanos. Enferm Infecc Microbiol Clin 30(1):33–39
198. Arabatzis M, Bruijnesteijn van Coppenraet LE, Kuijper EJ et al (2007) Diagnosis of common dermatophyte infections by a novel multiplex real-time polymerase chain reaction detection/identification scheme. Br J Dermatol 157(4):681–689

199. Dhib I, Fathallah A, Yaacoub A et al (2014) Multiplex PCR assay for the detection of common dermatophyte nail infections. Mycoses 57(1):19–26
200. Ebihara M, Makimura K, Sato K et al (2009) Molecular detection of dermatophytes and nondermatophytes in onychomycosis by nested polymerase chain reaction based on 28S ribosomal RNA gene sequences. Br J Dermatol 161(5):1038–1044
201. Gutzmer R, Mommert S, Küttler U et al (2004) Rapid identification and differentiation of fungal DNA in dermatological specimens by LightCycler PCR. J Med Microbiol 53(Pt 12):1207–1214
202. Kanbe T, Suzuki Y, Kamiya A et al (2003) PCR-based identification of common dermatophyte species using primer sets specific for the DNA topoisomerase II genes. J Dermatol Sci 32(2):151–161
203. Winter I, Uhrlaß S, Krüger C et al (2013) Molecular biological detection of dermatophytes in clinical samples when onychomycosis or *tinea pedis* is suspected. A prospective study comparing conventional dermatomycological diagnostics and polymerase chain reaction. Hautarzt 64(4):283–289
204. Kim JA, Takahashi Y, Tanaka R et al (2001) Identification and subtyping of *Trichophyton mentagrophytes* by random amplified polymorphic DNA. Mycoses 44(5):157–165
205. Ahmadi B, Mirhendi H, Shidfar MR et al (2015) A comparative study on morphological versus molecular identification of dermatophyte isolates. J Mycol Med 25(1):29–35
206. Bontems O, Hauser PM, Monod M (2009) Evaluation of a polymerase chain reaction-restriction fragment length polymorphism assay for dermatophyte and nondermatophyte identification in onychomycosis. Br J Dermatol 161(4):791–796
207. Elavarashi E, Kindo AJ, Kalyani J et al (2014) Application of PCR fingerprinting using (GACA)4 primer in the rapid discrimination of dermatophytes. Indian J Med Microbiol 32(3):236–239
208. Miao Z, Li S, Li D et al (2014) Rapid detection for rabbit-derived dermatophytes using microsatellite-primed polymerase chain reaction. J Mol Microbiol Biotechnol 24(1):53–58
209. Mehlig L, Garve C, Ritschel A et al (2014) Clinical evaluation of a novel commercial multiplex-based PCR diagnostic test for differential diagnosis of dermatomycoses. Mycoses 57(1):27–34
210. Vahidnia A, Bekers W, Bliekendaal H et al (2015) High throughput multiplex-PCR for direct detection and diagnosis of dermatophyte species, *Candida albicans* and *Candida parapsilosis* in clinical specimen. J Microbiol Methods 113:38–40
211. Nardoni S, Franceschi A, Mancianti F (2007) Identification of *Microsporum canis* from dermatophytic pseudomycetoma in paraffin-embedded veterinary specimens using a common PCR protocol. Mycoses 50(3):215–217
212. Gupta AK, Zaman M, Singh J (2008) Diagnosis of *Trichophyton rubrum* from onychomycotic nail samples using polymerase chain reaction and calcofluor white microscopy. J Am Podiatr Med Assoc 98(3):224–228
213. Uchida T, Makimura K, Ishihara K et al (2009) Comparative study of direct polymerase chain reaction, microscopic examination and culture-based morphological methods for detection and identification of dermatophytes in nail and skin samples. J Dermatol 36(4):202–208
214. Chandran NS, Pan JY, Pramono ZA et al (2013) Complementary role of a polymerase chain reaction test in the diagnosis of onychomycosis. Australas J Dermatol 54(2):105–108
215. Luk NM, Hui M, Cheng TS et al (2012) Evaluation of PCR for the diagnosis of dermatophytes in nail specimens from patients with suspected onychomycosis. Clin Exp Dermatol 37(3):230–234
216. Alexander CL, Shankland GS, Carman W et al (2011) Introduction of a dermatophyte polymerase chain reaction assay to the diagnostic mycology service in Scotland. Br J Dermatol 164(5):966–972
217. Ohst T, Kupsch C, Gräser Y (2016) Detection of common dermatophytes in clinical specimens using a simple quantitative real-time TaqMan polymerase chain reaction assay. Br J Dermatol 174(3):602–609

218. Paugam A, L'ollivier C, Viguié C et al (2013) Comparison of real-time PCR with conventional methods to detect dermatophytes in samples from patients with suspected dermatophytosis. J Microbiol Methods 95(2):218–222
219. Nimura K, Niwano Y, Ishiduka S et al (2003) Actin gene-targeted RT-PCR could be a useful method for evaluating *in vitro* fungicidal activity against dermatophytes. J Int Med Res 31(5):407–412
220. Iwanaga T, Anzawa K, Mochizuki T (2014) Quantification of dermatophyte viability for evaluation of antifungal effect by quantitative PCR. Mycopathologia 177(5–6):241–249
221. Worek M, Kwiatkowska A, Ciesielska A et al (2014) Identification of dermatophyte species using genomic *in situ* hybridization (GISH). J Microbiol Methods 100:32–41
222. L'Ollivier C, Cassagne C, Normand AC et al (2013) A MALDI-TOF MS procedure for clinical dermatophyte species identification in the routine laboratory. Med Mycol 51:713–720
223. Bader O (2013) MALDI-TOF-MS-based species identification and typing approaches in medical mycology. Proteomics 13(5):788–799
224. Karabıçak N, Karatuna O, İlkit M et al (2015) Evaluation of the Bruker matrix-assisted laser desorption-ionization time-of-flight mass spectrometry (MALDI-TOF MS) system for the identification of clinically important dermatophyte species. Mycopathologia 180(3–4):165–171
225. Nenoff P, Erhard M, Simon JC et al (2013) MALDI-TOF mass spectrometry - a rapid method for the identification of dermatophyte species. Med Mycol 51:17–24
226. Sánchez MJ, Pico AM, Tejedor FM et al (2014) Using a polymerase chain reaction as a complementary test to improve the detection of dermatophyte fungus in nails. J Am Podiatr Med Assoc 104(3):233–237
227. Friedman D, Friedman PC, Gill M (2015) Reflectance confocal microscopy: an effective diagnostic tool for dermatophytic infections. Cutis 95(2):93–97
228. Higashi Y, Miyoshi H, Takeda K et al (2012) Evaluation of a newly-developed immunochromatography strip test for diagnosing dermatophytosis. Int J Dermatol 51(4):406–409
229. Tsunemi Y, Hiruma M (2016) Clinical study of dermatophyte test strip, an immunochromatographic method, to detect *tinea unguium* dermatophytes. J Dermatol. doi:10.1111/1346-8138.13348
230. Smijs TG, Jachtenberg JW, Pavel S et al (2014) Detection and differentiation of causative organisms of onychomycosis in an *ex vivo* nail model by means of Raman spectroscopy. J Eur Acad Dermatol Venereol 28(11):1492–1499
231. White TJ, Bruns T, Taylor J (1990) Amplification and direct sequencing of fungal ribosomal RNA genes for phylogenetics. PCR Protocols: A Guid Methd Appl 18:315–322
232. Drummond AJ, Ashton B, Heled J, et al (2006) Geneious v2.5. 2006. http://www.geneious.com
233. Hall TA (1999) BioEdit: a user-friendly biological sequence alignment editor and analysis program for Windows 95/98/NT. Nucl Acids Symp Ser 41:95–98
234. Altschul SF, Madden TL, Schaffer AA et al (1997) Gapped BLAST and PSI-BLAST: a new generation of protein database search programs. Nucl Acids Res 25:3389–3402
235. Posada D, Crandall KA (1998) Modeltest: testing the model of DNA substitution. Bioinformatics 14:817–118
236. Goloboff PA, Farris JS, Nixon KC (2008) TNT, a free program for phylogenetic analysis. Cladistics 24:1–13
237. Huelsenbeck JP, Ronquist F (2001) MRBAYES: Bayesian inference of phylogenetic trees. Bioinform App Not 17(8):754–755
238. Kumar S, Stecher G, Tamura K (2016) MEGA 7: molecular evolutionary genetics analysis. Version 7.0 for bigger datasets. Mol Biol Evol 33(7):1870–1874
239. Elewski B (2009) Tinea capitis: a current perspective. J Am Acad Dermatol 42(1):1–20
240. Arenas R (2014) Micología Médica Ilustrada. McGraw-Hill, México, pp 67–98
241. Domínguez-Cherit J, Teixeira F, Arenas R (1999) Combined surgical and systemic treatment of onychomycosis. Br J Dermatol 140(4):778–780

242. Arenas R, Arce M, Leyva J (1999) Onicomicosis dermatofítica. Estudio abierto, comparativo y al azar entre pomada de bifonazol-urea combinada con un monopulso de itraconazol o terbinafina. Dermatol Rev Mex 43(4):149–156
243. Guarro J, Gené J, Stchigel AM (1999) Developments in fungal taxonomy. Clin Microbiol Rev 12(3):454–500
244. Gräser Y, Kuijpers AFA, Presber W et al (1999c) Molecular taxonomy of *Trichophyton mentagrophytes* and *T. tonsurans*. Med Mycol 37(5):315–330

The manufacturer's authorised representative in the EU is Springer Nature Customer Service Centre GmbH, Europaplatz 3, 69115 Heidelberg, Germany. If you have any concerns regarding our products, please contact ProductSafety@springernature.com

Printed and bound by CPI Group (UK) Ltd, Croydon, CR0 4YY
25/09/2026
02222851-0007